Lecture Notes in Physics

For information about Vols. 1–67, please contact your bookseller or Springer-Verlag.

Vol. 68: Y. V. Venkatesh, Energy Methods in Time-Varying System Stability and Instability Analyses. XII, 256 pages. 1977.

Vol. 69: K. Rohlfs, Lectures on Density Wave Theory. VI, 184 pages. 1977.

Vol. 70: Wave Propagation and Underwater Acoustics. Edited by J. Keller and J. Papadakis. VIII. 287 pages. 1977.

Vol. 71: Problems of Stellar Convection. Proceedings 1976. Edited by E. A. Spiegel and J. P. Zahn. VIII, 363 pages. 1977.

Vol. 72: Les instabilités hydrodynamiques en convection libre forcée et mixte. Edité par J. C. Legros et J. K. Platten. X, 202 pages. 1978.

Vol. 73: Invariant Wave Equations. Proceedings 1977. Edited by G. Velo and A. S. Wightman. VI, 416 pages. 1978.

Vol. 74: P. Collet and J.-P. Eckmann, A Renormalization Group Analysis of the Hierarchical Model in Statistical Mechanics. IV, 199 pages. 1978.

Vol. 75: Structure and Mechanisms of Turbulence I. Proceedings 1977. Edited by H. Fiedler. XX, 295 pages. 1978.

Vol. 76: Structure and Mechanisms of Turbulence II. Proceedings 1977. Edited by H. Fiedler. XX, 406 pages. 1978.

Vol. 77: Topics in Quantum Field Theory and Gauge Theories. Proceedings, Salamanca 1977. Edited by J. A. de Azcárraga. X, 378 pages 1978.

Vol. 78: Böhm, The Rigged Hilbert Space and Quantum Mechanics. IX, 70 pages. 1978.

Vol. 79: Group Theoretical Methods in Physics. Proceedings, 1977. Edited by P. Kramer and A. Rieckers. XVIII, 546 pages. 1978.

Vol. 80: Mathematical Problems in Theoretical Physics. Proceedings, 1977. Edited by G. Dell'Antonio, S. Doplicher and G. Jona-Lasinio. VI, 438 pages. 1978.

Vol. 81: MacGregor, The Nature of the Elementary Particle. XXII, 482 pages. 1978.

Vol. 82: Few Body Systems and Nuclear Forces I. Proceedings, 1978. Edited by H. Zingl, M. Haftel and H. Zankel. XIX, 442 pages. 1978.

Vol. 83: Experimental Methods in Heavy Ion Physics. Edited by K. Bethge. V, 251 pages. 1978.

Vol. 84: Stochastic Processes in Nonequilibrium Systems, Proceedings, 1978. Edited by L. Garrido, P. Seglar and P. J. Shepherd. XI, 355 pages. 1978

Vol. 85: Applied Inverse Problems. Edited by P. C. Sabatier. V, 425 pages. 1978.

Vol. 86: Few Body Systems and Electromagnetic Interaction. Proceedings 1978. Edited by C. Ciofi degli Atti and E. De Sanctis. VI, 352 pages. 1978.

Vol. 87: Few Body Systems and Nuclear Forces II, Proceedings, 1978. Edited by H. Zingl, M. Haftel, and H. Zankel. X, 545 pages. 1978.

Vol. 88: K. Hutter and A. A. F. van de Ven, Field Matter Interactions in Thermoelastic Solids. VIII, 231 pages. 1978.

Vol. 89: Microscopic Optical Potentials, Proceedings, 1978. Edited by H. V. von Geramb. XI, 481 pages. 1979.

Vol. 90: Sixth International Conference on Numerical Methods in Fluid Dynamics. Proceedings, 1978. Edited by H. Cabannes, M. Holt and V. Rusanov. VIII, 620 pages. 1979.

Vol. 91: Computing Methods in Applied Sciences and Engineering, 1977, II. Proceedings, 1977. Edited by R. Glowinski and J. L. Lions. VI, 359 pages. 1979.

Vol. 92: Nuclear Interactions. Proceedings, 1978. Edited by B. A. Robson. XXIV, 507 pages. 1979.

Vol. 93: Stochastic Behavior in Classical and Quantum Hamiltonian Systems. Proceedings, 1977. Edited by G. Casati and J. Ford. VI, 375 pages. 1979.

Vol. 94: Group Theoretical Methods in Physics. Proceedings, 1978. Edited by W. Beiglböck, A. Böhm and E. Takasugi. XIII, 540 pages. 1979.

Vol. 95: Quasi One-Dimensional Conductors I. Proceedings, 1978. Edited by S. Barišić, A. Bjeliš, J. R. Cooper and B. Leontić. X, 371 pages. 1979.

Vol. 96: Quasi One-Dimensional Conductors II. Proceedings 1978. Edited by S. Barišić, A. Bjeliš, J. R. Cooper and B. Leontić. XII, 461 pages. 1979.

Vol. 97: Hughston, Twistors and Particles. VIII, 153 pages. 1979.

Vol. 98: Nonlinear Problems in Theoretical Physics. Proceedings, 1978. Edited by A. F. Rañada. X, 216 pages. 1979.

Vol. 99: M. Drieschner, Voraussage – Wahrscheinlichkeit – Objekt. XI, 308 Seiten. 1979.

Vol. 100: Einstein Symposion Berlin. Proceedings 1979. Edited by H. Nelkowski et al. VIII, 550 pages. 1979.

Vol. 101: A. Martin-Löf, Statistical Mechanics and the Foundations of Thermodynamics. V, 120 pages. 1979.

Vol. 102: H. Hora, Nonlinear Plasma Dynamics at Laser Irradiation. VIII, 242 pages. 1979.

Vol. 103: P. A. Martin, Modèles en Mécanique Statistique des Processus Irréversibles. IV, 134 pages. 1979.

Vol. 104: Dynamical Critical Phenomena and Related Topics. Proceedings, 1979. Edited by Ch. P. Enz. XII, 390 pages. 1979.

Vol. 105: Dynamics and Instability of Fluid Interfaces. Proceedings, 1978. Edited by T. S. Sørensen. V, 315 pages. 1979.

Vol. 106: Feynman Path Integrals, Proceedings, 1978. Edited by S. Albeverio et al. XI, 451 pages. 1979.

Vol. 107: J. Kijowski, W. M. Tulczyjew, A Symplectic Framework for Field Theories. IV, 257 pages. 1979.

Vol. 108: Nuclear Physics with Electromagnetic Interactions. Proceedings, 1979. Edited by H. Arenhövel and D. Drechsel. IX, 509 pages. 1979.

Vol. 109: Physics of the Expanding Universe. Proceedings, 1978. Edited by M. Demiański. V, 210 pages. 1979.

Vol. 110: D. A. Park, Classical Dynamics and Its Quantum Analogues. VIII, 339 pages. 1979.

Vol. 111: H.-J. Schmidt, Axiomatic Characterization of Physical Geometry. V, 163 pages. 1979.

Vol. 112: Imaging Processes and Coherence in Physics. Proceedings, 1979. Edited by M. Schlenker et al. XIX, 577 pages. 1980.

Vol. 113: Recent Advances in the Quantum Theory of Polymers. Proceedings 1979. Edited by J.-M. André et al. V, 306 pages. 1980.

Lecture Notes in Physics

Edited by J. Ehlers, München K. Hepp, Zürich
R. Kippenhahn, München H. A. Weidenmüller, Heidelberg
and J. Zittartz, Köln
Managing Editor: W. Beiglböck, Heidelberg

137

From Collective States to Quarks in Nuclei

Proceedings of the Workshop
on Nuclear Physics with Real and Virtual Photons
Held in Bologna (Italy), November 25–28, 1980

Edited by H. Arenhövel and A. M. Saruis

Springer-Verlag
Berlin Heidelberg GmbH 1981

Editors

Hartmuth Arenhövel
Institut für Kernphysik Johannes Gutenberg-Universität
D-6500 Mainz

Anna Maria Saruis
Centro Studi e Ricerche
"E. Clementel"
C.N.E.N. via Mazzini 2
I-40138 Bologna

ISBN 978-3-540-10570-1   ISBN 978-3-540-38539-4 (eBook)
DOI 10.1007/978-3-540-38539-4

This work is subject to copyright. All rights are reserved, whether the whole or part of the material is concerned, specifically those of translation, reprinting, re-use of illustrations, broadcasting, reproduction by photocopying machine or similar means, and storage in data banks. Under § 54 of the German Copyright Law where copies are made for other than private use, a fee is payable to "Verwertungsgesellschaft Wort", Munich.

© by Springer-Verlag Berlin Heidelberg 1981
Originally published by Springer-Verlag Berlin Heidelberg New York in 1981

PREFACE

The workshop "FROM COLLECTIVE STATES TO QUARKS IN NUCLEI" on nuclear physics using real and virtual photons has been organized by C.N.E.N., Centro Studi e Ricerche "E. Clementel" in Bologna (Italy) on November 25-28, 1980, in the framework of the nuclear physics activities of the Institute.

The central idea of this workshop was to review recent developments in the field of probing nuclei with electromagnetic interaction for both experiment and theory, to discuss open problems and to try to look into the future of novel theoretical concepts and experimental techniques. The main topics were: i) the NN interaction and the many-body problem, ii) collective phenomena, iii) baryons and mesons in nuclei, iv) quarks in nuclei, and v) present and future experimental facilities.

All invited talks are contained in these proceedings and are organized following the program. In addition, 34 contributed papers have been published in a separate volume and distributed to the participants. About one-third of them has been presented orally.

The workshop has been sponsored by the Research Department RIT of the Italian Nuclear Energy Committee (C.N.E.N.). The generous collaboration of the CEN-Saclay Nuclear Physics Department, the Institute of Nuclear Physics of the Mainz University, the Max-Planck-Institute for Chemistry, Mainz, the Frascati National Laboratories, the Rome INFN Sanità and Physics Institute, and the Physics Institutes of the Bologna and Genova Universities is gratefully acknowledged. Particular thanks are due to Mrs. M. Accorsi, Dr. G. Bacchetta, Dr. S. Stipcich, friends, younger scientists and students for having kept the workshop running efficiently and cheerfully.

Mainz-Bologna, January 14, 1981   H. Arenhövel   A.M. Saruis

Detail from a relief by Iacopo della Quercia (15th century).
This photograph was taken from "S. Giacomo Maggiore in Bologna".
by Frederico Cruciani.

Organizing Committee:

H. Arenhövel, Mainz University
P.J. Carlos, CEN Saclay
C. Ciofi degli Atti, INFN Sanità, Roma
C. Coceva, CNEN Bologna
E. De Sanctis, INFN, Frascati
M. Giannini, Genova University
M. Rho, CEN Saclay
A.M. Saruis, CNEN Bologna
B. Schoch, Mainz University

Coordinator: ANNA MARIA SARUIS, CNEN Bologna
Secretariat: M. ANNOVI
Editorial Board: G. ABELLI, R. ABELLI, P. CENNI

Sponsored by
Comitato Nazionale Energia Nucleare, Research Department RIT

TABLE OF CONTENTS

I. THE NN INTERACTION AND THE MANY-BODY PROBLEM

The Paris Nucleon Nucleon Potential, New Developments 1
    R. VINH MAU

Recent Developments in the Bonn Potential 10
    K. HOLINDE

Variational Approach to Many-Body Problems in Finite Nuclei 20
    S. ROSATI

II. COLLECTIVE PHENOMENA

Magnetic Resonances and the Spin Dependence of the Particle-Hole Force 31
    S. KREWALD

A Mean Field Approach to the Description of Nuclear Structure:
Interpretations and Predictions 42
    B. GRAMMATICOS

Many-Body Aspects in Electron Scattering at Intermediate Energy 55
    B. FROIS

The Giant Dipole Resonance and the σ_{-1}, σ_{-2} Photonuclear Sum-Rules 65
    O. BOHIGAS

Electromagnetic Sum Rules 72
    G. ORLANDINI

Status of Nuclear Critical Opalescence 82
    J. DELORME

Inelastic Electron and Proton Scattering to Pion-Like Nuclear Excited
States 93
    H. TOKI and W. WEISE

The Nuclear Δ-Excitation 102
    K. KLINGENBECK

III. BARYONS AND MESONS IN NUCLEI

Three-Body Wave Functions and Electromagnetic Interactions 115
    C. CIOFI DEGLI ATTI, E. PACE and G. SALME

Electrodesintegration of Few-Body Systems 125
    I. SICK

Exchange Currents in the Deuteron 136
    H. ARENHÖVEL

Pion Production off Light Nuclei 148
    J.M. LAGET

Scaling Laws in High Energy Electron-Nuclear Processes 158
    M. CHEMTOB

Total Photonuclear Absorption Cross Section Measurements Below the Pion
Photoproduction Threshold 168
    P.J. CARLOS

Quasi-Deuteron Effects at Intermediate Energies 178
    B. SCHOCH

Direct Mechanism in Knockout Reactions with Real and Virtual Photons 186
    S. BOFFI

IV. QUARKS IN NUCLEI

A Model for Hadrons Based on the MIT Bag Model 196
  J.J. DE SWART

The Chiral Bag Model and the Little Bag 205
  V. VENTO

The Color Degree of Freedom and Multiquark States 212
  H. HØGAASEN

Topological Interpretation of Multiquark States 223
  B. NICOLESCU

Search for Dibaryonic Resonances of Small Mass ($Q_{DB} < 2.3$ GeV) 234
  G. TAMAS

Experimental Search for Dibaryon Resonances 243
  E. BOSCHITZ

V. PRESENT AND FUTURE EXPERIMENTS AND FACILITIES

Present Status of (ee'p) Experiments 251
  S. TURCK-CHIEZE

Future (e,e'p) Experiments at IKO 258
  C. DE VRIES, T. DE FOREST Jr., C.W. DE JAGER, E. JANS, J.H. KOCH,
  L. LAPIKAS, R. MAAS, H. DE VRIES and P.K.A. DE WITT HUBERTS

Saclay Activities in Electro- and Photonuclear Physics at Intermediate
Energies and Future Prospects 277
  C. SCHUHL

Coincidence Measurements with High Energy Electrons 286
  J.S. O'CONNELL

Polarization Experiments 296
  K.H. ALTHOFF

Experiments with Monochromatic and Polarized Photon-Beams 312
  L. FEDERICI, G. GIORDANO, G. MATONE, P. PICOZZA, R. CALOI, L. CASANO,
  M.P. DE PASCALE, M. MATTIOLI, E. POLDI, C. SCHAERF, P. PELFER,
  D. PROSPERI, S. FRULLANI and B. GIROLAMI

Photon Scattering 325
  B. Ziegler

Neutron Spectrometry and γ-Ray Transitions 339
  C. COCEVA

Photonuclear Physics with Synchrotron Radiation 348
  W.M. ALBERICO and A. MOLINARI

VI. THE FUTURE OF NUCLEAR PHYSICS WITH THE ELECTROMAGNETIC PROBE

Medium Energy Physics with C.W. Electron Accelerators 358
  D. DRECHSEL

Future Developments in Pion- and Kaon-Nuclear Physics and the EM Probe:
Examining Electromagnetic Entrails 368
  J.M. EISENBERG

Quarks in Nuclei 375
  C.W. WONG

Some Issues in Photonuclear Physics 385
  C. TZARA

Novel Techniques in Photonuclear Physics 393
  C. SCHAERF

Intermediate Perspectives 403
    T.E.O. ERICSON
List of Participants 410
Author Index 414

THE PARIS NUCLEON NUCLEON POTENTIAL
NEW DEVELOPMENTS

R. VINH MAU

Division de Physique Théorique[*], I.P.N. - Orsay 91406
and LPTPE, Université P. et M. Curie - Paris 75230

Introduction

It is now believed that hadrons are made of subhadronic constituents like quarks, gluons, etc... One is then entitled to demand that a theory of nuclear forces should be derived from the degrees of freedom of those fundamental constituents. During the last two years, several attemps [1] have been made in this direction. The available results, however, are still uncertain.

As a matter of fact, in a nucleus the following situation is somehow expected :
  i) for short internucleon separation distances where the overlap of the nucleons is important, the interaction between the nucleons is certainly dependent on the subhadronic degrees of freedom.
  ii) for large internucleon separation distances, because of <u>confinement</u>, only color-singlet objects, i.e. hadrons themselves, can be exchanged between the nucleons. The interaction is, then, dependent on the hadronic degrees of freedom. It is, in this case, unnecessary and uneconomical to take into account explicitly the quark and gluon degrees of freedom.

In view of these considerations, I think that a fruitful and reasonable approach to the problem of nuclear forces is to proceed from the outer fringe towards the inner core, namely one should
  i) try to determine carefully and accurately the long (LR) and medium range (MR) forces from the hadronic degrees of freedom,
  ii) try to determine the short range (SR) forces from a reliable model bringing the subhadronic (quarks, gluons, etc...) degrees of freedom into play.

The dividing line between the two regions may not be sharply defined and in principle there could be an intermediate region where the interplay of both types of degrees of freedom is significant. Moreover, this dividing line depends on the size of the domain where the quarks and gluons are confined in a hadron (big bag versus little bag).

---

[*] Laboratoire associé au C.N.R.S.

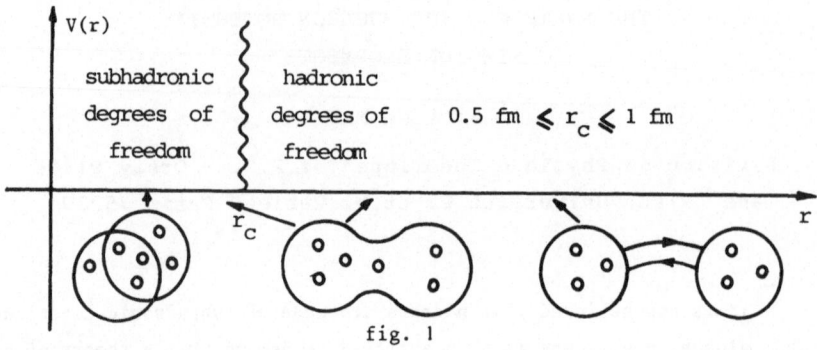

fig. 1

The Paris NN potential is constructed along these lines of thought, namely
   i) the LR + MR part are due to particle or multiparticle exchanges.
  ii) as no definite and reliable results, at present, are available from subhadronic models we prefer to determine, at least provisionally, the SR part from experiment, especially as the wealth of data -some of them of high precision- is now enormous.

The LR + MR part of the interaction

The derivation of the (LR + MR) parts of the Paris NN potential has been reported several times at various places [2]. Due to lack of time and space, I shall not repeat it here. Let me only recall that they include the one-pion, two-pion, and ω and A_1 mesons (as parts of the three-pion) exchange contributions. The two-pion exchange contribution was derived via dispersion relations from pion-nucleon phase shifts and pion-pion S- and P- wave amplitudes. In this way, the properties of mesons and their interactions both amongst themselves and with nucleons- i.e. the degrees of freedom of mesons and of isobars- are taken into account automatically and completely.

Before proceeding further in the determination of the SR forces, we have to check whether our calculation provides a realistic description of the actual (LR + MR) N-N forces. For this, we have compared
   i) the peripheral (J > 2) phase shifts calculated from these contributions with the experimental ones
  ii) the equivalent potential derived from this ($\pi + 2\pi + \omega + A_1$) exchange interaction with the phenomenological potentials.

These comparisons [3] show good agreement with phenomenology. However, they are second order comparisons and we believe that a direct comparison with data is even more meaningful. Recently, the polarization for low energy (6 - 20 MeV) pp and np scatterings has been measured with a high degree of accuracy [4]. As polarization depends only on the P and higher partial waves and as in turn these partial waves are mostly sensitive to the (LR + MR) forces, we have compared our predictions with these data [5]. The results are shown in Figures 2 and 3. The agreement between theory and experiment is very satisfactory.

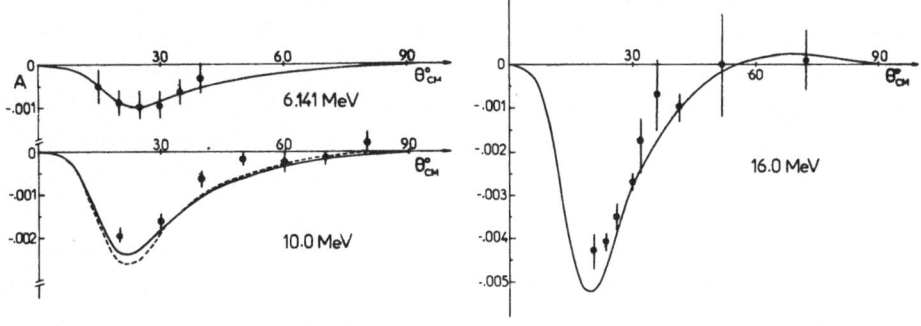

fig. 2

The analyzing power in pp scattering.
The solid lines refer to the Paris potential predictions. The dashed line to the MAW-X phase shift analysis.

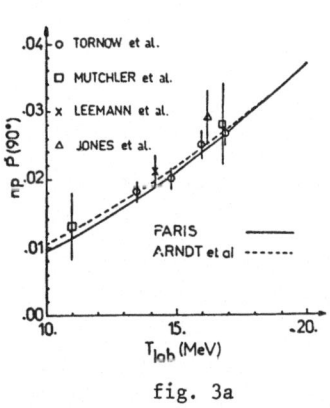

fig. 3a

np polarization at 90° c.m. angle as function of laboratory kinetic energy. The refs. for experimental points can be found in Tornow et al. [4].

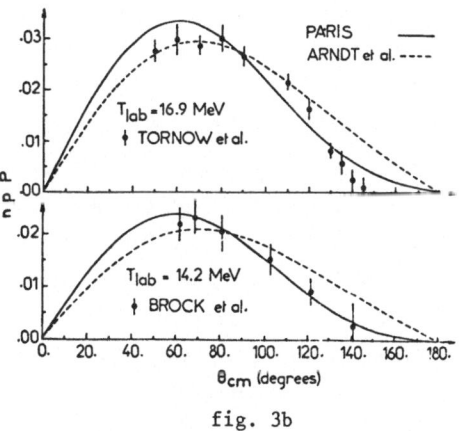

fig. 3b

np polarization at 14.2 and 16.9 MeV as function of c.m. angle.

This success in providing a good understanding of the two-nucleon interaction for long and medium ranges and at the same time a good quantitative fit of the data is noteworthy since the whole scheme is based on properties as fundamental as unitarity, analyticity and crossing.

The SR part of the interaction

i) A simple minded model [6]

To get a clear insight into the physics, we have first designed a very simple model. We wanted to demonstrate that once the LR + MR forces are accurately determined, the SR forces can be described by a model with a few parameters without affecting the LR + MR part. For this purpose, the LR + MR potential is cut off rather sharply at internucleon distances $r \sim 0.8$ fm and the SR ($r < 0.8$ fm) part is described simply by a constant soft core. This introduces the minimum number (five) of adjustable parameters corresponding to the five components (central, spin-spin, tensor, spin-orbit and quadratic spin-orbit) of the potential for each isospin state. We had already found that the central component of our LR + MR potential has a weak but significant energy dependence and that this energy dependence is, in a very good approximation, linear. Because of this, we expect also an energy dependence, presumably stronger, in the SR part. This reflects some kind of non-locality of the potential. Effectively, fitting the empirical phase shifts required an energy-dependent core for the central potential, the energy dependence being again linear, introducing therefore an additional parameter, the slope of the energy dependence. The proposed SR part is then determined by fitting all the known phase shifts (J<6) up to 350 MeV and the deuteron parameters. Although the number of free parameters is small (six in total for each isospin state) the quality of the fit is very good : The χ^2/degree of freedom values are 2.5 for pp scattering and 3.7 for np scattering, as good as those given by the best phenomenological potentials which contain many more free parameters.

ii) A more refined and practical parametrization [7]

The above SR model confirms also the idea that little freedom is allowed for the determination of the SR core once the LR + MR potential is well given. However the explicit expression of the resulting potential is not very convenient for practical use in many-body calculations : i) the energy dependence of the potential which can be treated naturally in the two-body scattering case, may be ill-defined in many-body systems, ii) the theoretical LR + MR potential presents itself as a dispersion integral ; iii) the presence of a sharp cutoff may cause troubles in numerical work. On the other hand, the SR part of the resulting potential was obtained in fitting the empirical phase shifts. A phase shift representation is useful to describe the overall properties of the interaction. It does not however lead necessarily to a unique solution, especially in the case of fixed-energy analysis. We therefore believe (and would like to emphasize) that a direct fit of experimental data rather than phase shifts is more meaningful for an accurate quantitative test of theoretical models.

Because of all these reasons, we have proposed [7] a parametrization of the potential which is convenient enough to facilitate its use in the many-body calculations. We adopted a unique analytical expression for the complete potential, namely a discrete sum of Yukawa-type terms which has the advantage that their forms are simple in both configuration and momentum spaces. The energy dependence has been transformed into a p^2 dependence which can be handled without ambiguity. The parameters of this parametrization are determined by fitting the shapes of the previous model, the phase shifts and the data themselves. The search on the core parameters was carried out with the best fit of the MAW phase shifts as a starting point. They are obtained by fitting a set of 913 data points between 3 and 330 MeV for pp scattering and a set of 2239 data between 13 and 350 MeV for np scattering. During the last two years, a great number of new measurements have been performed both at low and intermediate energies [8]. The new np data were used to improve the determination of the core parameters. The new pp data were not available when our fit was performed and hence not used for the determination of the core parameters. In this case, our results are predictions [9]. In Figures 4 and 5 are shown some examples of the fit.

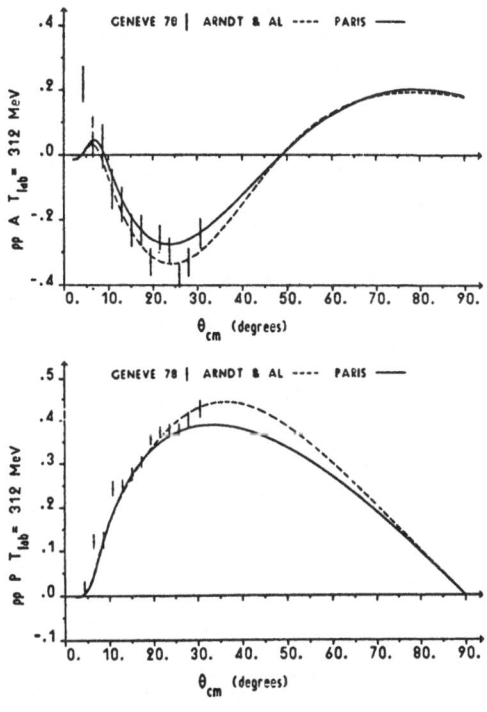

fig. 4

pp polarization and Wolfenstein parameter A.

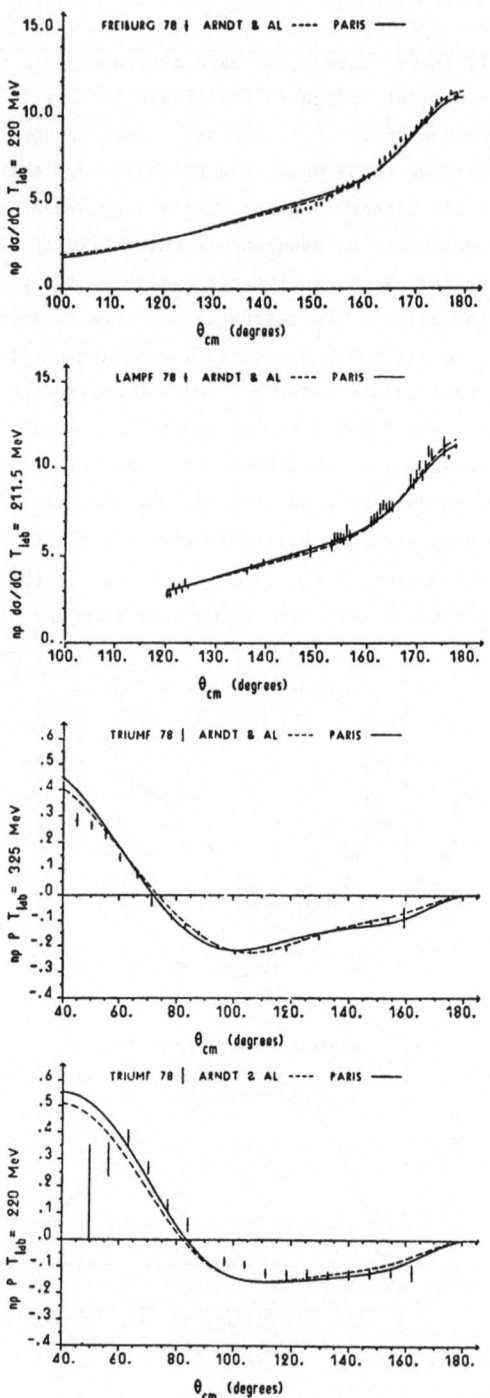

fig. 5
np differential cross section and polarization.

The χ^2/degree of freedom attains very satisfactory values : 1.99 for pp scattering and 2.17 for np scattering to be compared, for reference, with 1.33 and 1.80 for the Arndt et al. [10] phase shift analysis and with 4.76 and 9.99 for the Reid soft-core potential. This comparison is a kind of warning. It demonstrates clearly that two different potentials can fit equally well the phase shifts without fitting equally well the data.

We have restricted ourselves here to energies not too far above the π production threshold as we were using a non-relativistic potential picture. The analysis of higher energy data (e.g. 350 MeV to 800 MeV) is currently under investigation with the same dynamical input but in a relativistic scheme. In this regard, let me recall that our uncorrelated two pion-exchange provides inelasticity parameters that are expected to give a good description of the high partial waves.

Finally, comparison of the theoretical ($\pi + 2\pi + \omega + A_1$) exchange potential with the full Paris potential is shown in Figure 6 for two examples, the triplet central potential (T=1) and the tensor potential (T=0). Deviations of the theoretical (LR + MR) part from the full potential occur only for distances r < 1 fm.

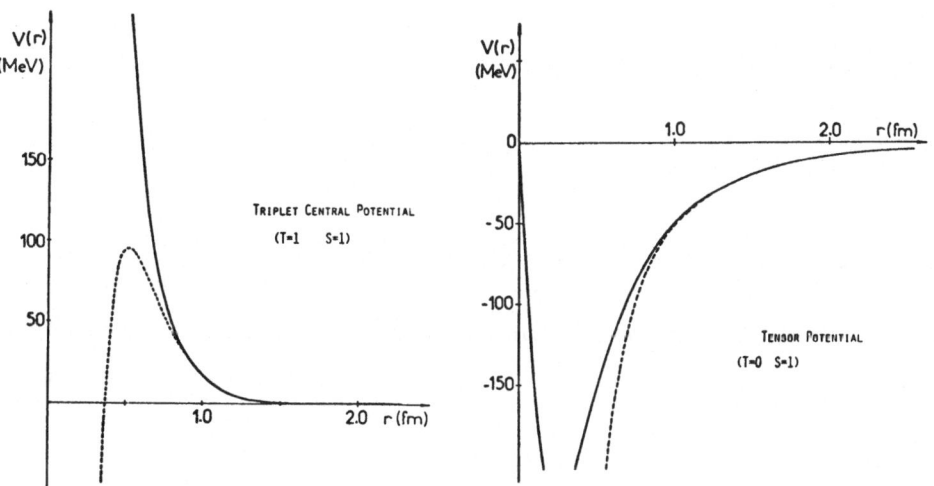

fig. 6

The solid lines refer to the complete potential, the dashed lines to the theoretical ($\pi + 2\pi + \omega + A_1$) potential.

This comparison along with that of our results with the Reid soft core potential results indicates what the two-pion exchange contribution (TPEC) is necessary for the medium range forces to get a good fit of the data. This fact is now recognized also by the advocates of the OBE models and recently inclusion of the TPEC in the OBE calculations have been carried out [11]. However, we believe that superimposing the TPEC on the OBE models may have shortcomings : besides double counting problems there is a conceptual inconsistency since in the OBE models by construction one should not have closed loops in the potential. Last but not least, the use of non covariant perturbation theory could lead to well known ambiguities encountered already in the early 50's.

In summary, I would like to emphasize again that a severe test of theoretical models can be achieved only by a direct comparison of the theoretical results with the experimental data rather than a comparison of phase-shifts. The ability of the Paris potential to provide a very good fit to the present NN world set of data up to 350 MeV in conjunction with the results shown in Figure 6 suggests that we have gained during the last few years a theoretical understanding of the NN interaction for interdistances larger than 0.8 - 1 fm. This also suggests that any ultimate theory of strong interaction should recover somehow the same results in that region.

The work described here has been carried out during several years by the Paris NN group. J. Côté, W.N. Cottingham, M. Lacombe, B. Loiseau, P. Pires, J.M. Richard and R. de Tourreil have contributed to this teamwork.

REFERENCES

1) C. de Tar, Phys. Rev. D17, 323 (1978). D. Liberman, Phys. Rev. D16, 1542 (1977). M. Harvey, Chalk River preprint CRNL-TP-80-JAN-1.

2) For a review, see for example R. Vinh Mau, The Paris N-N Potential in "Mesons in Nuclei" edited by M. Rho and D. Wilkinson (North-Holland, Amsterdam, 1979).

3) R. Vinh Mau, J.M. Richard, B. Loiseau, M. Lacombe and W.N. Cottingham, Phys. Lett. 44B, 1 (1973).
    W.N. Cottingham, M. Lacombe, B. Loiseau, J.M. Richard and R. Vinh Mau, Phys. Rev. D8, 800 (1973).

4) G. Bittner and W. Kretschner, Phys. Rev. Lett. 43, 330 (1979)
    W. Tornow, P.W. Lisowski, R.C. Byrd and R.L. Walter, Nucl. Phys. A340, 34 (1980)

5) J. Côté, P. Pirès, R. de Tourreil, M. Lacombe, B. Loiseau and R. Vinh Mau, Phys. Rev. Lett. 44, 1031 (1980) ;
    J. Côté, M. Lacombe, B. Loiseau, P. Pirès, R. de Tourreil and R. Vinh Mau, 5th Int. Symp. on Polarization Phenomena in Nuclear Physics, Santa Fe 1980, AIP Conference Proceedings, to be published.

6) M. Lacombe, B. Loiseau, J.M. Richard, R. Vinh Mau, P. Pirès and R. de Tourreil, Phys. Rev. D12, 1495 (1975).

7) M. Lacombe, B. Loiseau, J.M. Richard, R. Vinh Mau, P. Pirès and R. de Tourreil Phys. Rev. $\underline{C21}$, 861 (1980)

8) See reference [9] and references cited therein.

9) M. Lacombe, B. Loiseau, R. Vinh Mau, J. Côté, P. Pirès and R. de Tourreil, Preprint IPNO/TH 80-60 (1980).

10) R.A. Arndt, R.H. Hackman and L.D. Roper, Phys. Rev. $\underline{C15}$, 1002 (1977) and R.A. Arndt, private communication.

11) K. Holinde, Talk at this Symposium and references cited therein.

RECENT DEVELOPMENTS IN THE BONN POTENTIAL

K. Holinde

Institut für Theoretische Kernphysik der Universität Bonn
Nußallee 14-16, D-5300 Bonn, W.-Germany

The quality of various approximations, which are necessary to obtain an analytic expression of mesontheoretic NN potentials in r-space, is studied. First, we calculate the effect of such approximations on NN scattering phase shifts and the deuteron, starting from a one--boson-exchange version of the Bonn potential in momentum space. Second, new results for the noniterative 2π-exchange diagrams involving $N\Delta$- as well as $\Delta\Delta$ intermediate states are presented, derived from noncovariant perturbation theory. Current prescriptions for replacing all time-orderings of those diagrams by pion-range transition potentials (which can be handled in r-space) are tested.

INTRODUCTION

In the beginning of the development of mesontheoretical nucleon--nucleon potentials, i.e. in the sixties, such potentials [1-5] were usually represented in r-space, i.e. as an analytic function of the relative distance r of the two nucleons. The reason was essentially the following: one obtains in this way a representation in terms of central, spin-spin, tensor and spin-orbit contributions, suggested by phenomenological descriptions of the two-nucleon problem. In fact, historically, the existence of the ω-meson was predicted in order to account for the empirically observed strong spin-orbit term. Consequently, a direct comparison of these terms with those of phenomenological potentials [6] was possible, which gave a first idea of the quality of the theoretical model. For a detailed and quantitative analysis, of course, one has to evaluate NN scattering phase shifts or NN observables and compare them with the empirical values.

However, an analytic form in r-space can only be obtained after certain approximations: in meson theory, the nucleon-nucleon interaction originates from a superposition of relativistic meson-exchange contributions to the field-theoretic S-matrix, which is formulated in momentum space, as function of $\vec{q}$ and $\vec{q}'$, where $\vec{q}(-\vec{q})$ and $\vec{q}'(-\vec{q}')$ are the incoming and outgoing momenta of nucleon 1 (2) in the c.m. system. A direct Fouriertransformation would yield a function of $\vec{r}$ and $\vec{r}'$, i.e. a strongly nonlocal expression. Moreover, the transformation cannot be performed analytically.

In order to avoid the approximations which are necessary for going analytically from q-space to r-space, one alternatively, after 1970, constructed mesontheoretical potentials directly in momentum-space [7]. Some meson coupling constants, which were fixed by fitting to the NN scattering data, were different from those used in r-space models. However, this discrepancy is only partly due to the approximations made; another part comes from the use of different sets of mesons and different cutoff procedures. Definite conclusions about the quality of the approximations can, therefore, not be made by simply looking at the differences in the meson-nucleon coupling constants. What is required is a detailed numerical study of the effect of the approximations on NN scattering data in order to decide how much of the physics gets lost. This will be the subject of the first part of my talk. In the second part, I will start by presenting results we obtained recently for the noniterative 2π-exchange diagrams involving $N\Delta$- as well as $\Delta\Delta$ intermediate states, in the framework of noncovariant perturbation theory. Based on these results, the quality of current prescriptions for replacing all time-orderings (iterative + noniterative) of those diagrams by twice-iterated transition potentials (which can be handled in r-space) will be discussed.

Let me note finally that the topic of the present paper has nowadays become quite important again because there is a trend back to r-space models for the following reason: the range of validity of meson theory and a theory in terms of the basic constituents of nucleons and mesons, i.e. the quarks, is believed to be more suitably defined in r-space. Meson theory of the nucleon-nucleon interaction should break down for small distances (r < 1fm) and should be replaced, in this region, by results from quark calculations.

EFFECT OF APPROXIMATIONS IN MESONTHEORETIC POTENTIALS ON NN DATA
A mesontheoretic NN potential is built up by a superposition of contributions from exchanges of the various mesons and has the following structure in momentum space (omitting spin indices)

$$V^{(0)} = V(\vec{k},\vec{p};E',E)$$

where $\vec{k} = \vec{q}'-\vec{q}$, $\vec{p} = \frac{1}{2}(\vec{q}'+\vec{q})$ and $E = \sqrt{q^2+m^2}$, $E' = \sqrt{q'^2+m^2}$, m is the mass of the nucleon. In order to obtain an analytic expression in r-space, several approximations are usually done step by step:
1. on-shell approximation: $E' = E$
2. nonrelativistic approximation: $E = m + \frac{p^2}{2m} + \frac{k^2}{8m}$
   (keeping p^2-terms only in first order, i.e. in the central part).
3. neglect of all p^2-terms (which are complicated in r-space)
4. static limit: $E = m$
5. omission of $\Omega_{\sigma L} = \vec{\sigma}_1 \cdot \vec{k} \times \vec{p} \vec{\sigma}_2 \cdot \vec{k} \times \vec{p}$

This suggests to define $V^{(i)}(\vec{k},\vec{p})$, $i = 1,...,5$, i.e. the meson-theoretic potentials in the various approximation steps. Our numerical studies are based on a specific one-boson-exchange version (HM2 [8]) of the Bonn potential. All partial wave phase shifts are calculated using the potentials $V^{(i)}$, keeping the parameter set and the form factor the same. A typical example is the $^1D_2$ phase shift, shown in fig. 1. Whereas the first two approximations are very good, the omission of all p^2-terms and the $\Omega_{\sigma L}$-term has a drastic effect and must be considered to be a bad approximation to the original expression.

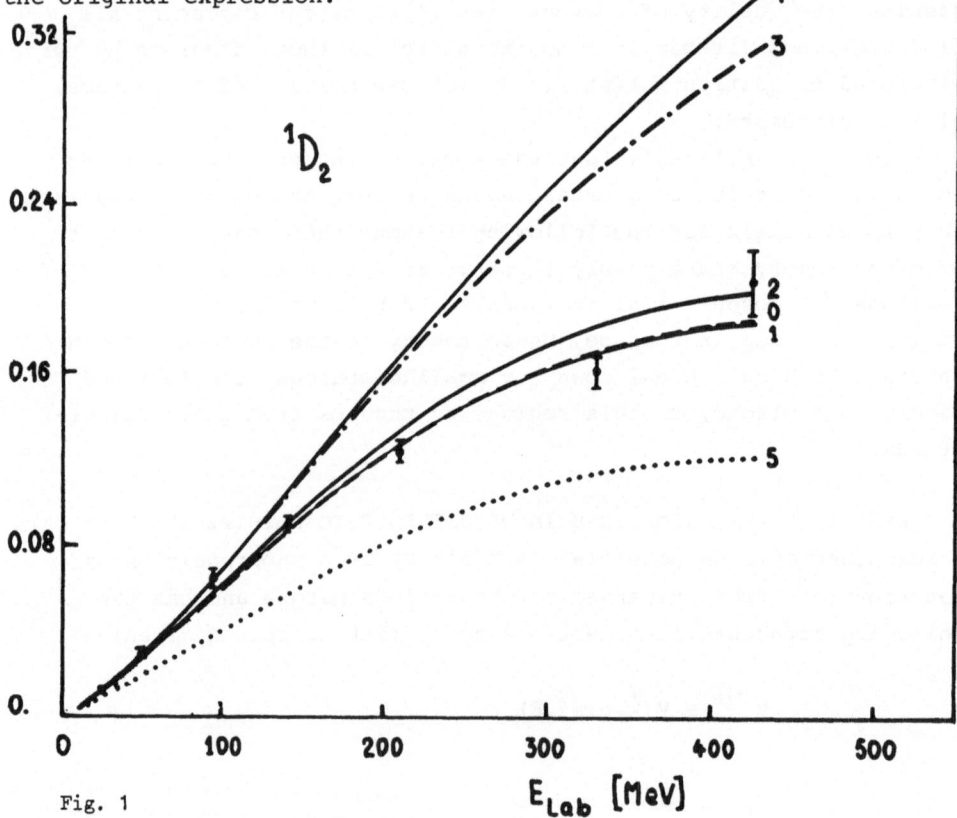

Fig. 1

For case 2 (and of course even more in case 1) it is possible to get again a good fit of the empirical phase shifts by a slight change ($\approx$ 10%) of the original meson parameters in $V^{(2)}$. This is not possible for the other cases.

It is encouraging that just the second approximation $V^{(2)}$, which can be transformed analytically into r-space (in contrast to $V^{(1)}$) is still a good approximation, in the sense that a quantitative fit of the phase shifts can be regained by a slight change of the parameters.

A study of the effect of different approximations of the quadrupole moment Q and the D-state probability P_D of the deuteron should show the influence of these approximations on the NN tensor force, a quantity which has enormous influence on all sorts of nuclear structure results and, therefore, must be determined as accurately as possible.

It turns out that the second approximation ($V^{(2)}$), being reasonable in NN scattering, leaves also the quadrupole moment unchanged, but considerably enhances the D-state probability. Thus, the full nonlocality structure of the momentum-space potential $V^{(0)}$ makes it possible to predict a relatively large quadrupole moment (which seems to be required by the empirical value inspite of possible meson-exchange-current corrections) and relatively small D-state probabilities (which also seems to be required by some few-body reactions [9]), i.e. to have a tensor force which is strongly suppressed in the inner part only. This flexibility is reduced by removing part of the nonlocal effects, i.e. by using $V^{(2)}$.

We feel that this is partly the reason why the (r-space) Paris-potential [10], which keeps p^2- and $\Omega_{\sigma L}$-terms and in this sense corresponds to $V^{(2)}$, has a relatively small Q (= 0.279 fm^2) together with a relatively large P_D (= 5.77%). For comparison, our (momentum-space) version HM2 has a larger Q, but on the other hand a considerably smaller P_D than the Paris-potential [10]. However, not all of the discrepancy is due to the inclusion of the full nonlocality structure in HM2, since the refitted $V^{(2)}$ ($V^{(2)}_{mod}$) has the same value for Q as the Paris potential, but still a much smaller value for P_D (= 4.44%). Obviously, $V^{(2)}_{mod}$ has a weaker tensor force than the Paris-potential, at least in the deuteron channel. This is demonstrated in fig. 2. The solid line shows the tensor force $V_T(r)$ of $V^{(2)}_{mod}$ (omitting the form factor) for isospin-zero states. The dashed-dot curve denotes the theoretical result of the (dispersion-

-theoretic) Paris-potential [10], which is derived from $(\pi+2\pi+\omega)$--exchange. A parametrization of this result in terms of simple Yukawa functions together with a regularization at the origin requires to modify the potential for $r \leq 1$fm (dashed curve), if a quantitative fit to the two-nucleon data is to be obtained, see ref. [10]. Obviously, the tensor force of $V_{mod}^{(2)}$ is much weaker in the inner region than both the 'theoretical' and 'empirical' curve of the Paris-potential.

Fig. 3 shows the analogous curves for the tensor force in isospin-one states. Here, the 'empirical' curve of the Paris--potential deviates from the theoretical result already at $r = 1.5$fm in order to obtain a good fit to the data. It is interesting, however, that the 'theoretical' OBE-potential $V_{mod}^{(2)}$ agrees with the empirical curve up to $r \simeq 1$fm (from outside).

Both figures (2 and 3) show that empirical NN data do not at all fix the potential inside 1fm, since $V_{mod}^{(2)}$ as well as the Paris--potential obtain a good fit to the NN phase shifts. This is not surprising, since e.g. the empirical curve should depend on the kind of parametrization of the Paris-potential. This implies that discrepancies inside 1fm between the theoretical and empirical curve in some components (central, tensor and so on), see ref. [10], do not prove unambiguously that meson theory predicts the wrong result inside 1fm, although, because of the quark structure, its basis is surely not so firm anymore in that region. We refer to the paper of Holinde and Mundelius [11] for further results.

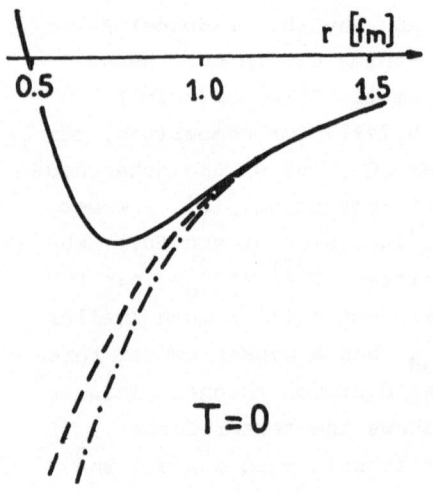

Fig. 2

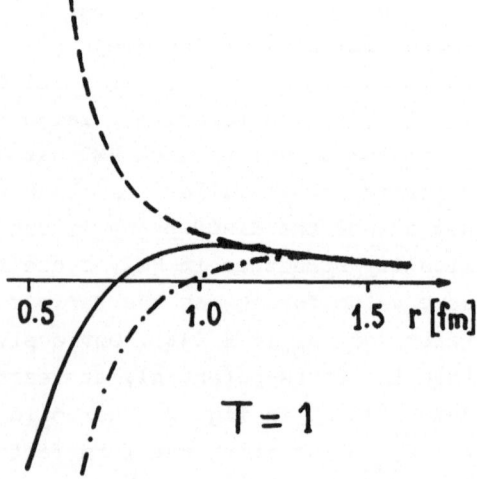

Fig. 3

ISOBAR CONTRIBUTIONS TO THE NN INTERACTION

In OBE models of the NN interaction, the intermediate-range attraction is described by the exchange of a scalar isoscalar σ-meson. This contribution effectively replaces the ($J^P=0^+$, I=0)- -part of the whole 2π-exchange minus the twice-iterated one-pion- -exchange (which is already included in the scattering amplitude by iterating OPEP). On the other hand, dispersion-theoretic methods obtain this contribution by using empirical πN- (and ππ) data and performing an analytic continuation.

However, both methods treat this contribution as part of the NN potential (of essentially scalar type), i.e. as a lowest-order contribution. Thus, the modification of the 2π-exchange in the medium, due to Pauli- and dispersion effects, which arise in a nuclear many-body theory, are suppressed, apart from the nucleon box diagram, which is treated as a second iteration of OPEP. These many-body effects should be important, especially in dense systems like nuclear matter or, even more, neutron stars.

A realistic treatment of such modifications suggests a much more explicit dynamical scheme, which starts from a field-theoretic Hamiltonian containing as basic ingredient not a potential, but NN- and NΔ-vertices, uses noncovariant perturbation theory and stays in momentum space, for details see ref. [12].

We have recently studied the isobar box diagrams (1-4 of fig. 4 which shows all time-orderings for the NΔ contribution) with NΔ- and ΔΔ-intermediate states, in NN scattering [13] and in nuclear matter [14]. We found that these diagrams provide roughly 30% of the inter- mediate-range attraction. Pauli- and dispersive effects reduce this contribution by as much as 30% in nuclear matter.

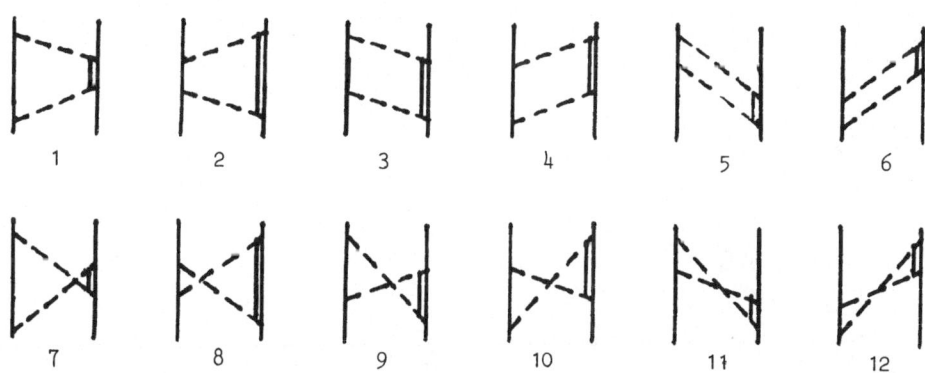

Fig. 4

Here I want to present some results we obtained recently for the noniterative, i.e. stretched-box (5,6 of fig. 4) and crossed-box (7-12 of fig. 4) diagrams, for NΔ- [15,16] as well as ΔΔ intermediate states [16]; namely, matrix elements $V(q',q|q_o)$ are shown for the $^1S_0$ partial wave and for $q = q_o = 250$ MeV as function of q'. Furthermore, $g^2_{NN\pi} = 14.4$ and $f^2_{N\Delta\pi} = 0.23$ suggested by the quark model. A monopole form factor is used at all NN and NΔ vertices with a cutoff-mass $\Lambda_\pi = 1$ GeV.

Fig. 5 shows the contributions of the (iterative) box (solid line), stretched-box (dashed line) and crossed-box diagrams (dash--dot line), omitting isospin factors. It is seen that the ΔΔ contribution is considerably larger than the NΔ contribution and can by no means be neglected. Due to its short range, however, its importance is reduced in higher partial waves. Furthermore, the noniterative diagrams are as important as the iterative ones.

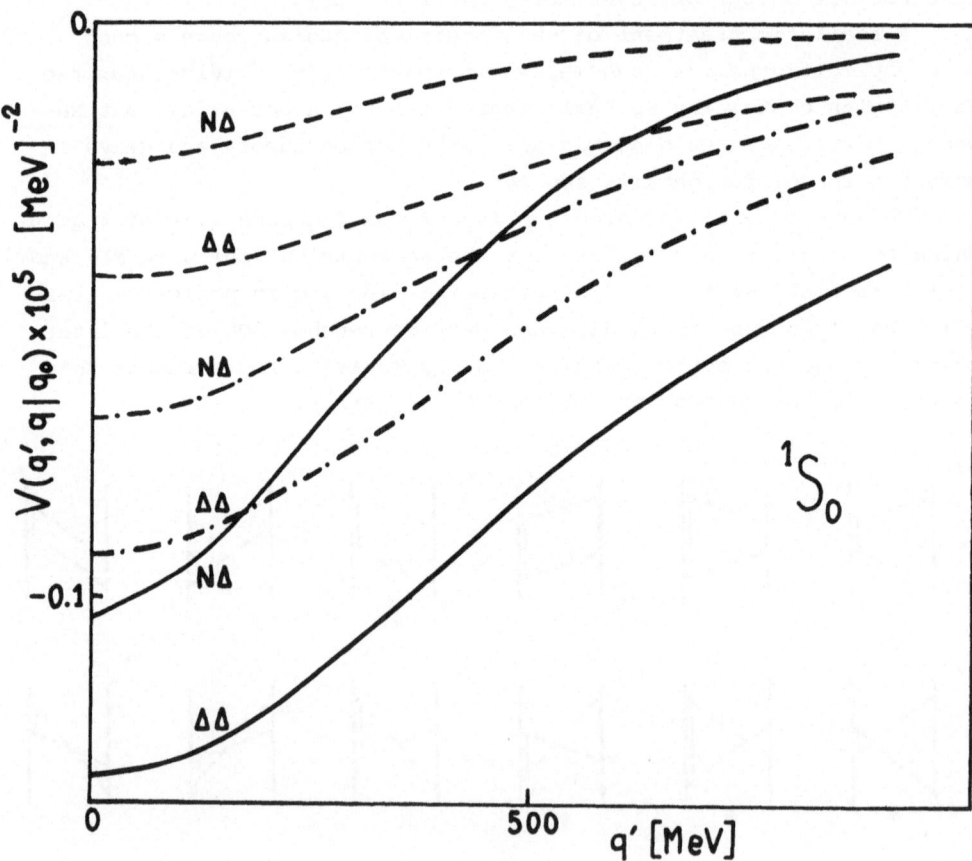

Fig. 5

Fig. 6 demonstrates that, in $^1S_0$, the use of transition potentials with simple pion-range propagators $((\vec{q}'-\vec{q})^2+m_\pi^2)$, dashed line) is a good approximation for the total NΔ contribution (diagrams 1-12 of fig. 4, solid line), but considerably overestimates the ΔΔ contribution. Again, isospin factors have been neglected. Note, however, that the approximation is poor also for NΔ in important higher partial wave states.

Including isospin factors $a \pm b \vec{\tau}_1 \cdot \vec{\tau}_2$ (+ for box, - for crossed-box diagrams, and $a = 2$, $b = \frac{2}{3}$ for NΔ-, $a = \frac{4}{3}$, $b = -\frac{2}{9}$ for ΔΔ-states), we can take over the above conclusions for the isoscalar piece of the sum of box and crossed-box diagrams, since there both terms have the same sign. On the other hand, box and crossed-box diagrams have opposite sign for the isovector $(\vec{\tau}_1 \cdot \vec{\tau}_2)$ piece and there is strong concellation. Consequently, the use of transition potentials of pion range also for the isovector piece, i.e. replacing the exact result by $(2 + \frac{2}{3}\vec{\tau}_1 \cdot \vec{\tau}_2)$ times the iterated contribution of pion range, (which is usually done in a coupled-channel treatment), overestimates the isovector exchange part enormously and leads, for example, to a vanishing NΔ contribution in isospin-zero channels.

Obviously, an improved approach would be to neglect the isovector piece, first proposed by the Stony Brook group [17], i.e. to approximate the exact result by a times the iterated pion-range contribution. This implies that the total contribution is treated

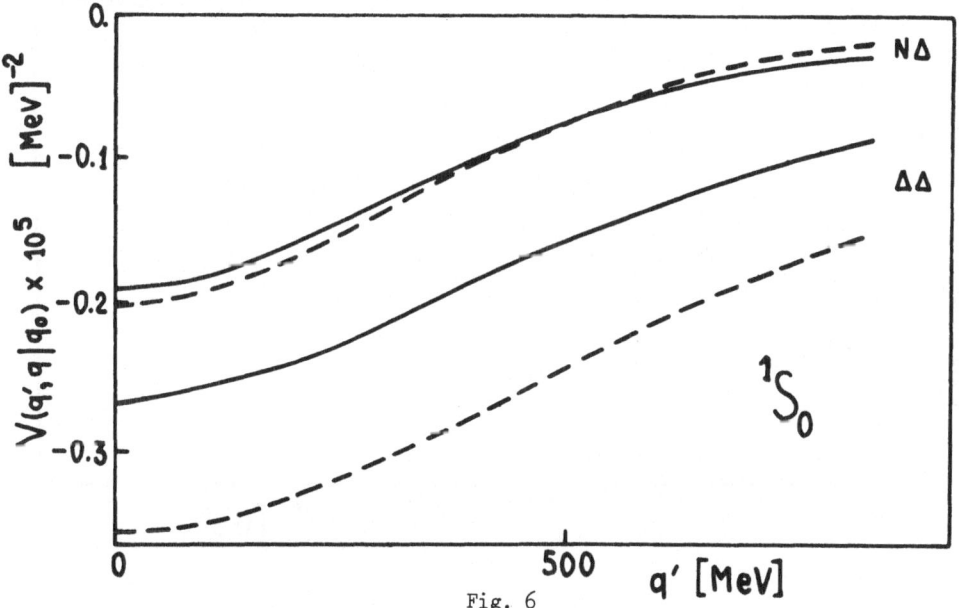

Fig. 6

to be exactly isoscalar. Fig. 7 shows that, in the $^1S_0$ state, this approximation (dashed line) is quite reasonable for NΔ but again strongly overestimates the ΔΔ contribution. Moreover, in $^3S_1$, also the NΔ contribution is considerably overestimated. In higher partial waves, the use of simple pion-range transition potentials overestimates the exact result, too. The reason is that the realistic contributions, being of shorter range due to the inclusion of recoil terms, are much more suppressed in these partial waves than contributions with pion range. Thus we believe that the prescription of dropping the isovector part in the transition potentials, though improving the situation, cannot account for the field-theoretic isobar contributions in a quantitative way. Moreover, such an approximation destroys the detailed structure of these terms, which plays an essential role in nuclear structure.

The same is true if we replace the sum of all NN, NΔ and ΔΔ diagrams (with isospin factors) by the corresponding sum of twice--iterated pion-range potentials (including here the 'iterative' isospin factor), as advocated by Smith and Pandharipande [18]. It turns out that this prescription overestimates the exact result considerably in all partial waves, at least by 20%.

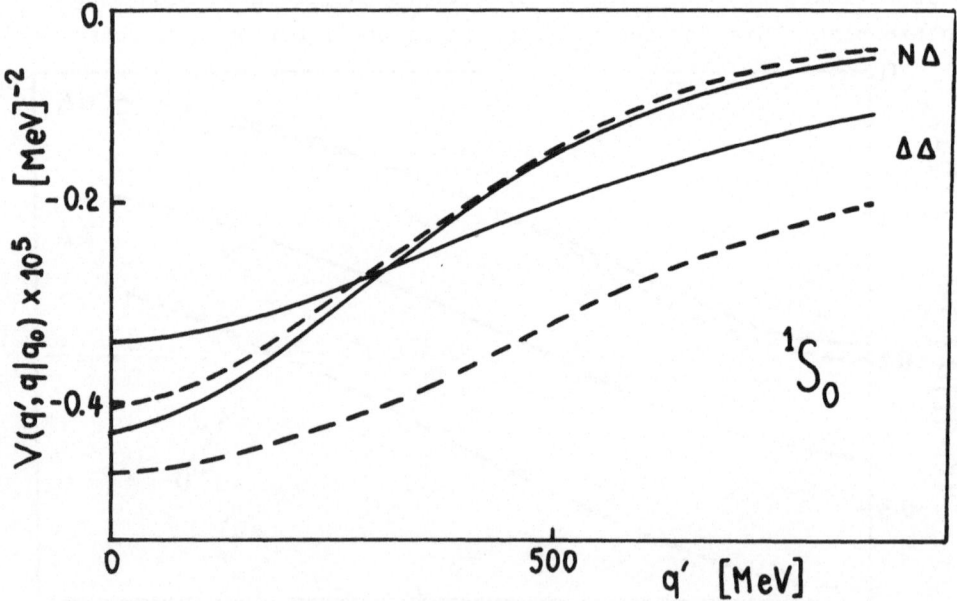

Fig. 7

CONCLUDING REMARKS

We have shown that, at least in the OBE frame, the nonrelativistic approximation of the NN potential, i.e. expanding E,E' and keeping nonlocal p^2-terms in the central part only, is, perhaps surprisingly, good regarding the fit of NN scattering data. However, the necessary approximations modify the tensor force in the medium-range region, leading to larger deuteron D-state probabilities, which has important consequences in nuclear structure.

Furthermore, realistic nuclear structure calculations require an explicit description of the intermediate-range attraction. It is shown here that the contributions of noniterative diagrams involving $N\Delta$ and $\Delta\Delta$ intermediate states are as important as the corresponding iterative diagrams. It turns out that the replacement of all time-orderings by pion-range transition potentials cannot be considered as a valid approximation to the true result.

Let me finish by making the following statement: I strongly feel that there is not the best potential. Its quality depends very much on what one wants to do with it.

REFERENCES

[1] R.A. Bryan and B.L. Scott, Phys. Rev. 135 (1964) B434
[2] R.A. Bryan and B.L. Scott, Phys. Rev. 164 (1967) 1215
[3] R.A. Bryan and B.L. Scott, Phys. Rev. 177 (1969) 1435
[4] T. Ueda and A.E.S. Green, Phys. Rev. 174 (1968) 1304
[5] K. Erkelenz, K. Holinde and K. Bleuler, Nucl. Phys. A139 (1969) 308
[6] T. Hamada and I. Johnston, Nucl. Phys. 34 (1962) 382
[7] For a review see K. Erkelenz, Phys. Reports 13C (1974) 191
[8] K. Holinde and R. Machleidt, Nucl. Phys. A256 (1976) 479
[9] H. Arenhövel and W. Fabian, Nucl. Phys. A282 (1977) 397
[10] M. Lacombe, B. Loiseau, J.M. Richard, R. Vinh Mau, J. Côté, P. Pirès and R. de Tourreil, Phys. Rev. C21 (1980) 861
[11] K. Holinde and H. Mundelius, Approximations in OBE-potentials and their Effect on Two-Nucleon Data, preprint, submitted for publication in Nucl. Phys.
[12] K. Holinde, Nucl. Phys. A328 (1979) 439; Physics Reports, in press
[13] K. Holinde, R. Machleidt, M.R. Anastasio, A. Fäßler and H. Müther, Phys. Rev. C18 (1978) 870
[14] M.R. Anastasio, A. Fäßler, H. Müther, K. Holinde and R. Machleidt, Phys. Rev. C18 (1978) 2416
R. Machleidt and K. Holinde, Role of the Single-Particle Potential in Nuclear Matter Calculations Including Mesonic and Isobar Degrees of Freedom, Nucl. Phys. A, in press
[15] K. Holinde, R. Machleidt, A. Fäßler, H. Müther, M.R. Anastasio, Noniterative Isobar Diagrams and Their Effect in NN Scattering, preprint, submitted for publication in Phys. Rev. C
[16] X. Bagnoud, K. Holinde and R. Machleidt, Isobar Contributions to the NN Interaction, in preparation
[17] J.W. Durso, M. Saarela, G.E. Brown and A.D. Jackson, Nucl. Phys. A278 (1977) 445
[18] R.A. Smith and V.R. Pandharipande, Nucl. Phys. A256 (1976) 327

VARIATIONAL APPROACH TO MANY-BODY PROBLEMS IN FINITE NUCLEI

S. Rosati

Istituto di Fisica dell'Università, Pisa, Italy
Istituto Nazionale di Fisica Nucleare, Sezione di Pisa, Italy

Abstract. The present status of the variational approach to finite nuclei, based on correlated wave functions is briefly sketched. A general outline of the problem, and some of the results recently obtained for ligth nuclei are discussed. The variational technique is applied to calculate the relevant coefficients of the nuclear mass formula and estimates of the ground state energies of ^{16}O and ^{40}Ca are also given.

1. Introduction

At present, three basically different theories are used in studying strongly interacting Fermi systems. The perturbative approach, either in the form of the BBG theory[1] or in the more recent version of the coupled cluster equations[2], is the most carefully investigated up to now. Nevertheless, the numerical approach[3] to the study of Fermi systems is presently undergoing a rapid development, and it has proved to be able to give the exact solution in some rather simplified situations. The variational theory has encountered an increasing attention in recent times. With respect to both perturbative and numerical approaches, the variational methods have the advantage that they are suitably applied to a wider class of problems. However, difficulties arise from the large sets of cluster terms which must be evaluated, and by the choice of the trial wave functions, which should be appropriately flexible. In the most of the cases, the variational studies of Fermi systems are based on a particle-particle interaction described by a (two-body) potential energy, and the non-relativistic limit is adopted. The Hamiltonian of the system is

$$H = -\frac{\hbar^2}{2m}\sum_{i=1}^{A}\vec{\nabla}_i^2 + \sum_{i<j=1}^{A} v(i,j) , \qquad (1.1)$$

where A is the number of particles. The trial wave function is written in the form

$$\Psi_V(1,..,A) = F(1,..,A)\, \Phi(1,..,A) \,. \tag{1.2}$$

2.1. <u>Choice of the model function Φ</u>. The model function Φ contains the statistics and excitation properties of the A-particle system and it may describe adequately the configurations with the particles at large or medium separations. The correlation factor $F(1,..,A)$ takes care of the important correlations when two or more particles are at small distances. However, the factor F can also include correlations of medium or long ranges so as to correct unsatisfactory behaviours of the model function Φ. For infinite, homogeneous and translationally invariant systems (for instance, nuclear matter) the model function is usually chosen to be a Slater determinant of free single particle wave functions

$$\Phi(1,..,A) = \det\left|\phi_{\alpha_i}(i)\right| \,, \tag{2.1}$$

with

$$\phi_{\alpha_i}(i) = \eta_{p(\alpha_i)}(i)\, \Omega^{-\frac{1}{2}}\, \exp(i\vec{k}_{\alpha_i}\cdot\vec{r}_i) \,. \tag{2.2}$$

The spin-isospin functions $\eta_p(i)$, $p=1,..,\nu$, constitute a complete orthonormalized set. The plane waves satisfy the periodicity conditions on the walls of a large cube of volume Ω which becomes infinite when the thermodynamic limit is performed. The allowed momenta fill a Fermi sphere of radius $k_F = (6\pi^2 \rho/\nu)^{1/3}$, where ρ is the density.

To describe the ground and the excited states of nuclei, the function Φ can be chosen as a shell-type wave function and it can be written as a superposition of Slater determinants of shell-model single particle wave functions. The theory which can then be developed is known as the Correlated Basis Functions method[4,5].

2.2. <u>Choice of the correlation factor F</u>. The simplest choice for F is the so called state-independent Jastrow ansatz

$$F(1,..,A) = \prod_{i<j=1}^{A} f(r_{ij}) \,. \tag{2.3}$$

The two-body correlation factor $f(r)$ heals to unity for large r values and it goes to zero when r approaches the hard-core radius. Moreover, f(r) must be conveniently structured at small and medium values of r. For the nuclear matter problem, a state-independent correlation factor allows for sufficiently accurate descriptions only for central interparticle interactions. In the case of nuclei, the most of the available approaches have been based on the form given in eq.(2.3). State-dependent two-body correlation factor can be written in the form

$$f(i,j) = \sum_p f^p(r_{ij}) \, O^p(i,j) , \qquad (2.4)$$

where $f^p(r_{ij})$ are functions of the interparticle distance r_{ij} only, and $O^p(i,j)$ are conveniently chosen operators. As an example, the operators which have been considered - all of them or only a part - in many nuclear matter calculations are

$$O^p(i,j) \equiv 1 , \, \vec{\sigma}_i \cdot \vec{\sigma}_j , \, \vec{\tau}_i \cdot \vec{\tau}_j , \, (\vec{\sigma}_i \cdot \vec{\sigma}_j)(\vec{\tau}_i \cdot \vec{\tau}_j) , \, S_{ij} ,$$
$$S_{ij}(\vec{\tau}_i \cdot \vec{\tau}_j) , \, (\vec{L} \cdot \vec{S})_{ij} , \, (\vec{L} \cdot \vec{S})_{ij}(\vec{\tau}_i \cdot \vec{\tau}_j) , \qquad (2.5)$$

where S_{ij} is the tensor operator, $\vec{L}$ the relative angular momentum and $\vec{S}$ the total spin of the pair. When r increases, the function $f^1(r)$ (see eq.(2.4)), goes to the unity, whilst the other functions $f^p(r)$, with p>1, go to zero. The expression (2.4) for f(i,j) should be well suited in the cases of NN potentials expressed in terms of the operators O^p. Since f(i,j), as given by eq.(2.4), does not commute in general with f(i,k), the correlation factor F is written in the form

$$F(1,..,A) = S \left(\prod_{i<j=1}^{A} f(i,j) \right) , \qquad (2.6)$$

where S is a symmetrizing operator.

2.3. <u>Cluster expansion</u>. Different cluster expansions can be devised for calculating the expectation values of the Hamiltonian and other important quantities. Two cluster expansions which have been widely studied are the so called <u>factor</u> and <u>sum-product</u> expansions[6,5]. The <u>power-series</u> expansion[7] for the expectation value of an operator Q is obtained by substituting $f_\alpha(i,j)$, defined by

$$f_\alpha(i,j) = \sum_p f_\alpha^p(r_{ij}) \, O^p(i,j) \, ,$$

$$f_\alpha^1(r_{ij}) = 1 + \alpha [f^1(r_{ij}) - 1] \, , \quad f_\alpha^p(r_{ij}) = \alpha f^p(r_{ij}) \, , \quad p > 1 \, , \tag{2.7}$$

for the generic two-body correlation factor $f(i,j)$ in the expression for $\langle Q \rangle$. The *associated function*

$$Q(\alpha) = \frac{\int d\tau \, \phi^* S(\prod_{i<j=1}^A f(i,j))^\dagger Q \, S(\prod_{i<j=1}^A f(i,j)) \phi}{\int d\tau \, |S(\prod_{i<j=1}^A f(i,j))\phi|^2} = \sum_{n=0}^\infty A_n \alpha^n \, , \tag{2.8}$$

is expanded in power series of α around the value $\alpha = 0$ and, finally, the value $\alpha = 1$ is considered.

In general, different expansion procedures give coincident results only in the limit where all the cluster contributions are summed up. At low orders of the cluster expansions, one expansion procedure can result more efficient than the others: for finite nuclei, the factor decomposition appears to give more accurate results[8]. All the cluster terms can be summed up by generalized Fermi hypernetted chain (FHNC) procedures. Let us consider the case of a state-independent Jastrow correlation factor. The FHNC equations for calculating the two-particle distribution function, can be derived for infinite homogeneous and translationally invariant systems[9] and for finite systems too[10]. The equations are easily solved for infinite systems but give rise to numerical problems for finite nuclei. The reason is that the unperturbed density matrix

$$\rho(\vec{r}_i, \vec{r}_j) = \sum_{\alpha=1}^A \phi_\alpha(i) \, \phi_\alpha^*(j) \, , \tag{2.9}$$

results to depend only on r_{ij} in the case of infinite systems, but it depends both on $\vec{r}_i$ and $\vec{r}_j$ for finite nuclei. As a consequence, one has to evaluate numerically convolution integrals of the form

$$\int d\vec{r}_3 \, a(\vec{r}_1, \vec{r}_3) \, b(\vec{r}_3, \vec{r}_2) \, , \tag{2.10}$$

which may be calculated by a Monte Carlo procedure. Alternatively, the functions entering in the FHNC equations can be expanded in the form

$$z(\vec{r}_1,\vec{r}_2) = \sum_{i,j} z_{ij}\, \phi_i(\vec{r}_1)\, \phi_j^*(\vec{r}_2) \ . \qquad (2.11)$$

The FHNC equations become in this way a set of algebraic equations for the expansion coefficients, and they can be solved if a (small) finite basis is considered as approximation to eq.(2.11). The <u>local density approximation</u> (LDA) in relation to the function $z(\vec{r}_1,\vec{r}_2)$ is accomplished through the position

$$z(\vec{r}_1,\vec{r}_2) = z(\vec{r}_{12},\vec{R}_{12}) \simeq z(r_{12},R_{12}) = Z(r_{12},\rho) \ , \qquad (2.12)$$

where ρ is the nuclear density at distance R_{12}. The function $Z(r_{12},\rho)$ is then evaluated by solving the nuclear matter problem at density ρ. Up to now, no numerical applications of the FHNC equations for finite nuclei have yet been made.

3. Results of variational studies on finite nuclei

3.1. Calculations on triton and ^{4}He nuclei.

A variational calculation with correlated wave functions of the form specified by eqs.(1.2), (2.3) has been recently performed[11] for the ground states of nuclei with A= 3, 4 and using the Reid soft-core potential[12]. The model function Φ is taken to be a pure spin-isospin function,

$$|\Phi(^3H)\rangle = A|n\uparrow n\downarrow p\uparrow\rangle \ , \qquad |\Phi(^4He)\rangle = A|n\uparrow n\downarrow p\uparrow p\downarrow\rangle \ , \qquad (3.1)$$

where A is the antisymmetrizing operator. The two-body correlation factor has the form

$$f(i,j) = f^c(r_{ij})\left[1 + u^\sigma(r_{ij})\,\vec{\sigma}_i\cdot\vec{\sigma}_j + \beta\, u^{t\tau}(r_{ij})\, S_{ij}\, \vec{\tau}_i\cdot\vec{\tau}_j\right] \ , \qquad (3.2)$$

where β is a variational parameter which is intended to vary the tensor correlation strength. The functions $f^c(r)$, $u^\sigma(r)$ and $u^{t\tau}(r)$ are determined by solving two-body Schroedinger-type equations with the particle-particle potential appropriate to the considered channel, plus an "induced" potential. The induced potential contains few trial parameters and it is such as to have satisfied the boundary conditions

$$f^c(r \to \infty) \to r^{-\alpha} \exp(-kr) , \quad rf^c(r)\big|_{r=0} = 0 ,$$
(3.3)

$$k = \sqrt{\frac{2m}{\hbar^2} \frac{(A-1)}{A} E_S} , \quad \alpha = \frac{1}{A-1} ,$$

where E_S, the separation energy of one particle from the nucleus, is taken as a variational parameter. A Monte Carlo method is used to calculate the expectation value of the Hamiltonian and the minimization with respect to the variational parameters is then accomplished. The upperbound obtained in ref.(11) for the triton is -6.86 ± 0.08 MeV, which compares well with the available evaluations obtained using either a complete set[13], or a different variational approach[14], or the Faddeev technique[15].

For ^{4}He (without coulomb repulsion), the upperbound for the energy is[11] -22.9 ± 0.5 MeV. This value is slightly higher than the -24.9 MeV obtained by KÜMMEL et al.[16] by the coupled cluster equations method, and it is slightly lower than the -20.5 MeV obtained by TJON[17] with the Yakubovsky equations.

3.2. Method of correlated basis functions for low levels of ^{16}O.

The ground state energy of ^{16}O has been calculated by MEAD and CLARK[18], at the second order of the cluster expansion, using a state-independent correlation factor. The NN model potentials considered are the so called OMY potential[19] and the KK potential adopted from KALLIO and KOLLTVEIT[20]. The two-body correlation factor used has the form

$$f(r) = 0 , \quad r \leqslant c ,$$
(3.4)
$$f(r) = \left\{ 1 - \exp[-\beta^2(r^2 - c^2)] \right\} \left\{ 1 + \gamma \exp[-\beta^2(r^2 - c^2)] \right\} , \quad r > c ,$$

where c is the hard-core radius. The parameter γ is fixed by satisfying the Pauli condition, which should produce a rapid cluster convergence. The correlation parameter β and the oscillator parameter $\hbar\omega$ are varied to minimize the energy. The results obtained[18] for the ground state of ^{16}O in the case of OMY potential are reported in Table I. No correction for motion of the center of mass has been included: a crude estimate of this correction is $-\frac{3}{4}\hbar\omega = -11.3$ MeV. The energy resulting for the ground state of ^{16}O lies somewhat above the experimental value (-127.6 MeV). However, a rather simplified and constrained two-body correlation factor has been adopted, so that the problem of the measure of possible energy improvement is still open.

In ref.(18), also the low-lying odd-parity levels of ^{16}O have been calculated by the method of correlated basis functions with a truncated basis of correlated states. In particular, the estimates for the KK potential are in rather good agreement with the results obtained by KALLIO and KOLLTVEIT[20] with a simplified Brueckner-Tamm-Dancoff method.

$\hbar\omega$ (MeV)	β (fm^{-1})	γ	$(H_{00})_2$ (MeV)	E_C (MeV)
14	1.114	2.070	-93.8	—
15	1.118	2.078	-94.85	15.6
16	1.122	2.087	-93.8	—

Table I. Results from minimization of the second order cluster approximation $(H_{00})_2$ to the energy of the ground state of ^{16}O and using OMY potential

3.3. <u>Calculations on light nuclei with correlated functions</u>. A careful study of the cluster convergence in the case of light nuclei is being pursued by GUARDIOLA and coworkers[8]. In a recent analysis[21], the potential S3 as given in ref.(22) has been considered to act in the even channels, while in the odd channels the potential is

$$V_{odd}(i,j) = V_0 \exp(-\lambda r_{ij}^2) , \quad V_0 = 1.000 \text{ MeV} , \quad \lambda = 3 \text{ fm}^{-2} . \quad (3.5)$$

The ground state energies of the nuclei ^{4}He, ^{16}O and ^{40}Ca have been evaluated in a variational approach, at the third order of the factor cluster expansion. The two-body correlation factor has been chosen of the form

$$f(r) = 1 + a \exp(-br^2) , \quad (3.6)$$

with a and b being free variational parameters so well as the oscillator parameter. In evaluating the energy per nucleon E/A, in ref.(22) the coulomb potential has not been included, but the effects due to motion of the center of mass were correctly taken into account. The results obtained for E/A and the r.m.s. radius $\langle r^2 \rangle^{\frac{1}{2}}$, are listed in

NUCLEUS	Variational approach		BHF theory		Experimental data	
	E/A (MeV)	$\langle r^2 \rangle^{\frac{1}{2}}$	E/A	$\langle r^2 \rangle^{\frac{1}{2}}$	E/A	$\langle r^2 \rangle^{\frac{1}{2}}$
^{4}He	-6.04	1.58	-6.29	1.69	-7.07	1.63÷1.71
^{16}O	-6.73	2.32	-7.41	2.55	-7.98	2.67÷2.72
^{40}Ca	-8.39	2.99	—	—	-8.55	3.43÷3.49

Table II. The variational and BHF theory estimates for the energy per nucleon (without coulomb repulsion) and the r.m.s. radius are listed for the nuclei ^{4}He, ^{16}O and ^{40}Ca. The experimental values of E/A, from ref.(23), and of $\langle r^2 \rangle^{\frac{1}{2}}$, from ref.(24), are also reported.

the second and third columns of Table I, while the fourth and fifth columns present the corresponding values calculated at the lowest order of the Brueckner-Hartree-Fock (BHF) theory. There is a rather satisfactory agreement among the variational and BHF results. However, the correlation factor can be generalized by allowing a flexibility larger than in eq.(3.6) and by including state dependences, with consequent improuvements in the variational estimates.

The extension of the variational treatment to non-closed shell nuclei is at present under investigation$^{(25)}$.

3.4. <u>The nuclear mass formula</u>. In principle, for a given potential it is possible to calculate the coeffiecients of the corresponding nuclear mass formula. As matter of fact, it is relatively simple to calculate the volume, symmetry and surface energy coefficients and the leading terms of the coulomb energy. On the other side, accurate evaluations appear quite difficult both for the rearrengement (namely non uniform bulk distributions) terms and for deformation terms. The available semiphenomenological mass formulas furnish rather small values for the rearrengement energies for spherical nuclei. In these cases, considering only volume, surface, symmetry and coulomb terms should represent a fairly accurate approximation. Let us consider the case of OMY potential. In correspondence to this potential, the problem of infinite nuclear matter has been extensively studied[26,27] to get the volume energy coefficient a_1, and, more recently[28], the symmetry and surface energy coefficients. In Table III, the second and the third lines report the values of a_1 and of the surface energy coefficient a_2 calculated at different density values: the techniques used to obtain a_1 and a_2 are discussed in refs.(27), (28), respectively. The coulomb energy for a nucleus with charge Z and bulk density ρ has been estimated through the formula

$$E_C = \frac{3}{5}\left(\frac{4\pi\rho}{3A}\right)^{1/3} Z^2 e^2 (1 + \eta_C A^{-2/3}) \quad , \quad (3.7)$$

where the first contribution is the bulk energy, while the second contribution represents the exchange-correction[28]. Table III presents the coulomb energies, calculated via eq.(3.7), and the energies per nucleon, obtained by adding the volume, surface and coulomb contributions, for the nuclei ^{16}O and ^{40}Ca. The estimates furnished for the quantities of interest by the mass formula of MYERS[29], are reported in the last column of Table III; the coulomb diffuseness-corrections and the rearrengement energies are -0.32 MeV and -0.49 MeV for ^{16}O, and -0.32 MeV and -0.21 MeV for ^{40}Ca, respectively.

		k_F (fm^{-1})	1.2	1.3	1.4	1.5	1.6	1.29
		a_1 (MeV)	-12.4	-14.5	-16.1	-16.5	-15.4	-16.0
		a_2 (MeV)	13.6	17.8	23.1	29.8	37.9	20.7
^{16}O	E_C/A (MeV)		0.84	0.90	0.96	1.02	1.08	0.94
	E/A (MeV)		-6.16	-6.54	-5.97	-3.68	0.67	-6.81
^{40}Ca	E_C/A (MeV)		1.75	1.88	2.09	2.16	2.30	1.92
	E/A (MeV)		-6.67	-7.42	-7.26	-5.66	-2.07	-7.99

Table III. Results obtained for OMY potential at different values of the Fermi momentum k_F. a_1 is the energy per particle in infinite symmetrical nuclear matter, a_2 is the surface energy coefficient. For 16O and 40Ca, the coulomb energies have been estimated by means of eq.(3.7), while E/A is the sum of the homogeneous bulk and nuclear surface energy contributions.

The OMY potential produces saturation for symmetrical infinite nuclear matter at too high density ($k_F \simeq 1.5$ fm^{-1}). However, the surface energy contribution is large and the minimum energy for ^{16}O and ^{40}Ca is obtained at smaller (bulk) density value. The diffuseness and rearrengement corrections, in analogy with the semiphenomenological previsions, should be rather small. In such a case, the energy per nucleon furnished by OMY potential should result about -7.5 MeV for ^{16}O and -8.5 MeV for ^{40}Ca. The diference between this estimate and the one given in sect. 3.2 for the ground state energy of ^{16}O is large.

Acknowledgement. The author thanks Prof. R. Guardiola, for sending his results prior to publication.

References

(1) B. Day, "The Meson Theory of Nuclear Forces and Nuclear Matter", eds. D.Shütte, K. Holinde, K. Bleuler, (Wissenschaftverlag, 1980)
(2) K.H. Lührmann and H. Kümmel, Nucl. Phys. A194 (255) 1972
(3) D.M. Ceperley and M.H. Kalos, "Monte Carlo Methods in Statistical Physics", Edited by K. Binder, Springer (1978)
(4) E. Feenberg, "Theory of Quantum Fluids", Academic, New York (1969)
(5) J.W. Clark, in: Progress in Particle and Nuclear Physics, vol. 2, Edited by D. Wilkinson, Pergamon, Oxford (1979)
(6) J.W. Clark and P. Westhaus, Jour. Math. Phys. 9 (1968) 131; 9 (1968) 149
(7) S. Fantoni and S. Rosati, Nuovo Cim. 20 (1974) 179
(8) R. Guardiola, Nucl. Phys. A328 (1979) 490
R. Guardiola and A. Polls, Nucl. Phys. A342 (1980) 385
R. Guardiola, A. Polls and J. Ros, Nuovo Cim., to be published
(9) S. Fantoni and S. Rosati, Nuovo Cim. 25 (1975) 593
(10) S. Fantoni and S. Rosati, Nucl. Phys. A328 (1979) 478
(11) J. Lomnitz-Adler, V.R. Pandharipande and R.A. Smith, preprint (1980)
(12) R. V. Reid, Ann. of Phys. 50 (1968) 411
(13) P. Numberg, D. Prosperi and E. Pace, Nucl. Phys. A285 (1977) 58
(14) M.A. Hennel and L.M. Delves, Nucl. Phys. A246 (1975) 490
(15) A. Laverne and C. Gignoux, Nucl. Phys. A2 3 (1973) 597
R.A. Brandeburg, Y.E. Kim and A. Tubis, Phys. Rev. C12 (1975) 1368
W. Clockle and R. Offermann, Phys. Rev. C16 (1977) 2039
(16) H. Kümmel, K.H. Lührmann and J.G. Zabolitzky, Phys. Rep. C36 (1978) 1
(17) J.A. Tjon, Phys. Rev. Lett. 40 (1978) 1239
(18) L.R. Mead and J.W. Clark, Phys. Lett. 90B (1980) 331
(19) T. Ohmura, M. Morita and M. Yamada, Progr. Theor. Phys. 15 (1956) 222
(20) A. Kallio and K. Kolltveit, Nucl. Phys. 53 (1964) 87

(21) A. Faessler, R. Guardiola, H. Müther and A. Polls, to be published
(22) I.R. Afnan and Y.C. Tang, Phys. Rev. <u>175</u> (1968) 1337
(23) A.H. Wapstra and K. Bos, Atomic Data and Nuclear Data Tables, <u>19</u> (1977) 177
(24) C.W. de Jager, H. de Vries and C. de Vries, Atomic Data and Nuclear Data Tables, <u>14</u> (1974) 479
(25) R. Guardiola, private communication
(26) E. Campani, S. Fantoni and S. Rosati, Nuovo Cim. Lett. <u>15</u> (1976) 217
(27) S. Rosati, in:"From Nuclei to Particles", Varenna Summer School (1980), in press; see also references quoted therein
(28) S. Fantoni and S. Rosati, Nuovo Cimento, in press
(29) W.D. Myers, Atomic Data and Nuclear Data Tables, <u>17</u> (1976) 411

MAGNETIC RESONANCES AND THE SPIN DEPENDENCE OF THE PARTICLE-HOLE FORCE

S. Krewald
Institut für Kernphysik, KFA Jülich, D-5170 Jülich, West Germany

Abstract: Spin-isospin modes in nuclei are shown to be an excellent tool to investigate the spin-isospin dependent part of the particle-hole interaction. Many-body effects in connection with the recently discovered magnetic high spin states and Gamow-Teller resonances are discussed. The interaction deduced from the present analysis does not imply precritical phenomena attributed to pion condensation.

1. Theory of Collective States and the Spin-Isospin Dependent Particle-Hole Interaction

Among the few many-body theories which attempt to describe excited states in heavy nuclei, the concept of "interacting Fermi systems" due to Landau and Migdal has been very successful[1,2]. Starting from the density-density correlation function, a closed system of equations can be derived which allows to calculate the excitation energies and the transition amplitudes of collective states. These equations are in principle exact, provided one uses a highly renormalized particle-hole (ph) interaction. This, of course, is the place where the difficulties of the many-body problem are hidden. Landau and Migdal suggested not to derive this interaction, but to expand it in terms of Legendre polynomials and to extract the free parameters introduced this way from experiment.

At this stage, the connection to the bare nucleon-nucleon interaction is no longer obvious[3] and, therefore, this theory can only relate different experimental facts with each other. One example is the link between isotope shifts and the excitation energy of the breathing mode. From the experimental knowledge of isotope shifts in the lead region, the corresponding interaction parameter (f_0 in eq. (2)) was deduced. Now one was able to predict the excitation energy of the breathing mode, which later was confirmed experimentally[4].

In even-even nuclei, the excitation energies and transition probabilities are obtained from the following random-phase equation:

$$x^\mu_{\nu_1\nu_2} = \frac{n_{\nu_1}-n_{\nu_2}}{\varepsilon_{\nu_1}-\varepsilon_{\nu_2}-\Omega_\mu} \sum_{\nu_3\nu_4} F^{ph}_{\nu_1\nu_3,\nu_2\nu_4} x^\mu_{\nu_3\nu_4} \qquad (1)$$

Here ε_ν are the single particle energies, Ω_μ is the excitation energy of the collective state $|\mu\rangle$, n_{ν_1} are occupation probabilitis (= 0 or 1), F^{ph} is the effective ph interaction and x^μ are the transition amplitudes which determine transition probabilities and scattering cross sections. In practical applications, only the zeroth order of the expansion of F^{ph} is taken and tensor invariants are neglected which leads to the following ansatz:

$$F_0^{ph}(r_1 r_2) = C_0 \{f_0 + f_0' \vec{\tau}_1 \cdot \vec{\tau}_2 + g_0 \vec{\sigma}_1 \cdot \vec{\sigma}_2 + g_0' \vec{\sigma}_1 \cdot \vec{\sigma}_2 \vec{\tau}_1 \cdot \vec{\tau}_2\} \delta(r_1 - r_2) \tag{2}$$

The constant C_0 is the inverse of the density of states at the Fermi surface.

$$C_0 = \frac{\pi^2 \hbar^2}{k_F m^*} = 302 \frac{m}{m^*} \left[\text{MeV fm}^2\right] \tag{3}$$

This restriction to zeroth order seems to be justified from the investigations done so far on electric properties of nuclei.

The spin-dependent part of the interaction which describes the magnetic properties of nuclei is much less well known. The major physical reason is the lack of really collective magnetic states in nuclei which would allow to test the ph-interaction. Within the ansatz (2), the parameters g_0 and g_0' were deduced from magnetic moments and transition probabilities in odd-mass nuclei. A strongly repulsive interaction in these channels was found to be required. This result, however, leads to serious problems in even-even nuclei. Here it is well known that the energies of unnatural parity ("magnetic") states are rather close to the unperturbed particle-hole energies which one deduces from the mass differences of neighbouring nuclei. Therefore the spin-dependent ph-interaction should be weakly repulsive only. In theoretical investigations of pion condensation one even expects a strongly attractive force in the spin-isospin channel. Clearly the ansatz (2) is too narrow to explain these phenomena simultaneously. We claim that the different behaviour of magnetic moments in odd-even nuclei and excited states in even-even nuclei is due to the momentum transfer involved in the various processes. Therefore the ansatz (2) has to be generalized because it is valid only in the limit of small momentum transfer.

Now one-pion exchange which is the longest range component of the nuclear force is expected to play a special role in the spin-isospin channel[5,6]. Therefore we do not have to introduce an arbitrary q-dependence but can start from a physical motivation as a guideline. We choose a generalized ph-force[7] which includes in addition to the zero range part in eq. (2) contributions due to pion-(π) and rho-meson (ρ) exchange:

$$F_{\sigma\tau}^{ph}(q) = g_0' \sigma \cdot \sigma' \tau \cdot \tau' - \frac{4\pi f_\pi^2}{m_\pi^2} \tau \cdot \tau' \frac{\vec{\sigma} \cdot \vec{q} \vec{\sigma} \cdot \vec{q}'}{q^2 + m_\pi^2} - \frac{4\pi f_\rho^2}{m_\rho^2} \frac{(\vec{\sigma} \times \vec{q}) \cdot (\vec{\sigma} \times \vec{q})}{q^2 + m_\rho^2} \tag{4}$$

The rho-meson has been added because it is the only other meson with isospin one. Since the ρ-mass is large and therefore the range of the ρ-interaction is small, one might argue that the effects of the ρ-meson should be included in the zero-range parameters. This would lead to an additional zero-range tensor parameter in eq. (2). This complication can be avoided by using the ρ-meson explicitly. Furthermore, the tensor force from the ρ-exchange potential cuts off the tensor force of the π-exchange potential at small distances or high momentum transfers which may have important physical consequences.

Since the interaction due to ρ-exchange is short-ranged, this part of the nucleon-nucleon interaction gets appreciably modified by the short-range repulsive correlations of the ω-exchange potential. Therefore one should in principle fold the ρ-exchange potential with an appropriate correlation function. The result of such a calculation, however, can be approximated by multiplying the ρ-coupling constant by a factor 0.4 [8]. This factor is not applied to the tensor contribution of the ρ-meson, because the tensor couples only relative D with relative S-states, so that the tensor contribution is only little influenced by short-range correlations. In fig. 1a, a spin-isospin average[6] of the effective particle-hole interaction is shown to

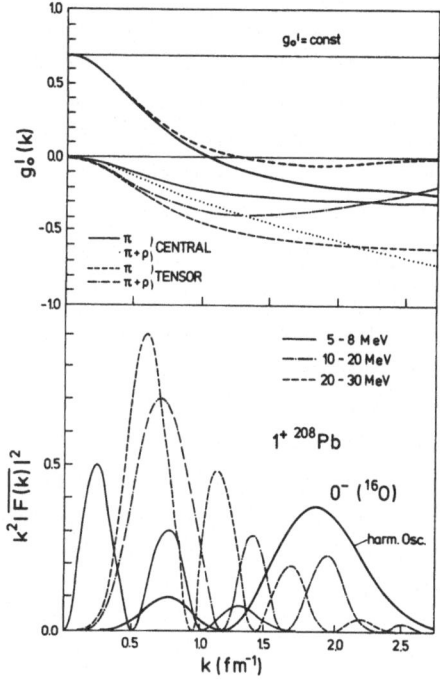

Fig. 1: (a) Spin-isospin averaged particle-hole interaction in momentum space, consisting of a zero range piece, π- and ρ-exchange. The finite range contributions are broken up into central and tensor pieces. Thick full line: sum of all contributions. Thick dashed line: sum of all contributions including reduction factor for short-range correlations.
(b) Momentum distribution of 1^+ states in ^{208}Pb and 0^- state in ^{16}O.

illustrate the q-dependence of this force. The π- and ρ-exchange have been separated into a central and a tensor part. The thick full line is the sum of all contributions, the thick dashed line is again the sum of all contributions, however, with the central part of the ρ-exchange multiplied by a reduction factor 0.4. Free meson coupling constants $f_\pi^2 = 0.081$ and $f_\rho^2 = 4.86$ are used, but the constant g_0' is taken as a free parameter. In order to test the q-dependence of this interaction, the unnatural parity states in ^{16}O are especially appropriate, since their momentum distributions have maxima in the momentum region between 1 fm^{-1} and 3 fm^{-1}, as shown in the lower part of fig. 1. From the analysis of the excitation energies and transition probabilities of unnatural parity states in ^{16}O and ^{12}C a Landau parameter $g_0' = 0.75$ was deduced[7].

2. Microscopic Structure of the Magnetic High-Spin States in ^{208}Pb [10])

Recently magnetic high-spin states have been discovered in ^{208}Pb by inelastic electron scattering at backward angles[9]. These states are of considerable physical interest because, as a consequence of the high multipolarity, the cross sections are peaked at a momentum transfer of approximately $q = 2$ fm^{-1}. Therefore these states are very sensitive to the high-momentum components of the generalized interaction (4). The number of 1p1h excitations which can contribute to these states is severely restricted by the high multipolarity. Since the experimental excitation energies are close to the shell-model ph-energies, the 12⁻ state at 6.43 MeV and the 14⁻ state at 6.74 MeV were tentatively interpreted as pure $\nu(1j_{15/2}, 1i_{13/2}^{-1})$ ph-configurations, while the 12⁻ state at 7.06 MeV was assumed to be a pure $\pi(1i_{13/2}, 1h_{11/2}^{-1})$ excitation. This simple interpretation faces one problem, however, because the experimental cross section is only 50 % of the 1p1h-prediction. Here we show that (i) the effects of the π- and ρ-exchange potential give rise to a very weak interaction in this momentum transfer region, and (ii) that the fragmentation of the single-particle strength due to the phonon coupling is mainly responsible for the reduction of the cross section.

From the investigations in part 1, we obtained a Landau parameter $g_0' = 0.75$. Since we analyzed predominantly isovector states, the value of g_0 remained ambiguous. The magnetic high-spin states, however, are very sensitive to both g_0 and g_0'. In the limit of a δ-force, the interaction between the two 12⁻ configurations is given by the proton-neutron interaction $g_0 - g_0'$. We find that $g_0 = 0.25$ has to be used in order to reproduce the excitation energies of the 12⁻ and 14⁻ states. These Landau parameters are very close to the values $g_0 = 0.22$ and $g_0' = 0.65$ determined by Bäckman et al.[11] from the Reid soft-core potential.

J^π	12⁻		12⁻		14⁻	
model	E (MeV)	$\frac{BML}{BML_1}$	E (MeV)	$\frac{BML}{BML_2}$	E (MeV)	$\frac{BML}{BML_3}$
1p1h-a	6.49	1.00	7.18	1.00	6.49	1.00
1p1h-b	6.77	0.44	7.86	0.91	7.14	0.82
1p1h-c	6.60	1.18	7.52	0.94	6.68	0.96
2p2h	6.55	0.60	7.37	0.54	6.65	0.45
exp	6.43		7.06		6.75	

Table 1: Excitation energies and B(M,L) values for the magnetic high-spin states in ^{208}Pb obtained with the following models: 1p1h-a = a single particle-hole configuration; 1p1h-b = RPA with zero range interaction; 1p1h-c = RPA with finite range interaction; 2p2h = two-particle two-hole calculation described in the text. The experimental data are from ref. 9). The B(M,L) values are divided by the shell model values (1p1h-a) $BML_1 = 0.57 \cdot 10^{23} \mu^2 fm^{22}$; $BML_2 = 0.68 \cdot 10^{23} \mu^2 fm^{22}$; $BML_3 = 1.33 \cdot 10^{27} \mu^2 fm^{26}$.

In table 1, the energies and B(M,L) values of the magnetic high-spin states are shown. Using a zero-range force, the B(M,L) values are reduced, but the excitation energies are pushed far above the shell model values. Only the π- and ρ-exchange provide a mechanism to reduce the excitation energy.

It is well known that the coupling to phonons may modify the single-particle states appreciably[12]. The $\nu j_{15/2}$ state comes at an excitation energy of 1.42 MeV relative to the ground state of ^{209}Pb, which is only 1.2 MeV below the $(3^- \times \nu 2g_{9/2})_{15/2^-}$ configuration. Therefore a considerable mixing of these configurations has to be expected, which strongly reduces the single-particle strength

$$Z_m^{(\alpha)} = |<A+1,\alpha|a_m^+|A,0>|^2 \tag{5}$$

of the $|\alpha> = |15/2^->$ state at 1.42 MeV. This effect is especially large for the so-called spin-orbit partners which are shifted into the next lower major shell and therefore are embedded in single-particle states of opposite parity. All the dominant configurations of the high-spin states are of that special type.

We evaluate the single-particle strength by taking into account explicitly the coupling of phonons in ^{208}Pb to single-particle states, thus obtaining quasiparticle states in the neighbour nuclei. These quasi-particle and quasi-hole states are used to construct a core-coupling random phase wave function containing the most relevant 2p2h configurations. Details are given in ref. 13). We mention, however, that we include in the present case also the interaction between the 2p2h configurations. Including as phonons only the 3^- state at 2.61 MeV and the 5^- state at 3.19 MeV, the single-particle strength is reduced to $Z = 0.55$ for the $\nu j_{15/2} i_{13/2}^{-1}$ and to $Z = 0.58$ for the $\pi i_{13/2}^{-1} h_{11/2}$ configuration. These single-particle strengths were not further reduced, when more phonons were taken into account. The results of the 2p2h calculation with a finite-range interaction are given in table 1. The 2p2h components of the wave function strongly reduce the B(M,L) values and even produce a further decrease of the excitation energy. The fraction of normalization of the wave function carried by the 2p2h components is 52.6 % for the 14^- state and 47.9 % and 48.4 % for the 12^- states. The mixing between proton and neutron configurations in the case of the 12^- states is of the order of 0.03 in the amplitudes.

In fig. 2, the inelastic electron scattering cross sections at $\vartheta = 90°$ and $\vartheta = 160°$ are shown for the three magnetic high-spin states. The calculations were performed in Distorted Wave Born Approximation, using the code HEIMAG by J. Heisenberg[14]. The Woods-Saxon functions employed as single-particle wave functions in the present calculation were obtained from a potential[15] fitted to the neutron rms-radius given by Negele's Hartree-Fock calculation[16]. The inclusion of 2p2h-configurations reduces the cross sections, as expected, so that both shapes and absolute magnitudes of the cross sections are now in good though not perfect agreement with the experimental data.

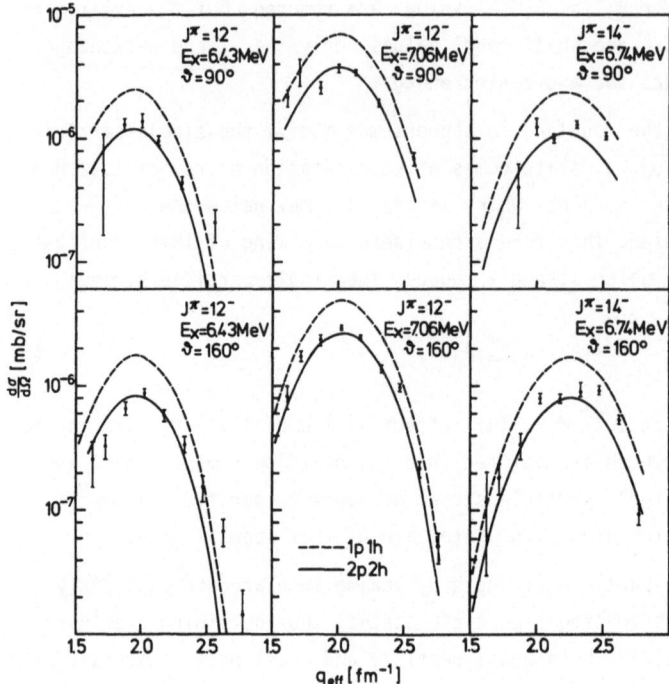

Fig. 2: Experimental cross sections for inelastic electron scattering of ref. 9), compared with an RPA calculation (dashed) and a 2p2h-calculation (solid). Both calculations were performed with a finite-range interaction including one-pion and rho-meson exchange.

The single-particle strength missing in the three low-lying states is fragmented into many states at higher energies in this approach. The next 12^- states come at 8.10, 8.23, 8.38 and 8.65 MeV, while 14^- states are expected at 8.12, 8.34 and 8.70 MeV. The maximal inelastic electron scattering cross section of each state is less than 10 % of the experimental strength of the known high spin states, however.

A possible discrepancy between the magnitude of (e,e') and (p,p') cross sections suggested by Lindgren et al.[17] could be resolved with our more realistic wave functions, except for the first 12^- state. Here one should look for a possibly unresolved 11^- or 13^- state.

3. Nuclear Structure Effects Connected with Charge-Exchange Reactions

Using a high energetic proton beam Goodman et al.[18] recently obtained detailed experimental results on the Gamow-Teller resonances (GTR) in many nuclei. The structure of these states is very similar to the well known isobaric analog states (IAR): both resonances can be described as a superposition of proton-particle neutron-hole states, coupled to 0^+ in the case of the IAR and to 1^+ in the GTR case. Since the levels excited by a (p,n)-reaction are connected with a change in isospin, they allow a selective investigation of the isospin-dependent part of the particle-hole interaction. The interaction (4) is strongly repulsive for small momentum transfer, but as the momentum transfer increases, the one-pion exchange cancels the repulsive components. Therefore this interaction can build collective magnetic states only, if the Fourier components of the wave function are concentrated at low momentum transfers. This is the case for the GTR the energy of which is strongly pushed up above the unperturbed ph-energies. Therefore this state promises to be a useful tool to investigate the ph-interaction in the spin-isospin channel. It turned out, however, that in a straightforward RPA calculation (using experimental single-particle energies ε_ν), the energy of the GTR comes 2.7 MeV below the experimental value, whereas the IAS is nicely reproduced.

A similar effect happens in the case of the giant dipole resonance (GDR) which is several MeV too low if an interaction consistent with the symmetry energy is used[15]. Now it is well known that the phonon contributions to the single-particle energies give rise to a compression of the Brueckner-Hartree-Fock spectrum in odd-mass nuclei[12]. The coupling of phonons to a given collective state with energy h can be expressed in terms of a selfenergy $\Sigma(\hbar\omega_\lambda)$, the real part of which can be written as a principle value integral

$$\Sigma(\hbar\omega_\lambda) = P \int dE_i \, \frac{|M_\lambda(E_i)|\rho(E_i)}{\hbar\omega_\lambda - E_i} \tag{6}$$

where $M(E_i)$ is the matrix element which couples the given collective state $\hbar\omega_\lambda$ to 2p2h states. Brown and Speth pointed out that this coupling may vanish in special situations[19]. In fig. 3, two limiting situations are shown. In the GDR case (fig. 1a) the repulsive ph-interaction pushes this resonance right in between the corresponding 2p2h states. Therefore the principle value (eq. (6)) is small and the effect of the phonons on the energy of the GDR is small. Therefore the compression of the single-particle spectrum due to phonons must be removed.

The situation of the IAR is shown schematically in fig. 3b. Since the IAR is the $T_>$ state the corresponding 2p2h $T_>$ states are much higher in energy (the coupling to the $T_<$ states is weak). Therefore the principle value integral (3) is large which means that the phonon coupling is <u>not</u> removed and one has to use the experimental single-particle energies.

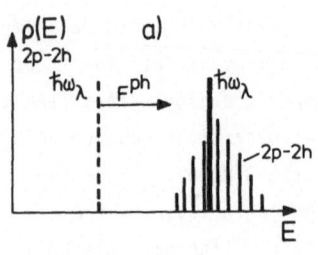 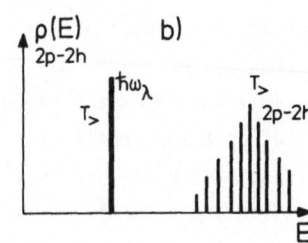

Fig. 3: Two limiting cases included in eq. (6): (a) the collective resonance $\hbar\omega_\lambda$ is shifted due to the ph-interaction into the corresponding 2p2h states (e.g. GDR, GTR); (b) the relevant 2p2h states are higher in energy than the given collective resonance (e.g. IAR).

Within the context of a 1p1h calculation, one can approximate the "dynamical theory" of Brown and Speth[19] by using an appropriate single-particle spectrum. Using the phenomenological formula given by Brown and Speth[19] which interpolates between the single-particle spectrum calculated within the Brueckner-Hartree-Fock approach ($m^*/m \sim 0.65$) and the experimental one ($m^*/m \sim 1$) we obtain an effective mass of $m^*/m = 0.75$. Since this is very close to the effective mass of the Skyrme III interaction, we used the corresponding single-particle spectrum in the calculation of the GDR which now comes out at the observed energy (see table 2).

	GDR (^{208}Pb)		IAR (^{208}Bi)		GTR (^{208}Bi)	
	theory	experiment	theory	experiment	theory	experiment[18]
E [MeV]	13.7	13.7	17.0	18.0	18.9	18.4±0.2
s.p. spectrum	SK III		exp. energies		SK III	

Table 2: Comparison between theoretical and experimental energies of the electric dipole resonance (GDR), isobaric analogue resonance (IAR) and Gamow-Teller resonance, respectively. The parameters $f_0'^{in} = 0.60$, $f_0'^{ex} = 1.8$ and $g_0' = 0.65$ were used. The energies refer to the ground state of ^{208}Pb.

The behaviour of the GTR is similar to the GDR because the GTR is the $T_<$ state. The GTR couples strongly to the 2p2h ($T_<$) states which are in the same energy region, i.e. the selfenergy (eq. (6)) is small. Therefore we used also in this case the single-particle spectrum calculated with Skyrme III. In the present context we used $g_0' = 0.65$ suggested from the Reid soft core[11]. Bearing in mind the uncertainties connected with the single-particle spectrum our theoretical results shown in table 2 are in fair agreement with experiment. The total Gamow-Teller strength concentrated in the GTR at 18.9 MeV is about 82 %. Using the same parameters and Skyrme III single-particle energies which fit the GDR, the energy of the IAR is 2 MeV too high compared with experiment, as expected from the dynamical theory. Therefor we conclude that the comparison of the GDR and IAR provides an excellent example of the importance of the "dynamical theory of collective states". In addition we would like to point out that the GTR in heavy nuclei so far is the best example of a collective magnetic state. An open problem remains in the absolute (p,n) cross sections shown in fig. 4. The effective projectile target nucleon interaction and the optical model parameters are taken from Love and Petrovich[20].

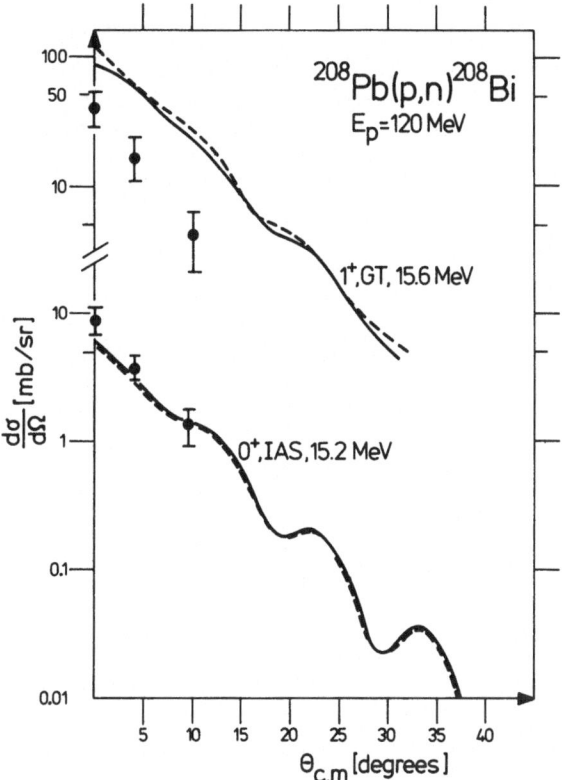

Fig. 4: Differential (p,n) cross section for the reaction ^{208}Pb(p,n)^{208}Bi at E_p = 120 MeV. The dashed line represents results where the dynamical theory has been taken into account, whereas the full lines follow from a conventional RPA calculation.

Whereas the theoretical IAR cross section agrees with the experimental one, the theoretical GTR cross section is about a factor of two too large. Oset and Rho suggest that the Lorentz-Lorenz effect might at least partially explain this reduction[21]. This effect is used by Knüpfer, Dillig and Richter to account for the missing M1 strength in heavy nuclei[22]. Certainly more quantitative investigations in finite nuclei are highly desirable.

4. Application to the (e,e') Cross Section of the 1$^+$(T=1, E=15.1 MeV) State in ^{12}C [23]

Since Migdal's suggestion[5] of the possible existence of a phase transition in nuclei to a condensed pion phase, this phenomenon has been subject to various theoretical considerations. Though this effect seems not to exist in finite nuclei, nor to give rise to large observable manifestations in heavy ion scattering[24], data do not exclude the possibility of being close to the onset of pion condensation. The indication of such a precursor phenomenon would be an enhanced (e,e') or (p,p') cross section at momentum transfers near the critical momentum for pion condensa-

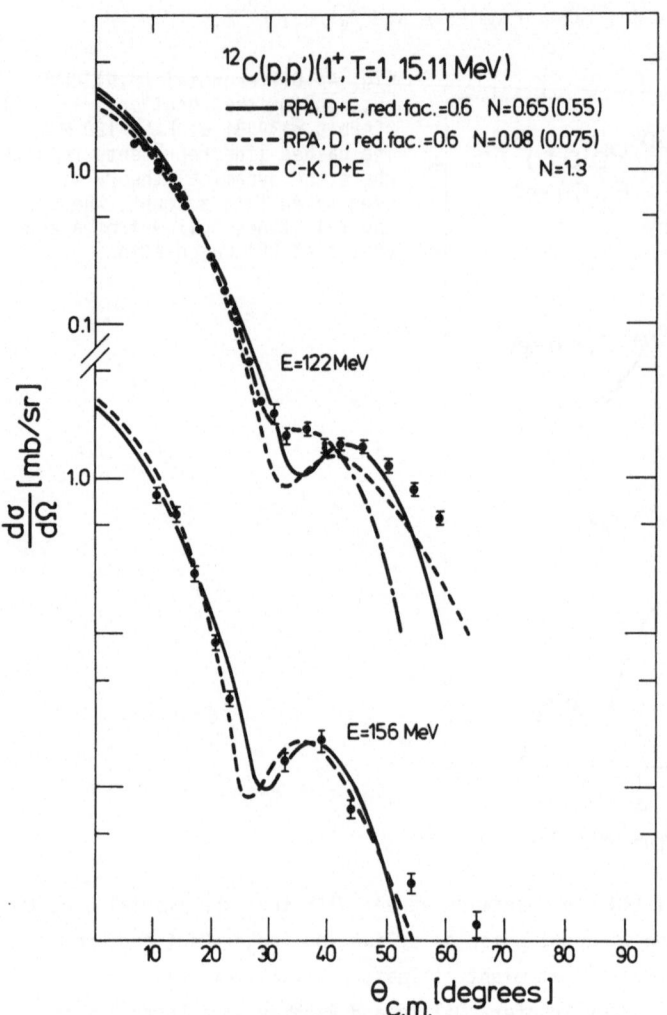

Fig. 5: (p,p') cross sections calculated in DWBA. Here D and D+E denote the results where direct and direct plus exchange amplitudes have been taken into account. N denotes the normalization factor used for the coupling potential.

tion[25,26]. In particular, Ericson and Delorme and Toki and Weise have focussed attention on the second maximum of the (e,e') form factor, which is a factor of five smaller than experiment if derived from Cohen-Kurath wave functions.

The differences between the work of our group[23] and the one of Ericson[25] and Weise[26] are: (i) the q dependence of the effective interaction. Here Ericson used a very small g_0' parameter whereas Weise includes the full 2π-exchange contribution instead of the resonant part (ρ-meson) only. Though they can fit (e,e') data, their precritical enhancement of the second maximum of the (p,p') cross sections is a factor of 3 larger than found in a recent experiment[27]. (ii) Instead of harmonic oscillators, we used Woods-Saxon functions which are more realistic for unoccupied levels. (iii) We used $6\hbar\omega$ and $10\hbar\omega$ spaces in order to check the convergence in a configuration space expansion. Corrections for many-particle many-hole effects are included in all approaches. With our force introduced in chapter 1, the (e,e') cross section can be reproduced reasonable well[23] and so can the (p,p') cross section as shown in fig. 5.

The results discussed in this contribution have been obtained in collaboration with G.E. Brown, V. Klemt, F. Osterfeld, J. Speth, T. Suzuki, and J. Wambach.

References

1. A.B. Migdal, Theory of finite Fermi systems (Wiley, New York, 1967).
2. J. Speth, E. Werner, and W. Wild, Phys. Rep. $\underline{33}$ (1977) 127.
3. G.E. Brown, Rev. Mod. Phys. $\underline{43}$ (1971)1.
4. J. Speth, L. Zamick, and P. Ring, Nucl. Phys. $\underline{A232}$ (1974) 1.
5. A.B. Migdal, Rev. Mod. Phys. $\underline{50}$ (1978) 107.
6. G.E. Brown and W. Weise, Phys. Rep. $\underline{C27}$ (1976) 1.
7. J. Speth, V. Klemt, J. Wambach, and G.E. Brown, Nucl. Phys. $\underline{A343}$ (1980) 382.
8. M.R. Anastasio and G.E. Brown, Nucl. Phys. $\underline{A285}$ (1977) 516.
9. J. Lichtenstadt, J. Heisenberg, C.N. Papanicolas, C.P. Sargent, A.N. Courtemanche, and J.S. McCarthy, Phys. Rev. $\underline{C20}$ (1979) 497.
10. S. Krewald and J. Speth, Phys. Rev. Lett. $\underline{45}$ (1980) 417.
11. S.O. Bäckman, O. Sjöberg, and A.D. Jackson, Nucl. Phys. $\underline{A321}$ (1979) 10.
12. I. Hamamoto, Phys. Rep. $\underline{10}$ (1974) 63; P. Ring and E. Werner, Nucl. Phys. $\underline{A211}$ (1973) 198.
13. J.S. Dehesa, S. Krewald, J. Speth, and A. Faessler, Phys. Rev. $\underline{C15}$ (1977) 1858.
14. J. Heisenberg, private communication.
15. G.A. Rinker and J. Speth, Nucl. Phys. $\underline{A306}$ (1978) 360.
16. J.W. Negele and D. Vautherin, Phys. Rev. $\underline{C5}$ (1972) 1472.
17. R.A. Lindgren, W.J. Gerace, A.D. Bacher, W.G. Love, and F. Petrovich, Phys. Rev. Lett. $\underline{42}$ (1979) 1524.
18. C.D. Goodman et al., Phys. Rev. Lett. 44 (1980) 1755; C.D. Bainum et al., Phys. Rev. Lett. $\underline{44}$ (1980) 1751.
19. G.E. Brown, J.S. Dehesa, and J. Speth, Nucl. Phys. $\underline{A330}$ (1979) 290.
20. W.G. Love, F. Petrovich, in "The (p,n) reaction and the nucleon-nucleon force", ed. C.D. Goodman et al. (Plenum, New York, 1980) p. 30.
21. E. Oset and M. Rho, Phys. Rev. Lett. $\underline{42}$ (1979) 47.
22. W. Knüpfer, M. Dillig, and A. Richter, Phys. Lett. $\underline{95B}$ (1980) 349.
23. T. Suzuki, F. Osterfeld, and J. Speth, submitted to Phys. Lett.
24. S. Krewald and J.W. Negele, Phys. Rev. $\underline{C21}$ (1980) 2385.
25. M. Ericson and J. Delorme, Phys. Lett. $\underline{76B}$ (1978) 182.
26. H. Toki and W. Weise, Phys. Rev. Lett. $\underline{42}$ (1979) 1034; Phys. Lett. $\underline{92B}$ (1980) 265.
27. M. Haji-Saeid et al., Phys. Rev. Lett. $\underline{45}$ (1980) 880.

A MEAN FIELD APPROACH TO THE DESCRIPTION OF NUCLEAR

STRUCTURE : INTERPRETATIONS AND PREDICTIONS

B. Grammaticos

Service de Physique Nucléaire Théorique

C.R.N. de Strasbourg BP 20 CR

67037 STRASBOURG Cedex France

Abstract : We present a unified description of the nuclear ground and excited states based on the self consistent mean field approach. Typical results on spherical and deformed nuclei, ranging from static properties to the effect of dynamical correlations are given. The success of the theory used in the interpretation of the nuclear structure seems to guarantee it usefulness as a predictive tool.

I - INTRODUCTION

The abundance of high quality experimental data on nuclear structure presents the theorist with the challenge of providing an accurate interpretation . In many cases the experimental situation is such that the theorist must resort to a high degree of sophistication in order to be able to, quantitatively, account for the experimental data. Furthermore with high energy projectiles being used as nuclear probes a simple non-relativistic approach is often inadequate and mesonic degrees of freedom must somehow be taken into account. In the case of the data we are going to review such a situation does not, hopefully, prerail and we can attempt a simple non relativistic approach.

Even limiting ourselves to the non-relativistic case we are still faced with a formidable task [1]. An " ab initio " solution to the nuclear structure problem would consist in an adequate definition of a realistic nucleon-nucleon interaction (a problem which is not yet quite solved despite the substantial recent progress [2]) and a subsequent solution of the many particle Schrödinger equation. Such a " realistic " approach, is, apparently, extremely tedious and in the few cases where it has been materialized, it has not given quite satisfactory results. Fortunately enough, there exists a different approach, or rather a whole family of them based on the nuclear mean field assumption. This notion while retaining most of the desirable features of the more realistic approaches, leads to not excessively lengthy calculations. The main assumption in nuclear mean field theories is that the nucleus can be described as a system of independent quasi-particles. Once this basic assumption is made one is lead naturally to the replacement of the " true " nuclear force by an effective one. This effective interaction must, in the spirit of pseudopotentials in scattering theory, account for all the complexity left out of the nuclear wave-function. The calculation of the effective interaction starting from first principles is, indeed, possible,- though cumbersome. In most cases one proceeds by postulating a simple analytic form for the interaction and subsequently adjusts the free parameters on pertinent data. Of course some of the features of the effective force, as, for example, the density dependence, can be predicted on quite general ground. Once the effective interaction is given and a space of wave-functions is specified one can calculate the energy of the nucleus and obtain, through the application of a variational principle , in a self-consistent way, ground state properties. This method has been used with great success in the past few years in approximations such as the Hartree-Fock [3] or the Hartree Bogoliubov [4] one. Whenever one is concerned with the description of the excited states of the system a time dependent extension of the mean-field approximation is in order. Although this method has been substantiated by some most ingenious calculations [5] it suffices in most cases to limit oneself to approximations thereof. Namely the small amplitude or random phase approximation (RPA) and the small velocity or adiabatic approximation.

This paper deals with the application of the mean field approach to the description of the nuclear structure. Our main point is that such calculations can actually be performed and, provided, that they are carried through in an internally consistent way, that they can yield quite reliable results. In this way one can use the mean field approach as a tool for interpretations and predictions in nuclear structure. At the same time the comparison with experiment enables us to assess the validity and limitations of the mean field assumption.

II - STATIC RESULTS

a) Spherical nuclei.

The totality of the results we are going to present have been obtained with Gogny's D1 effective interaction [4]. The latter is a finite range density dependent interaction whose parameters have been adjusted so as to reproduce at the Hartree-Fock approximation the properties of ^{16}O and ^{90}Zr, as well as some chosen pairing matrix elements. Its exact form is the following :

$$U = \sum_{i=1,2} (W_i + B_i P_\sigma - H_i P_\tau - M_i P_\sigma P_\tau) e^{-r^2/\mu_i^2} + t_o(1+P_\sigma)\rho^{1/3}(r)\delta(\vec{r}_1-\vec{r}_2) + iW_{LS}\vec{\nabla}\times\delta\vec{\nabla}\cdot(\vec{\sigma}_1+\vec{\sigma}_2)$$

with $\mu_1 = 0.7$ fm $W_1 = -402.4$ MeV $B_1 = -100$ MeV $H_1 = -496.2$ MeV $M_1 = -23.56$ MeV

$\mu_2 = 1.2$ fm $W_2 = -21.3$ MeV $B_2 = -11.77$ MeV $H_2 = 37.27$ MeV $M_2 = -68.81$ MeV

$t_o = 1350$ MeV fm^4 $W_{LS} = -115$ MeV fm^5 (or $W_{LS} = -130$ MeV fm^5 for D_1')

The Hartree Bogoliubov (HFB) version of the mean field approximation has been used throughout. In this way a fully self consistent description of the pairing correlations can be achieved. The latter turn out to be quite important for nuclei other than the doubly magic and the lightest N=Z ones. A striking example of the importance of the pairing correlations is presented by the charge densities of ^{90}Zr and ^{140}Ce (Fig. 1) (although these correlations are small for the former nucleus due to its near magicity). The effect of pairing is a smoothing of the density oscillations in the interior of the nucleus, bringing thus the charge density closer to the experimental one which presents systematically less oscillations than the Hartree-Fock one.

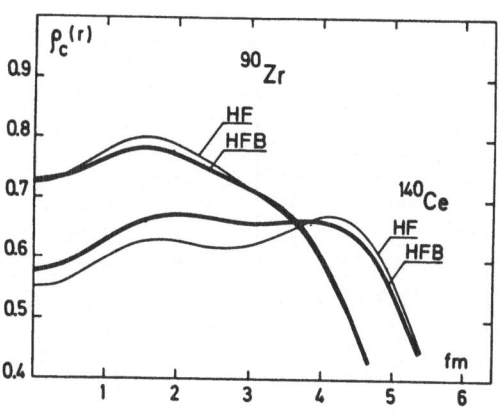

Fig. 1

The quality of the results obtained with Gogny's interaction is clearly depicted

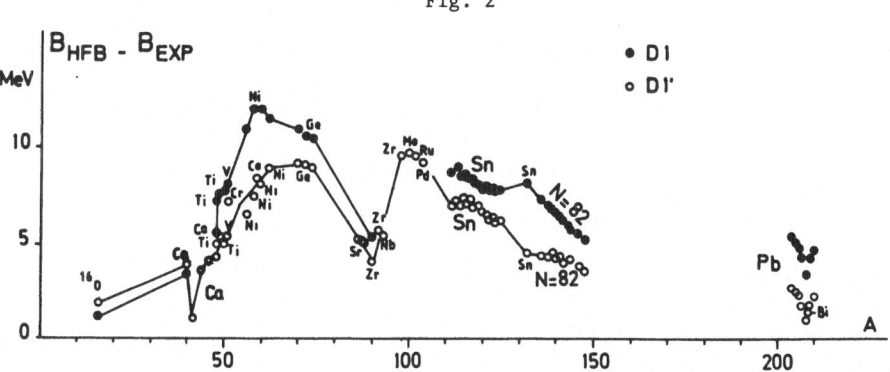

Fig. 2

in Fig. 2 where we plot the difference between experimental and theoretical binding energies for nuclei all over the periodic table. We remark that the differences are smaller than 10 MeV and much better than that in most cases, especially with D_1'. In Table I we present the first few moments of the charge distribution of selected nuclei [6]. The overall agreement with experiment is quite satisfactory. Some minor problems seem to exist in the case of ^{208}Pb. This can be also clearly seen in the case of the charge density distribution of the said nucleus which presents very

Table I

		$\langle r_{ch}^k \rangle^{1/k}$		
		k =2	4	6
^{40}Ca	EXP	3.48		
	HFB	3.44	3.76	4.03
^{48}Ca	EXP	3.48	3.78	4.05
	HFB	3.47	3.76	3.99
^{58}Ni	EXP	3.77	4.06	4.31
	HFB	3.77	4.08	4.34
^{90}Zr	EXP	4.25	4.58	4.81
	HFB	4.23	4.55	4.81
^{116}Sn	EXP	4.62		
	HFB	4.56	4.90	5.17
^{124}Sn	EXP	4.67		
	HFB	4.64	4.96	5.22
^{208}Pb	EXP	5.50	5.85	6.13
	HFB	5.44	5.78	6.05

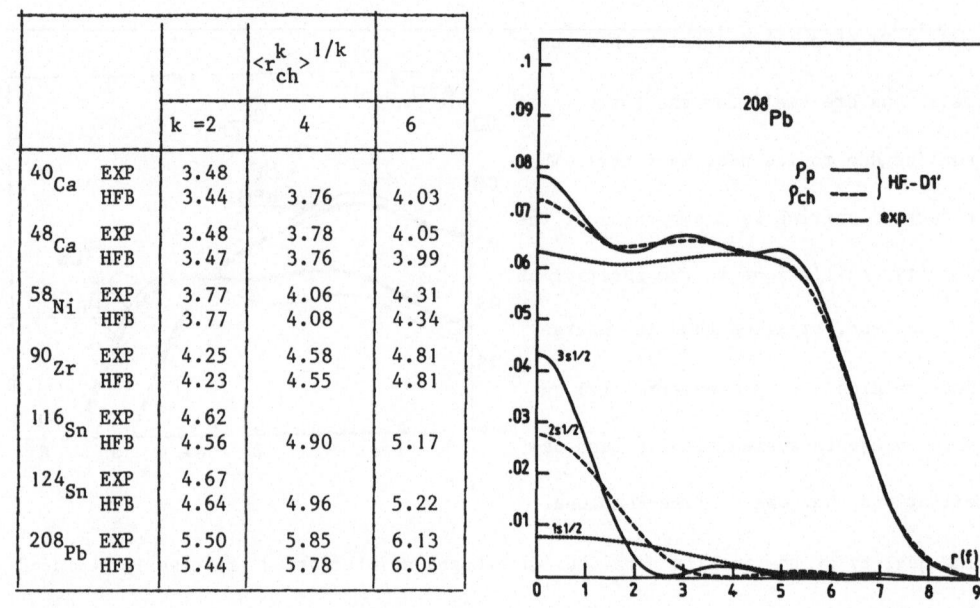

Fig. 3

strong oscillations in the nuclear interior (Fig. 3). An analysis of there oscillations attributes them mainly to the $3S_{1/2}$ level. As we will see in the next section a change in the occupation probability of this level will result to a substantial smoothing of the charge density.

b) Deformed nuclei.

The success of the study of spherical nuclei with Gogny's interaction and the HFB approach has spurred the extension of this method to the domain of deformed nuclei [7,8]. The generalization of the formalism is quite straightforward although the practical implementation presents considerable difficulties and leads to lengthy calculations.

One of the most interesting results of the study of deformed nuclei is the existence of shape transitions in families of isotopes. In Fig. 4 we display the results for the energy versus deformations for three such families. The choice of the quantity $Q_{20}A^{-5/3}$ as deformation parameter (with Q_{20} being the usual quadrupole moment)

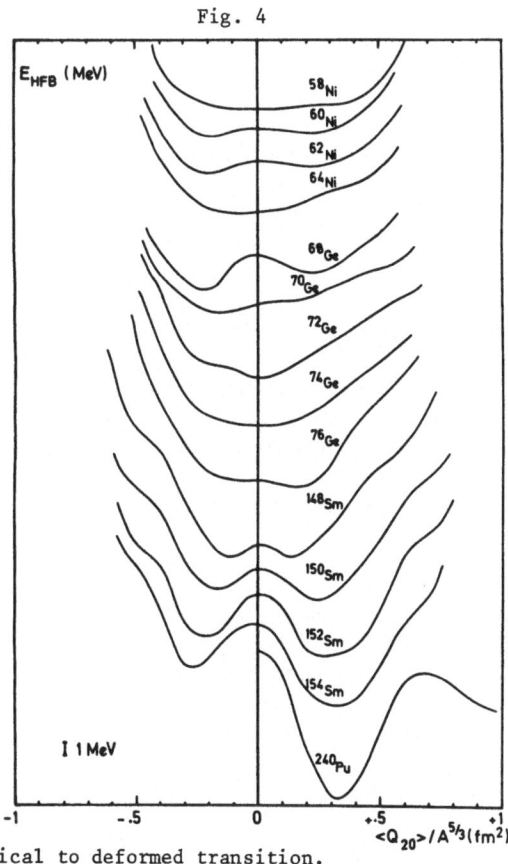

Fig. 4

ensures that the deformations remain comparable for all mass numbers. This is most clearly seen in the case of ^{240}Pn, whose ground state deformation in not much different with the ones encountered in ^{58}Ni.

One clearly sees in the figure the various transitions from spherical to deformed the spherical for the Ni isotopes, although a great softness characterizes the whole family. In the Ge case we encounter an oblate to prolate transition, while for the Sm isotopes we are in presence of a spherical to deformed transition.

In some cases it may even happen in a particular family of transitional nuclei that the transition proceeds through ellipsoidal (triaxial) deformations. In this case a member of the family may posseses a triaxial ground state. This is the case for example for the nucleus ^{134}Ce, as can be seen in Fig. 5, where we plot the energy versus the β,γ Bohr deformation parameters. However in all triaxial nuclei encountered a great softness with respect to the γ deformation was observed.

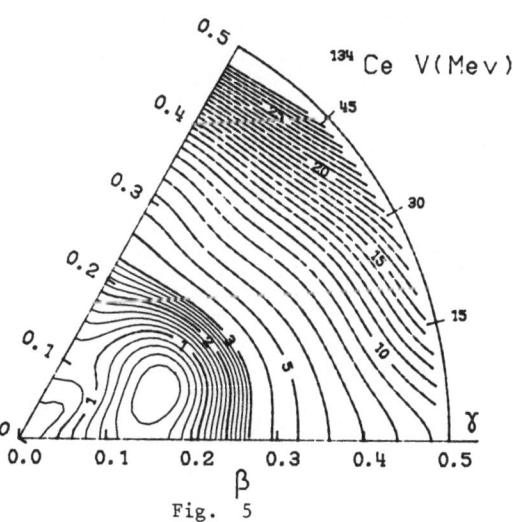

Fig. 5

III - DYNAMICAL CORRELATIONS.

a) Small amplitude vibrations of spherical nuclei.

In order to study the importance of dynamical correlations in the ground state of nuclei we distinguish the spherical from the deformed ones. The former, being particularly stiff, execute only small amplitude collective motions. If one starts from an energy functional calculated an a one-body density matrix (according to the mean field hypothesis) the small amplitude assumption leads to an expansion around the Hartree-Fock equilibrium density matrix. This leads to the well known RPA equations. One obtains thus the energy of the excited states of the system together with their amplitude (X^N, Y^N).

In ref. [9] have been presented RPA results concerning RPA excitation energies and transition strengths for various spherical nuclei. A major result of these calculations has been the prediction of the existence of a giant monopole resonance in medium and heavy nuclei [10]. Its analysis, based on experimental data has made possible the determination of the value of the nuclear matter compression modulus.

Here we limit ourselves to the presentation of some selected transition densities for ^{208}Pb (Fig. 6). The 3^- state has been obtained through a very precise measurement performed up to very high transfer in Saclay [11]. The rest of the states come from the extensive MIT experiment [12]. The agreement between theory and experiment is quite satisfactory and would encourage detailed comparisons in other regions of the periodic table as well.

What is also most important is the investigation of the effect of dynamical correlations on the ground state wave-functions. This can be carried through in the framework of the RPA provided one makes the assumption that the ground state of the system is approximated by the vacuum of quasi-bosons θ_N^+ associated to the RPA eigenvectors (X^N, Y^N).

We have thus $\theta_N^+ = \sum_{ph} X_{ph}^N A_{ph}^+ + Y_{ph}^N A_{ph}$
where $A_{ph}^+ = a_p^+ a_h$ defined on the Hartree-Fock basis. So the ground states is given by $\theta_N |GS\rangle = 0$. The explicit calculation (which can be found in ref. [13]) gives

$|GS\rangle = \exp\left[-\frac{1}{2} \sum_i \omega_i (b_i^+ b_i^+ - b_i b_i)\right] |HF\rangle$

The ω_i's are related to the eigenvalues r_i of the matrix $\tilde{Y}Y$ through $\omega_i = \text{argsh } r_i^{1/2}$ and the quasi-bosons b_i are deduced from the initial ones A^+ via the matrices D which diagonalizes $\tilde{Y}Y$ i.e. $b^+ = DA^+$.

It is clear that the ω_i are a measure of the correlations in the ground state. Once the form of the ground state wave-function is known one can calculate the occupation probabi-

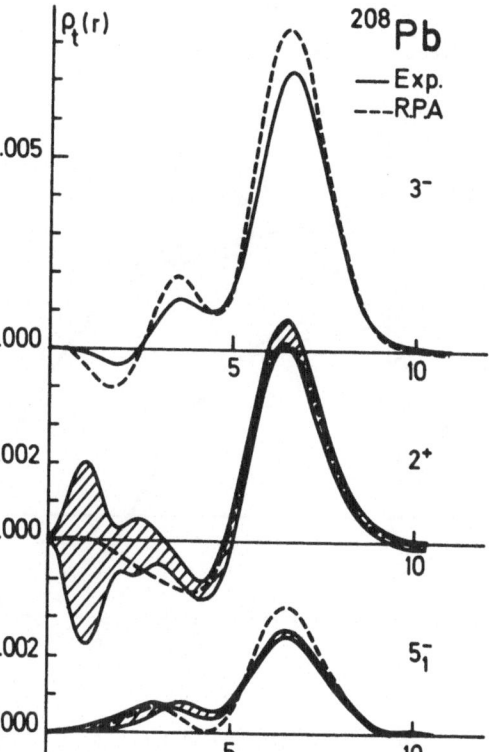

Fig. 6

lities of the particle and hole states. The change in a particle-particle element of the density matrix for example is given graphically by :

$\delta \rho_{pp} = \sum_{hN} \begin{pmatrix} p & h \\ & N \end{pmatrix}$

where $\begin{pmatrix} N \end{pmatrix}$ is the RPA propagator of the excitation N. As is well known this procedure leads to a double counting in the expectation value for all one or two body operators. However it can be most easily corrected provided we substract from the $\delta \rho_{pp}$ previously defined the quantity :

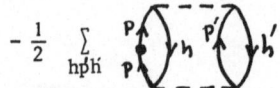

The effect of these long range correlations on the charge densities of ^{40}Ca, ^{48}Ca,

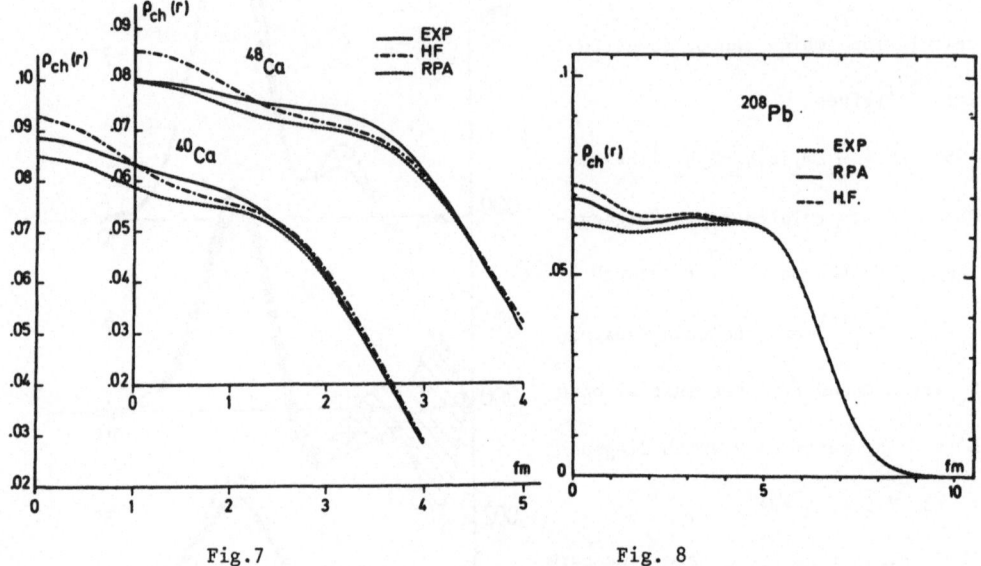

Fig.7 Fig. 8

^{208}Pb is shown in figures 7 and 8. It is clear that the inclusion of this effect smoothes out the Hartree-Fock oscillations and brings it closer to the experimental one. However, at least in the case of ^{208}Pb, there remain still some significant differences. This is far from being negative result as we will see in section IV.

b) Adiabatic vibration of soft (deformed) nuclei.

The existence of rigid deformed nuclei is undeniable. The observation, experimentally, of rotational spectra spectra extending up to considerably high spin constitutes a substantial evidence. However there exists also a large class which occupy the space between the rigid spherical and the rigid deformed ones. These nuclei have a main common feature : they are solft with respect to deformations. This means that, independently of their ground state deformation, their description through a single Slater determinant (or HFB wave-function) is inadequate. The amplitude of the zero-point motion of the nucleus is sufficiently large to mix quite different shapes in the ground state wave-function. Another common feature of the collective motion of such nuclei is that it is associated to small velocities i.e. It is adiabatic (at least the modes on which we are going to focus our interest). One way to take into account the ground state correlations induced by the large amplitude collective motion is through the generator coordinate method. In this method one starts from a constrained mean-field calculation which generates functions with various deformations. They are subsquently projected in good angular momentum and mixed. The weight functions which determine the mixing are determined variationally. However, this approach can be substantially simplified once ones makes use of the adiabaticity and retains just quadratic (in velocity) terms in the nuclear collective Hamiltonian. A restriction of the collective degrees of freedom to the quadrupole ones (which are the most important for the low lying excitation spectrum of soft nuclei) leads then to a Hamiltonian of the Bohr-Mottelson type in its Kumar-Baranger version. Once the collective Hamiltonian is solved (and such a program has actually been carried through on a microscopic basis [14]) one obtains, apart from the energy of excited states, an expression for the nuclear wave-function which incorporates the correlations induced by the collective motion of the system. The correlated ground state as well as the transition densities can then be easily calculated.

As an illustration of this method we present the results obtained for the ground state and first transition charge densities of ^{58}Ni. The choice of this nucleus has been dictated by the fact that it is a most typical soft nucleus as can be seen in figure 9 where we plot its energy surface. In figure 10 we give the uncorrelated ground state density obtained through a HFB calculation (as well as the one obtained with Skyrme interaction SIII). They both present large oscillations. Once we include the dynamically correlated ground states the correlations are strongly damped

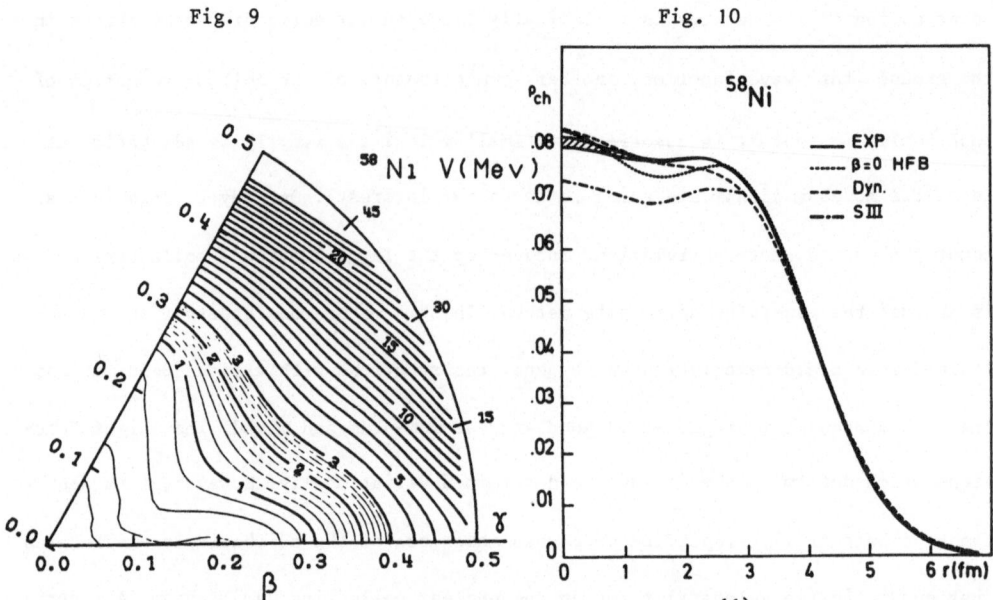

Fig. 9 Fig. 10

and the agreement with the model independent analysis result [11] substantially improved.

In the case of the transition density we have tried a more simplified approach in order to assess the importance of triaxiality on its spatial distribution. We have thus performed calculations at various fixed triaxialities, using the Davydov-Filippov model, for $\beta = 0.2$, and also at $\beta = 0.3$ for $\gamma = 0$. The results for $\beta = 0.2$ are contained in the dashed region of fig. 11. This deformation corresponds roughly to the mean dynamical deformation of the nucleus and allows one to predict the shape of the transition density, at least with a simple minded collective wave-function. A comparison with the experimental result reveals thus clearly

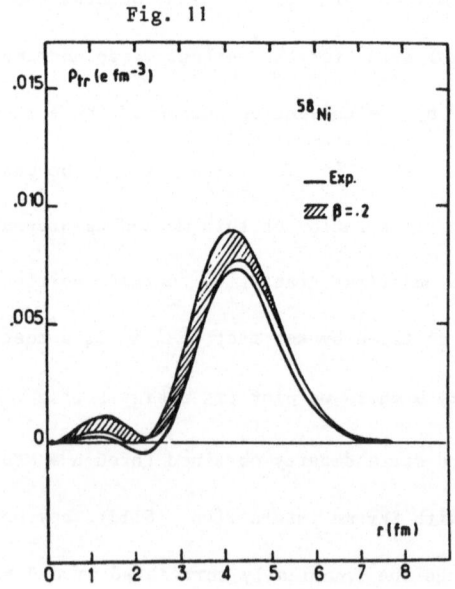

Fig. 11

the necessity for a more sophisticated collective wave-function, taking into account vibrations in the β-γ plane as well as rotations. Such a calculation, using Kumar-Baranger's version of the Bohr Hamiltonian is actually in progress.

IV - CONCLUSION

In the previous sections we have presented a brief summary of results obtained in the framework of the mean-field approximation. We have shown that starting from an adequately defined effective nuclear interaction one can obtain a unified microscopic description of the nuclear ground and existed states. This internal consistency is achieved through the systematic use of the mean field assumption in the calculations of the static properties as well as the dynamical ones.

The quality of the results, as we have seen, is quite satisfactory. In some cases, we have in mind the charge density of ^{208}Pb in particular, some discrepancies still exist. This simply implies that there exist in the nucleus correlations that cannot be represented as superposition of simple particle-hole ones. It is already most gratifying that an independent quasi-particle picture of the nucleus, based on the mean field assumption, corrected for some dynamical correlations, can attain such a degree of accuracy.

AKNOWLEDGEMENTS

The involvement of the author in the study of the most interesting aspects of nuclear structure has been made possible mainly through his collaboration with M. Girod, to whom he wishes to express his gratitude. He aknowledges also most interesting discussions with D. Gogny, J. Déchargé and B. Frois who kindly made their results available to him.

References

1) Proceedings of the International Conference on recent progress in many-body theories, Nucl. Phys. A328 1.2 (1979)

2) R. Vinh Mau, this Conference

3) D. Vautherin, D.M. Brink, Phys. Rev. C5 (1972) 626

   P. Quentin, H. Flocard, Ann. Rev. Nucl. Sci. 28 (1978) 523

4) J. Déchargé, D. Gogny, Phys. Rev. C21 (1980) 1568

5) P. Bonche, S.E. Koonin, J.W. Negele, Phys. Rev. C13 (1976) 1226

6) D. Gogny in " Nuclear Physics with Electromagnetic Interactions", (Springer Verlag 1979) p. 88

7) J. Déchargé, M. Girod, D. Gogny, Phys. Lett. 55B (1976) 361

8) M. Girod, B. Grammaticos, Phys. Rev. Lett. 40 (1978) 361

9) J.P. Blaizot, D. Gogny, Nucl. Phys. A284 (1977) 429

10) J.P. Blaizot, D. Gogny, B. Grammaticos, Nucl. Phys. A265 (1976) 315

11) B. Frois, private communication

12) J. Heisenberg in " Nuclear Physics with Electromagnetic Interactions ", (Springer Verlag 1979) p. 33

13) J. Déchargé, D. Gogny in 5è Session d'Etudes Biennale, Aussois 1979, LYCEN 7902, C.12

14) M. Girod, K. Kumar, B. Grammaticos, P. Aguer, Phys. Rev. Lett. 41 (1978) 1765.

MANY-BODY ASPECTS OF ELECTRON SCATTERING AT INTERMEDIATE ENERGY

B. Frois
DPh-N/HE, CEN Saclay, 91191 Gif-sur-Yvette Cedex, France

I. INTRODUCTION

The many-body problem plays a special role in nuclear physics. Most nuclei have a number of constituents which is neither small enough to permit the use of techniques developed for few body problems, nor large enough to approach the limit where statistical methods become applicable. Thus, it should be possible to obtain a solution to this problem in which all the fundamental interactions between all the nuclear constituents appear explicitly. It is a very difficult task and this complexity is a clear challenge to our ability to develop refined theoretical techniques since it is not possible to solve the nuclear many body problem in the general case with the largest computers available. The experimental determination of the spatial distributions of nuclear constituents provides a solid foundation for further progress. It severely tests theoretical approaches by providing direct information about the nuclear wave functions which are the solutions of the problem. The spatial distribution of nucleons constitutes a touchstone of any microscopic calculations. For a long time only integral properties of the wave functions have been measured, such as transition probabilities $B(E\lambda)$. Now precise experimental densities can be obtained by scattering of intermediate energy electrons. It is precisely in this area that the impact of the latest generation of electron accelerators has been felt in nuclear physics. There are no other available experimental probes having the advantages enjoyed by the electron. The weakly interacting electron penetrates the nucleus with a well understood mechanism, the electromagnetic interaction, and is able by varying the momentum transfer to map out all the multipoles of nuclear densities throughout the nuclear volume without ambiguity. The previous generation of experiments had been limited by experimental considerations to maximum momentum transfers of about 2.7 fm^{-1}. Because of the inverse relationship between momentum transfer and wavelength such experiments were insensitive to the density in the central region of the nucleus. The extension of electron scattering techniques to momentum transfers adequate to probe the interior of the nucleus has required experiments which are at the frontier of present techniques. The cross sections that have to be measured are very small, typically between 10^{-35} cm^2/sr and 10^{-38} cm^2/sr. In order to isolate discrete excited states at high momentum transfers one needs exceedingly good energy resolution. $\Delta E/E = 10^{-4}$ is already insufficient for the study of some weak excitations that are very interesting such as 0^+ states or $0^+ \to 0^-$ transitions. These constraints are very severe and the feasibility of high momentum transfer experiments for deformed nuclei is very recent. This explains why there is much less electron scattering data than available from other spectroscopic tools. Nevertheless, we are gradually seeing the emergence of a new generation of electron scattering data that provide landmarks in the history of nuclear structure.

In this talk, I would like to make a brief review of some recent electron scattering experiments from medium and heavy nuclei. Examples have been chosen to illustrate the power of electron scattering experiments at intermediate energy. No attempt has been made to provide a comprehensive survey of all the experimental work which is being carried out throughout the world in this field. The experiments I have chosen to present are directly interpretable and give both qualitative and quantitative information in a simple way to help us understand various aspects of the nuclear many body problem.

II. THE GROUND STATE CHARGE DISTRIBUTION OF CLOSED SHELL NUCLEI

About ten years ago, a major advance was achieved by nuclear theorists with the introduction of density dependent terms in the nucleon-nucleon interaction. This cured one of the major difficulties associated with the short range repulsion in the NN interaction. Mean field theories are now correctly predicting the saturation properties of nuclei. The average shapes of nuclear charge distributions are also reasonably well reproduced. It is only in the interior of the nucleus that significant differences remain between experiment and theory. A reliable comparison in the nuclear interior requires form factors measured up to $q \sim 3.5$ fm^{-1}. Such experiments have been carried out at Saclay for ^{40}Ca, ^{58}Ni, ^{116,124}Sn and ^{208}Pb. The cross sections decrease by a factor larger than 10^{10} in this range of momentum transfer. Fig. 1 shows the charge densities deduced from a simultaneous fit to elastic electron scattering and muonic X rays data[1]). The experimental uncertainty in the density, which is of the order of 1 %, is represented by the shaded area in the figure. The dashed curves are mean field predictions of Dechargé and Gogny[2]). In the case of ^{208}Pb and ^{40}Ca the theoretical calculations include ground state correlations calculated in a self-consistent way with the same finite range force.

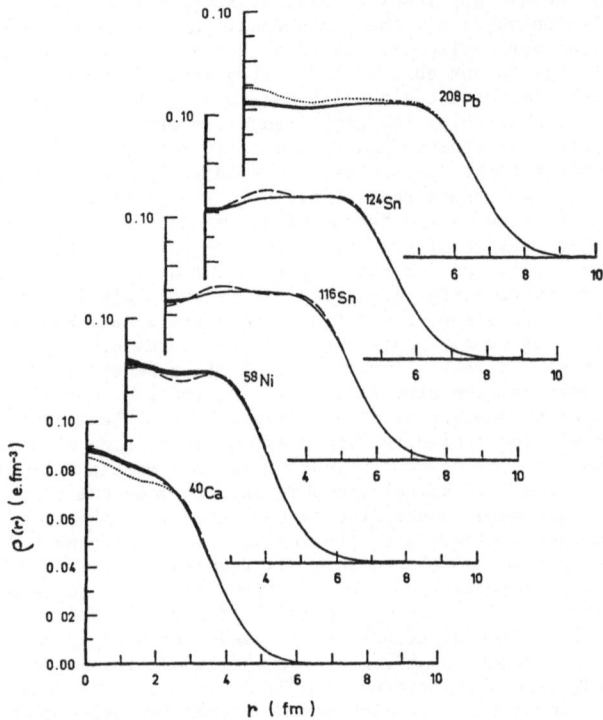

Fig. 1 - *Charge densities of the ground state of ^{40}Ca, ^{58}Ni, ^{116}Sn, ^{124}Sn, ^{208}Pb. The solid lines in black are the experimental results. The dashed lines are mean field predictions. The dotted lines are the same calculations including long range correlations.*

Some systematic trends can be extracted from the comparison of the theoretical and experimental charge densities. The theory predicts too much structure, except in the case of ^{40}Ca where the inclusion of ground state correlations has considerably improved the shape of the theoretical density. However, there is a lack of balance in the distribution of the charge in the nucleus. There is not enough charge in the center of ^{40}Ca while there is too much charge in the center of ^{208}Pb. It should be noted that the experimental uncertainty in the density is imperceptible. This comes from the combination of very precise muonic X rays measurements that fix the absolute normalization and precise elastic scattering data over a variation of 10^{10} in cross section. A 1 % uncertainty in {r} space does not mean that all uncertainties are 1 % in cross sections. The correlation is much more indirect and requires a deeper look into Fourier analysis. The best example was shown by Negele and Riska[3]) with calculations of meson exchange corrections for charge distribution. A 0.5 % effect in the charge distribution is associated with a 30 % variation in the cross section at some values of the momentum transfer. It is also reassuring that measurements at different laboratories, and reanalysis of these data by various world experts are yielding the same results.

An interesting effect has been observed in the measurement of charge differences between neighbouring nuclei[1]. The charge difference between the isotopes of heavy nuclei is surprisingly well reproduced by all mean field calculations even though all these calculations predict too much structure for the charge density of each nucleus. This implies that both the wave function of the last nucleon and the polarisation induced in the nucleus by the addition of the last nucleons are correctly predicted by mean field calculations for a heavy nucleus. An experiment has recently been completed at Saclay in which the charge differences between neighbouring nuclei in the region of lead were measured, extending to higher momentum transfers previous data from Mainz[4]. Fig. 2 shows the ratio of cross sections of ^{205}Tl to ^{206}Pb. The highest momentum transfer data measured at Mainz have been omitted just for the clarity of the figure since their statistical errors are much larger than those of the present experiment. However these data are compatible with our new results. The striking variation observed in the cross section ratio is mostly due to the effect of the $3s_{1/2}$ proton, whose wave function in momentum space has a peak in the vicinity of 2 fm^{-1}. There is a good agreement between the experiment and the theoretical prediction of Décharge and Gogny[2] although some differences in detail still remain.

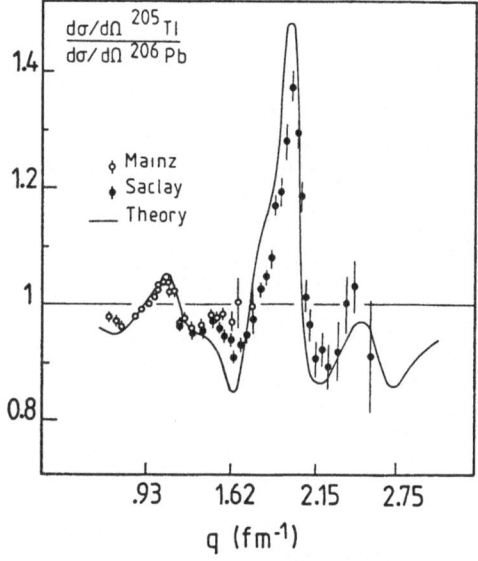

Fig. 2 - *The ratio of cross sections ^{205}Tl to ^{206}Pb. The open circles are data from Mainz. The black points are preliminary data from Saclay.*

All these experiments clearly show the progress accomplished by nuclear theory. But systematically something is missing to reproduce correctly the shape of the charge distribution of closed shell nuclei. The inclusion of ground state correlations has improved the agreement with experiment but they are not sufficient. The difficulty of extending such calculations arises from the nature of effective interactions. Some correlations are already included and there appears at present no clear way to disentangle the effects of the higher order corrections in the theory.

III. THE TRANSITION CHARGE DENSITY OF THE FIRST EXCITED STATE OF ^{208}Pb

The first excited state of ^{208}Pb is a very collective octupole vibration whose energy and transition probability can be understood within the framework of the random phase approximation. Recently, new measurements by Goutte et al.[5] have determined the transition charge density of this level to an accuracy comparable to the measurements of the ground state density. The result is shown in Fig. 3 together with various theoretical predictions. It is striking that all the theories fail to reproduce the shape of the oscillations in the interior of the nucleus, even when the strength of the transition has been adjusted so that the surface peak is correctly reproduced. This effect is closely related to the central depression observed in ^{208}Pb. Correlations tend to remove particles from inside of the nucleus, smoothing the fluctuations both in the charge distribution of the ground state and in the tran-

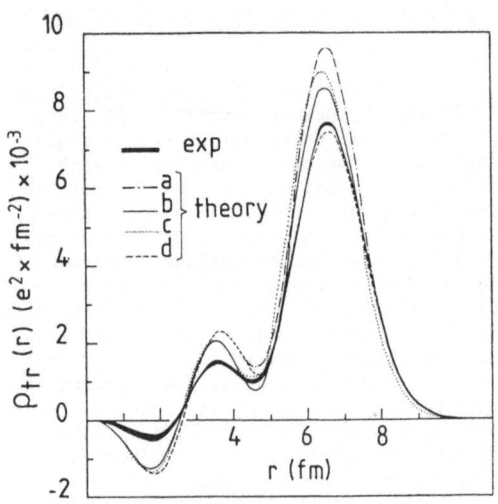

sition charge density of the 3⁻ state. This can be seen in the negative lobe of the transition charge density which is almost entirely due to the $[2f_{7/2} 3s_{1/2}^{-1}]$ particle hole component. A strong depletion of the proton $3s_{1/2}$ orbit would improve the agreement between theory and experiment for the charge distribution of the ground state. This would also reduce the amplitude of the $[2f_{7/2} 3s_{1/2}^{-1}]$ ph configuration and thereby reduce the amplitude of the negative lobe in the 3⁻ transition charge density.

Fig. 3 - The heavy line is the transition charge density that corresponds to the best fit to all existing data. The othen curves are RPA calculations : a) Hamamoto ; b) Dechargé and Gogny ; c) Bertsch and Tsai ; d) Heisenberg[6]).

IV. HIGH MULTIPOLARITY MAGNETIC TRANSITIONS TO PARTICLE HOLE STATES

We have just discussed the example of a very collective excitation which includes many particle hole configurations. It is quite obvious that one would also like to isolate simple particle hole excitations to have additional information on the wave functions and the occupation probabilities of the different orbits. An excellent technique for isolating single particle-hole wave functions is the observation of high multipolarity magnetic transitions to "stretched" configurations. Such configurations can be produced by very few particle hole configurations, because of their parity and of the restrictions due to the high angular momentum involved both in the initial and in the final state. By carefully choosing the kinematic conditions it is also possible to experimentally isolate these transitions quite well. Because of the properties of the magnetic operator, each contributing multipolarity can be preferentially excited in a certain range of momentum transfer. The transverse nature of the magnetic transition strongly enhances the cross sections at backward angle, while simultaneously all the longitudinal excitations are considerably reduced. Another advantage of the study of these transitions is that the orbital component of the angular momentum is identically zero. These transitions are induced only by the spin density so they probe a very specific part of the nucleon-nucleon interaction. One might be puzzled by the question : "if these transitions are so interesting and are so easy to isolate, how can it be that they have not been measured before?" The answer is quite simple, it was technically impossible before the present generation of experimental facilities which include magnetic spectrometers of very high energy resolution and excellent background rejection together with electron accelerators of very high intensity.

Fig. 4 is a summary of the magnetic strength of the known 1 hw stretched states which have been measured to date mostly by a combined effort of different groups at MIT Bates laboratory[7]. The strengths are presented as functions of mass number A in units of the pure single particle strength. All magnetic transitions are seen to be reduced by about 50 % from the pure single particle estimate. For closed shell nuclei such as ^{40}Ca, ^{90}Zr, ^{208}Pb, only one state of the highest multipolarity is observed, while for non closed shell nuclei such as ^{54}Fe, the strength is fragmented in many states (Fig. 5).

Recent attempts have been made to understand the structure of the M12 and M14 transitions in ^{208}Pb [ref.[8]]. Both the work of Hamamoto et al. and Krewald et al. have shown that the 50 % reduction observed in the strength can be explained by configuration mixing. Recently, the work of Dechargé, Sips and Gogny has determined the M12 and M14 form factors in a fully self consistent RPA calculation with the same

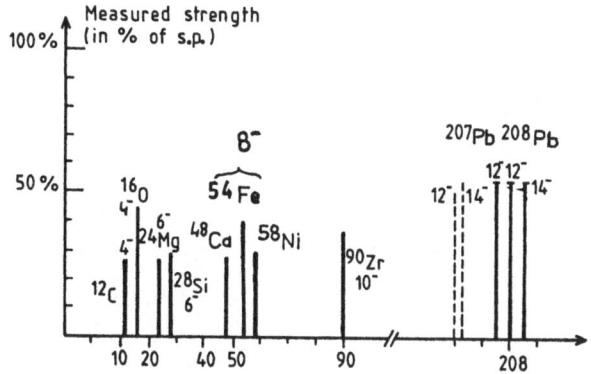

Fig. 4 - Measured magnetic strength in single particle units.

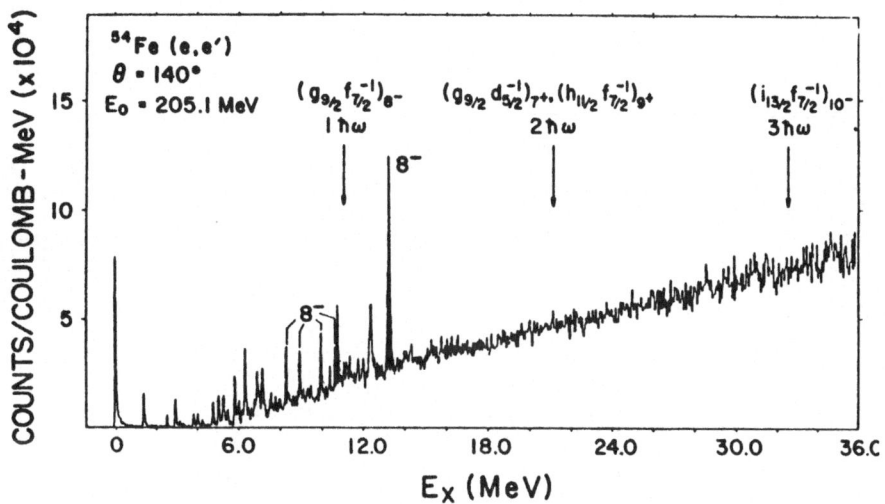

Fig. 5 - Inelastic electron spectrum from ^{54}Fe measured by R. Lindgren et al.[7]). The M8 Strength is fragmented in many states.

finite range force D1 that was used to describe the 3⁻ state of ^{208}Pb (Fig. 3) and the ground state charge densities of Fig. 1. In this work, there are no free parameters because the effective force has been fixed by matching known properties of ^{16}O, ^{90}Zr and Sn isotopes. Their RPA calculation which includes correlations in the ground state, confirms the result of Hamamoto et al. and of Krewald et al. (Fig. 6). However for the 12⁻ at 6.43 MeV which is a neutron state, this calculation gives an unexpected and disastrous result (Fig. 7), which is not seen in the 12⁻ proton state (Fig. 8). Dechargé et al. have investigated this effect. They have found that the cause is a destructive interference between the two 12⁻ states, the difference predicted by the theory being too small $\Delta E \sim 300$ keV. By increasing the energy of the proton state by 500 keV more reasonable result is obtained (curve a). They have also found a better agreement by neglecting the proton 12⁻ state in the calculation of the neutron 12⁻ state and vice-versa. This result is quite interesting because it shows that these experimental results are very sensitive to the various theoretical approximations. It is clear that the HF energies used in the calculation of Dechargé et al. are not sufficient and that one should definitely include higher order corrections such as particle vibration coupling in order to obtain a fully self-consistent description for not only these stretched states but also the ground state and the collective excitations such as the first 3⁻.

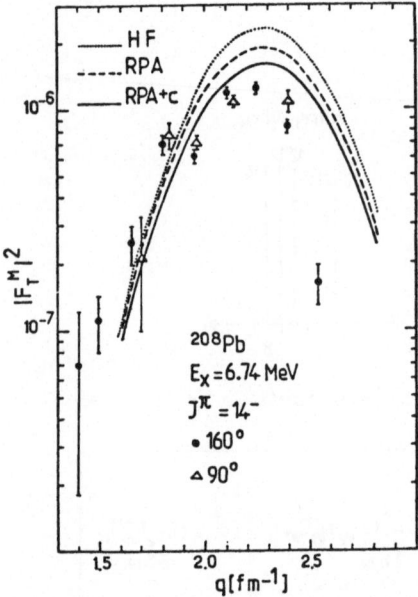
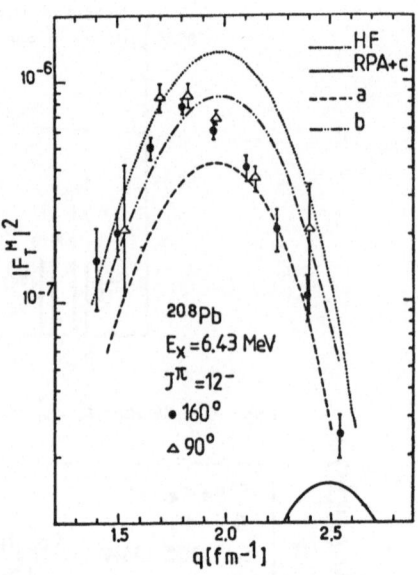

Fig. 6 – Form factor of the M14 magnetic transition in ^{208}Pb. The data have been obtained by Lichtenstadt et al.[7]. The theoretical curves are from Dechargé et al.[8].

Fig. 7 – M12 form factor of the neutron state. Curve a correspond to an artificial increase of the difference in energy between the 12$^-$ proton and neutron states of 500 keV. Curve b neglects the proton state.

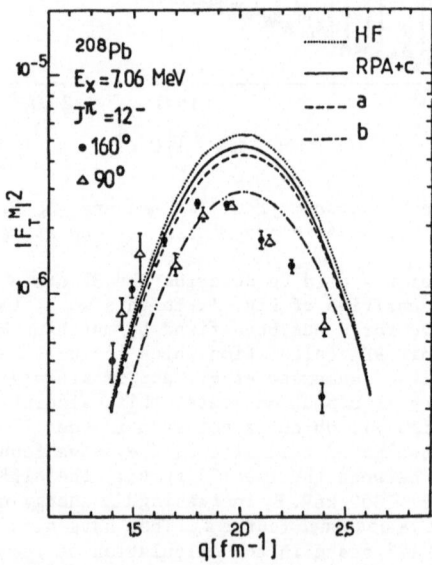

Fig. 8 – M12 from factor of the proton state. Curve a correspond to an artificial increase of the difference in energy between the 12$^-$ proton and neutron state of 500 keV. Curve b neglects the neutron state.

V. THE DETERMINATION OF THE SHAPE OF VALENCE ORBITS BY HIGH MULTIPOLARITY MAGNETIC ELECTRON SCATTERING

For odd-even nuclei with a ground state of angular momentum equal to the maximum angular momentum of all occupied shells, the magnetic form factor at high momentum transfer is almost entirely due to the intrinsic magnetization density of the impaired nucleon. The highest magnetic multipole of the form factor can be isolated by elastic scattering at backward angles. The interpretation of the scattering magnetic data can be directly related to the wave function of the valence nucleon. In particular, it is possible to determine the shape of the valence orbit in the interior of the nucleus. In order to determine accurately the radial part of the wave function it has been necessary to measure very small cross sections, of the order of 10^{-37} cm^2/sr.

Such measurements have been done for many nuclei, from ^{17}O to ^{209}Bi. In this talk, I would like to discuss only the case of the $f_{7/2}$ and $g_{9/2}$ shells that was studied in great detail by the electron scattering group at Saclay[9].

Fig. 9 shows the magnetic form factor of ^{93}Nb which has been mapped out in the entire region of momentum transfer to the highest multipolarity by experiments at IKO[15], MIT Bates[16] and Saclay[9]. At high momentum transfer, the form factor is almost a pure M9 multipole. The solid curve is a theoretical prediction of Decharge and Gogny[2] which corresponds to the magnetization of a pure $g_{9/2}$ orbit. Configuration mixing effects would not change the shape of the M9 form factor. This can be easily understood. Configuration mixing depletes the $g_{9/2}$ orbit but has a very small

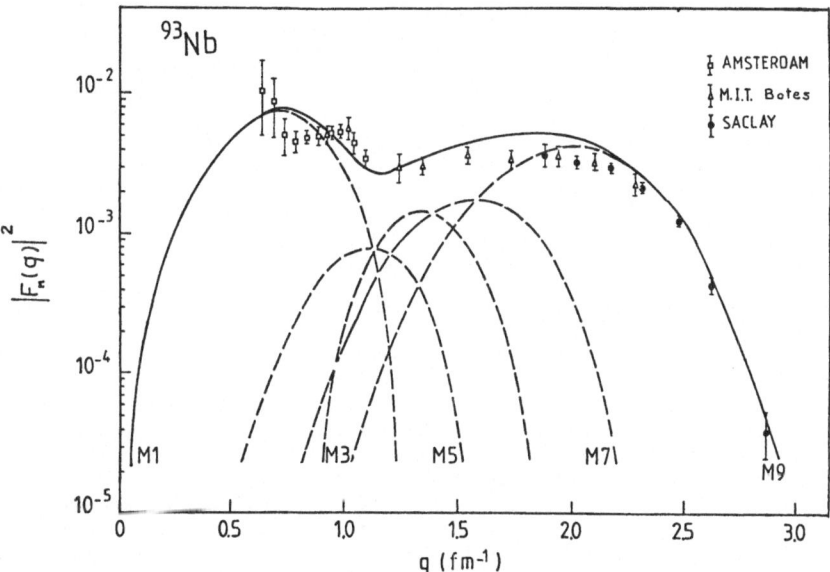

Fig. 9 - Magnetic form factor of the $1g_{9/2}$ proton orbit in ^{93}Nb. The solid curve is a theoretical calculation[2] and the dashed curves are the main contributions of the individual multipoles.

probability of populating shells of the same parity as ^{93}Nb ground state whose angular momentum is higher than or equal to 9/2. A second possible source of ambiguity in the interpretation of these experiments would be meson exchange corrections. Recent calculations by Desplanques and Mathiot[10] have shown that these corrections are small in the M9 region due to partial cancellations of the different terms in the calculation. Here again, the magnetic high multipolarity transition is a very selective probe and can be interpreted unambiguously.

Fig. 10 is a comparison of the radii deduced from these experiments for the $f_{7/2}$ shell proton (^{51}V) neutron (^{49}Ti) and the $g_{9/2}$ shell proton (^{93}Nb) neutron (^{87}Sr), with theoretical mean-field predictions from Negele and Vautherin (DME)[11], Campi and Sprung (DDHF)[12], and Decharge and Gogny (DDHFB)[2].

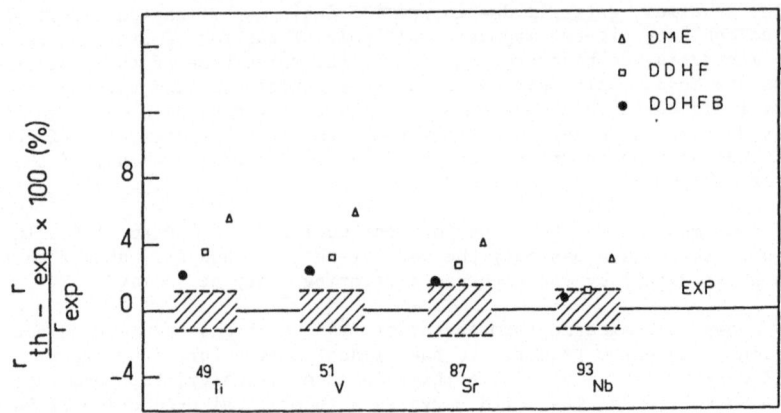

Fig. 10 - Percentage deviations between the values of the rms radii of the $1f_{7/2}$ and $1g_{9/2}$ shells found by magnetic elastic electron scattering and three different mean field calcualtions described in the text. The shaded area corresponds to the experimental uncertainty.

The difference between experiment and theory is not very large, but significantly all the calculations predict radii that are larger than the experiment.

The calculation of Decharge and Gogny gives the best agreement. It is evident from the small differences that the approximations employed are good though they are not sufficient to reproduce the experimental results. It should be emphasized at this point that by no mean this disagreement can be ignored. Because of the great sensitivity of the method a 1 % difference in rms radius may correspond to a 50 % difference in the magnetic cross section at high momentum transfer.

VI. A TEST OF THE INTERACTING BOSON MODEL IN ^{154}Gd

The interacting boson model has been remarkably successful in the description of energy levels and transition probabilities for deformed nuclei. It is striking to see how well the evolution from one nucleus to the next in an open shell is represented by a very limited number of parameters. What makes this model so appealing is its simplicity, it involves only s and d bosons made of nucleon pairs. Electron scattering is an especially attractive probe for testing this model because it is the only probe that can determine the radial structure functions which are needed to predict the variation of the form factors. A direct implication of this model is that there are only two linearly independent form factors for quadrupole transitions and all quadrupole transitions in the same nucleus should be a linear combination of these two form factors. A good example is the case of ^{154}Gd where it is possible to measure the form factors of the 2^+ rotation and of the 2^+ β and γ vibrations. An experiment is being carried out by Hersman et al.[13] at MIT Bates laboratory to study this problem. A spectrum of inelastically scattered electrons from ^{154}Gd is shown in Fig. 11. It shows that the excitation of the 2^+β is very small compared to the excitation of the 2^+γ. This experiment is an excellent example of the high resolution required to make these measurements. 10 keV is a remarkable result at 100 MeV, but at 500 MeV it is not possible at present to obtain such an energy resolution, so the momentum transfer range of such excitations will be difficult to extend beyond 2 fm^{-1}. For the moment, cross sections have been taken only at low momentum transfer (Fig.12) for the 2^+ rotation and the 2^+ beta and gamma vibrations. A first analysis does not support the interacting boson prediction for the transition probabilities and would

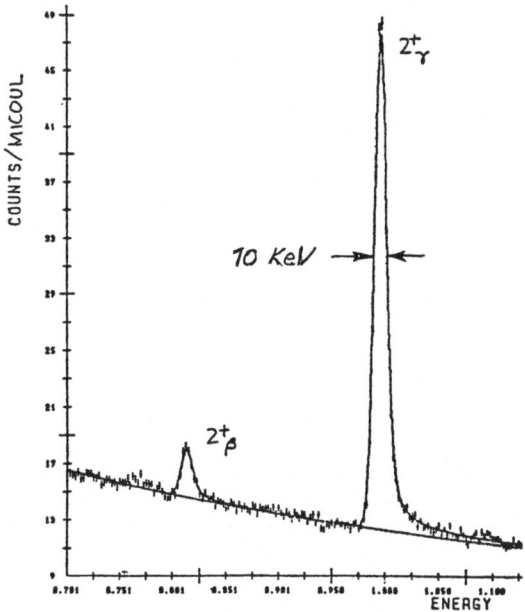

Fig. 11 - *Inelastic spectrum of electrons scattered from ^{154}Gd at 100 MeV.*

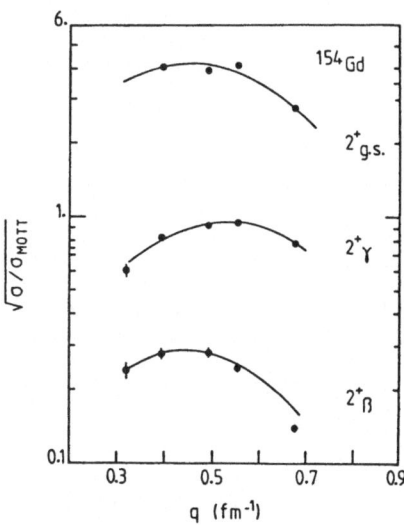

Fig. 12 - *Inelastic form factors of the 2^+ g.s., β and γ vibrations in ^{154}Gd measured by Hersman et al. at MIT Bates laboratory.*

be much closer to mean field predictions. A possible explanation of the discrepancy is in the weakness of the $2^+\beta$ excitation, since the approximations of the interacting boson model would be probably more justified for excitations of approximately the same order of magnitude.

Additional measurements at higher momentum transfer will check whether or not the three form factors are linearly independent. They will also indicate if other bosons are needed by the model to reproduce the experimental data.

An interesting study will be the measurement of the variation of the 2^+ and 4^+ form factors in a chain of isotopes. Such a study was described by Dieperink recently[14]), but to date the available data are not sufficient to be conclusive.

CONCLUSIONS

Modern electron scattering experiments are giving very detailed information on the structure of nuclei, providing the shape of the nuclear charge and magnetization distributions with an unprecedented accuracy. They now represent the most stringent experimental tests of nuclear structure calculations.

We have seen that the bulk properties of nuclei are fairly well reproduced by the mean field theory. But systematically there are significant differences between the theory and the experimental data which provide quantitative estimates of the higher order corrections that should be added to the present many body calculations. A consistent picture seems to emerge from the data available now. Correlations tend to modify the relative occupations but do not seem to modify dramatically the shape of the wave functions.

The study of deformed nuclei is still in its infancy, because very few experiments have reached sufficiently high momentum transfers to provide a reliable measurement of the fluctuations of the transition charge densities. No model presently available gives a satisfactory description of the results observed.

The solution of these problems presents a clear challenge to theorists in the years to come. One hopes that the forthcoming generation of electron scattering data will help to disentangle different effects in the nuclear many body problem.

Some experiments are already at the frontier of experimental techniques. We are now fully aware of the necessity of high momentum transfer data for the reconstruction of the transition charge densities in configuration space. Because of this fact, it will be necessary to improve the presently available energy resolution if one wants to isolate specific nuclear excitations at high q. Higher beam energies would also be useful for isolating the transverse contribution to nuclear excitations because one can then take the data for a given q value at more forward scattering angles. But the main advantage of a higher beam energy coupled with a spectrometer having sufficient energy resolution would be an increase in the number of possible experiments due to the reduction in the counting time, which varies approximately as the square of the incident energy for fixed momentum transfer. An accelerator of variable energy between 500 MeV and 2 GeV with an energy resolution of the order of 10^{-5} would be an ideal tool.

It is a pleasure for me to thank all my colleagues at Saclay who have contributed to this work. I would like to thank J. Heisenberg, W. Hersman, R.A. Lindgren, C. Papanicolas, K. Seth, C. Williamson and J. Wise for communication of their results before publication. I am very undebted to L.S. Cardman for his so helpful comments and his careful reading of this manuscript.

REFERENCES

[1] J.M. Cavedon, Thèse de doctorat d'état, Université de Paris, 1980 and references therein.
[2] J. Decharge and D. Gogny, Phys. Rev. C21 (1980) 1568.
[3] J. Negele and D. Riska, Phys. Rev. Lett. 40 (1978) 1005.
[4] H. Euteneuer et al., Nucl. Phys. A298 (1978) 452.
[5] D. Goutte et al., Phys. Rev. Lett. 45 (1980) 1618.
[6] J. Heisenberg, Advances in Nucl. Phys. to be published.
[7] T.W. Donnelly et al., Phys. Rev. Lett. 21 (1968) 1196.
  I. Sick et al., Phys. Rev. Lett. 23 (1969) 1117.
  T.W. Donnelly et al., Phys. Lett. 32B (1970) 545.
  R.A. Lindgren et al., Phys. Rev. Lett. 40 (1978) 504.
  J. Lichtenstadt et al., Phys. Rev. C20 (1979)
  J. Heisenberg, private communication (^{90}Zr).
  R.A. Lindgren, private communication (^{54}Fe).
  C. Papanicolas, private communication (^{207}Pb).
  K. Seth and C. Williamson, private communication (^{40}Ca).
  J. Wise, private communication (^{48}Ca).
[8] I. Hamamoto et al., Phys. Lett. 93B (1980) 213.
  S. Krewald and J. Speth, Phys. Rev. Lett. 45 (1980) 417.
  J. Decharge et al., Phys. Lett., in press.
  W. Knüpfer et al., Phys. Lett., in press.
[9] S.K. Platchkov et al., to be published. This paper will be a review of all the results from Saclay, some results were published by P. De Witt Huberts et al. Phys. Lett. 71B (1977) 317 and Phys. Lett. 60B (1976) 157.
  I. Sick et al., Phys. Rev. Lett. 38 (1977) 1259.
  S.K. Platchkov et al., Phys. Lett. 86B (1979) 1.
[10] B. Desplanques and J.F. Mathiot, to be published.
[11] J.W. Negele and D. Vautherin, Phys. Rev. C5 (1972) 1472.
[12] X. Campi and D. Sprung, Nucl. Phys. A194 (1972) 401.
[13] W. Hersman et al., private communication.
[14] A. Dieperink, Proceedings of the symposium on perspectives in electro and photo-nuclear physics, Saclay 1980, to be published in Nuclear Physics.
[15] G. Box, Thesis, University of Amsterdam, 1976.
[16] R.C. York et al., Phys. Rev. C19 (1979) 574.

THE GIANT DIPOLE RESONANCE AND σ_{-1}, σ_{-2} PHOTONUCLEAR SUM-RULES

Oriol BOHIGAS

Division de Physique Théorique*, Institut de Physique Nucléaire
F-91406 Orsay Cedex

1. INTRODUCTION

The nuclear photo-effect, discovered in the late 40's, is by now very well known experimentally and constitutes one of the most prominent examples of collective motion in nuclei. It has the following two main characteristics : i) the photoabsorption cross-section shows, all over the periodic table, a broad peak that takes a large part of the integrated photo cross-section, ii) the variation of the peak energy is a smooth function of the mass number A. The purpose of the present talk is to discuss what can be learned from the study of the integrated photo cross-sections

$$\sigma_{-1} = \int \frac{\sigma(\omega)}{\omega} d\omega \quad \text{(bremsstrahlung weighted)} \tag{1a}$$

and

$$\sigma_{-2} = \int \frac{\sigma(\omega)}{\omega^2} d\omega \tag{1b}$$

where $\sigma(\omega)$ is the total photoabsorption cross-section.

Notice the particular energy weighting in eqs.(1a,1b) which facilitates the comparison with finite energy evaluations and insures that the values of σ_{-1} and σ_{-2} are dominated by the nuclear photo-effect (long range correlations) in contrast to, for instance, $\sigma_0 = \int \sigma(\omega) d\omega$, for which medium and short range correlations play also an important role. All the way we shall consider only dipole contributions to $\sigma(\omega)$ and finite wavelength modifications will be ignored.

2. EXPERIMENTAL DATA

A great deal of information on the dipole strength comes from photonuclear data [1-4]. Two main sources are available :
  i) neutron emission cross-sections (γ,xn). The sum of all the partial neutron cross-sections gives practically the total cross-section $\sigma(\omega)$ for heavy nuclei because in that case the (γ,$\dot{p}$) contribution is very small due to the Coulomb barrier. For light nuclei, however, photo-neutron cross-sections will give only a fraction of the total cross-section.

  ii) Total absorption cross-sections. One must extract from the data the nuclear contribution out of the total measured cross-section. The non-nuclear part (mainly Compton cross-section and cross-sections for electron pair production) increasing rapidly with Z, this method has been used only for light nuclei (A ≤ 40).

For those reasons both methods are complementary but unfortunately it has not been possible to make a direct test on the consistency of data extracted by the two techniques.

As is well known the data on the peak energy E_D of the dipole resonance are rather well reproduced by

$$E_D = 79 \, A^{-1/3} \text{ MeV} \tag{2}$$

---
* Laboratoire associé au C.N.R.S.

Some data on σ_{-1} and σ_{-2} are reproduced on table 1, the values for light nuclei coming from total cross-section measurements, for medium and heavy nuclei from photoneutron cross-sections. The empirical values for $A \geqslant 100$ are well reproduced by

$$\sigma_{-1} = (0.22 \pm 0.02) \; A^{4/3} \; \text{mb} \tag{3a}$$
$$\sigma_{-2} = (2.7 \pm 0.2) \; A^{5/3} \; \mu\text{b/MeV} \tag{3b}$$

Notice that for light nuclei the values of σ_{-1} and σ_{-2} are significantly larger than the ones provided by eqs.(3).

	$\sigma_{-1} \; A^{-4/3}$ (mb)	$\sigma_{-2} \; A^{-5/3}$ (μb.MeV^{-1})
^{7}Li	0.42	10.6
^{12}C	0.32	5.0
^{16}O	0.36	5.8
^{40}Ca	0.33	4.8
^{90}Zr	0.18	2.3
Sn	0.20	2.7 $\pm$ 0.2
Ce	0.22	2.5 $\pm$ 0.2
Sm	0.21	2.8 $\pm$ 0.2
^{197}Au	0.21	2.6 $\pm$ 0.2
^{208}Pb	0.22	2.6 $\pm$ 0.2

Table 1 : Experimental values of σ_{-1} and σ_{-2}.

3. GENERAL REMARKS AND SIMPLE MODELS [5-7]

The integrated photoabsorption cross-sections σ_p are related to the sum-rules m_p

$$\sigma_p = 4\pi^2 \; \frac{e^2}{\hbar c} \; m_{p+1}(D_z) \qquad (p = 0, -1, -2) \tag{4}$$

where

$$m_p(D_z) = \sum_n E_n^p \; |<n|D_z|0>|^2 \; ; \tag{5}$$

σ_{-2} is related to the static dipole polarizability α_D

$$\sigma_{-2} = 2\pi^2 \; \frac{e^2}{\hbar c} \; \alpha_D \tag{6}$$

In eq.(5) E_n is the excitation energy of the eigenstate $|n\rangle$ and D_z is the z-component of the dipole operator $\vec{D}$ referred to the center of mass $\vec{R}$

$$\vec{D} = \sum_{i=1}^{Z} (\vec{r}_i - \vec{R}) \tag{7}$$

It can also be written

$$\vec{D} = \frac{NZ}{A} \vec{r}_{ZN} \tag{8}$$

where $\vec{r}_{ZN} = \vec{R}_Z - \vec{R}_N$ is the relative coordinate of the center of mass of protons with respect to the center of mass of neutrons. Consequently, photoabsorption is only sensitive to the relative motion of protons as a whole with respect to neutrons as a whole. Let us consider

$$\sigma_{-1} = \begin{cases} = \frac{4}{3} \pi^2 \frac{e^2}{\hbar c} \left(\frac{NZ}{A}\right)^2 \langle 0|\vec{r}_{ZN}^2|0\rangle & (9) \\ = 4\pi^2 \frac{e^2}{\hbar c} \langle 0|D_z^2|0\rangle & (10) \end{cases}$$

Eq.(9) tells that σ_{-1} provides a measure of the Goldhaber-Teller zero-point motion. Eq.(10) that it provides a measure of the ground state expectation value of a two-body operator (D^2) giving thus <u>direct information on two-body correlations.</u>

Let us now assume that the motion of the coordinate $\vec{r}_{ZN}$ is decoupled (as it is in the harmonic oscillator independent particle model-HOSM) and that its Hamiltonian H_{ZN} is

$$H_{ZN} = \frac{p^2}{2\mu} + \frac{1}{2} \mu \Omega^2 r_{ZN}^2 \tag{11}$$

where $\mu = (NZ/A)m$. One has $E_0 = \hbar\Omega$, $\sigma_0 = 60(NZ/A)$MeV.mb, $\sigma_{-1} = \sigma_0/\hbar\Omega$ and $\sigma_{-2} = \sigma_0/(\hbar\Omega)^2$. The HOSM gives $\hbar\Omega = \hbar\omega$ where $\hbar\omega$ is the frequency of the HOSM. Taking $\hbar\omega \simeq 41 A^{-1/3}$ MeV and $(NZ)/A \simeq A/4$ one gets

$$E_D = 41 A^{-1/3} \text{ MeV}, \quad \sigma_{-1} = 0.37 A^{4/3} \text{ mb}, \quad \sigma_{-2} = 8.92 A^{5/3} \text{ μb.MeV}^{-1} \tag{12}$$

in disagreement with the empirical values. If one takes $\hbar\Omega = 79 A^{-1/3}$ MeV in order to reproduce the observed giant resonance energy one has

$$E_D = 79 A^{-1/3} \text{ MeV}, \quad \sigma_{-1} = 0.19 A^{4/3} \text{ mb}, \quad \sigma_{-2} = 2.4 A^{5/3} \text{ μb.MeV}^{-1} \tag{13}$$

in good agreement with the data for heavy nuclei. Thus, roughly doubling the independent particle model value of the frequency of the relative motion of protons with respect to neutrons is adequate to reproduce the empirical values of E_D, σ_{-1} and σ_{-2}.

Let us now extract from the data the amplitude of the r_{ZN}-motion in the ground state (see table 2). From the experimental knowledge of σ_{-1} (first column) is obtained the value of $\langle r_{ZN}^2 \rangle^{1/2}$ (second column) by use of (9). The ratio $\mathcal{R}$ of the zero-point GT root mean square radius to the experimental charge root mean square radius r_c (third column)

$$\mathcal{R} = \left(\frac{\langle 0|r_{ZN}^2|0\rangle}{r_c^2}\right)^{1/2} \tag{14}$$

is given in the fourth column. The fifth column gives an estimation $\bar{\mathcal{R}}$ of $\mathcal{R}$ obtained as follows : take (11), with $\hbar\Omega = 79\ A^{-1/3}$ MeV, $\mu \simeq (A/4)m$, the h.o. 1s state for

	σ_{-1} (exp.) (mb)	$\langle 0\vert r^2_{ZN}\vert 0\rangle^{1/2}$ (fm)	$\langle r^2_c \rangle^{1/2}$ (exp.) (fm)	$\mathcal{R}$	$\bar{\mathcal{R}}$
^{7}Li	4.6	1.28	2.41	0.53	0.51
^{9}Be	5.2	1.05	2.51	0.42	0.43
^{12}C	8.8	1.01	2.45	0.41	0.36
^{16}O	14.5	0.97	2.72	0.36	0.29
^{40}Ca	45.5	0.69	3.48	0.20	0.16
^{90}Zr	70.6	0.39	4.28	0.09	0.09
^{208}Pb	270.	0.34	5.50	0.06	0.05

Table 2 : First column : experimental values of σ_{-1}. Third column : experimental charge root mean square radii. See text for further explanation.

the zero-point motion and $r^2_c \simeq 0.9\ A^{2/3}$ fm^2. One obtains $\bar{\mathcal{R}} = 1.87\ A^{-2/3}$, in very good agreement with the value $\mathcal{R}$ extracted from experiment. For light nuclei, where $\mathcal{R}$ is large, chances are that once the nucleus has absorbed a γ-ray the proton and neutron distributions will have a small enough overlap such that studies of the deexcitation of the GDR may provide information on neutron-neutron correlations. In particular, the possibility to observe multineutron bound states in this way should be explored (for ^{7}Li, the 3n threshold is not far from the giant resonance region).

Let us now consider what are the predictions concerning E_D, σ_{-1} and σ_{-2} of usual microscopic theories of collective motion, like TDA or RPA. Before describing selfconsistent treatments which, in particular, take exchange effects into account, we consider the simplest model, i.e. the degenerate schematic model. Both TDA and RPA produce an upward collective shift of a single state $\vert c\rangle$ (the collective one) which exhausts all the strength. But TDA (which contains no ground state correlations) and RPA (which contains ground state correlations) predict different strengths :

$$\frac{\vert \langle c\vert D\vert 0\rangle\vert^2_{TDA}}{\vert \langle c\vert D\vert 0\rangle\vert^2_{RPA}} = \left(\frac{\Delta\varepsilon}{\hbar\Omega}\right)^{-1} \simeq \frac{79}{41} , \qquad (15)$$

where $\Delta\varepsilon$ is the particle-hole energy and $\hbar\Omega$ is the energy of the collective state. The TDA value of σ_{-1} corresponds to an independent particle evaluation, should therefore be identified to the one given by eq.(12) and will be in disagreement with the data. On the contrary, σ_{-1}(RPA) will coïncide with eq.(13) and will agree with the data. So the schematic model indicates that the ground state long range correlations included in RPA (and neglected in TDA) produce the effect of doubling the frequences of the relative proton-neutron motion with respect to the independent particle model value, in which only Pauli correlations have been included.

4. SELFCONSISTENT DESCRIPTION [8]

We shall now discuss results obtained within a selfconsistent HF-RPA scheme, using effective interactions of the Skyrme type. Force-parameters have been fitted to observed bulk properties of ground states given by the Hartree-Fock method. These forces contain momentum dependent terms that induce a non-locality of the average field. Like more realistic effective forces, they contain a two-body density dependent term. For isovector operators (like $\vec{D}$), the velocity dependent terms contribute to the commutator $[V,\vec{D}]$ (exchange effects). As is well known, the Time Dependent Hartree-Fock derivation of the RPA specifies how to construct the response function and the RPA matrices from the effective interaction used in determining the ground state. The RPA calculations discussed in what follows are selfconsistent because they have been performed using this procedure.

On table 3 are reproduced the computed values of σ_{-1} and σ_{-2}. The following remarks can be made :
i) There is an overall general agreement between the computed RPA values and the experimental ones.
ii) This agreement is achieved for light as well as for heavy nuclei.
iii) There is an important reduction of the σ_{-1}-value when going from the uncorrelated (TDA) to the correlated (RPA) ground state evaluation, as qualitatively predicted by simple models.
iv) The computed values do not critically depend on the interaction used.

Also included in table 3 are values of σ_{-2} obtained using Migdal's estimate

$$\sigma_{-2} = 2\pi^2 \frac{e^2}{\hbar c} \frac{R^2 A}{40 a_\tau} \tag{16}$$

where R is the sharp nuclear radius and a_τ is the volume symmetry energy coefficient appearing in the Weizsäcker formula. As can be seen, Migdal's estimation of the polarizability is very rough and can only be used for heavy nuclei. In the last column of table 3 are reproduced lower bounds of σ_{-2} obtained by the relation [9,10]

$$\alpha_D \gtrsim \frac{A^2}{16m^2} <HF|[[D_z,T],[H,[T,D_z]]]|HF>^{-1} \tag{17}$$

where H is the total Hamiltonian H = T + V. By comparing results of the last two columns, which are obtained with the same force, one can see that the bounds are 20-35 % lower than the exact RPA values and provide an efficient and simplified way to estimate the dipole polarizability.

To end up, let us draw the following conclusions :
- The general agreement between theory and experiment provides a very strong indication that measurements of total photoabsorption cross-sections in light nuclei and of photoneutron cross-sections in heavy nuclei are <u>consistent</u>.

- Dipole long range correlations in the ground state reduce considerably (from 25 to 40 %) the uncorrelated values of σ_{-1} and bring them to agreement with the data.

- The static dipole polarizability α_D is correctly described in a self-consistent RPA scheme.

	$\sigma_{-1} A^{-4/3}$ (mb)				$\sigma_{-2} A^{-5/3}$ (μb.MeV^{-1})				
	Exp	S III [11]	Sk M [12]	S II [11]	Exp	S III [11]	Sk M [12]	S II [11]	S II.Ref [9,10]
^{16}O	0.36	0.33 (0.45)	0.31 (0.46)	0.32 (0.45)	5.8	5.7	6.5	5.5	4.8
^{40}Ca	0.33	0.25 (0.40)	0.27 (0.40)	0.28 (0.40)	4.8	4.4	4.8	4.2	3.4
^{208}Pb	0.22	0.22 (0.33)	0.21 (0.34)	0.22 (0.35)	2.6	3.0	3.3	2.9	2.1
						1.9	1.7	1.6	← Migdal Eq. (16)
						28.1	31.0	34.1	← a_τ (MeV)

Table 3 - Comparison between experimental and calculated values of σ_{-1} and σ_{-2} for different forces using selfconsistent RPA. In parenthesis are given the values of σ_{-1} calculated with the H.F. uncorrelated ground-state, i.e. the Tamm-Dancoff evaluation. In the last column are reproduced the lower bounds of σ_{-2} given by sum-rule techniques in ref |9,10| . In the bottom lines are reproduced values of σ_{-2} given by eq. (16) and the volume symmetry energy corresponding to each force.

- The zero point Goldhaber-Teller motion can be extracted directly from the data and evaluated in a simple model.

The author acknowledges D. Vautherin and Nguyen Van Giai for permission to present results prior to publication.

REFERENCES

[1] "Electro- and Photonuclear Reactions", ed. by S. Costa and C. Schaerf, Lecture Notes in Physics, Vols. 61 and 62, Springer-Verlag 1977.
[2] B.L. Berman and S.C. Fultz, Rev. Mod. Phys. 47 (1975) 713.
[3] R. Bergère in ref.[1].
[4] J. Ahrens et al., Nucl. phys. A251 (1975) 479 ; Nuov. Cim. 32 (1976) 364.
[5] A. Dellafiore and D.M. Brink, Nucl. Phys. A286 (1977) 474.
[6] A.M. Lane and A.Z. Mekjian, Phys. Rev. C8 (1973) 1981.
[7] O. Bohigas in "Theory and applications of moment methods in Many-Fermion systems", ed. by B.J. Dalton et al., Plenum Press 1980, p.499.
[8] O. Bohigas, Nguyen Van Giai and D. Vautherin, Proceedings of the International Conference on Nuclear Physics, Berkeley 1980 ; and to be published.
[9] E. Lipparini, G. Orlandini, S. Stringari and M. Traini, Nuov. Cim. 42A (1977) 296.
[10] D.M. Brink and R. Leonardi, Nucl. Phys. A258 (1976) 285.
[11] M. Beiner, H. Flocard, N.V. Giai and P. Quentin, Nucl. Phys. A238 (1975) 29.
[12] H. Krivine, J. Treiner and O. Bohigas, Nucl. Phys. A336 (1980) 155.

ELECTROMAGNETIC SUM RULES

Giuseppina Orlandini[*]
Dipartimento di Matematica e Fisica
Università di Trento
I-38050 Povo (Trento), Italy

Abstract: The sum rules (SR) are introduced showing their connection with the linear response of a nucleus to an external electromegnetic field. Their relation to some macroscopic "collective" aspects of the nucleus are briefly outlined. Energy weighted sum rules (EWSR) for electron scattering at constant momentum transfer are studied for light and medium weight nuclei. The q-dependence of the different contributions to these SR permits to discuss the rôle of the nuclear potential and of correlations in the electroexcitation process.

1. Introduction

Since the beginning of quantum mechanics, the SR have been successfully used as a method to discuss general features of excitation mechanism. Various aspects and applications of these techniques in nuclear physics can be found in the reviews of refs. 1-6. It is interesting to notice how the SR are connected to a fundamental quantity in an excitation process as the response function[7].

Let us suppose that a nucleus whose Hamiltonian is H_0 is exposed to a general probe (real photon, virtual photon, meson...). If $-\lambda F \cos \omega t$ is the perturbing interaction, then the state of the nucleus is governed by

$$H = H_0 - \lambda F \cos \omega t$$

In general, F can depend on the coordinates, spins and isospins of the nucleons, on the momentum transferred to the nucleus and it can have a definite multipole character as well. The quantity describing the linear response of the nucleus in the limit of small perturbations is the dynamical polarizability of the probe

$$\alpha(\omega) = \lim_{\lambda \to 0} \frac{\langle\psi(t)|F|\psi(t)\rangle - \langle 0|F|0\rangle}{\lambda \cos \omega t} = \sum_n 2\omega_{no} \frac{|\langle n|F|0\rangle|^2}{\omega_{no}^2 - \omega^2} \tag{1}$$

which can be studied in two limiting cases

$$\alpha(\omega)\Big|_{\omega \to \infty} = -\frac{2}{\omega^2} \sum_n \omega_{no}|\langle n|F|0\rangle|^2 - \frac{2}{\omega^4} \sum_n \omega_{no}^3|\langle n|F|0\rangle|^2 \ldots = -\frac{2}{\omega^2} S_1 - \frac{2}{\omega^4} S_3 \ldots \tag{2}$$

$$\alpha(\omega)\Big|_{\omega \to 0} = 2 \sum_n \frac{|\langle n|F|0\rangle|^2}{\omega_{no}} + 2\omega^2 \sum_n \frac{|\langle n|F|0\rangle|^2}{\omega_{no}^3} \ldots = 2S_{-1} + 2\omega^2 S_{-3} + \ldots \tag{3}$$

---

[*] Supported by the Deutsche Forschungsgemeinschaft, contract no. Eh 4/11

S_m are the moments of the strength distribution or structure function connected to the differential cross section of the process in the first order approximation. It is well known that - under a suitable hypothesis of convergence - for m > 0, S_m represent the m-energy-weighted sum rules. They can be written in the form[8]

$$S_{m_{odd}} = \frac{1}{2}(-)^b <0|[F_a, F_b]|0> \; ; \quad S_{m_{even}} = \frac{1}{2}(-)^b <0|\{F_a, F_b\}|0>$$

$$F_a = [H_o,[H_o...[H_o, F]...]] \quad \text{a times}$$

$$a + b = m$$

The connection between $\alpha(\omega)$ and S_m can be very useful in some cases, in fact, it has been demonstrated[7] that the electric dynamical polarizability evaluated in the TDHF framework is related to RPA energies and matrix elements, and a method has been developed to obtain SR for RPA from the TDHF dynamical polarizability from the various terms of the expasnsions of eqs. (2) and (3).

2. Applications of sum rule techniques

In many cases, the SR permit to connect microscopic and macroscopic aspects of the nucleus and allow, from the comparison with experiments, to understand which features play the biggest rôle in a certain excitation process. The most famous sum rule is $S_1(E1)$ for photoabsorption processes with F equal to the dipole operator, which is proportional to the integrated photoabsorption cross section

$$\int \sigma_{Abs}(\omega)d\omega = 4\pi^2 \, S_1^{TRK}(E1) = 2\pi^2 <0|[D_z,[H_o,D_z]]|0> = \frac{2\pi^2 \, e^2}{m} \frac{NZ}{A}(1+k)$$

The comparison between the experimental[9] and theoretical[10] values of k has permitted to understand the rôle of the exchange potential in this excitation and the importance of tensor correlations in the region of energy between the giant dipole resonance and the pion threshold.

More recently other giant resonances of different multipolarity have been measured in electromagnetic processes. The sharper they are, the more they justify a SR approach to be studied; for example, the resonance energy $\omega_R = S_1/S_0$ can be given in this case with good approximation also by other ratios of different SR

$$\omega_R = (S_3/S_1)^{1/2} \; ; \; (S_1/S_{-1})^{1/2} \; ; \; (S_{-1}/S_{-3})^{1/2}$$

The strength distribution is directly given by S_0. In the limit case of a δ-resonance all the ratios would give the same values of ω_R; for a more spread resonance they would slightly differ from the first. However, it often happens that it is possible to evaluate SR with m ≠ 0 with more accuracy than S_0 so that one can give better predictions of ω_R using S_1 and S_3 or S_{-1}. This has been done for example in refs. 11-14. To study such resonances, one can also use a collective approach constructing a collective Hamiltonian whose parameters (spring constant K and mass M) define ω_R and the strength

$$H_{coll} = \frac{1}{2} K Q^2 + \frac{M}{2} \left(\frac{dQ}{dt}\right)^2 \qquad \omega_R = \sqrt{K/M}$$

It is often possible to identify the macroscopic parameters K and M with some SR, which are evaluated on a microscopic level[6,15]. For example, in the framework of TDHF theory a collective Hamiltonian can be constructed whose M and K are given by[7] $M = 1/2S_1^{RPA}$; $K = S_3^{RPA}/2(S_1^{RPA})^2$; so that $\omega_R = (S_3^{RPA}/S_1^{RPA})^{1/2}$. Explicit evaluation of S_3^{RPA} for isovector modes (by using eq. (2))[16] has permitted to predict ω_R for isovector monopole, dipole and quadrupole excitations in double closed nuclei. For the same nuclei an ATDHF model has been applied to the description of isoscalar giant monopole resonances in terms of collective masses and potentials[17] leading to $\omega_R = (S_{-1}/S_{-3})^{1/2}$.

Another example of the connection between microscopical and macroscopical features is given by the link between the stiffness parameters (nuclear compressibilities) and $S_3(F)$ in a scaling description of the deformed nuclear state[12]. Furthermore, very recently it has been shown how for the dipole excitation operator ($F = D_z$) S_3^{RPA} is connected to typical ingredients of nuclear structure like the symmetry energy potential and that it is proportional to the nuclear stiffness parameter corresponding to Goldhaber-Teller deformations[18]. As to the inverse energy-weighted dipole SR, S_{-1} is proportional to the static polarizability and Migdal[19] related it, in the case of E1 operator, to the symmetry energy coefficient of the semiempirical mass formula. Much later Marshalek and Da Providencia related this SR defined in the RPA scheme to the static HF polarizability[20].

3. Electronuclear sum rules

In the case of electroexcitation (virtual photons, $F = F(q)$), one has several possibilities of defining SR because both momentum and energy transfer can vary independently within the space-like region ($q^2 - \omega^2 > 0$) so that one may choose different curves $q = q(\omega)$ along which one integrates the structure functions[4]. In the case q = const. a simple evaluation in terms of ground state expectation values is often possible, since the charge and current operators do not depend on ω. The variation of q allows then the study of the spatial distribution of such basic phenomena as for example exchange forces and tensor correlations in nuclei. In the one-photon approximation, neglecting the rest mass of the electron[21]

$$\frac{d\sigma}{d\Omega\, d\omega} = \sigma_M \sum_n \int d\omega_n \delta(\omega_n + q^2/2mA - \omega_0 - \omega) \cdot [\frac{q_\mu^4}{q^4} F_C(\vec{q},\omega_n) + (tg^2\frac{\theta}{2} - \frac{q_\mu^2}{2q^2}) F_T(\vec{q},\omega_n)]$$

where F_C and F_T are the Coulomb and transverse structure functions.

$$F_C = \sum_n |\langle n|\rho(\vec{q})|0\rangle|^2 = \frac{4\pi}{2J_0+1} \sum_{n,J\geq 0} |\langle n||M_J(\vec{q})||0\rangle|^2$$

$$F_T = \sum_n |\langle n|\vec{J}_T(\vec{q})|0\rangle|^2 = \frac{4\pi}{2J_0+1} \sum_{n,J\geq 1} \{|\langle n||T_J^{el}||0\rangle|^2 + |\rangle n||T_J^{Mag}(q)||0\rangle|^2\}$$

The charge and transverse current operators are in the non-relativistic impulse approximation

$$\rho(\vec{q}) = \sum_{K=1}^{A} e_K(q)\, e^{i\vec{q}\cdot\vec{r}_K}$$

$$\vec{J}_T(\vec{q}) = \vec{J}_T^{conv} + \vec{J}_T^{spin} = \sum_{K=1}^{A} \frac{e_K(q)}{m} \vec{P}_{KT} e^{i\vec{q}\cdot\vec{r}_K} + i\vec{q}\times \sum_{K=1}^{A} \frac{\mu_K(q)}{2m} \vec{\sigma}_K\, e^{i\vec{q}\cdot\vec{r}_K}$$

e_K and μ_K indicate the charge and magnetic form factors. EWSR at q = const. are so constructed[22]

$$S_N^C = \sum_n \omega_{no}^N\, |\langle n|\rho(\vec{q})|0\rangle|^2$$

$$S_N^T = \sum_n \omega_{no}^N\, |\langle n|\vec{J}_T(\vec{q})|0\rangle|^2$$

So, under particular kinematic conditions, playing with the kinematic factors before the structure functions, one may hope to relate the integrated weighted cross section to the Coulomb or transverse SR. In some cases, it may be interesting to decompose the SR into a multipole series projecting out transitions of a particular multipolarity.

$$S^{\lambda,N} = \sum_J S_J^{\lambda,N}$$

with
$$S_J^{\lambda,N} = \frac{4\pi}{2J_0+1} \sum_n \omega_{no}^N\, |\langle n||O_J^\lambda||0\rangle|^2$$

where $O_J^\lambda = M_J$, T_J^{el} or T_J^{mag} for λ = C, el or mag. Similarly to the case of the total one, the multipole transverse SR is the incoherent sum of electric and magnetic contributions for an unoriented nucleus.

Most of the theoretical studies have concentrated on S_0^C and S_0^T with rather good agreement with the few existing experimental data[23]. But already Drell and Schwartz[24] pointed out that S_1 is a much more sensitive test for exchange forces. Further studies on S_1 have been performed comparing theory and experiments for light nuclei[25]. Even S_3 has been evaluated and used to study the momentum dependence of the mean excitation energy of the Coulomb operator, which for low q, reduces to the monopole[26]. RPA results are given for ^{16}O and ^{208}Pb.

A systematic study of both Coulomb and transverse parts of S_1 for light and medium weight nuclei is done in ref. 22. But before going into details about the results of this study, some words should be said about the convergence and the comparison with the experimental data. In this study the question of convergence of this SR is discussed in the framework of analytical models and the conclusions are that, at least within the framework of a potential model, this SR should exist. At the same time, an explicit calculation has been done for the deuteron, where the form factors can be numerically evaluated[27]. The results of this investigation may be summarized as follows: The Coulomb SR converges rather rapidly and is exhausted by the contribution up to the photon point more and more as q increases (fig. 1). Table 1 lists the values of the integration up to the photon point together with the results of the

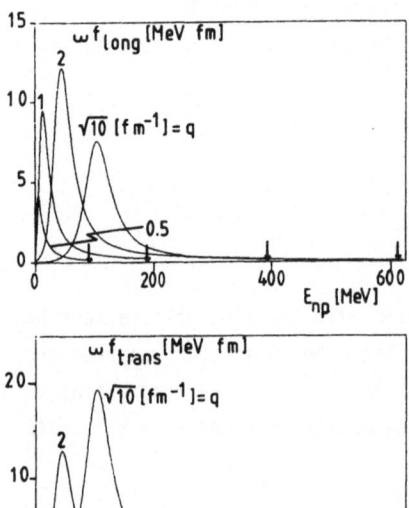

Fig. 1: Energy weighted form factors for various constant momentum transfers. Arrows indicate photon points.

Table 1

q fm^{-1}	photon point MeV^2(fm)	DC MeV^2(fm)
0.5	4.20	5.25
1	15.3	17.3
2	33.5	35.4
$\sqrt{10}$	29.6	31.0

double commutator. The agreement is satisfactory and shows that with increasing momentum transfer the contribution from the time-like region decreases from 20% at $q = 0.5$ fm^{-1} to 4% at $q = \sqrt{10}$ fm^{-1}. The transverse SR converges for the Reid soft-core potential (fig. 1) but it does not for hard-core potentials like for example the Hamada-Johnston. However, the spin part is always convergent and is dominant with respect to the convection SR for not too small q (fig. 2). It also almost exhausts the SR in the space-like region. On the contrary, the convection SR receives the bigger contribution from the time-like and dominates in the low q region. Of course, these conclusions are valid only for the deuteron and one is not allowed to extrapolate them sic et simpliciter to heavier nuclei. Nevertheless, some results are encouraging, especially as to the problem of the comparison with the experiments like for example the exhaustion of the Coulomb SR and for high q the dominance of the spin SR in the physical region.

On the other hand, the comparison of the theoretical SR with experimental determinations remains still rather problematic: on the side of the theory one must mention the difficulties in the choice of the right operators describing the process, e.g., exchange currents and at high q the problem of the presence of pion production, isobar excitations and relativistic effects, on the side of the experiments difficulties arise in separating the longitudinal and transverse form factors and in the necessity to subtract radiative corrections and contributions from pair produced electrons and pions from the data.

Recently[28], longitudinal and transverse inelastic form factors in ^{56}Fe have been separated for both discrete states and the continuum over a significant range of the q-ω plane. The results of a Fermi gas calculation agree reasonably well with the magnitude and shape of the transverse component but large disagreement is observed for the longitudinal form factor for $q > 1.5$ fm^{-1}. The problems in the measurement and in the comparison with the theoretical results are discussed.

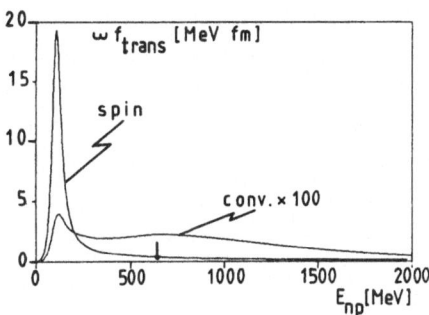

Fig. 2: Transversal energy weighted form factor for $q = \sqrt{10}$ fm^{-1}. The arrow indicates the photon point.

We look forward to an increasing effort of both theorists and experimentalists in solving the problem associated with this comparison.

Coming back to the study of the EWSR S_1^C and S_1^T and their multipole terms have been evaluated using the following ingredients: for ^{2}H, ^{3}H and ^{3}He "exact" non-relativistic ground state wave functions obtained as numerical solutions for realistic two-nucleon potentials are used. For ^{4}He the wave functions are those of the harmonic oscillator modified by short range correlations obtained for the RSC potential by a perturbative method. For heavier nuclei $|0>$ is the HF ground state as obtained with an effective two-body potential. Because of the Thouless theorem, the values of this EWSR in these nuclei can be considered to have RPA accuracy.

The energy weighted Coulomb sum rule is given by (see also ref. 29)

$$S_1^C = S_{c1}(q)(1 + \Delta_{exc}(q) + \Delta_{corr}(q))$$

where

$$S_{c1} = \frac{NZ}{2Am} e^2 q^2$$

$$\Delta_{exc} = <0|[\rho^+(q),[V,\rho(q)]]|0>/S_{c1}(q)$$

$$\Delta_{corr} = \frac{4}{e^2 NZ} < | \sum_{K<l} e_K e_l \sin^2(\vec{q}\cdot(\vec{r}_K - \vec{r}_l)/2)0>$$

Δ_{corr} embodies the p-p correlation and is a consequence of the centre of mass corrections to S_1^C; it is quite essential, in fact, to use the commutator relations for relative coordinates and momenta to obtain it. Δ_{exc} is the Fourier transform of the exchange interaction weighted with the two-particle density. The limits for $q \to 0$ are $\Delta_{exc}(0) = k^{TRK}$ and $\Delta_{corr}(0) = 0$.

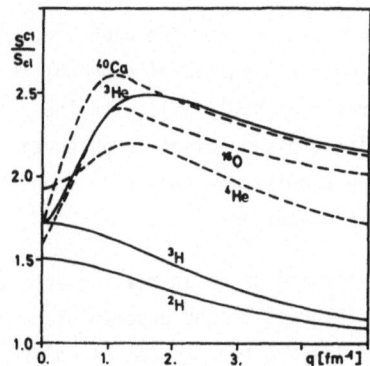

Fig. 3: S_1^C/S_{cl} as a function of q for various nuclei.

In fig. 3 S_1^C is shown for different nuclei as a function of q. The initial increase of the SR (for nuclei with Z > 1) is due to Δ_{corr}. In fig. 4a Δ_{exc} normalized to unity for q = 0, shows as function of q, a very similar shape for various nuclei. Fig. 4b, where the influence of the potential and of dynamical correlations is studied for ^{4}He, shows that inclusion of tensor correlations lowers the curve. Fig. 5 shows the multipole decomposition of $S_1^C(q)$. At low q the dipole dominates and the potential terms in the double commutator are

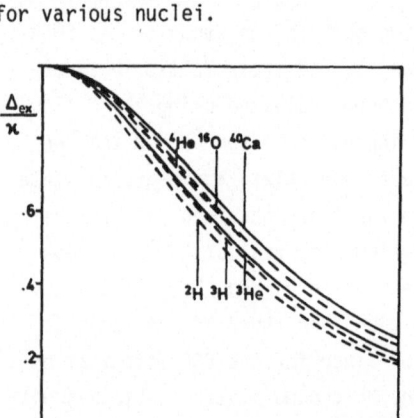
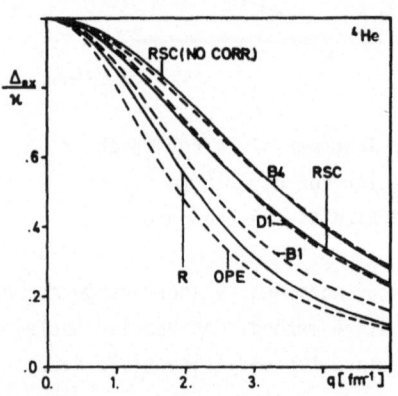

Fig. 4 Normalized Coulomb enhancement $\Delta_{exc}(q)/K^{TRK}$ as a function of q. a) For various nuclei and b) for ^{4}He with various potentials and correlations.

particularly important for it, contributing about 50%. For the higher multipoles, which become more important with increasing momentum transfer, the potential contribution is much smaller.

Fig. 6 shows that the spin current gives the biggest contribution at not small q, and one can see that the potential contributes at low q, while the kinetic energy dominates at high q, where the transverse SR is essentially determined by the magnetic moments of the bound nucleons.

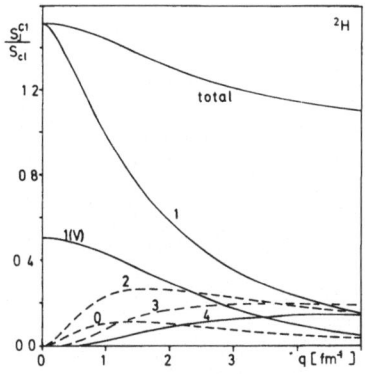

Fig. 5: Partial wave decomposition of $\overline{S_1^C \cdot S_J^{C1}}/S_{c1}$ as function of q for ^{2}H. The sum of all multipoles ("total") and the potential energy contribution for the dipole (1(V)) are also shown.

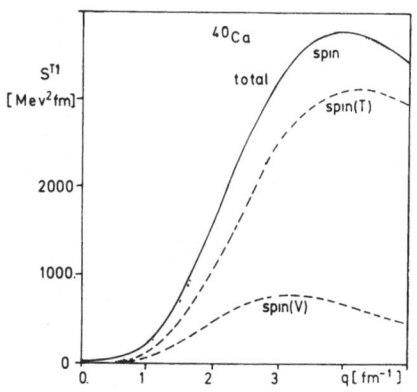

Fig. 6: S_1^T as a function of q for ^{40}Ca (solid curve). Separately the spin current SR (dotted curve) and the contribution to it of kinetic and potential energy term are shown.

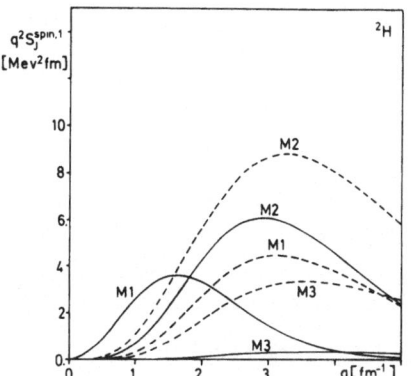

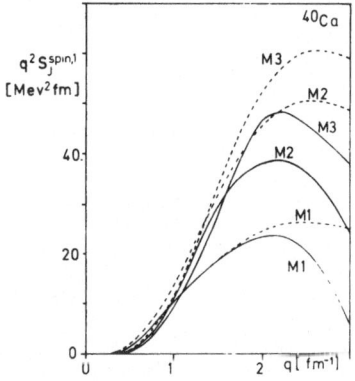

Fig. 7: Potential (solid curves) and kinetic (dashed curves) energy contributions to the magnetic multipoles of the spin current sum rule for ^{2}H and ^{40}Ca.

The multipole decomposition of the spin SR is shown in fig. 7. The potential contributes appreciably only for the lowest multipoles. The limit for $q \to 0$ reproduces the results of M. Traini[30], who pointed out the rôle of the exchange and spin orbit potentials, generalizing the Kurath SR to higher multipoles.

Finally, I will mention another interesting application of electronuclear SR to the description of the quasi-free peak[31]. In the region of rather high energies and momentum transfers, the quasi-free scattering dominates and the structure function given by

$$R(\vec{q}, \omega) = \delta(\omega - \frac{q^2}{2mA} - \omega_{no})|<0|F(\vec{q})|n>|^2$$

represents a rather smooth shape, which can be characterized by few moments, namely few SR for q = const. In fact, it may be argued that high energy transfer ω corresponds roughly to small interaction times t. Position, width and shape of the quasi-elastic peak can be described by the first few terms in the "cumulant" expansion of its Fourier transform.

$$Q(t) = \int R(\vec{q}, \omega) e^{i\omega t} d\omega$$

$$= \exp\left[it \frac{S_1}{S_0} + \frac{(it)^2}{2!} \left(\frac{S_2}{S_0} - \left(\frac{S_1}{S_0}\right)^2\right) + \frac{(it)^3}{3!} \cdot \left(\frac{S_3}{S_0} - 3\frac{S_2 S_1}{S_0^2} - 2\left(\frac{S_3}{S_0}\right)^2\right) + \ldots \right]$$

It has been shown that the observed shift of the quasi-elastic peak is related to exchange parts of the two-body interaction. In particular, calculations for ^{12}C show that hard-core potentials come close to the experimental value whereas the effective interactions all give too low results.

References:

1. R. Leonardi and M. Rosa-Clot, Nuovo Cim. 1 (1971) 1
2. J.S. O'Connell, Proc. Int. Conf. on Photonuclear physics and applications, Asilomar/Calif., 1973, ed. by B.L. Berman
3. W. Weise, Phys. Rep. 13C (1974) 53 and Int. School on Electro- and photonuclear reactions, Erice, 1976, in Lecture Notes in Physics, Vol. 61 (Springer-Verlag, 1977) p. 484
4. D. Drechsel, Proc. IVth Seminar on Electromagnetic interactions of nuclei at low and medium energies, Moscow, 1977
5. H. Arenhövel, Proc. Int. Conf on Nuclear Physics with electromagnetic interactions, Mainz, 1979, eds. H. Arenhövel and D. Drechsel, in Lecture Notes in Physics, Vol. 108 (Springer-Verlag) p. 159
6. O. Bohigas, A.M. Lane and J. Martorell, Phys. Rep. 51 (1979) 267
7. S. Stringari, E. Lipparini, G. Orlandini, M. Traini and R. Leonardi, Nucl. Phys. A309(1978) 177
8. R. Leonardi and M. Rosa-Clot, Nuovo Cim. 69A (1970) 1
9. B. Ziegler, Proc. Int. Conf. on Few body systems and electromagnetic interactions, Frascati, 1978, eds. C. Ciofi degli Atti and E. De Sanctis, in Lecture Notes in Physics, Vol. 86 (Springer-Verlag, 1978) p. 100
   A. Leprêtre et al., Phys. Lett. 79B (1978) 43
   R. Bergère, Mainz Conference, 1979, (see ref. 5) p. 138
10. A. Arima, G.E. Brown, H. Hyuga and M. Ichimura, Nucl. Phys. A205 (1973) 27
    W.T. Weng, T.T.S. Kuo and G.E. Brown, Phys. Lett. 46B (1973) 329
    M. Fink, M. Gari and H. Hebach, Phys. Lett. 49B (1974) 20

11. E. Lipparini, G. Orlandini and R. Leonardi, Phys. Rev. Lett. 36 (1976) 660
12. J. Martorell, O. Bohigas, S. Fallieros and A.M. Lane, Phys. Lett. 60B (1976) 313
13. R. Leonardi, E. Lipparini and G. Orlandini, Phys. Lett. 64B (1976) 21
14. D.M. Brink and R. Leonardi, Nucl. Phys. A258 (1976) 285
15. K. Goeke, A.M. Lane and J. Martorell, Nucl. Phys. A296 (1978) 109 and references therein
16. S. Stringari, E. Lipparini, G. Orlandini, M. Traini and R. Leonardi, Nucl. Phys. A309 (1978) 189
17. K. Goeke, B. Castel, Phys. Rev. C19 (1979) 201
18. E. Lipparini and S. Stringari, to be published
19. A. Migdal, Journ. Phys. USSR 8 (1944) 331; JEPT 15 (1945) 81
20. R. Marshalek and J. Da Providencia, Phys. Rev. C7 (1973) 2281
21. T. De Forest and J.D. Walecka, Adv. Phys. 15 (1966) 1
22. V. Tornow, G. Orlandini, M. Traini, D. Drechsel and H. Arenhövel, Nucl. Phys. A348 (1980) 157
23. V.D. Efros, Yad. Fiz. 18 (1973) 1184 (Sov. Journ. Nucl. Phys. 18 (1974) 607) and references therein
24. S.D. Drell and C.L. Schwartz, Phys. Rev. 112 (1958) 568
25. A.Yu. Buki, N.G. Shevchenko and A.V. Mitrofanova, Yad. Fiz. 24 (1976) 457 (Sov. Journ. Nucl. Phys. 24 (1976) 237)
E.L. Kuplennikov, V.A. Goldstein, N.G. Afanasev, V.G. Vlasenko and V.I. Startsev, Yad. Fiz. 24 (1976) 22 (Sov. Journ. Nucl. Phys. 24 (1976) 11)
26. E. Lipparini, G. Orlandini and R. Leonardi, Phys. Rev. C16 (1977) 812
27. W. Leidemann, diploma thesis, Mainz, 1980
H. Arenhövel, Invited talk at the workshop "Intermediate energy nuclear physics with monochromatic and polarized photons", Frascati, 1980
28. R. Altemus, A. Cafolla, D. Day. J.S. Mc Carthy, R.R. Whitney and J.E. Wise, Phys. Rev. Lett. 44 (1980) 965
29. A. Dellafiore and M. Traini, Nucl. Phys. A344 (1980) 509
30. M. Traini, Phys. Rev. Lett. 41 (1978) 1535
31. R. Rosenfelder, Phys. Lett. 79B (1978) 15; Ann. of Phys. 128 (1980) 188

STATUS OF NUCLEAR CRITICAL OPALESCENCE

J.Delorme
Institut de Physique Nucléaire de Lyon (et IN2P3)
Université Claude Bernard-Lyon 1
69622 Villeurbanne Cedex, France

1. Introduction :

During the recent period the extensive efforts devoted to the detection of pion condensation in nuclear systems have been disappointed. A common belief is that the threshold density is largely higher than that of nuclear matter. There has been consequently a growing interest in the search for critical effects or precursors which can occur till rather far from the critical point. They could thus be observable under usual nuclear conditions and provide informations on the proximity of the transition itself. Their signature would be a strong increase of specific spin correlations in a range of energy and momentum characteristic of pion condensation ($\omega \approx 0$, $q \approx 2$ to $3\ m_\pi$). The name of critical opalescence which has been proposed to christen these phenomena is reminiscent of critical scattering on magnetic substances which also originates from strong spin fluctuations [1,2]. The present experimental situation seems at best unconclusive though some hope of substantial effects has been sustained sometime by a tentative analysis of anomalies in magnetic form factors [3,4]. The available data are briefly reviewed below. More emphasis is put on the necessity of more complete information and on the criteria of selectivity which should be fulfilled by future experiments. Such a discussion is better introduced by the consideration of the general features of critical opalescence which dictate the choice of the best probe.

2. General features of nuclear critical opalescence :

The topic of nuclear critical opalescence presents aspects relevant to both classical nuclear physics and mesonic degrees of freedom. Indeed it can be viewed as a collective phenomenon in the spin-isospin channel, i.e. the channel of states which have the quantum numbers of the pion (the so-called pion-like states $T = 1, J^\pi = 0^-, 1^+, 2^-...$), or as a manifestation of pion degrees of freedom. These two descriptions represent both sides of the same physical reality. Each has advantages and shortcomings. It should be clear that, though technically different they can be conducted to practically equivalent results. In the collective approach, an effective particle-hole interaction is diagonalized in the RPA scheme (fig.1). The prominent long range component of the p-h force in the $\sigma\tau$ channel is given by the attractive one pion exchange

$$V_\pi = -f_\pi^2 \vec{\tau}_1 \cdot \vec{\tau}_2 \; \vec{\sigma}_1 \cdot \vec{q} \; \vec{\sigma}_2 \cdot \vec{q} / (\vec{q}^2 + m_\pi^2)$$

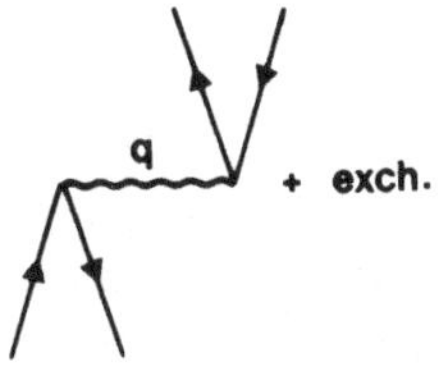

There is a large uncertainty for the shorter range part. It is usually very schematically represented by a contact repulsive interaction $g' \vec{\tau}_1 \cdot \vec{\tau}_2 \; \vec{\sigma}_1 \cdot \vec{\sigma}_2$ governed by the Landau-Migdal parameter g'. The isobar degrees of freedom (essentially Δ excitation) can be introduced by a renormalization of the interaction which has the effect of increasing the OPE attraction. A logical procedure would be to follow the development of the collective phenomenon as a function of the density with a fixed interaction. In view of the uncertainty on the force, it is more convenient to keep the usual nuclear conditions and to vary g', a decrease of g' producing the same effect as an increase of the density. As an illustration the evolution of the energies of the lowest 2^- isovector states of ^{16}O is represented in fig. 2 which is borrowed from a recent work of Meyer-ter-Vehn[5]. It is seen that when the repulsive component is low, one state is pushed considerably downwards by the OPE attraction. Ultimately, beyond the transition point (attained for null excitation energy in a RPA theory), the nucleus would present a permanent spin-isospin deformation with a typical layered structure in the same manner as strong quadrupole coupling leads to permanent shape deformation. This is

Fig.1 The schematic p-h interaction in the $\sigma\tau$ channel

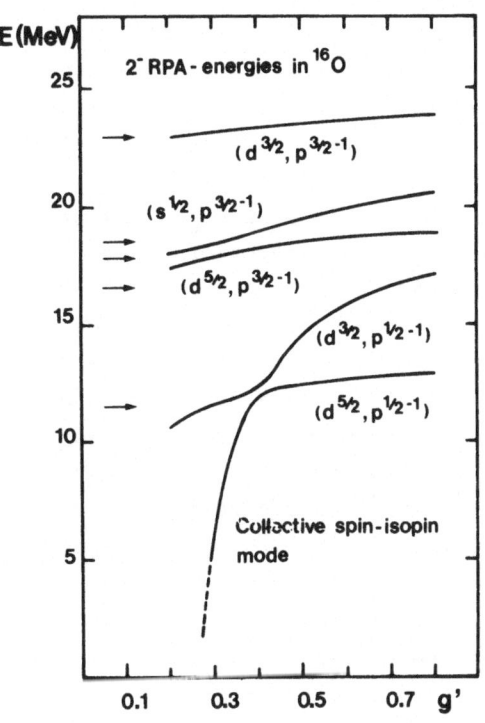

Fig. 2 The development of the spin isospin collective mode (according to ref.5)

the way pion condensation is realized in finite systems (see e.g. the reviews of Migdal[6] and Meyer-ter-Vehn[5] for more details). One has noted the well-known fact that with the pion force alone, nuclei would be already in the "condensed phase" i.e. spin-isospin deformed. Concurrently as the energy of the collective state drops to zero, the longitudinal spin form factor for its excitation (i.e. the matrix element of the $\vec{\sigma}.\vec{q}$ operator) shows a very spectacular increase for momenta in the critical range, the more pronounced the nearer the transition point. This is the critical opalescence phenomenon which signs the importance of the longitudinal spin fluctuations announcing the change of structure of the deformed phase.

The second description singles out the pion field component and its renormalization in the nuclear medium through the p-wave pion-nucleon interaction (fig.3). In a mean field theory (hence the equivalence with the RPA), the free pion propagator $D(q,\omega) = [\vec{q}^2 - \omega^2 + m_\pi^2]^{-1}$ is changed into the renormalized one

$$\widetilde{D}(q,\omega) = [\vec{q}^2 - \omega^2 + m_\pi^2 + \vec{q}^2 \alpha_0(q,\omega)]^{-1}$$

by the nuclear axial polarizability α_0 (defined < 0) representing nucleon-hole and Δ-hole excitations by $\vec{\sigma}.\vec{q}\,\tau$ operators. The frequency and momentum dependence of the pion self-energy $\vec{q}^2 \alpha_0(q,\omega)$ is such that for a sufficiently large value of the polarizability (which is a function of the density) a singularity of $\widetilde{D}$ occurs at $\omega = 0$, $q = q_c$ the critical momentum (2 to 3 m_π).

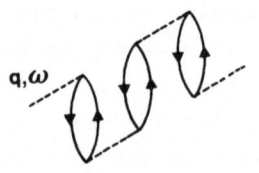

Fig.3 Renormalization of the pion propagator by the medium polarization.

This is the onset of pion condensation in nuclear matter. At lower densities, the pion field, though non singular, still presents an enhancement $\widetilde{D}/D$ which is maximum around $\omega = 0$ and $q = q_c$. This is again the critical opalescence phenomenon. The description becomes somewhat more complicated in a finite nucleus but the main feature remains : the pion field form factors, i.e. by PCAC the $\langle \vec{\sigma}.\vec{q}\,\tau \rangle$ matrix elements undergo large enhancements in the characteristic momentum range at the approach of the transition point. For commodity we will keep throughout the denomination "critical opalescence" to designate the medium enhancement of the pion field irrespective of its amplitude. It is clear that there exists no precise boundary of the critical domain and that even for moderate "opalescence" the underlying physical phenomenon remains the same, i.e. the alignment of the spins through the pion exchange force.

In the two descriptions we have presented, a crucial role is played by the (repulsive) forces other than the OPE which act in the $\sigma\tau$ channel. In the RPA approach, collectivity appears if the overall particle-hole force is sufficiently large in the momentum range relevant to pion condensation to drive strength from many high lying excitations and build the collective mode. On the contrary, if the repulsion is important as given for instance by a value of g' as big as 0.8 in the schematic representation chosen above, the OPE attraction would be cancelled and the critical effects would disappear. In the field approach, the inclusion of forces other than OPE amounts to replace the polarizability by an effective lower one α (at the usual density α_0 would be large enough to cause pion condensation). The relevant quantity is indeed the irreducible pion self-energy $\vec{q}^2\alpha$ obtained by summing the graphs of fig.4. In the case of a contact interaction, α_0 is replaced by

$$\alpha = \alpha_0/(1-g'\alpha_0)$$

which can thus be considerably lower than α_0 for large values of g' (Lorentz-Lorentz effect[7]) and reduces accordingly the amplitude of critical opalescence. The size of the non-pion components of the $\sigma\tau$ force at the relevant transfers is thus the key which governs critical phenomena. There is still a large uncertainty in theoretical estimations [8,9] of this short range part but the calculations converge towards a rather large repulsion (g' = 0.6 to 0.9 with a more or less marked momentum dependence) which is very unfavourable to important amplification of the pion field. Phenomenological determinations come essentially from the position of pion-like levels. We will not consider here the information from the Lorentz-Lorenz renormalization of Gamow-Teller transitions [10] which is not exactly relevant since it concerns only the zero-transfer region (note however that it is consistent with the overall picture of large repulsion).

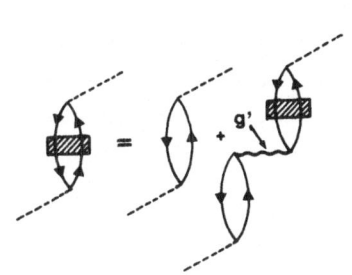

Fig.4 The equation for the irreducible self-energy (shaded blob).

3. The pion-like levels :

A general observation about the $\sigma\tau$ channel is that it does not show prominent collective features. The energies of the known pion-like states have no marked deviation from their unperturbed p-h positions. This very remark about the 0^- (T = 1) state of ^{16}O was at the origin of early scepticism about Migdal's suggestion of pion condensates in nuclei[11] (earlier considerations against $\sigma\tau$ instability can be found in ref.12). A RPA calculation of the relevant levels of ^{16}O with a schematic (OPE + contact) interaction showed indeed that the observed energies demanded a relatively large repulsion corresponding to g' = 0.7 to counterbalance the OPE attraction [5,13]. More recently a detailed analysis has been performed over a wide range of selected states from C to Pb with different momentum content so as to span the momentum dependence of the force [14]. A ρ meson exchange component was added to the interaction because its tensor part has the well known property of cancelling the tensor OPE at large momenta. The resulting fit gave g' = 0.55 with such a q dependence that an extra repulsion of 0.15 is produced at q = 2.5m_π. These numbers can even be considered as lower limits since the interaction was not renormalized to allow for the attractive contribution of the Δ isobars. The general information from energy levels is thus in favour of an overall weak force at the relevant momenta which would push the condensation threshold to very large densities and would produce only small pionic opalescence. One can make however two sorts of objections which have more concern with the precision of the result rather than with the general conclusion of a large repulsive component.

First the fitted forces have a very schematic character and their determination is probably not unique. Two different examples have been quoted above. Indeed the consideration of the $\sigma\tau$ channel is not sufficient to sign the pion-channel : there are two degrees of freedom, one transverse ($\vec{\sigma}\times\vec{q}$ mode) illustrated by rho meson coupling and one longitudinal ($\vec{\sigma}\cdot\vec{q}$ mode) which is sole relevant to pion properties. It is very important to realize that in the present problem this distinction is much more relevant than the usual scalar/tensor separation. To fix the ideas, an interaction containing a g' contact term and $\pi + \rho$ exchanges has the following longitudinal and transverse components V_L and V_T (up to possible nucleon form factors) :

$$V_L = \left(g' - f_\pi^2 \frac{q^2}{q^2+m_\pi^2}\right)\vec{\sigma}_1\cdot\hat{q}\,\vec{\sigma}_2\cdot\hat{q} \quad V_T = \left(g' - f_\rho^2 \frac{q^2}{q^2+m_\rho^2}\right)(\vec{\sigma}_1\times\hat{q})\cdot(\vec{\sigma}_2\times\hat{q}) \quad (1)$$

This is to be compared to the force used in ref.14 where the central part of the ρ exchange is decreased by the effect of short range correlations :

$$V_L = \left(g' + 0.4 f_\rho^2 \frac{q^2}{q^2+m_\rho^2} - f_\pi^2 \frac{q^2}{q^2+m_\pi^2}\right)\vec{\sigma}_1\cdot\hat{q}\,\vec{\sigma}_2\cdot\hat{q} \quad V_T = \left(g' - 0.6 f_\rho^2 \frac{q^2}{q^2+m_\rho^2}\right)(\vec{\sigma}_1\times\hat{q})\cdot(\vec{\sigma}_2\times\hat{q}) \quad (2)$$

This force is thus more repulsive in both channels. One should not be surprised by the appearance of a ρ contribution in the longitudinal part because the presence of short range correlations means that the exchange is not that of a pure ρ-meson (recall that a great part of g' is believed to arise from ρ exchange between correlated nucleons[15]). One would thus need in principle an independent determination of the force in the two channels. This is not an easy task since their relative weight for a given state strongly depends on the structure of the p-h excitation to begin with. An exception is constituted by the 0^- levels which are dominated by the longitudinal interaction. They are however very scarce and not easily excited. One should add that there is little doubt that the overall picture of a relatively weak force at the relevant momenta will not be significantly altered. It might well happen however that the exact balance between V_T and V_L could be slightly modified with a little more repulsion in one channel compensated by more attraction in the other one.

A second and connected comment should be made about the rather low sensitivity of many of the pion like states to the precise value of the force. For instance it can be seen on fig.2 that the curves representing the evolution of the 2^- states of ^{16}O are very flat in the region g' > 0.5 : a variation of 0.1 corresponds to a shift of only 200 keV. The only and very important exception arises from the state which would become collective with sufficient attraction. In the case of ^{16}O, this state has precisely not yet been seen, one possible reason being its very weak excitability at low transfers (see the spin form factors shown in ref.5). The sensitivity of such excitations that I would call "pion favoured" or "true" pion like is due to the peculiar structure of their $\langle \vec{\sigma}.\vec{q} \rangle$ (or equivalently pion field) form factor which is in phase with the polarizability α, i.e. it peaks at $q \approx p_F$, the Fermi momentum, and has no node in the momentum range where α is large. These states are then able to benefit best from the pion attraction and to drive maximum strength from the high excitations represented in the polarizability. In the same way one can probably find states with maximum sensitivity to the transverse spin mode. It appears thus that some possibilities remain open for a much more selective determination of the $\sigma\tau$ force. Of particular interest would be a systematic study of the "true" pion-like or collective states as emphasized by Meyer-ter-Vehn [5]. It seems worth to devote a systematic program to the identification of these specific excitations. We will return later to this important question.

4. The form factors of the longitudinal spin mode :

The simple consideration of the energy of the appropriate levels does not fully exploit the specific property of critical opalescence namely the enhancement of the pion field form factors in the momentum range 2-3 m_π. Furthermore the scale of the amplification varies much more rapidly with g' than the energies. For instance calculations in ^{12}C show that the average renormalization factors are 1.5, 2, 3 and 4.5 for g' = 0.7, 0.6, 0.5 and 0.4 respectively[3]. This increased sensitivity is not without shortcomings. It is likely that at the considered momenta other types of correlations or meson exchanges can seriously affect the form factors. Below a certain level, say a factor 2, the pion field amplification would be very difficult, if not impossible, to identify among all sources of "nuclear noise". With this reservation in mind, one can use the previous discussion as a basis to establish the criteria for best detection of the searched phenomenon. As a spin-isospin ordering effect it should be looked for in isovector form factors of the spin type. It is also clear that the specific information on the pion channel is essentially contained in the

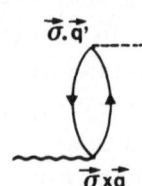

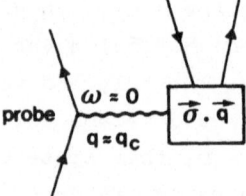

Fig.5 The conversion between transverse (wavy line) and longitudinal (dashed line) couplings through the nuclear polarizability.

Fig. 6 The probe for pion opalescence.

form factors of the $\langle \vec{\sigma}\cdot\vec{q} \rangle$ mode (or longitudinal spin form factors). One can add incidentally that the $\langle \vec{\sigma}\times\vec{q} \rangle$ form factors are not completely unsensitive to pion opalescence. There is actually some coupling between the transverse and longitudinal modes because the momentum transfer $|\vec{q}-\vec{q}'|$ (of the order of the inverse nuclear radius) prevents complete orthogonality between the two vertices (fig.5). This conversion of a transverse probe into a pion is however relatively weak so that only a very large pion field enhancement would have a marked influence on the transverse form factors.

The conditions for a selective detection clearly take shape : are sould choose a proble which couples to the longitudinal spin mode $\langle \vec{\sigma}.\vec{q} \rangle$ (fig.6) and look for an increase of the corresponding form factors at the relevant momenta. Hence the crucial prime requirement should be a good knowledge of the driving interaction (wavy line on fig. 6) leading to the spin excitation so as to permit a clean separation of the longitudinal and transverse spin information. A weakly interacting probe would be preferable to avoid final state interaction (and connected) problems which would produce some mixing of the two modes and blur the information extracted from the data. Unfortunately photon and neutrino reactions present major drawbacks. The coupling of the (virtual) photon to the magnetization is indeed transverse so that (e,e') reactions are purely informative on the ($\vec{\sigma} \times \vec{q}$) mode. As for lepton production by neutrino reactions, it has both components. The longitudinal contribution has however the unpleasant feature of being proportional to the squared lepton mass at zero energy transfer. It has thus generally weak cross sections and would be difficult to separate from the transverse component except in special cases. It appears that in contradistinction to the transverse mode there is not clean specific probe of the spin longitudinal degree of freedom. One has to turn to reactions involving strongly interacting particles with their already mentionned inherent uncertainties. Among them, the photopion reactions (γ,π), (e,e'π), ($\pi,2\gamma$) and the inverse processes[1,16-18] are the best understood mainly because they are connected by current algebra to the weak currents. Other probes have been proposed like (p,p')[19,20] and ($\pi,2\pi$)[21] but the interpretation of the experiments strongly relies on our knowledge of the elementary reaction on nucleons which is not always sufficient at the needed energies and momenta.

The easiest and very informative experiments would be inclusive measurements of the inelastic response function (fig.7). They would not only probe the spatial (longitudinal) spin correlations with their typical momentum dependence but also their lifetime which should increase at the approach of the phase transition[22]. This feature appears as a softening of the quasi-elastic peak, i.e. a shift towards the low energy transfers. Theoretical calculations have been limited up to now to

Fig.7 The spin longitudinal response (the hatched circle stands for the nuclear complexity).

the case of infinite nuclear matter. On the experimental side, only the (e,e') response is presently available[23]. As we have already stressed, it is not properly adequate to the searched pion degrees of freedom and does not anyway show outstanding anomalies (see however ref.24). The use of other probes depends on the possibility of extracting the spin longitudinal information form the transverse and non-spin degrees of freedom and occasionnally dominating the final state interaction complications. This involves more specific experiments with e.g. polarized particles. There exist yet no systematic investigation of the various possibilities. Among the best known processes we have seen that neutrino reactions do not appear very promising because of the dominance of the charge and spin transverse modes. Inclusive photopion reactions offer interesting perspectives when working with polarized photons at specific angles (scattering angle $\theta = \pi/2$, $\phi = 0$). Preliminary studies have also been presented for $(e,e'\pi)$[17] and $(\pi, 2\pi)$[21] production near threshold.

The difficulties encountered in the interpretation of inclusive reactions are also present for more exclusive processes where one looks at the excitation of specific pion-like states. These necessitate stringent resolution requirements. The available experiments are still very scarce though they will develop in the near future. Most efforts has been concentrated up to now on the excitation of the 1^+ (15.11MeV) state of ^{12}C. Anomalies in the form factor measured by (e,e') scattering [25,26] have been tentatively interpreted [3] as a signal for strong pion opalescence (4.5 enhancement) and this analysis was confirmed[4] by recent measurements of the M1 form factor of ^{13}C[27]. In view of the low sensitivity of this transverse information to critical phenomena (an enhancement of $|<\vec{\sigma}\cdot\vec{q}>|^2$ by a factor 20 would be necessary to produce a factor 2 in the squared $<\vec{\sigma}\times\vec{q}>$ form factor) and the rather complicated structure of these open shell nuclei, it seems that reasonable alternative explanations are possible though a definitive interpretation is yet to be found [28-30]. The same state has been explored by (p,p') reactions at 800 MeV (Los Alamos)[31] and 400 MeV (Saturne)[32]. The interpretation of the data is somewhat obscured at the first energy by uncertainties in the driving amplitude. The situation at the lower energy is in principle under better control. No definitive analysis has yet been produced. It seems however that the interaction contains much more transverse component than first believed. Nevertheless the large opalescence advocated from the (e,e') data seems excluded. Preliminary calculations indicate that the data at both energies can be accounted for with a rather moderate pion field enhancement (a factor 2) corresponding to g' = $0.55^{33)}$. This value can also produce agreement with the M1 form factor. One will remark that it is sensibly lower than that extracted from the energy levels. Other yet unexplored reactions can show an interesting sensitivity. Lepton processes leading to 0^- states would be a very selective probe of the $\langle \vec{\sigma}\cdot\vec{q}\rangle$ mode with an unfortunately very small cross section ; leptonic excitation of $J \neq 0$ states would be predominantly transverse. The photopion reactions (γ,π) and (π,γ) have been shown to present promising features at angles around $\pi/2^{16)}$. They have however the drawback of strongly interacting probes and one has not yet enough experience to insure that the pion-nucleus dynamics is under control. Such experiments are nevertheless currently planned or performed in the appropriate energy and momentum range $^{34,35)}$.

5. Conclusion :

The present information on the pion mode in nuclei is rather adverse to strong critical opalescence phenomena. Energy levels are interpretable in terms of a rather weak particle-hole force where the pion attraction is counterbalanced by short range repulsion. Form factor measurements permit more direct exploration of the longitudinal spin-isospin correlations. The scarce data seem consistent with moderate pion field opalescence. It is clear that complementary information is necessary for a more precise statement. Though theoretical investigations are not yet completed, it seems that there exist no perfectly selective experiment for the pion channel with the exception of the hardly feasible excitation of 0^- states. One need therefore a variety of measurements making use of the whole arsenal of spin sensitive probes. The investigations should be pursued in several nuclei especially those with closed shell structure which are the natural field of application of RPA theory. Special attention is to be devoted to the search and the study of the "pion favoured" states already pointed out by Meyer-ter-Vehn. Comparison of $\langle\vec{\sigma}\cdot\vec{q}\rangle$ and $\langle\vec{\sigma}\times\vec{q}\rangle$ form factors in the critical momentum region would be especially instructive. Such an effort is deserved if we want to understand the rather elusive pion mode in nuclei and its eventual extension to pion condensation.

References

1. M.Ericson and J.Delorme, Phys.Lett. 76D (1978) 182
2. M.Gyulassi and W.Greiner, Ann.Phys.(NY) 109 (1977) 485
3. J.Delorme, M.Ericson, A.Figureau and N.Giraud, Phys.Lett. 89B (1980) 327
   J.Delorme, A.Figureau and N.Giraud, Phys.Lett. 91B (1980) 328
4. J.Delorme, A.Figureau and P.Guichon, Report LYCEN 8023,Lyon (1980)

5. J.Meyer-ter-Vehn, Report PLF 34, Garching (1980)
6. A.B.Migdal, Rev.Mod.Phys. 50 (1978) 107
7. M.Ericson and T.E.O.Ericson, Ann.Phys. (NY) 36 (1966) 323
8. G.E.Brown, S.O.Bäckman, E.Oset and W.Weise, Nucl.Phys. A 286 (1977) 191
9. W.Dickhoff, A.Faessler, J.Meyer-ter-Vehn, H.Müther, preprint (1980)
10. E.Oset and M.Rho, Phys.Rev.Lett. 42 (1979)47
11. S.Barshay and G.E.Brown, Phys.Lett. 47B (1973) 107
12. W.Burr, D.Schütte and K.Bleuler, Nucl.Phys. A133 (1969) 581
13. J.Meyer-ter-Vehn, unpublished report (1979)
14. J.Speth, V.Klemt, J.Wambach and G.E.Brown, Nucl.Phys. A 343 (1980) 382
15. G.Baym and G.E.Brown, Nucl.Phys. A 247 (1975) 395
16. J.Delorme, to appear in Journal of Physics G
17. J.M.Eisenberg, preprint (1980)
18. T.E.O. Ericson and C.Wilkin, Phys.Lett. 57B (1976) 1
19. E.E.Saperstein, S.V.Tolokonnikov and S.A.Fayans, JETP Lett. 25 (1977) 513
20. H.Toki and W.Weise, Phys.Rev. Lett. 42 (1979) 1034
21. J.M.Eisenberg, Phys.Lett. 93B (1980) 12
22. W.Alberico, M.Ericson and A.Molinari, Phys.Lett. 92B (1980) 153
23. P.Barreau et al., contribution to the Symposium on "Perspectives in Electro -and Photo- Nuclear Physics, Saclay (1980)
24. Y.Kawazoe et al., preprint (1980)
25. J.B.Flanz et al., Phys.Rev.Lett. 43 (1979) 1922
26. R.Neuhausen, private communication
27. R.S.Hicks et al., private communication
28. H.Toki and W.Weise, Phys.Lett. 92B (1980) 265
29. T.Suzuki, H.Hyuga, A.Arima and K.Yazaki, preprint (1980)
30. T.Suzuki, F.Osterfeld and J.Speth, preprint (1980)
31. M.Haji-Saeid et al., Phys.Rev.Lett. 45 (1980) 880
32. J.L.Escudié , private communication
33. H.Toki and W.Weise, contribution to the Symposium on "Perspectives in Electro-and Photo- Nuclear Physics", Saclay (1980)
    W.Weise, private communication
34. J.Deutsch and P.Truöl, private communication
35. B.A.Mecking, private communication.

INELASTIC ELECTRON AND PROTON SCATTERING
TO PION-LIKE NUCLEAR EXCITED STATES [+)]

H. Toki [++)] and W. Weise
Institute of Theoretical Physics
University of Regensburg
D-8400 Regensburg, Germany

1. Introduction and Motivation

The motivation to investigate the structure of pion-like nuclear excited states (i.e. states with unnatural parity, $J^\pi = 0^-, 1^+, 2^-, ...$), especially at high momentum transfers, derives from their close relationship to the problem of pion condensation [1,2,3]. The issue has been summarized in a recent review by Meyer-ter-Vehn [4], and a survey has been given by Delorme [5] at this conference.

We would like to look at this question from the point of view of determining the strength of the virtual pion field in a nuclear environment. Consider the static one-pion exchange (OPE) interaction in free space,

$$V_\pi(\vec{q}) = - \frac{f^2}{m_\pi^2} \frac{\vec{\sigma}_1 \cdot \vec{q}\, \vec{\sigma}_2 \cdot \vec{q}}{\vec{q}^2 + m_\pi^2} \vec{\tau}_1 \cdot \vec{\tau}_2 \,. \tag{1}$$

where $f^2/4\pi = 0.08$, $\vec{q}$ is the momentum transfer carried by the pion. If the vacuum is replaced by a Fermi sea filled with nucleons, the pion field will polarize the medium, as shown in Fig. 1, by virtual excitation of nucleon-hole and Δ-hole pairs carrying pion quantum numbers. The polarization strength is controlled essentially by three

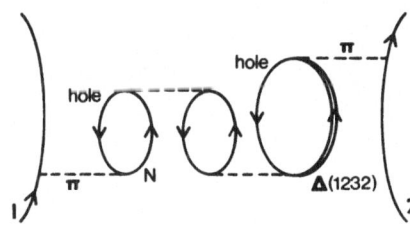

Figure 1: One-pion exchange between nucleon 1 and 2 is modified in a nuclear medium by mesic polarization effects via the virtual excitation of high-lying nucleon-hole and $\Delta(1232)$-hole states.

[+)] Invited talk presented by W. Weise; work supported in part by Deutsche Forschungsgemeinschaft.

[++)] Now at Department of Physics, Michigan State Univ., East Lansing, Michigan, USA

factors: the strength of the p-wave pion-nucleon coupling, the density of the medium and, very importantly, the strength of repulsive short range correlations accompanying the driving OPE interaction. The modified OPE interaction can then be written

$$\tilde{V}_\pi(\vec{q}) = \frac{V_\pi(\vec{q})}{\varepsilon(q)} \qquad (2)$$

where the mesonic polarization effects are summarized in terms of the diamesic function $\varepsilon(q)$. If, in a much simplified picture, the repulsive short range correlations are parametrized by $g' \vec{\sigma}_1 \cdot \vec{\sigma}_2 \; \vec{\tau}_1 \cdot \vec{\tau}_2$ times a delta function in r-space, then

$$\varepsilon(q) = 1 + [g' - \frac{f^2}{m_\pi^2} \frac{q^2}{q^2 + m_\pi^2}] U(q,\rho) \qquad (3)$$

where $U(q,\rho)$ contains the information about each individual nucleon-hole or Δ-hole excitation, represented by a single "bubble" in Fig. 1, including possible vertex factors.

The size of the diamesic function is essentially determined by the competition between the attraction from OPE (2/3 of which comes from the tensor part) and the short-range repulsion controlled by g'. The characteristic features of the inverse diamesic function $1/\varepsilon(q)$ are shown in Fig. 2.

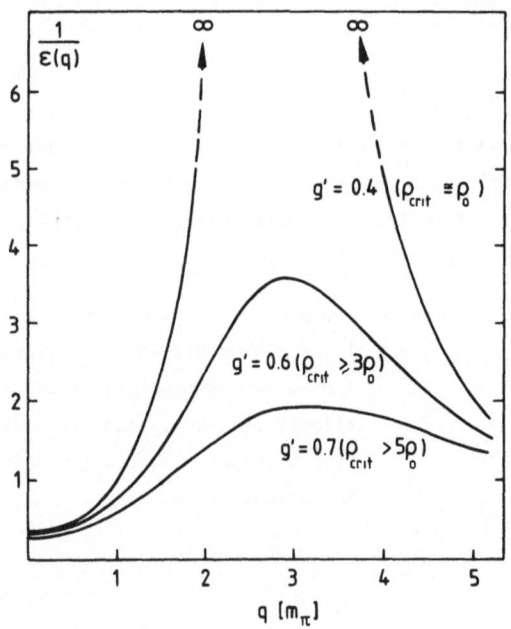

Figure 2:

The inverse static diamesic function at normal nuclear matter density, $\rho_o = 0.17$ fm^{-3}, shown as a function of momentum transfer q. Note the strong dependence on the parameter g' which measures repulsive short range correlations accompanying OPE. To each g', the corresponding critical density ρ_{crit} for pion condensation has been assigned.

With increasing momentum transfer q, the attraction from OPE increases, such that the medium acts to amplify the pion field. The maximum enhancement appears at $q \sim 3m_\pi$, since the medium cannot easily accomodate momentum transfers larger than twice the Fermi momentum. The degree of amplification depends sensitively on g'. In fact, critical amplification ($1/\varepsilon \to \infty$) occurs for $g' \simeq 0.4$ at the density of normal nuclear matter, $\rho_0 = 0.17$ fm^{-3}. This is the threshold for pion condensation. Even if this threshold is far from being reached, pronounced amplification phenomena are expected unless g' is large enough to prohibit any critical effects.

Guided by these observations, the following questions emerge, all of which are strongly related to each other:

(a) How far away is the pion condensate [1-4]?

(b) Does the nucleus act as a "short-wavelength amplifier" for the pion field [6-9]?

(c) What is the nature of spin-isospin correlations in nuclei [10]?

Figure 3: Probing the virtual pion field in nuclei by selective excitation of pion-like excited states.

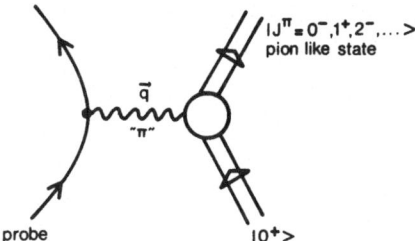

The idea how to provide answers to some of these questions by detailed experiments is illustrated in Fig. 3: in the ideal case, a probing particle acts as a source for a pion field of well defined momentum $\vec{q}$ which induces a transition to a pionlike state. In Born approximation, the cross section $d\sigma/d\Omega$ is roughly proportional to the square of the inverse diamesic function (or rather its analogue for a finite nucleus, to be discussed later). Hence pionic enhancements, if existent at all, should show up in $d\sigma/d\Omega$ under appropriate angles.

As a sideremark, we mention that in the long wavelength ($q \to 0$) limit, the diamesic function, eq. (3), causes quenching rather than amplification. This quenching is systematically observed in Gamow-Teller [11] and magnetic multipole transitions [12-13]. It puts important constraints on repulsive short-range correlations described by g'. A large fraction of this quenching is actually due to virtual Δ-hole excitations [11,13].

2. Case 1: Inelastic electron scattering into $^{12}C(1^+, T=1)$

The 1^+ state in ^{12}C at 15.1 MeV has so far been the only pionlike state whose high momentum transfer properties have been explored with sufficient accuracy. We consider the electron scattering case first, although this is not an optimum probe for the nuclear pion field, because of the transverse ($\vec{\sigma} \times \vec{q}$) nature of the M1 operator, as opposed to the longitudinal ($\vec{\sigma}\cdot\vec{q}$) coupling of the pion source. For small nuclei, this restriction is, however, not so severe, because of surface effects.

The long wavelength properties of the 1^+ state in question are supposed to be described accurately by the Cohen-Kurath (CK) model which exploits many-particle many-hole configurations within the p-shell. It is known for example that 4p4h-components are important. The well-known failure to reproduce the M1 formfactor with the CK model is illustrated in Fig. 5; this is not surprising: a pure p-shell model space is simply to small to accomodate the high-momentum transfer properties of the 1^+ state. Core polarization effects are important, as pointed out in ref. [14]. Further reaching conclusions have been drawn in ref. [15], where the required enhancement beyond the CK result has been interpreted as a signature of pionic critical opalescence. We have examined [15] this in some detail using the following model:

(a) The p-shell structure of the 1^+ state is supposed to be described appropriately by the CK model.

(b) Virtual nucleon-hole excitations beyond the p-shell, as well as Δ-hole excitations, are incorporated by diamesic function techniques explained in detail in ref. [17].

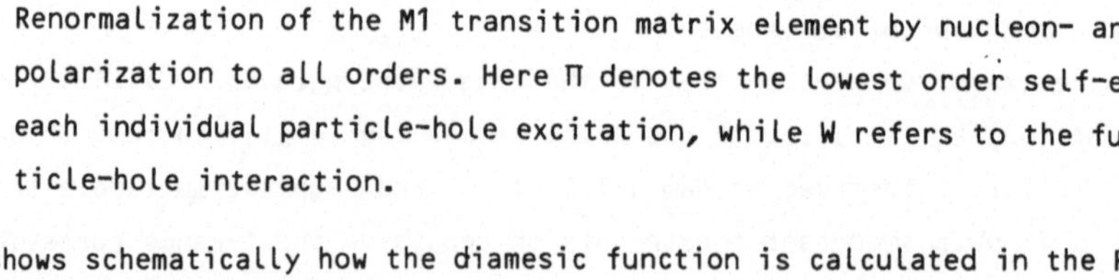

Figure 4: Renormalization of the M1 transition matrix element by nucleon- and Δ-hole polarization to all orders. Here Π denotes the lowest order self-energy for each individual particle-hole excitation, while W refers to the full particle-hole interaction.

Fig. 4 shows schematically how the diamesic function is calculated in the RPA approximation for a finite nucleus. If $\Pi_J(q, q')$ is the pion self-energy representing the relevant nucleon- and Δ-hole excitations then the iteration of Π_J to all orders yields the response function

$$R_J(q,q') = \Pi_J(q,q') + \int_0^\infty dk\ \Pi_J(q,k)\ W_J(k)\ R_J(k,q'). \tag{4}$$

Actually Π_J is a matrix, and $W_J(k)$ refers to matrix elements of the full nucleon- or Δ hole interaction (see ref. [17] for further details). The properties of this particle-hole interaction are an important issue in our development, so that a somewhat more detailed discussion will be devoted to this question.

Our point of view [16] is that the particle-hole interaction operating in spin-isospin excitation channels can be split into the leading one-pion exchange interaction, the spin-isospin dependent part of two-pion exchange interaction including ρ-meson exchange, and short-range correlations accompanying these. Therefore,

$$W(\vec{q}) = V_\pi(\vec{q}) + V_{2\pi}(\vec{q}) + [g'(q) \vec{\sigma}_1 \cdot \vec{\sigma}_2 + h'(q) S_{12}(\hat{q})] \vec{\tau}_1 \cdot \vec{\tau}_2. \qquad (5)$$

The two-pion exchange is taken from the Paris NN interaction, with addition of iterated one-pion exchange, so that the important second order tensor force is automatically included. Thus $V_{2\pi}$ is the sum of ρ exchange and non-resonant 2π exchange background, their strength being determined by the square of the relevant $\pi\pi \to N\bar{N}$ helicity ampltitudes. We note that $V_{2\pi}$ is proportional to $(\vec{\sigma}_1 \times \vec{q}) \cdot (\vec{\sigma}_2 \times \vec{q})$, hence it does not contribute to pion-like excitations in infinite nuclear matter. It is of great importance, however, in the discussion of M1 excitations of pion-like states in small nuclei, since the isovector (2π) system carries photon quantum numbers.

The additional correlations accompanying V_π and $V_{2\pi}$ are described by g' (for the spin-spin part) and h' (for the tensor part), respectively. Now, h' turns out to be small of order $(\lambda q)^2$, where λ is a typical distance over which short-range correlations act. The important quantity is g'(q). Note that the long wavelength limit of the particle-hole interaction leads to

$$W'(q \to o) = g'(o) \vec{\sigma}_1 \cdot \vec{\sigma}_2 \vec{\tau}_1 \cdot \vec{\tau}_2,$$

which is the limit of Landau-Migdal theory. Phenomenological determinations within that framework lead to $g'(o) \simeq 0.6$. We are interested, however, in the high momentum transfer region, $q \sim 2-3\, m_\pi$. A reliable theory of g'(q) at large momentum transfers does not exist, although attempts are being made to perform calculations in this direction [18]. Assuming a smooth q-dependence of g'(q) because of the short range of the underlying interactions, we prefer to treat g' as a parameter.

The actual calculation of the diamesic function proceeds in a 20 $\hbar\omega$ harmonic oscillator space. Fig. 5 shows the influence of mesic polarization on the M1 formfactor [16,23], illustrating the significance of different pieces of the particle-hole interaction operating in the diamesic function. In the absence of $V_{2\pi}$, the repulsive g' moves the minimum to the right place, while V_π acts such as to produce only a moderate enhancement in the form factor for g' = 0.55. The ρ exchange part of $V_{2\pi}$, because of its alignment with the M1 operator, increases the formfactor at high q, but not by a sufficient amount. At this level Delorme et al. [15] raised the form factor by reducing g' to a value smaller than 0.4, suggesting that critical conditions are very closely approached. On the other hand, once the full $V_{2\pi}$ is employed by adding non-resonant pieces to ρ-exchange, there seems to be no need for a drastic reduction of g', as Fig. 5 shows. (A $g' = 0.55$ would raise the critical density in nuclear matter considerably beyond three times nuclear matter density.)

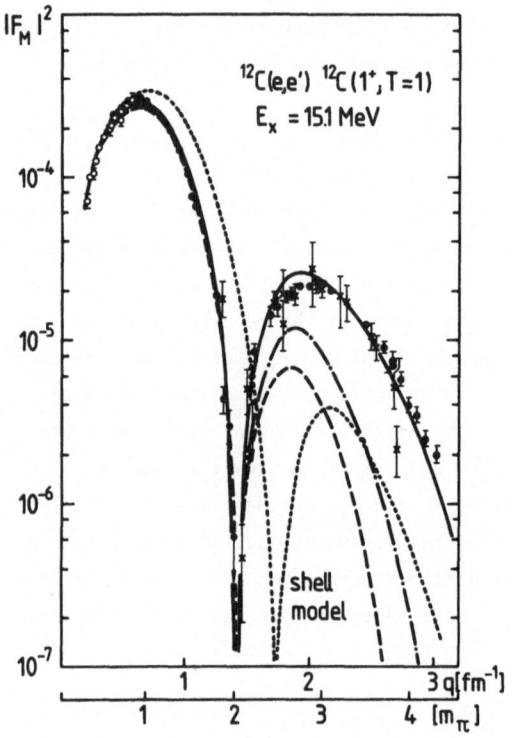

Figure 5:

M1 form factor in ^{12}C. Short-dashed curve: Cohen-Kurath shell model result; dashed curve: mesonic polarization incorporated with V_π and g' only ($V_{2\pi} = 0$) for $g' = 0.55$; dash-dot curve: same calculation with V_ρ added to the particle-hole interaction, but omitting non-resonant parts of $V_{2\pi}$; solid curve: full calculation including V_π, $V_{2\pi}$ and $g' = 0.55$, showing the importance of background parts in $V_{2\pi}$ in addition to ρ exchange.
(experimental data: ref. [25])

An alternative approach of the Tokyo group [19] emphasizes the role of genuine two-body mechanisms in addition to 1p1h core polarization processes. It has to be sorted out whether parts of such terms can be interpreted as exchange corrections (vertex corrections in the language of Landau-Migdal theory) already incorporated phenomenologically in the definition of g'. Those parts related to the tensor force turn out to be small. The Jülich group [20] has replaced the diamesic function approach by an explicit $6\,\hbar\omega$ 1p1h RPA calculation, which appears to be sufficient to reproduce what is otherwise treated as a polarization effect on top of a small (p-shell) model space. This procedure requires an arbitrary reduction factor to reproduce the effect of many-particle-many-hole configurations in the p-shell. Also, Δ-isobars are omitted, which account for a large fraction of the polarization effect at large q. On the other hand, convection current contributions are incorporated, which raise the second maximum of the form factor. There remain some questions about the uniqueness of the particle-hole interaction used in this work.

The qualitative conclusions are, however consistent with those drawn before, namely that relatively large values of g' are favoured, such that critical conditions are far from being met.

3. Case 2: Inelastic proton scattering into $^{12}C(1^+, T=1)$

One might expect that inelastic proton scattering would be a better way to study the presence or absence of pionic amplification effects, because the scattered proton naturally provides a pion source [16]. Experiments have extended recently into the high momentum transfer region at Indiana (120 - 150 MeV), Saclay (400 MeV) and Los Alamos (800 MeV). We wish to concentrate here first on the highest energy, where problems due to reaction mechanism are minimal and the scattering can be described appropriately within Glauber theory. Problems still exist, however, at the level of the spin-isospin dependent two-nucleon amplitudes $f_{\sigma\tau}$ at high energies. Omitting spin-orbit parts which are small in isovector channels, we write

$$f_{\sigma\tau}(q, E) = [A(q, E)\vec{\sigma}_1 \cdot \hat{q}\, \vec{\sigma}_2 \cdot \hat{q} + B(q, E)\, \vec{\sigma}_1 \cdot \vec{\sigma}_2]\, \vec{\tau}_1 \cdot \vec{\tau}_2 \qquad (7)$$

and assume that the real part of A can be identified with that obtained from π and 2π exchange; thus A is proportional to $V_\pi - V_{2\pi}$. Again, a large transverse coupling implied by $V_{2\pi}$ acts to reduce the pion source related to V_π. This effect seems to be substantiated also by recent developments [21] to obtain a reliable parametrization of $f_{\sigma\tau}$. The parameter B is supposed to be a smoothly varying complex function of E and q, which is fit to low q data.

Fig. 6 shows $d\sigma/d\Omega$ at 800 MeV (with data from ref. [22]), employing $f_{\sigma\tau}$ together with the diamesic function model used in the description of the M1 form factor. The steep negative slope in $d\sigma/d\Omega$ at large q indicates that (a) a strong transverse component in $f_{\sigma\tau}$ of a 2π exchange range seems indeed to be required and (b) the non-observation of strong enhancements around $q \sim 3\, m_\pi$ is consistent with the relatively large g' (between 0.5 and 0.6, rather than 0.4) suggested also by the M1 formfactor analysis. The calculations shown in Fig. 6 exhibit too deep minima in $d\sigma/d\Omega$; these are probably enforced by too small values of Im A (see eq. (7)) in our calculation. In fact, the most recent parametrizations of $f_{\sigma\tau}$ [21] indicate relatively large imaginary parts in the $\vec{\sigma}\cdot\vec{q}\,\vec{\sigma}\cdot\vec{q}$ part of the amplitude. Refined calculations are in progress [23].

The situation at 400 MeV appears to be very similar. In fact, the data [24] are almost identical to those at 800 MeV. While $f_{\sigma\tau}$ is supposed to be better known at 400 MeV, distortion effects are also more important and require a full DWIA calculation [23].

Figure 6:

Glauber model calculation of inelastic proton scattering at 800 MeV leading to $^{12}C(1^+, T=1)$. The elementary two-body amplitude has been used as in eq. (7), with $A = -\frac{M}{2\pi}[V_\pi - V_{2\pi}]$, $B = -\frac{M}{2\pi} \cdot 0.3$ otherwise, the model is the same as that used in the calculation of the M1 formfactor. Dashed curve: Cohen-Kurath shell model result; solid curves: mesic polarization included, employing the full particle-hole interaction with V_π, $V_{2\pi}$ and g', for different values of g'.

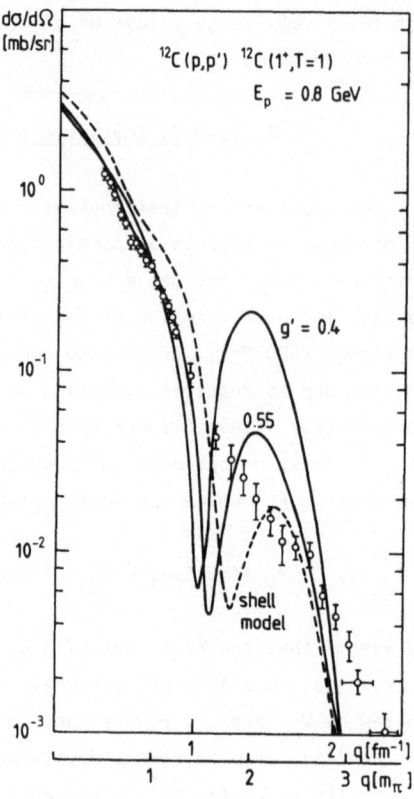

4. Conclusions

So far, there is no evidence for a strong amplification of the pion field inside a nucleus. The repulsive short-range correlations accompanying one-pion exchange seem to be sufficiently strong such as to prevent the nucleus from approaching critical conditions. However, such conclusions need to be established more systematically by investigations of pion-like states in a larger variety of nuclei.

According to our present understanding, nuclei are far from being pion-condensed. It is nevertheless worth noting that the issue of pion-condensation has had considerable impact on recent experimental proposals to perform detailed investigations of the short-wavelength properties of nuclear spin-isospin correlations.

References

1. A. B. Migdal, Rev. Mod. Phys. 50 (1978) 107

2. G.E. Brown and W. Weise, Phys. Reports 27 (1976) 1

3. See contributions of: A.B. Migdal, R.F. Sawyer, G. Baym and D.K. Campbell, S.-O. Bäckman and W. Weise, in: Mesons in Nuclei, Vol. III, M. Rho and D. H. Wilkinson, eds., North-Holland (1979)

4. J. Meyer-ter-Vehn, Phys. Reports (1980), in print

5. J. Delorme, lecture presented at this conference

6. M. Ericson and J. Delorme, Phys. Lett. 76 B (1978) 241;

7. M. Gyulassi and W. Greiner, Ann. of Phys. 109 (1977) 485

8. S. A. Fayans, E. E. Saperstein and V. E. Tolokonnikov, Nucl. Phys. A 326 (1979) 463

9. H. Toki and W. Weise, Phys. Rev. Lett. 42 (1979) 1034

10. J. Speth, V. Klemt, J. Wambach and G.E. Brown, Nucl. Phys. A 343 (1980) 382

11. E. Oset and M. Rho, Phys. Rev. Lett. 42 (1979) 47;
    I.S. Towner and F.C. Khanna, Phys. Rev. Lett. 42 (1979) 51

12. W. Knüpfer, M. Dillig and A. Richter, Phys. Lett.

13. H. Toki and W. Weise, Phys. Lett. 47 B (1980) 12

14. H. Sagawa, T. Suzuki, H. Hyuga and A. Arima, Nucl. Phys. A 322 (1979) 361

15. J. Delorme, M. Ericson, A. Figureau and N. Giraud, Phys. Lett. 89 B (1980) 327;

16. H. Toki and W. Weise, Phys. Lett. 92 B (1980) 265

17. H. Toki and W. Weise, Z. Phys. A 292 (1979) 389, A 295 (1980) 187

18. W. H. Dickhoff, J. Meyer-ter-Vehn, A. Faessler, H. Müther, preprint (to be published)

19. T. Suzuki, H. Hyuga, A. Arima and K. Yazaki, preprint (submitted to Phys. Lett.)

20. T. Suzuki, F. Osterfeld and J. Speth, preprint (submitted to Phys. Lett.)

21. W. G. Love, private communication

22. J. M. Moss et al., Phys. Rev. Lett. 44 (1980) 1189;
    M. Haji-Saeid et al., Phys. Rev. Lett. 45 (1980) 880

23. H. Toki and W. Weise, in preparation

24. J. L. Escudié et al., preprint (submitted to Phys. Rev. Lett.)

25. J. Flanz et al., Phys. Rev. Lett. 43 (1979) 1922;
    J. Neuhausen, private communication

THE NUCLEAR Δ-EXCITATION

K. Klingenbeck

Institute for Theoretical Physics, University of Erlangen-Nürnberg, Erlangen

I. Introduction

In low energy nuclear physics one usually treats the nucleus as a collection of inert particles, the nucleons as the basic constituents of a nucleus. On the other hand we are all aware of the fact that the nucleon has an internal structure: there is a whole spectrum of excited states, the various baryons characterized by their spin-, isospin- and strangeness. Certainly the corresponding internal (subnucleonic) degrees of freedom of the nucleon do also show up in a complex nucleus if the necessary excitation energy ω (roughly 200 MeV $<\omega<$ 1 GeV) is transfered to the system. Consequentely this excitation energy is not available to the external motion of nucleons, thermalization is suppressed and the energy is stored inside of the nucleonic quark system.

Therefore entering this energy domain of nuclear physics, we have in one way or another to extend the conventional nuclear picture to account for those internal degrees of freedom of bound nucleons. From this point of view the propagation of excited baryons and their coupling to the surrounding nuclear medium becomes one of the challenging and fascinating questions to be investigated with nuclear reactions at the corresponding excitation energies.

The way we will approach this problem is to introduce the baryon as the basic nuclear constituent and to treat the nucleus as a Many Baryon System including both, the external as well as the internal degrees of freedom. Consequently, a complex nucleus is expected to exhibit a nuclear excitation spectrum in the region of the elementary baryon resonances, reflecting the excitation strength of the corresponding many baryon eigenmodes. This will be discussed in detail later on for the Δ(3,3) isobar. Clearly, conventional nuclear physics is still contained in such a scheme with the excitation energy approaching zero. Here all the baryons are essentially in their nucleonic ground state, the excited baryon states are only felt as relatively small dispersive effects via virtual isobar admixtures (for a review see ref. 1). This paper will however exclusively deal with the real excitation, i. e. with nuclear excitations right in the region of elementary baryon resonances.

II. The Nucleus as a Many Baryon System

In this chapter we shall briefly outline the formal development of such a many baryon picture[2] which can necessarily be done only on a very sketchy way. To begin with we have to comment on how to introduce the baryon as an elementary object.

II.1 The free baryon

Introducing the concept of bound states embedded in the continuum (BSEC) we start out from stable or bare baryons which may be viewed as bound states of the corresponding quark bag. However those bare states are embedded in and coupled to real decay continua, giving rise to a mass shift and some width of the physical baryon. This is graphically demonstrated in fig. 1 for the Δ(3,3) isobar. It is in this sense that the baryon may be called the strong interaction analogue of the Fano resonances[3] of atomic physics.

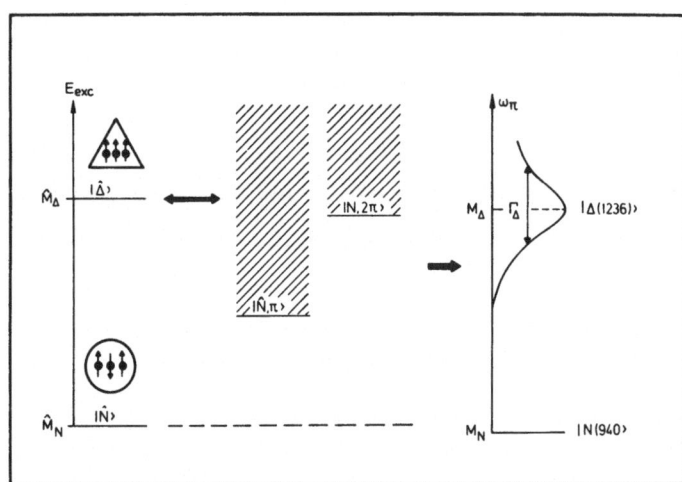

Fig. 1: The Δ(3,3) resonance as a BSEC.

More formally this concept may be written down in a Hamiltonian formalism:

$$h = \hat{h}_B + h_m + h_c \tag{1a}$$

where $\hat{h}_B$ describes bare baryons $|\hat{B}_\nu\rangle$:

$$\hat{h}_B |\hat{B}_\nu\rangle = \hat{M}_\nu |\hat{B}_\nu\rangle \tag{1b}$$

and where h_m describes the various meson fields ($\pi, \varrho, k, \ldots$). In a underlying quark picture the derivation of the coupling operator h_c would be equivalent to solve the problem of how one quark bag manages to decay into asymptotically free quark clusters of baryons and mesons. It is here where we introduce this coupling on the basis of phenomenological transition operators $\mathcal{L}$:

$$h_c = \mathcal{L} + \mathcal{L}^+ \quad ; \quad \mathcal{L} = \sum_{m,\nu,\nu'} \mathcal{L}_{m\nu\nu'} \tag{1c}$$

where $\mathcal{L}_{m\nu\nu'}$ transforms $|\hat{B}_\nu\rangle$ to $|\hat{B}_{\nu'}\rangle$ via the meson field $|m\rangle$. Eliminating the meson fields by projecting the hamiltonian of eq. (1a) onto the one baryon subspace, the stable part $\hat{h}_B$ is complemented by an additional term, which reflects the influence of mesonic decay continua and generates the mass shift and width:

$$h_B = \hat{h}_B + \delta h_B \quad ; \quad \delta h_B = \mathcal{L}^+ (E^+ - \hat{h}_B - h_m - h_c)^{-1} \mathcal{L} \tag{2a}$$

The physical baryons are then the eigenstates of an energy dependend and complex, i. e. nonhermitean, Hamiltonian:

$$h_B |B_\nu\rangle = (M_\nu + i \Gamma_\nu/2) |B_\nu\rangle \tag{2b}$$

With an expansion of δh_B in powers the meson-baryon coupling one obtains the well known diagrammatic series for the self-energy of fig. 2.

II.2 The Many Baryon System

To describe the Many Baryon System we similarly start out from A stable (bare) baryons and meson fields, with each individual baryon being coupled to the meson field:

$$H = \hat{H}_B(1,...,A) + h_m + H_C(1,...,A) \qquad (3a)$$

$$\hat{H}_B = \sum_{i=1}^{A} \hat{h}_B(i) \quad ; \quad H_C = \sum_{i=1}^{A} h_c(i) \qquad (3b)$$

Again the elimination of the meson field, generates an additional A-body part of the Many Baryon Hamiltonian, $\mathcal{H}_B$, containing implicitly the influence of meson continua:

$$\mathcal{H}_B(1,...,A) = \hat{H}_B(1,...,A) + \delta\mathcal{H}_B(1,...,A) \qquad (4a)$$

with

$$\delta\mathcal{H}_B(1,...,A) = \sum_{i,k} \mathcal{L}^+(i)\left(E^+ - \hat{H}_B - h_m - H_C\right)^{-1} \mathcal{L}(k) \qquad (4b)$$

Obviously this part can be split up into two contributions:

(i) single particle of selfenergy terms, which by expansion exhibit a subseries very similar to the free selfenergy, however modified by the presence of the Many Baryon propagator; in addition other dynamical corrections do appear (see fig.3).

(ii) interactions, due to the fact that one baryon can couple to the meson field which in turn couples back to another baryon. Clearly from fig. 4 we obtain the baryon-baryon interaction as a generalization of the NN-interaction, the OBE-terms, many boson exchanges and many body forces.

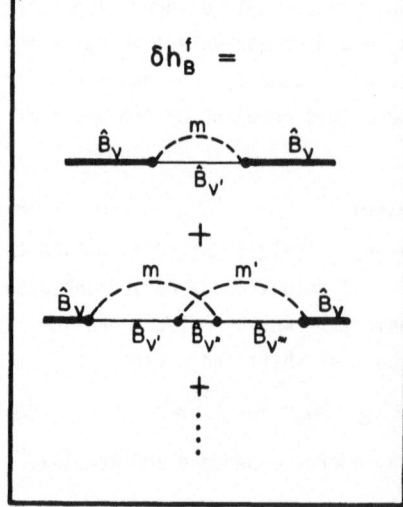

Fig. 2:
The free selfenergy of a baryon.

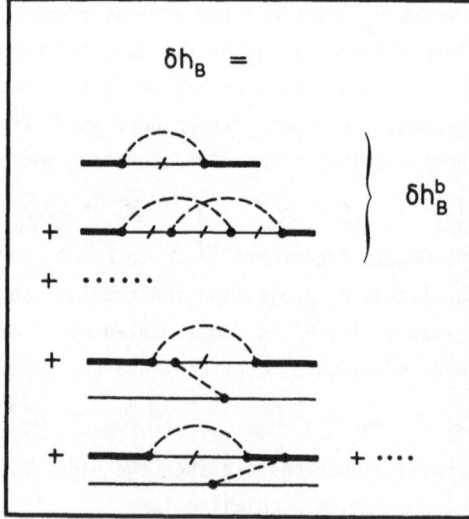

Fig. 3: Selfenergy contributions for a bound baryon.

Fig. 4: Some contributions to the baryon-baryon interaction.

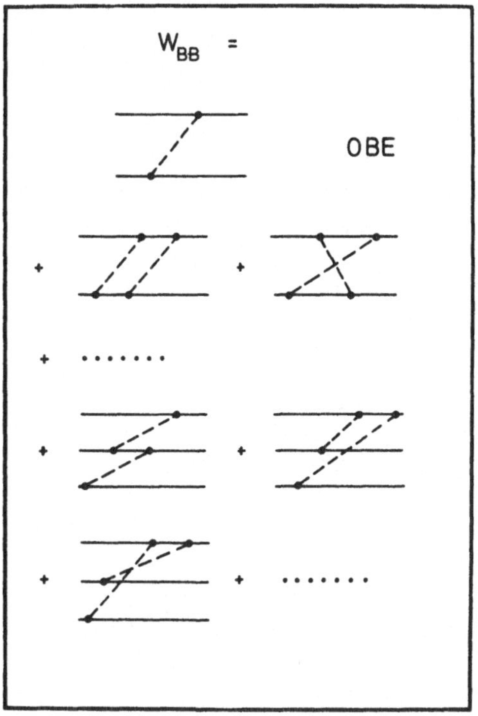

So far for the formal development which is certainly very useful as a systematic guide in actual problems. However clearly, the Hamiltonian problem of eq. (4) is at least as complicated as conventional nuclear many body physics. Consequently one can only attempt a solution within a certain model and one needs some test from a comparison with experimental data for reactions which probe those subnuclear excitations of nuclei.

III. The Reaction Mechanism

For the discussion of reactions let us assume that we have solved the structure problem of finding the eigenmodes and eigenenergies of a particular system, described by the Hamiltonian of eq. (4):

$$\mathcal{H}|\mu\rangle = \mathcal{E}_\mu |\mu\rangle \qquad (5)$$

Since $\mathcal{H}$ is generally nonhermitean the eigenenergies are complex with the eigenmodes $|\mu\rangle$ acting as the poles of the transition amplitude of (pion or photon induced) nuclear reactions in the corresponding range of excitation energy:

$$\hat{T}^{res}_{fi} = \mathcal{L}^+_f (\omega - \mathcal{H})^{-1} \mathcal{L}_i \qquad (6a)$$

$$T^{res}_{fi} = \sum_\mu \langle f|\mathcal{L}^+_f|\mu\rangle \frac{1}{\omega - \mathcal{E}_\mu} \langle \mu|\mathcal{L}_i|i\rangle \qquad (6b)$$

According to this Resonance Fluorescence Mechanism the resonant part of the reaction proceeds via the excitation of all those intermediate resonances. They may be considered as the nuclear analogues of the elementary baryon resonances if we restrict the discussion to the excitation of only one baryon.

For energies in the region of the $\Delta(3,3)$ isobar a corresponding graphical representation is given in fig. 5. Here the eigenmodes of the corresponding system, the $(\Delta,(A-1)N)$ system, are excited which then can decay into various final channels. It is this nuclear Δ-excitation which shall be discussed more detailed in the following.

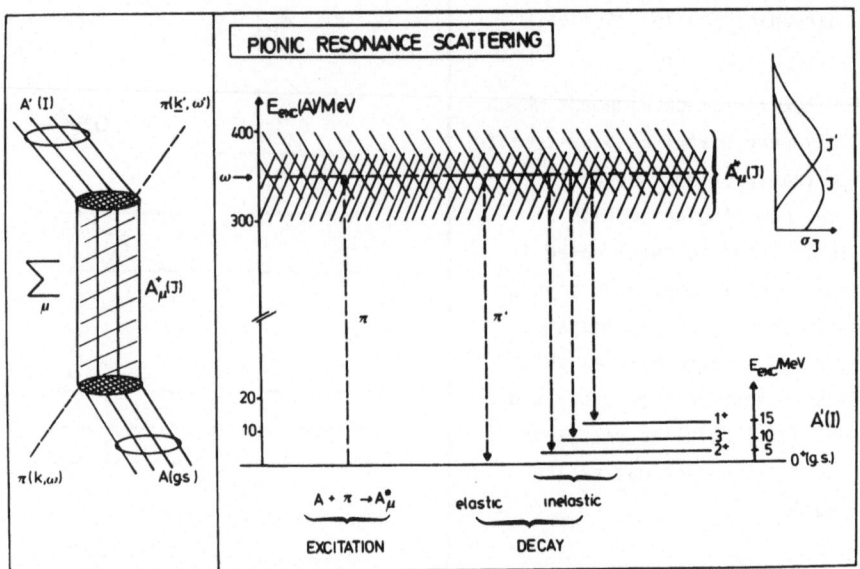

Fig. 5: Schematic representation of the reaction mechanism of eq. (6).

IV. The Nuclear Δ-Excitation

In this chapter we restrict the discussion to the particular system of one Δ and (A-1) nucleons. For any meson- or photon-induced reaction in the Δ-energy range, the Δ-hole ($\Delta\bar{N}$) configurations play the role of doorway states, a common starting point for all Δ-hole or Isobar-Doorway models[4-9]. However with any such ph-configuration the system is not yet in an eigenstate, there is a damping (due to the Δ-selfenergy terms) and most importantly a coupling (due to the ΔN interaction) of different ph basis states. Correspondingly by diagonalization we will expand the eigenmodes in a $\Delta\bar{N}$ basis, using harmonic oscillator single particle states:

$$|J^{\pi},T,\mu\rangle = \sum_{\beta,\alpha} c_{\beta\alpha}^{J,T,\mu} \, |[\Delta_{\beta} \otimes \bar{N}_{\alpha}]^{J,T}\rangle \qquad (7)$$

IV.1 The One Boson Exchange ΔN Interaction (OBE)

The OBE-interaction is considered to be relatively well determined and is constructed from π- and correlated ρ-exchange for the direct and the exchange term of the ΔN interaction[10-14]. The corresponding ph-diagrams are shown in fig. 6a. We would like to stress, however, one particular feature of the one pion-contribution to the ΔN-exchange force: it is strongly energy dependend and complex, since the exchanged pion can propagate on-shell. In the ph-space it describes the elastic rescattering of pions through intermediate $\Delta\bar{N}$ excitations thereby building up a strong coherence in the different spin-isospin states of eq. (7).

Fig. 6:
(a) OBE-contribution to the ΔN interaction
    (a1,(a2),pp(ph) diagrams)
(b) as (a) for the TBE processes

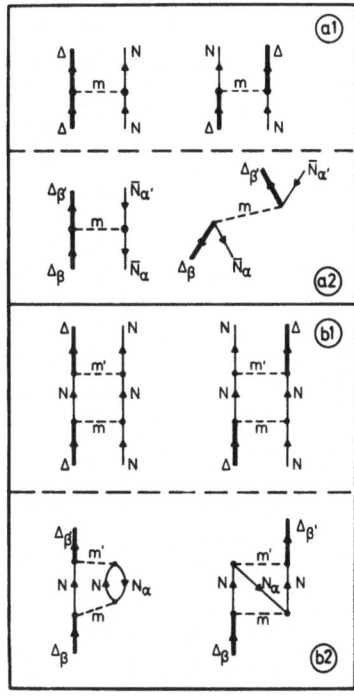

IV.2 The Two Boson Exchange ΔN interaction (TBE)

Clearly the TBE-forces of fig. 6b correspond to the π-absorption in the two-nucleon channel for both the direct and the exchange diagram. In the Hartree approximation they give rise to a Δ selfenergy via this absorptive channel. In the present context several investigations on those terms have been carried through recently, for the direct term[6] and in addition the exchange as well as residual interaction terms[7]. Since only the direct and the exchange diagram have been found to be important and of comparable strength let us concentrate on this Hartree part for the following discussion. In an r-space representation it may be represented by a nonlocal selfenergy-operator of scalar and tensor character. It is therefore not easy to keep the connection to the parametrization of an optical potential for the Δ introduced in ref. 5. However it might well be that the complex Δ spin-orbit interaction is related to an incomplete representation of the absorptive selfenergy by a local and central parametrization.

IV.3 The Quasifree Width

The width of a nuclear Δ in a specific (ΔN̄)-configuration is certainly different from the free width essentially from two reasons: Firstly Pauli blocking[5,6] reduces the phase space for the Δ → Nπ decay. Secondly the off-shell situation for a Δ in some single particle state introduces an additional formfactor. Both effects lead to sizeable reduction of the free Δ-width as compared to the quasifree width.

IV.4 The Spin-Orbit Interaction

Both in the quark modell[15] and in meson exchange theories, the Δ is expected to experience a spin-orbit force similar to the nucleons. Its importance in the framework of Δh-models has been pointed out in ref. 5. Therefore we discuss the influence of a spin-orbit force assuming the same strength for the Δ and the N:

$$V_{\ell s}(r) = \kappa \, \frac{1}{r} \, \frac{\partial}{\partial r} w(r) \qquad (8)$$

with a Wood-Saxon shape w(r); note however that κ is taken to be real.

For the actual applications, we will use and compare the following two models:

(i) The OBE-terms are treated explicitly; the quasifree width is approximated by the free width and is combined with a phenomenological absorptive width to produce the damping of $\Delta \bar{N}$ states:

(ii) the quasifree and the absorptive mechanisms (direct and exchange) are taken into account explicitly in the Hartree approximation to the TPE on the same microscopic level as the OBE contributions.

Clearly the corresponding eigenmodes, to be used in eq. (6) are then obtained by a TDA-diagonalization within a certain $\Delta \bar{N}$ basis.

IV.5 Applications: π-nucleus scattering, π^0-photoproduction and the (e,e') reaction

In fig. 7 elastic π-^{12}C scattering calculations are presented together with the experimental data; a conservative estimate of the present theoretical uncertainties is indicated which are due to different treatments of the damping mechanisms and the spin-orbit interaction. Clearly there may be a substantial improvement in the forward direction however with the structure being rather stable up to moderate scattering angles. Both the theoretical uncertainty as well as the sensitivity increase with scattering angle, opening up the possibility to learn about details but only if enough data for a systematic and consistent analysis are available.

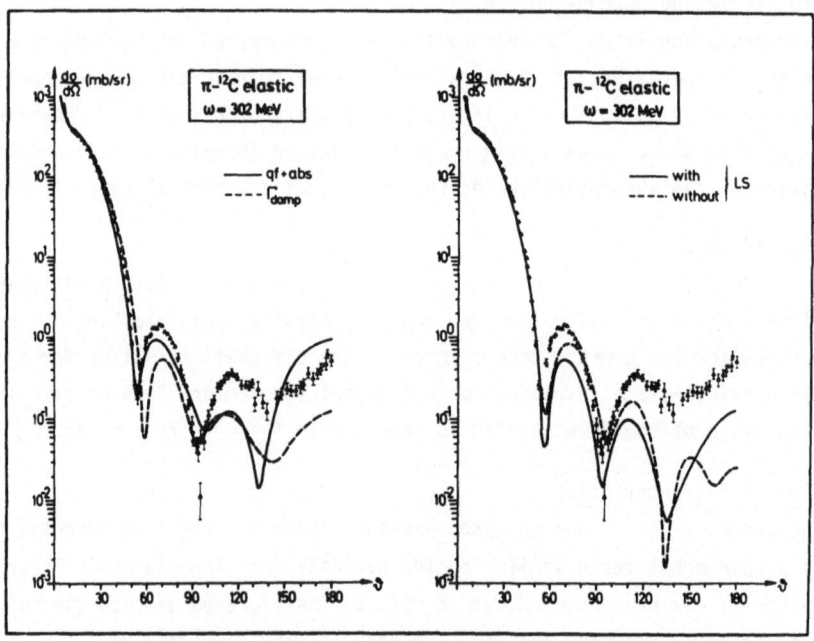

Fig. 7: Differential cross sections for π-^{12}C elastic scattering, with different treatments of the damping mechanisms (see text) and with/without spin-orbit force. The shaded area indicates the resulting uncertainties. Experimental results from refs. 26, 27.

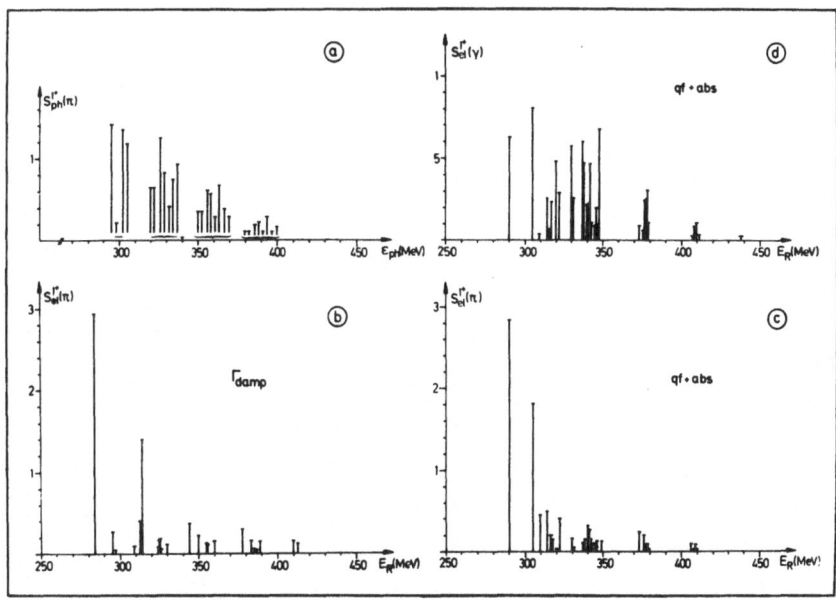

Fig. 8: (a) pionic transition strength into Δ-hole states; (b) pionic transition strength into diagonalized states, assuming constant damping; (c) as b) with explicit treatment of damping; (d) as c) for the photoexcitations. All four figures are drawn for 1$^+$-states and for an excitation energy of ω = 300 MeV.

There is an interesting feature of the intermediate eigenmodes which is illustrated in fig. 8 for the case of the M1 eigenmodes. Due to the coherence, introduced by the OPE-ΔN exchange interaction there are one (or two) outstanding resonances which collect the pionic ph-strength very much along the lines of an extended schematic model. This phenomenon appears to be independend on the specific model, thereby introducing some stabilizing effect on the influence of the damping mechanism. We should add however that this phenomenon is connected with the elastic pion channel; it reflects the importance of the coherent or elastic pion propagation leading to a large elastic width (see ref. 16 for the connection of the multiple scattering to the Isobar Doorway approach). The results discussed here are representative for a wide range of energies around the Δ(3,3) resonange. Using the simplified description of a constant damping width elastic πA-scattering[8] has been successfully described, particularly the structure of the differential cross sections and the variation of this structure with excitation energy. In addition also inelastic reactions are determined from the same spectrum of intermediate resonances. Also for those weak channels the experimental data can be reproduced reasonably well without readjustment of the parameters of the theory[8].

Let us briefly summarize some conclusions which can be drawn on the nuclear excita-

Fig. 9:
Pionic excitation strength distribution of ^{12}C for the dominant multipolarities.

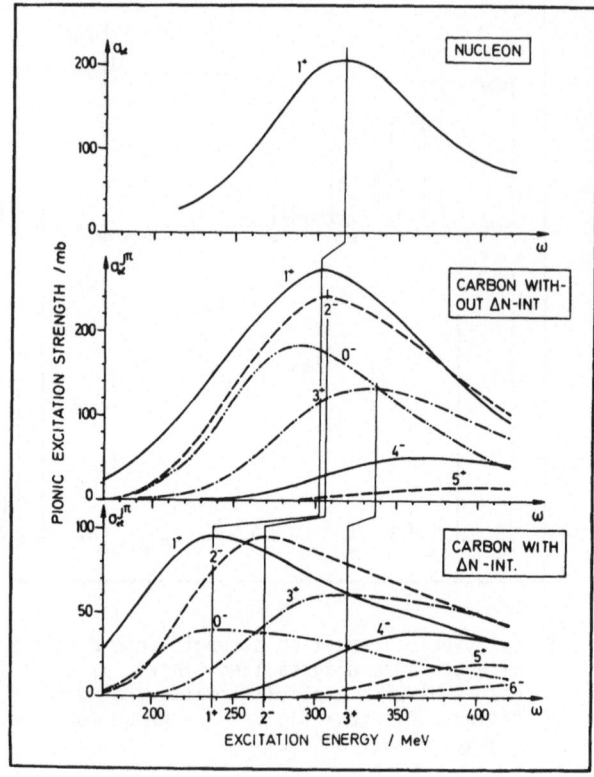

tion spectrum observed with pions. The pionic excitation strength distribution is characterized by broad and overlapping multipoles of magnetic type (see fig. 9). Compared to the elementary Δ, a pure M1 transition out of the nucleon, the nuclear Δ and the corresponding hole can be produced in various angular momentum states leading to different multipole excitations. In a pure particle-hole model the important multipoles are almost degenerate; however due to the interaction they are dynamically discriminated from each other, they are shifted and split up. And it is this relative strength distribution which reflects the underlying dynamics and which finally determines the structure of the differential cross sections. Furthermore if we look at different π-induced reaction channels we observe that the various multipole resonances contribute differently. For example in fig. 10 the M1-multipole is seen to be strong in elastic scattering, even stronger for inelastic scattering into the 1^+ state of ^{12}C with a rapid fall off and rather unimportant for the radiative pion capture into ^{12}N(g.s.), the same, i. e. the isobaric analogue state to the 1^+ above. Consequently a specific final channel acts as a sort of multipole filter. From those arguments it is by no means trivial to describe consistently a whole set of reactions from the same intermediate spectrum.

As a final point, we will discuss the role the photon with an emphasis on the complementarity to the pionic probe. To begin with, we compare in fig. 8 the excitation of 1^+ resonances with pions and photons respectively. We have seen already previously that the pion likes to excite only a few states strongly, but the photon behaves very differently. Again this effect is rather general and does also show up

Fig. 10:
Contributions of the various multipolarities to different reactions.

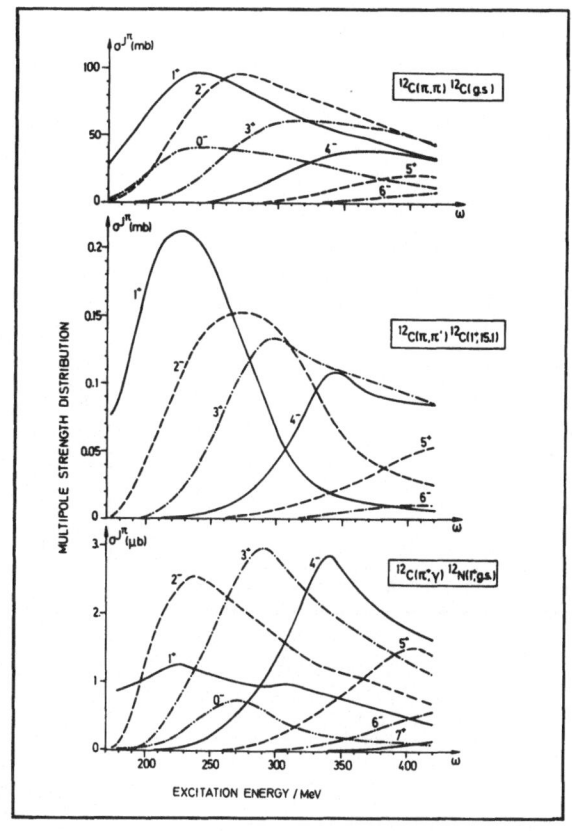

with the explicit inclusion of absorptive terms. This means that the photon predominantly excites those modes where the coherent pion propagation is suppressed and whose properties are dominated by the damping mechanisms[17].
As an application the coherent π^0-photoproduction[18-20] is shown in fig. 11; obviously with the explicit treatment of the damping mechanisms the structure of the cross sections becomes more pronounced, we obtain sharper minima and again a strong modification at large angles.

There is however another and importantly unique property of the photon. In contrast to a pion the photon can also excite electric states, $E\lambda$ resonances, as they show up in π^0-photoproduction into excited target states. This electric strength is appreciable and it interferes strongly with the magnetic strength (see fig. 12). Evidently from the final angular distribution, this interference may be constructive or destructive, depending on the energy and the angle. Now in this case the coherent π-propagation is suppressed from both sides, the photon in the initial channel and the excited state in the final channel. Consequently we should have here a good candidate to study many baryon aspects beyond the OPE.

And clearly replacing the real photon by a virtual photon, that means studying deep inelastic electron scattering, we can carry over all those arguments. Experimentally the nuclear response function in the Δ region has been measured at the two laboratories, at Stanford[21] and more recently at Saclay[22,23]. Clearly from fig. 13 there is a significant underestimation of the strength function in the so called dip region, using quasifree models for the Δ excitation[24,25]. On the other hand introducting the dynamics, i. e. using the nuclear eigenmodes, provides a mechanism to

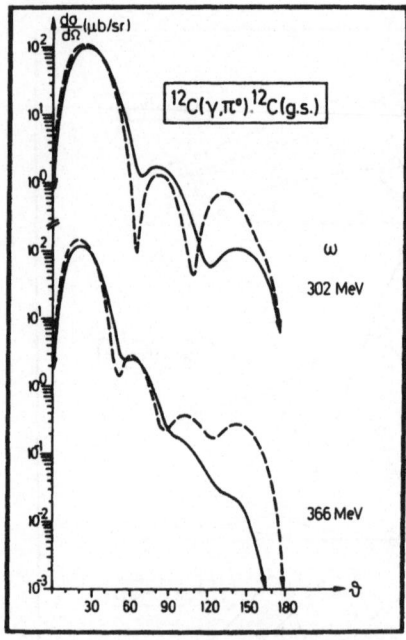

Fig. 11: Coherent π^0-photoproduction on ^{12}C. Solid line (dashed line) for constant (explicit) damping (see text).

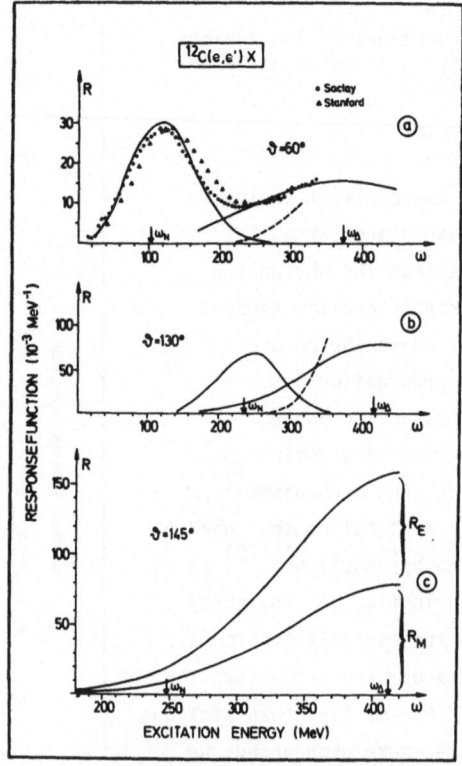

Fig. 13: Nuclear response function for different electron scattering angles; experimental results from refs. 21, 22.

Fig. 12: π^0-photoproduction into excited target states; dashed (dot-dashed) curve represents the contribution of $M\lambda$ ($E\lambda$) intermediate states in eq. (6). The solid curve gives the total result.

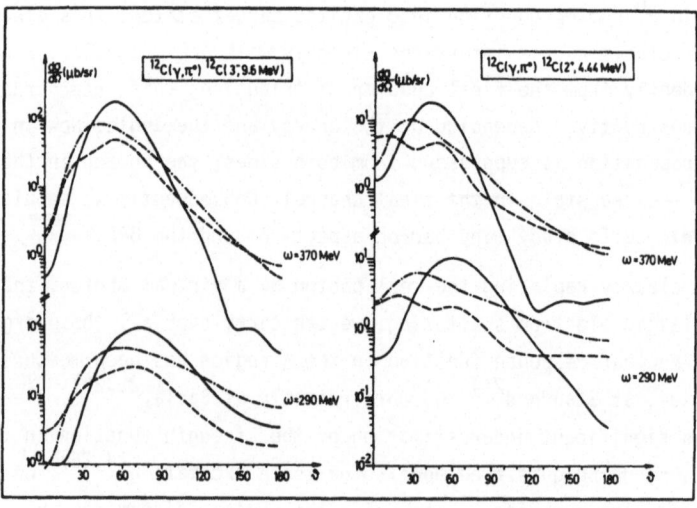

fill up this dip. We should however mention that due to the different kinematics of the virtual photon (compared to pions or real photons), i. e. due to the rather large momentum transfer, predominantly those Δ-hole states are excited, with the Δ in high oscillator configurations. Consequently the strong resonances of pion scattering (see fig. 8) are suppressed and the details of the quasifree and absorption mechanism are expected to be more important in this case. From such considerations it is plausible that a few corrections to the quasifree reaction[25] mechanism might already be sufficient to include the main effects.

V. Summary and Outlook

Above pion threshold the internal modes of motion of a nucleon start to play a dominant role. Consequently also the conventional picture of a nucleus must be extended to include those subnucleonic degrees of freedom. As an immediate consequence the nucleus exhibits new modes of excitation, which have been studied in the framework of a Many Baryon Model for the example of the Δ-isobar.

Here we find that the Nuclear Δ-excitations are concentrated in the elementary resonance region, however in detail they strongly depend on the ΔN dynamics. This spectrum is very complicated simply due to the fact that both, the internal and the external degrees of freedom, are subject to the strong interaction and are coupled with each other by the same strong mechanisms. However the same spectrum shows up in a whole variety of different reactions; thus a systematic and complete investigation offers the oportunity to pin down present uncertainties in more detail.

In this connection we would like to stress again the role of the photon, as it allows to suppress the influence of the strong but rather well determined, OPE-interaction, or equivalently the coherent π-rescattering mechanism, usually dominating π-induced processes.

In this paper we addressed ourselves to the nuclear Δ excitation, but a similar analogy does also appear for the strange baryons. The hypernuclear states can be considered as nuclear analogues of the elementary Λ- or Σ-particles, although quantitatively the phenomena are different. The Λ and Σ are narrow compared to Δ and their interaction with nucleons is relatively weak. Therefore those nuclear Λ-states are to a good approximation simple $\Lambda \bar{N}$ states.

Although very little is known about the excitation of higher resonances in nuclei, strange as well as nonstrange and although an extension from our present understanding is certainly speculative it is quite conceivable that a similar analogue between the elementary and the nuclear excitations might exist. Consequently the nucleus might exhibit a rich spectrum of subnuclear excitations also in other energy regimes, opening up a wide and fascinating field of nuclear research at medium energies.

References

1) H. J. Weber and H. Arenhövel, Phys. Rep. 36 (1978) 277
2) M. G. Huber and K. Klingenbeck, Nukleonika, Vol. 24 (1979) 95 and preprint Erlangen 1980
3) U. Fano, Nuovo Cimento 12 (1935) 156; Phys. Rev. 124 (1961) 1866
4) L. S. Kisslinger and W. L. Wang, Phys. Rev. Lett. 30 (1973) 1071 and Ann. Phys. (N.Y.) 99 (1976) 374
   L. S. Kisslinger, Phys. Rev. C 22 (1980) 1202
5) F. Lenz and E. J. Moniz, Phys. Rev. C 12 (1975) 909
   M. Hirata, J. H. Koch, F. Lenz and E. J. Moniz, Phys. Lett. 70 B (1977) 281
   M. Hirata, F. Lenz and K. Yazaki, Ann. Phys. (N.Y.) 108 (1977) 116
   M. Hirata, J. H. Koch, F. Lenz and E. J. Moniz, Ann. Phys. (N.Y.) 120 (1979) 205
   F. Lenz and M. Thies, in: Studies in High Energy Physics, Vol. 2, Harwood Acade. Publishers, London
6) W. Weise, Nucl. Phys. A 278 (1977) 402
   E. Oset and W. Weise, Phys. Lett. 77 B (1978) 159; Nucl. Phys. A 319 (1979) 477; Nucl. Phys. A 329 (1979) 365
7) H. M. Hofmann, Z. Phys. A 289 (1979) 273
8) M. Dillig and M. G. Huber, Phys. Lett. 48 B (1974) 417
   K. Klingenbeck, M. Dillig and M. G. Huber, Phys. Rev. Lett. 41 (1978) 387
   M. G. Huber and K. Klingenbeck, in: Studies in High Energy Physics, Vol. 2, Harwood Academic Publishers, London
   K. Klingenbeck, Phys. Lett. 85 B (1979) 21
   K. Klingenbeck and M. G. Huber, Phys. Rev. C 22 (1980) 681
9) M. Danos and H. T. Williams, Phys. Lett. 89 B (1980) 169
10) H. Sugawara and F. von Hippel, Phys. Rev. 172 (1968) 1764
11) S. Jena and L. S. Kisslinger, Ann. Phys. (N.Y.) 85 (1974) 251
12) H. Arenhövel, Nucl. Phys. A 247 (1975) 473
13) M. Dillig, Phys. Rev. D 13 (1976) 179
14) G. E. Brown and W. Weise, Phys. Rep. 22 (1975) 909
15) H. J. Pirner, Phys. Lett. 85 B (1979) 190
16) W. A. Friedman, Phys. Rev. C 12 (1975) 1294
17) E. Oset and W. Weise, Phys. Lett. 94 B (1980) 19
    W. Weise, invited talk at the Symposium on Perspectives in Electro- and Photo-nuclear Physics, Saclay, Sept. 29 - Oct. 3, 1980
18) R. M. Woloshyn, Phys. Rev. C 18 (1978) 1056
19) J. H. Koch and E. J. Moniz, Phys. Rev. C 20 (1979) 235
20) K. Klingenbeck and M. G. Huber, J. Nucl. Phys. G 6 (1980) 961
21) R. R. Whitney et al., Phys. Rev. C 9 (1974) 2230
22) J. Mougey et al., Phys. Rev. Lett. 41 (1978) 1645
23) J. Morgenstern et al., contrib. to the Symposium on Perspectives in Electro- and photonuclear Physics, Saclay, Drpz. 29 - Oct. 3, 1980 and contribution to this workshop
24) G. Do Dang, Phys. Lett. 69 B (1977) 425 and Z. Phys. A 294 (1980) 377
25) J. M. Laget, contrib. to the Symposium on Perspectives in Electro- and Photo-nuclear Physics, Saclay, Sept. 29 - Oct. 3, 1980 and contribution to this workshop
26) J. Piffaretti et al., Phys. Lett. 67 B (1977) 289 and 71 B (1977) 324
27) B. Chabloz et al., Phys. Lett. 81 B (1979) 143

THREE-BODY WAVE FUNCTIONS AND ELECTROMAGNETIC INTERACTIONS[*]

C.Ciofi degli Atti[+], E.Pace[+o] and G.Salmè[+]

[+]Istituto Nazionale di Fisica Nucleare, Sezione Sanità
Istituto Superiore di Sanità, Radiation Laboratory
V. Regina Elena 299, I-00161, Rome, Italy

[o]Istituto di Fisica, Università di Roma
I-00161, Rome, Italy

Abstract: The interpretation of various electromagnetic properties of three-body systems in terms of different types of three-body non-relativistic wave functions is presented. The possibility of obtaining consistent results independently of the computational method adopted in the solution of the three-body problem is thoroughly discussed.

The three-body wave function. Two types of non-relativistic three-body wave functions have been widely used in the calculation of a large body of experimental data (the ground state energy and radius, the charge form factor and the charge density, the quasi-elastic cross sections and the scaling function F(y)): the Faddeev wave function by Kim et al[1] and the variational wave function by Nunberg et al[2].

The general form for the three-body wave function is

$$\Psi_{1/2M}(\vec{a},\vec{b}) = \sum_k C_k \,|(\ell_a \ell_b)L,(S_{12}\,1/2)S;1/2\,M\rangle \qquad (1)$$

with $\quad k \equiv \{\ell_a \ell_b L S_{12} S\}$.

In eqn (1) $\vec{a}$ and $\vec{b}$ are the intrinsic variables describing, respectively, the relative motion of particles 1 and 2 and the relative motion of particle 3 with respect to the pair 1,2; ℓ_a and ℓ_b are the corresponding angular momenta. The wave function components $|(\ell_a \ell_b)L,(S_{12}\,1/2)S;1/2\,M\rangle$ have been obtained in Ref.1 by means of Faddeev technique using the restricted form ($^1S_0 + ^3S_1 - ^3D_1$) of the Reid Soft Core (RSC) interaction and $\ell_a + \ell_b \leq 4$; in Ref.2 the full RSC interaction has been used, all wave function components with $\ell_a + \ell_b \leq 28$ were considered and the wave function components have been obtained variationally by using a harmonic oscillator basis expansion. The values of the ground state energy and radius are listed in Table 1; as we have previously pointed out[3] there is a disturbing difference in the values of the mean kinetic energy predicted by the two approaches.

___
[*] Presented by C. Ciofi degli Atti

	Interaction	$\ell_a+\ell_b \leq$	$<T>$	E_3	R
A	RSC ($^1S_0+^3S_1-^3D_1$)	4	35.9	-6.98	2.25
B	RSC (all waves)	28	52.1	-7.3	1.92
Experiment				-8.48	1.88±0.05

Table 1 - Triton mean kinetic energy $<T>$ and binding energy E_3, and rms radius R of ^{3}He, corresponding to the Faddeev (A) and variational (B) wave functions of Refs.1 and 2, respectively. Energies are in MeV and radii in fm. (After Ref.3a).

<u>Elastic Electron scattering</u>. The charge form factor of ^{3}He is given by the expression

$$F_{ch}(q^2) = F_p(q^2) G_E^p(q^2) + 0.5 F_n(q^2) G_E^n(q^2) \qquad (2)$$

where

$$F_{p(n)}(q^2) = \frac{1}{4\pi} <\Psi_{1/2M}|\int d\Omega_{\hat{q}}\, e^{i\vec{q}\cdot\vec{r}_{p(n)}}|\Psi_{1/2M}> \qquad (3)$$

is the proton (neutron) body form factor and G_E^N is the nucleon electromagnetic form factor (in eqn (2), $\vec{r}_{p(n)}$ is measured from the center of mass of the nucleus). In Fig.1 the theoretical charge form factors are compared with the experimental data in a region of momentum transfer where relativistic effects are expected to be negligible.

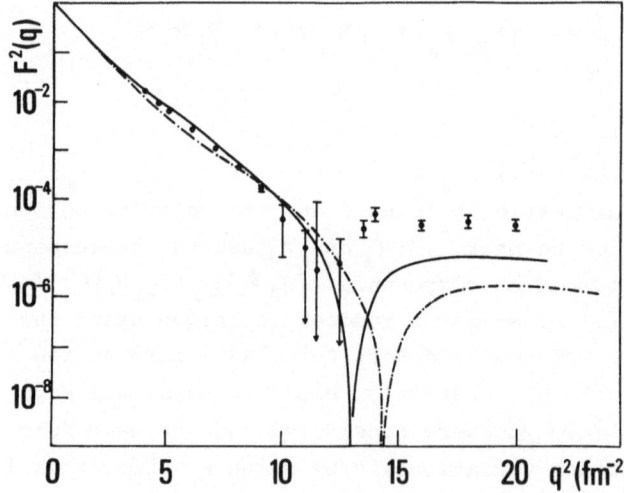

Fig.1 - Charge form factor of ^{3}He. The experimental points are from Ref.4. The theoretical curves correspond to the wave functions of Ref.1 (dot-dashed line) and Ref.2 (full line).

The theoretical point charge density can be computed directly by Fourier transforming F_p. However, Sick[5] has obtained an "experimental" density by assuming $F_p=F_n$ and by defining an "experimental" proton body form factor

$$F_P^{"exp"}(q^2) = \frac{F_{ch}^{exp}(q^2)}{G_E^P(q^2) + 0.5\, G_E^n(q^2)} \qquad (4)$$

from which the "experimental" point density has been obtained, after correcting for MEC and relativistic effects. For this reason we have calculated, with the wave function of Ref.2, the point charge density in two different ways, namely by Fourier transforming the body proton form factor, i.e.

$$\rho_p^{th}(r) = \frac{1}{(2\pi)^3} \int e^{i\vec{q}\cdot\vec{r}} F_p(q^2)\, d^3q \qquad (5)$$

and by the expression

$$\rho_p^{"th"}(r) = \frac{1}{(2\pi)^3} \int e^{i\vec{q}\cdot\vec{r}} F_p^{"th"}(q^2)\, d^3q \qquad (6)$$

with

$$F_p^{"th"}(q^2) = \frac{F_{ch}^{th}(q^2)}{G_E^P(q^2) + 0.5\, G_E^n(q^2)} \qquad (7)$$

Eqn (6) is the quantity which should be directly compared with Sick's density; if $F_p = F_n$, then $\rho_p^{th} = \rho_p^{"th"}$. The two densities ρ_p^{th} and $\rho_p^{"th"}$ are compared in Fig.2 with Sick's experimental density and with the Faddeev den-

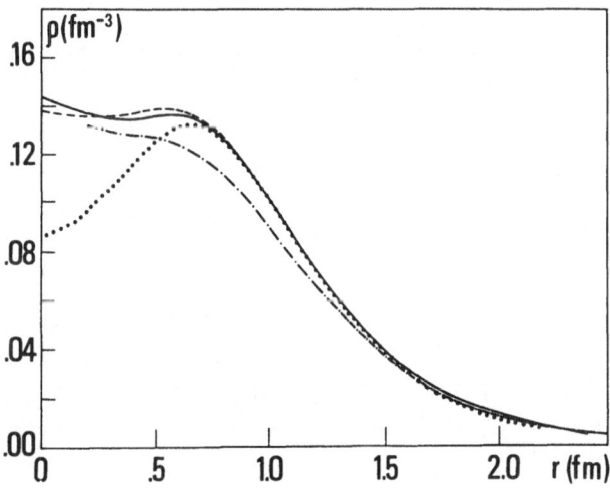

Fig.2 - Point charge density of ^{3}He. The experimental density (dots) is from Ref.5. The theoretical curves correspond to the wave function of Ref.1 (dot-dashed line) and to the wave function of Ref.2 using eqn(5) (full line) and eqn(6) (dashed line).

sity corresponding to the wave function of Ref.1 and reported in Ref.5. From this figure it can be concluded that $\rho_p^{th} \neq \rho_m^{th}$. Actual calculations of these quantities from the wave functions of Ref.2 do show that they appreciably differ[3b].

Inclusive quasi-elastic scattering. Assuming non relativistic kinematics, the momentum and energy conservations in a quasi-elastic process read[6]

$$\vec{q} - \vec{k}_N = \vec{k}_R \equiv -\vec{k} \qquad (8)$$

$$\omega - \frac{k_N^2}{2M} = \frac{k_R^2}{2(A-1)M} + E \qquad (9)$$

where $\vec{q}$ is the momentum transfer, ω the energy tranfer, $\vec{k}_N$ the momentum of the emitted nucleon, $\vec{k}_R$ the momentum of the recoiling nucleus and $\vec{k}$ the internal momentum of the nucleon; $E = M_{A-1} + M - M_A + E_{A-1}^f$ is the removal energy and E_{A-1}^f is the intrinsic excitation energy of the (A-1) system (for A=3, $E_2^f = 0$ for the two-body channel and $E_2^f = t^2/M$ for the three-body channel, t being the relative momentum of the recoiling two-nucleon pair). In PWIA the inclusive quasi-elastic cross section for ^{3}He can be expressed in terms of the Spectral Function P by

$$\frac{d^2\sigma}{dE_2 d\Omega_2} = \int_{k'_{min}}^{k'_{max}} dk \int_0^{2\pi} d\varphi\, \sigma_{ep} P(k) \frac{Mk}{q} +$$

$$+ \int_0^{E_{2\,max}^f} dE_2^f \int_{k_{min}}^{k_{max}} dk \int_0^{2\pi} d\varphi\, \sigma_{ep} P_p(k, E_2^f) \frac{Mk}{q} + \qquad (10)$$

$$+ \int_0^{E_{2\,max}^f} dE_2^f \int_{k_{min}}^{k_{max}} dk \int_0^{2\pi} d\varphi\, \sigma_{en} P_n(k, E_2^f) \frac{Mk}{q}$$

where $\sigma_{eN} = (d\sigma/d\Omega)_{eN}$.
The first term in the rhs of eqn (10) describes the two-body channel $^3\text{He} \rightarrow p+d$, whereas the other terms describe the three-body channels $^3\text{He} \rightarrow p+(np)$, $^3\text{He} \rightarrow n+(pp)$, respectively.

The theoretical quasi-elastic cross sections calculated with Faddeev and variational wave functions, using relativistic kinematics, are compared in Fig.3 with the experimental data by Day et al[7]. It should be

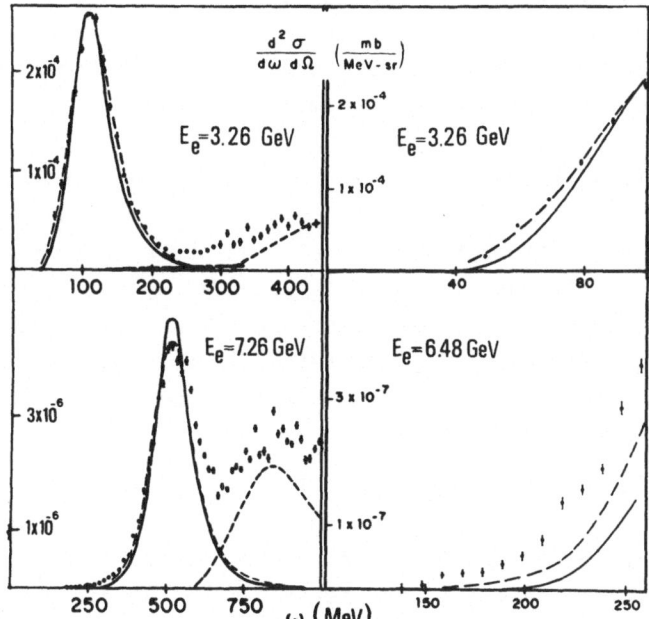

Fig.3 - Inclusive quasi-elastic cross sections for ^{3}He. The experimental data are from Ref.7. The full curve has been obtained in Ref.7 using the wave function of Ref.1 and the dashed curve has been obtained in Ref.3a using the wave function of Ref.2. The right figures show, expanded, the left wing of the peak. In the left figures, the dashed line at high ω represents the contribution from N^* excitations and the long dashed line is the contribution from MEC; references to the original works can be found in Ref.7.

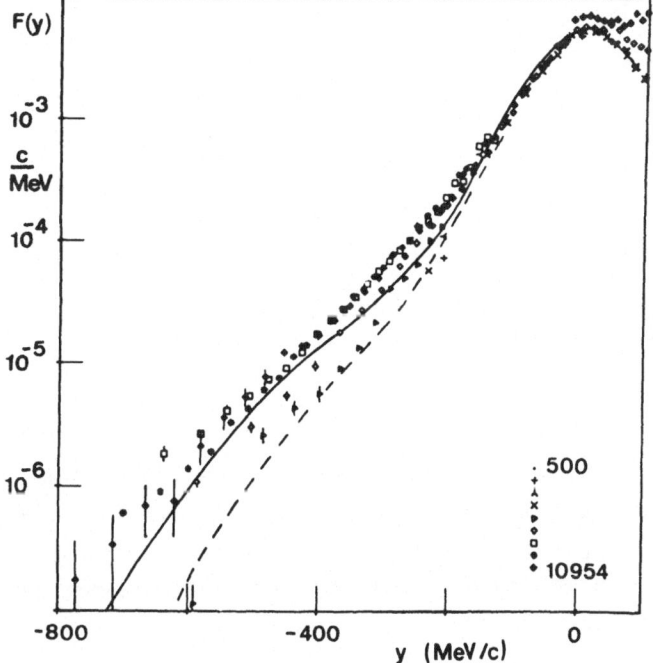

Fig.4 - Scaling function $F(y)$ (eqn(16)) calculated with the variational wave function of Ref.2 (full line; preliminary results were reported in Ref.3a). The dashed curve has been obtained in Ref.12 using the wave function of Ref.1. The experimental points are from Ref.12.

pointed out that the experimental data are performed in such kinematical conditions that the left wing of the peak ($\omega < q^2/2M$) is very sensitive to the nucleon high momentum components without being affected by MEC and N^* production effects; moreover the final state interaction is negligible in the processes at very high energy and momentum transfer[7,8].

The Scaling Function F(y). A general discussion of scaling laws in electron scattering is given by West[9]. Here only the so called y scaling[10-12] will be considered. Although actual calculations have been performed with relativistic kinematics, as it is required by the high energies involved in the experiments, non-relativistic kinematics will be adopted in the following for ease of presentation. From eqn (9), the component of the internal momentum parallel to the momentum transfer will be

$$k_{\parallel} = \vec{q}\cdot\vec{k}/q = (2M\omega - q^2)/2q \equiv y \tag{11}$$

if q is very large. In this case E_2^f and $k_\perp$ can be disregarded in σ_{eN}. If it is further assumed that the proton and neutron momentum distributions are the same, i.e.

$$P_p(k) = P_n(k) = \int_0^\infty P_n(k, E_2^f) dE_2^f \equiv P(k) \tag{12}$$

then one obtains from eqn (10)

$$\frac{d^2\sigma}{dE_2 d\Omega_2} = (2\sigma_{ep} + \sigma_{en}) \frac{dy}{d\omega} F(y) \tag{13}$$

where the Scaling Function F(y) is

$$F(y) = 2\pi \int_y^\infty p(k) k \, dk \tag{14}$$

Thus the quantity

$$\frac{d^2\sigma}{dE_2 d\Omega_2} (2\sigma_{ep} + \sigma_{en})^{-1} \frac{d\omega}{dy} = F(y) \tag{15}$$

will scale in y, i.e. the lhs measured for fixed y will be independent of ω and q. An analytical derivation of F(y) using relativistic kinematics is under way and preliminary results show that the relation between F(y) and $P(k, E_2^f)$ is not as simple as in eqn (14)[13]. The calculation of F(y) using relativistic kinematics has been performed as proposed in Ref.12, that is by computing the expression

$$F(y) = \left(\frac{d^2\sigma}{d\mathcal{E}_2 d\Omega_2}\right)^{th} (2\sigma_{ep}^{th} + \sigma_{en}^{th})^{-1} \frac{d\omega}{dy} \qquad (16)$$

for values of q large enough that F(y) becomes independent of q. A comparison between theoretical and experimental F(y) is shown in Fig.4. The agreement between the experimental data and the variational F(y) is fair although the theoretical curve is lower than the experimental data, particularly in the region -200 MeV/c $\lesssim$ y $\lesssim$ -400 MeV/c; such a disagreement is not surprising in view of the disagreement found in the charge form factor and in the quasi-elastic cross sections.

Elastic and quasi-elastic electron scattering: summary. The analysis of elastic and quasi-elastic electron scattering by the three-body systems in terms of realistic wave functions shows that:
1) the Faddeev wave function of Ref.1 and the variational wave function of Ref.2 lead to different elastic and quasi-elastic cross sections and scaling functions F(y);
2) there is no agreement between theoretical and experimental charge form factors and quasi-elastic cross sections for q $\gtrsim$ 3 fm^{-1} and k $\gtrsim$ 2 fm^{-1}, respectively; since one has roughly q$\sim$3k/2, it can reasonably be surmised that the origin of the disagreement is the same in both processes, namely a lack of high momentum components (k $\gtrsim$ 2fm^{-1}) in the available three-body wave functions;
3) MEC effects do not resolve the discrepancy between experimental data and theoretical calculations since, from one side, they do not affect the left wing of the quasi-elastic peak and, from the other, they only slightly improve the situation in elastic scattering, as shown in Fig.5;
4) it has recently been shown[18] that N* components in the three-body wave function have large effects on the high momentum part of the charge form factor. In this region, unfortunately, also the uncertainties on the nucleon electromagnetic form factor have large effects (see e.g. Ref.14);
5) three-body forces[19-21] might be one of the candidates to resolve the discrepancy between experimental data and theoretical calculations. However, before introducing three-body forces or other exotic effects, the apparent disagreement between Faddeev and variational calculations should be understood. Progress along this direction occurred recently.

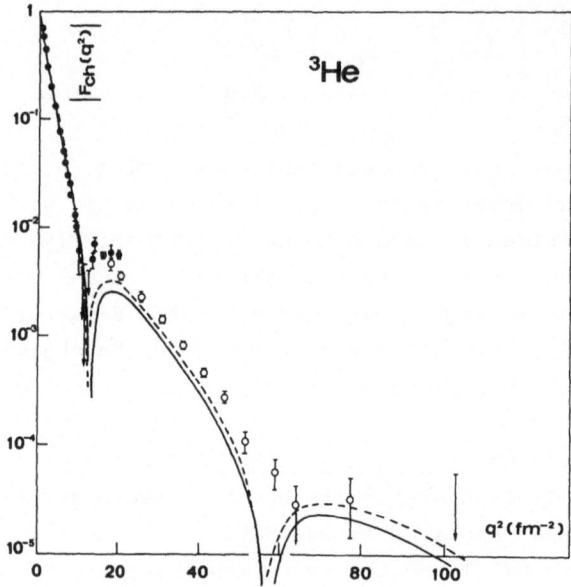

Fig.5 - The charge form factor of ^{3}He calculated with the wave function of Ref.2 without (full line) and with MEC$^{(15)}$ and relativistic$^{(16)}$ corrections (dashed line). Open dots: Ref.17; full dots: Ref.4. (After Ref.14)

Towards convergent results between Faddeev and variational approaches.
The origin of the discrepancy between the predictions of the wave functions of Refs.1 and 2 might be twofold:
1) the different interactions used in the two calculations. Hajduk and Sauer[22] have investigated this problem and found that the restricted form ($^1S_0 + ^3S_1 - ^3D_1$) and the full RSC interactions yield almost the same results, as far as the ground state energy is concerned: for example the values of the mean kinetic energy are $\langle T \rangle$ =49.022 MeV and $\langle T \rangle$ =49.925 MeV corresponding to the restricted and full interactions respectively;
2) the different number of high angular momentum components included in the wave functions. The results presented in Table 2 clearly show that Faddeev type calculations which include high angular momentum components are in better agreement with variational calculations.

$\ell_a + \ell_b \leq$	A	B	C
4	35.9	45.7	-
8	-	50.2	49.925
28	-	52.1	-

Table 2 - Triton mean kinetic energy $\langle T \rangle$ in MeV as a function of the maximum value of $\ell_a + \ell_b$. Case A corresponds to the Faddeev wave function of Ref.1, case B to the variational wave function of Ref.2 and case C to the Faddeev wave function of Ref.22.

Conclusions and future perspectives. From the analysis we have exhibited it appears that:
1) Faddeev and variational approaches to the three-body systems lead to consistent results if in the former approach wave function components with $\ell_a + \ell_b > 4$ are considered;
2) the discrepancy between theoretical calculations and experimental data involving high momentum components has not yet been resolved;
3) the true "conventional" high momentum background resulting from a correct non-relativistic treatment is a prerequisite for obtaining unambiguous information about "non conventional" effects which produce high momentum components;

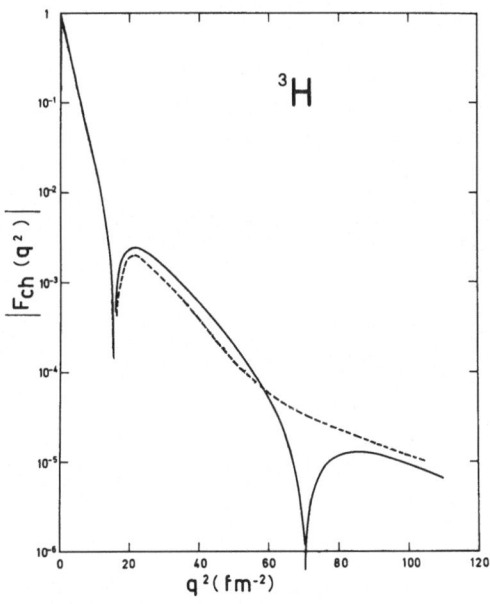

Fig.6 - The charge form factor of ^{3}H calculated with the wave function of Ref.2 using neutron charge form factor from Ref.24 (full line) and from Ref.25 (dashed line). (After Ref.14)

4) as for the future developments we insist on the necessity of a measurement of the charge factor of ^{3}H up to $q \sim 10$ fm^{-1}. This would represent an invaluable check of MEC effects on the charge density operator (see e.g. Refs.23,6) and of the high momentum behaviour of the neutron electromagnetic form factor, as shown in Fig.6.

References

(1) R.A. Brandenburg, Y.E. Kim and A. Tubis, Phys.Rev. C12 (1975) 1368
(2) P. Nunberg, E. Pace and D. Prosperi, Nucl.Phys. A285 (1977) 58
(3) a) C. Ciofi degli Atti, E. Pace and G. Salmè, "Perspectives in Electro and Photo-Nuclear Physics" (Nuclear Physics, in press)

b) C. Ciofi, E. Pace and G. Salmè, "Electron and Pion Interactions with Nuclei at Intermediate Energies", eds. W.Bertozzi, S. Costa and C. Schaerf (Harwood Academic, New York, 1980) p. 369.
(4) J.S.McCarthy, I. Sick and R. R. Whitney, Phys. Rev. C15 (1977)1396
(5) I.Sick,"Few Body Systems and Electromagnetic Interactions", eds. C. Ciofi degli Atti and E. De Sanctis, Lecture Notes in Physics, Vol. 86 (Springer, Berlin, 1978) p. 300.
(6) C. Ciofi degli Atti, Progress in Particle and Nuclear Physics, ed. D. Wilkinson, Vol. 3 (Pergamon Press, Oxford, 1980) p. 163
(7) D.Day et al, Phys.Rev. Lett. 43 (1979) 1143
(8) I. Sick, this Workshop
(9) G.B. West, Phys.Rep. 18C (1975) 264
G.B. West, "Electron and Pion Interactions with Nuclei at Intermediate Energies", eds. W. Bertozzi, S. Costa and C. Schaerf (Harwood Academic, New York, 1980) p. 417
(10) Y.Kawazoe, G. Takeda and H. Matsuzaki, Progr.Theor. Phys. 54 (1975) 1394
(11) P.D. Zimmerman et al, Phys. Rev. C19 (1979) 279
(12) I.Sick, D.Day and J.S.McCarthy, Phys.Rev.Lett. 45 (1980) 871
(13) C. Ciofi degli Atti, E. Pace and G.Salmè, to be published
(14) E.Pace, Lett.Nuovo Cimento, 26 (1979) 615; Errata 27 (1980) 544
(15) J.Borysowicz and D.O. Riska, Nucl.Phys. A254 (1975) 301
(16) J.L.Friar,Ann. Phys. (N.Y.) 81 (1973) 332
(17) R.G.Arnold et al, Phys. Rev. Lett. 40 (1978) 1429
(18) Ch. Hajduk, P.Sauer and W.Strueve, this Workshop
(19) J.L.Ballot and M.Fabre de la Ripelle, "Electron and Pion Interactions with Nuclei at Intermediate Energies", eds. W.Bertozzi, S.Costa and C. Schaerf(Harwood Academic, New York, 1980) p. 391
(20) Y.E.Kim, Muslim and T.Ueda, Proc.9th Int.Conf. on the Few Body Problem, Eugene, 1980, Vol.1, p. 48
(21) S.A.Coon and W.Glöckle, Proc. 9th Int. Conf. on the Few Body Problem, Eugene, 1980, Vol.1, p. 36
(22) Ch.Hajduk and P.Sauer, private communication
(23) J.A.Tjon, Phys.Rev.Lett.40 (1978) 1239
(24) S.Blatnik and N.Zovko, Acta Phys.Austr. 39 (1974) 62
(25) F. Iachello, A.D. Jackson and A. Lande, Phys.Lett. 43B (1973)191

ELECTRODESINTEGRATION OF FEW-BODY SYSTEMS

Ingo Sick

Dept. of Physics, University of Basel,

Switzerland

I. Introduction

In the past, the investigation of nuclei by electrodesintegration has been a tool very little in use. Only in a few places electrodesintegration has been contributing to our understanding of nuclei. Quasielastic electron scattering was used to determine nucleon Fermi momenta in nuclei[1], electrodesintegration of light nuclei was used to check "exact" wave function calculations[2], quasielastic scattering off the deuteron was employed to determine neutron form factors[3].

The limited interest in electrodesintegration has several reasons, some of which are not relevant, others that are. To cite an example for each: As a function of electron energy loss ω the inclusive spectrum of scattered electrons represents a rather wide and featureless bump; according to a widespread feeling we have never learned much from such bumps. In order to understand in detail the relevant nuclear properties and the reaction mechanism, one needs electron desintegration data over a large region of ω and momentum transfer q; such data are hard to take. As a consequence electrodesintegration is rarely studied, and even some elementary questions like the applicability of sumrules have not yet been answered in a satisfactory way. During the past few years, the interest in electrodesintegration has increased considerably, to the point where the study of (e,e') has become quite fashionable. A more quantitative description of the inclusive scattering process is being developed[4], and experiments are producing data over a large q- and ω-range for both the longitudinal and the transverse response function[5]. From these developments, an improved understanding of electrodesintegration in terms of the reaction mechanism and nuclear properties can be expected.

The usefulness of electrodesintegration as a tool to investigate nuclei has been enhanced greatly by looking at kinematical domains not usually considered in the past. This is true in particular for threshold desintegration of very light nuclei, the topic of this talk. Basically two types of experiments and related physics questions should be mentioned here: First, the electrodesintegration of the deuteron, The transverse part of the cross section for the excitation of the (unbound) singlet -s state of the deuteron shows in the impulse approximation (IA) a pronounced interference minimum at $q^2 = 12$ fm^{-2}. This usually dominating IA-contribution being very small, the presence of meson exchange current (MEC) contributions appears in a very prominent way. As a consequence, electroexcitation of the ^{1}s-state of the deuteron has become a subject of intense study[6], and H. Arenhövel will certainly discuss this subject in his talk.

Secondly, the electrodesintegration at very large q but comparatively small ω (50 MeV<ω<<$q^2/2m$) has been introduced as a most promising tool. With this type of experiment a new kinematical region of electrodesintegration, and a new domain of physics - the long sought for high momentum components - have become accessible. It is this topic I am going to concentrate upon here.

II. Electrodesintegration at large q, "low" ω

To discuss the electrodesintegration for this region I will, for the moment, postulate the validity of the impulse approximation. The scattered electron is assumed to transfer momentum $\vec{q} = \vec{e} - \vec{e}'$ and energy ω = e - e' to one nucleon. This nucleon, having initial momentum $\vec{k}$ and separation energy E, recoils with momentum $\vec{k} + \vec{q}$. Nucleon final state interaction, and the contribution of different reaction mechanisms like MEC will for the moment be ignored; to be precise, we will assume that they take place sufficiently far from the electron-nucleon vertex to not influence the inclusive electron energy spectrum. I will show below that for the kinematical region discussed this assumption on the reaction mechanism is appropriate indeed.

The cross section for inclusive scattering of electrons is dominated by the quasielastic peak, a bump that occurs at ω ≃ $q^2/2m$ with a width given by the average momentum of nucleons within the nucleus. To properly describe the cross section dσ/dΩdω the knowledge of integral quantities - Fermi momentum k_F, average separation energy, eventually some higher moments - in general is sufficient. The corresponding cross sections experimentally are easily measurable; the perhaps biggest problem concerns the radiative corrections for large ω.

The kinematical situation I am interested in here, large q and comparatively very small or very large ω, is quite different. Here the cross sections are very small, and within the IA only high-momentum or large separation energy components of the wave function contribute. From the point of view of the physics investigated this region is of highest interest.

At very large energy loss ω, reaction mechanisms other than quasielastic scattering dominate. In particular the Δ-excitation process and the contribution of MEC dominate the cross section. Suggestions[7] to exploit this region for the study of high-k or high-E components therefore will be very difficult to realize.

As pointed out in ref.[8] the low-ω large-q region looks much more promising. Here the real or virtual excitation of the Δ is negligible (see below), and we can easily discuss the properties of the nuclear spectral function S(k,E) that matter. If experimentally we impose a large q, and a small $\omega \sim (\vec{k} + \vec{q})^2/2m$ (using nonrelativistic kinematics for the sake of the discussion) then $\vec{k}$ must roughly equal $-\vec{q}$, and be large as well. A more detailed discussion[8] tells us that at small ω and large q the electrodesintegration cross section is sensitive to values of S(k,E) near $k = q \cdot (A-1)/A$; the k-resolution S(k,E) is measured with is given approximately by $\sqrt{2m(\omega-\omega_{el})}$, with $\omega_{el} = q^2/2mA$. If we can study (e,e') at large q and small ω, then we can have access to high momentum components.

The subject of high momentum components in nuclear wave functions for a long time has been of great interest, for reasons I hardly need to explain. Many types of experiments have been carried out with the search for high-k components in mind. To cite a few, let me mention π-absorption, the (π,p) reaction and its inverse, the (γ,p) reaction, the (X,p) inclusive spectra, aso. The interpretation of the experimental results has produced rather disappointing conclusions discussed in more detail in ref[9]. The reaction mechanism in general is not the simple IA process initially imagined. When dealing with small wave function components and strongly interacting probes (or reaction products) two-step processes that depend on the low-k components dominate the cross section. The contribution of the large-k components thus is very difficult to extract. Reactions trying to exploit the large momentum mismatch between initial and final state are particularly prone to this limitation[10].

The best way I can think of to avoid the known difficulties of these past attempts is inclusive electron scattering at large q and comparatively small ω. The elementary condition, an IA cross section depending on high k components only, is trivially fulfilled. The improvements over previous attempts, the absence of strongly interacting hadrons subject to multistep reactions, is obvious. The interaction of the electron is much too weak to make multistep reactions likely. The knocked out nucleon, in traversing the (A-1) nucleus, certainly is subject to additional reactions; but this is of little concern since only the inclusive electron scattering cross section, without consideration of the ultimate fate of the recoil nucleon, is of interest. By detecting the lepton only, and by ignoring the properties of the final hadronic state, the major stumbling block of past attempts is removed.

The non-importance of the recoil-nucleon final state interaction (FSI) in inclusive electron scattering might not be familiar to everybody, and some comment is in order. A crude treatment of the recoil nucleon FSI can be performed by including a (real) energy dependent optical potential to describe the nucleon in the nuclear medium. Such a calculation shows[7] that that FSI is negligible once the energy of the recoil nucleon (which amounts to $\omega-\omega_{el}$, roughly) is larger than ~50 MeV. In this sense, "threshold" in the present context means 50 MeV $<\omega-\omega_{el} \ll q^2/2m$. Similar arguments apply to the absorptive part of the optical potential which leads[4], roughly speaking, to a smearing of $\sigma(\omega,q)$ over a small ω-region.

The unimportance of FSI is entirely analogous to the situation occuring in deep inelastic electron scattering off the nucleon, the famous experiment that detected the presence of partons in nucleons. In these experiments the pointlike constituents - the quarks - were found to be asymptotically free, despite the fact that a strong FSI - confinement - obliges quarks to recombine into observable elementary particles. The FSI is weakly affecting the inclusive cross section if it occurs at distances $\gg q^{-1}$ of the scattering vertex.

III. ^{3}He electrodesintegration

The preceeding general remarks on (e,e') and past attempts to determine high-momentum components explain why I propose (e,e') to be a most promising, new tool. These ideas have motivated the study[11] of (e,e') at large q and low ω by an experiment we performed 2 years ago at SLAC.

This experiment[12] was carried out on ^{3}He, a nucleus that is predestined for such an investigation. Only for ^{3}He the calculations of nuclear wave functions include the short-range correlations between nu-

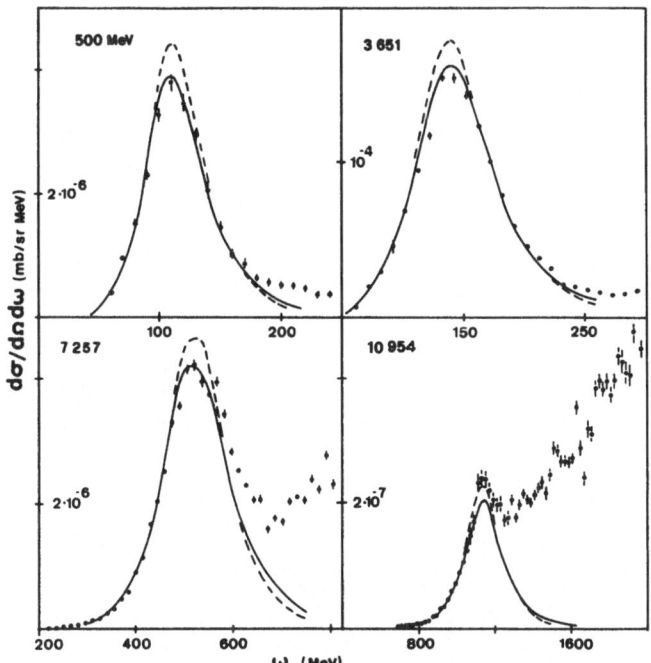

Fig.1 Experimental cross sections and predictions employing Faddeev (dashed), variational (dotted) and modified Faddeev (solid) spectral functions.

cleons in a sufficiently realistic way to justify, to a degree, the qualifier "exact" calculation. For ^{3}He we therefore can make a significant comparison between experiment and theoretical prediction at large momenta k.

The results of the experiment are shown in figs. 1,2. The data have been taken at electron energies from 3-10 GeV and at a scattering angle of 8°. The corresponding momentum transfer q varies between $2 \div 7$ fm^{-1}. Figure 1 shows the data that cover the complete quasielastic peak, figure 2 shows the "threshold" data of particular interest here.

The dashed curves in figs. 1,2 represent the theoretical predictions calculated using the program QUASI[13] and the spectral function of ^{3}He derived[2] from the Faddeev wave function[14] calculated using the RSC nucleon-nucleon interaction. While the main quasielastic peak (sensitive to k < 200 MeV/c) is properly reproduced, the cross sections in the "threshold" region (fig.2) are vastly underestimated. This shows that the Faddeev wave function has high-momentum components of too small an amplitude.

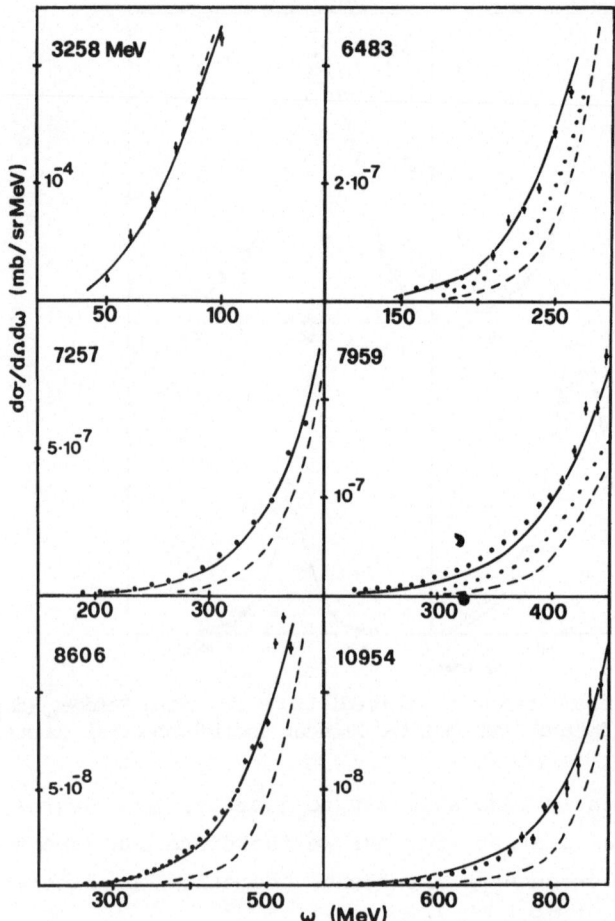

Fig.2 Experimental cross sections and predictions employing Faddeev (dashed), variational (dotted) and modified Faddeev (solid) spectral functions.

A somewhat better agreement with the data is obtained when using the recent variational calculation[15] for the 3-body wave function. This calculation gives (for reasons not yet entirely understood) high-k components of considerably larger amplitude. Accordingly, the agreement with the data at large q, low ω is better (see dotted curves in figure 2).

The difference between calculation and experiment cannot be accounted for by nucleon final state interaction; FSI has been taken into account, but is very small. Additional reaction channels like π-production[16] or Δ-excitation[17] give small contributions as well. The two-step processes of importance for electron scattering, MEC, also yield a contribution[18] to $\sigma(\omega,q)$ much smaller than the experimental cross sections. These MEC processes are of importance at large ω and large scattering angles, and at small ω for the "pathological" case of the IA diffraction minimum of the deuteron transverse M1 form factor. For the low-ω, small angle regime studied here the calculations[18] yield a negligible contribution of MEC. We consequently can assume the reaction mechanism (IA) to be the correct one and assign the deviations appearing in fig. 2 to properties of the ³He spectral function at large k. A detailed analysis shows that the Faddeev momentum space density is low by a factor 2÷10 in the k-region of 200÷600 MeV/c.

The calculations performed above leave open one point[10]: the interplay of initial versus final state correlation in the short range nucleon-nucleon interaction responsible for the occurence of high momentum components. Qualitatively, we expect the final state correlation to be of little importance since the nucleon-nucleon interaction in the high relative angular momentum state (produced after transfering a large q to one nucleon) is weak compared to the interaction in relative $\ell=0$ states (mainly responsible for the initial-state short range correlation). A quantitative treatment of this point is still lacking, however.

In order to see whether a phenomenological momentum distribution can fit the data correctly, I have multiplied the Faddeev spectral function $S(k,E)$ with a factor $f(k) \propto (1 + (k/\bar{k})^n)$. Using n=2.5 and $\bar{k}$=285 MeV/c gives a near-perfect fit of the data (see solid line, figs 1,2). This function $f(k)$ quantitatively describes the amount by which the Faddeev calculation underestimates the high momentum components in the region k = 200÷600 MeV/c.

Summarizing this section, we can say that the IA cross section at large q, low ω is sensitive to the high momentum components, and that one can make a quantitative comparison between experiment and the most realistic wave functions available. We have also verified that the reaction, which a priori is the cleanest one ever used in the search to determine high momentum components, receives no appreciable contribution from reaction mechanism other than the one described by the impulse approximation.

The big question nevertheless concerns the reaction mechanism. Is it indeed the IA process, or could there anything be wrong with the calculation that pretend that the non-IA processes give small contributions?

This question is a serious one given the bad experience we made with
the reaction mechanism of processes used to "measure" large k in the
past. With respect to this question, the full beauty of the inclusive
(e,e') experiment I am advocating really comes to bear: The scaling
property of the (e,e') cross sections provides us with an <u>experimental</u>
determination of the reaction mechanism too !

IV. Scaling

The occurence of scaling in (e,e') implies that at large q $\sigma(\omega,q)$
becomes a function of a single variable only, and no longer depends on
the physically independent variables ω and q.

The scaling property of deep inelastic electron-nucleon scattering
[19] has led to the most convincing evidence for the presence of quarks
in nucleons. The fact that scaling does occur could only be explained
by postulating that the electron scatters off quasifree, pointlike constituents. For the electro-desintegration of nuclei, the occurence of
this Bjorken-type scaling would be of lesser interest. It could be used
to prove that electrons scatter from nucleons, but it would not inform
us about the quantitative nuclear properties; the scaling function derived from $\sigma(q,\omega)$ converges in the large-q limit to a δ-function in the
scaling variable x.

For nuclei, the concept of y-scaling as proposed by G. West is more
fruitful. This type of scaling is derived using as the basic assumption
that the electron scatters off a quasifree nucleon having before the reaction a momentum component $k_{//} = y$ parallel to $\vec{q}$. If the data does show scaling,
then the assumed reaction mechanism is confirmed; in this respect x- and
y-scaling are entirely analogous. However, the function $\sigma(\omega,q)$ scales to,
F(y), is not trivial since it represents the momentum distribution of
nucleons in nuclei. Again, scaling is a concept that only applies in the
limit of large q ($q \gg k_F$), and does not make sense for the $q < 2\text{fm}^{-1}$ where
most quasielastic electron scattering data have been taken at in the
past.

Here, I do not want to derive scaling properly, but refer the reader to the paper of G. West[20]. I should perhaps give a simple qualitative argument, however, why scaling could occur. Writing down energy
and momentum conservation for the quasifree scattering process introduces as variables ω, $\vec{q}^2$, E, $\vec{k}^2$ and $(\vec{k} + \vec{q})^2$. In the limit of large q,
only the terms containing ω, q^2, $\vec{k} \cdot \vec{q}$ survive, while E, k_i^2 become negli-

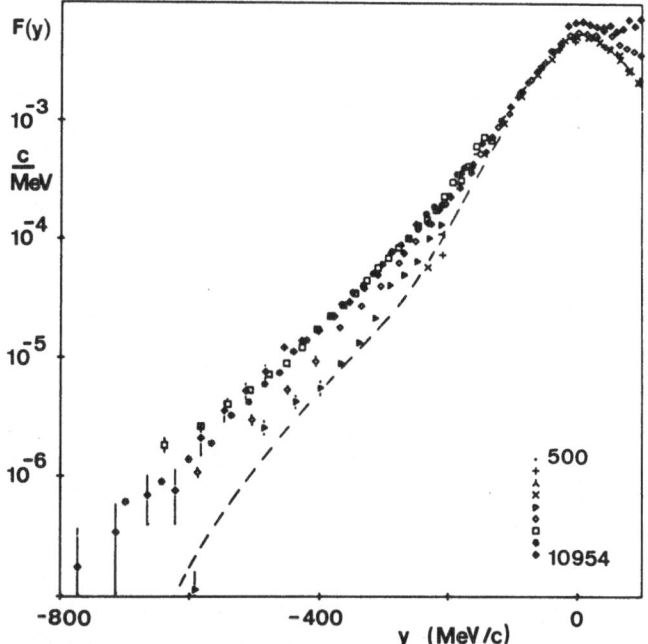

Fig.3 Experimental scaling function F(y).

gible. The equation obtained thus allows to compute from the experimental variables q,ω the quantity $k_{\parallel} = \vec{k}\cdot\vec{q}/q = y(\omega,q)$. The cross section, after dividing out the elementary electron-nucleon cross section, becomes a function F(y) of y only.

Figure 3 shows the values of F(y) for ^{3}He calculated from the experimental data covering an energy region of 0.5 to 10 GeV. The data points show a most striking scaling behaviour: as the electron energy, i.e. momentum transfer, increases the different sets of data merge into a unique function of y which, for q → ∞, represents the momentum distribution of nucleons in ^{3}He.

The scaling property has a most remarkable consequence: it proves that the reaction mechanism is indeed the one assumed. To compute F(y) explicit use was made of the fact that the electron scatters in a quasi-free way off an object having the mass and the form factor of the nucleon. If the data do scale as expected in this model, then the assumed mechanism is the correct one. For any other process - meson exchange current contributions, Δ-excitation, knockout plus short range final state correlations - no scaling is expected. For instance, no scaling is found if for Δ-excitation the theoretical predictions or the experimental data (y>0) are analyzed. If the cross sections for these processes might at all scale, then only if the appropriate parameters of the scattering system - Δ-form factor, mass of the two nucleon system, etc. - were used.

V. Conclusion

The scaling property thus takes away the biggest worry we had concerning the use of the tool of electrodesintegration at large q, small ω. The reaction mechanism basically is the simple one we considered, and we can avoid the most important difficulty (multistep reactions) previous attempts to see high momentum components have encountered. The quantitative comparison of experiment and theory as performed in section III (independent of the asymptotic-q limit) can be expected to become a valid way to determine high components in nuclear wave functions.

References

1) E. Moniz et al., Phys. Rev. Lett. 26 (71) 445
   R.R. Whitney et al., Phys. Rev. C9 (74) 2230
2) A.E.L. Dieperink et al., Phys. Lett. 63B (76) 261
   G.G. Simon et al., Nucl. Phys. A 324 (79) 277
3) E.B. Hughes et al., Phys. Rev. 139 (65) B458
4) Y. Horikawa et al., to be publ. in Phys. Rev. C
   R. Rosenfelder Ann. Phys. 128 (80) 188
5) J. Mougey et al., Phys. Rev. Lett 41 (78) 1645
   R. Altemus et al., Phys. Rev. Lett. 44 (80) 965
6) M. Bernheim et al., to be publ. in Phys. Rev. Lett. and references therein.
7) W. Czyz et al., Ann Phys. 21 (63) 47
8) I. Sick Proc. Conf. Mod. Trends in Elastic Electron Scattering, Amsterdam, 1978, p.155
9) For review see Proc. Workshop Program Options in Intermediate Energy Physics, LAMPF, 1979, p.99
10) R.D. Amado Phys. Rev. C9 (79) 1473

11) I. Sick et al., Phys. Rev. Lett. $\underline{45}$ (80) 871
12) D. Day et al., Phys. Rev. Lett. $\underline{43}$ (79) 1143
13) I. Sick, to be publ.
14) R.A. Brandenburg et al., Phys. Rev. $\underline{C12}$ (75) 1368
15) P. Nunberg et al., Nucl. Phys. $\underline{A285}$ (77) 58
16) F. Borie, Z. Phys. $\underline{A285}$ (78) 129, and priv. com.
17) G. Do Dang, Phys. Lett. $\underline{69B}$ (77) 425, and priv. com.
18) T.W. Donnelly et al., Phys. Lett. $\underline{76B}$ (78) 383, and priv. com.
19) W.R. Francis et al., Phys. Rep. $\underline{54C}$ (79) 308
20) G.B. West, Phys. Rep. $\underline{18C}$ (75) 264

EXCHANGE CURRENTS IN THE DEUTERON

Hartmuth Arenhövel
Institut für Kernphysik
Johannes Gutenberg-Universität
D-6500 Mainz, Federal Republic of Germany

Abstract: The present status of exchange currents and their contribution to deuteron disintegration by real and virtual photons is reviewed. The importance of the Siegert theorem for electric transitions is stressed. Various experiments in photoabsorption, electron and photon scattering with respect to influences of exchange currents are discussed.

1. Introduction

Throughout the history of exchange currents (EC) in nuclear physics the deuteron has played a very important rôle. One reason is that the thermal n-p capture forming a deuteron showed a clear signature of a serious failure of the classical concept. The discrepancy between experiment and theory was soon interpreted as an indication for the presence of exchange currents[1]. However, from this early recognition to the present quantitative evaluation of EC, there was still a long way to go. And here again, the deuteron has played a key rôle. Because of its simple structure as a two-body system it allowed clear cut calculations without the usual approximations associated with the conventional many-body problem of more complex nuclei. Therefore, I think, it is justified to concentrate for the purpose of this workshop on the deuteron even though many interesting EC effects have been investigated in three-body systems and heavier nuclei. In the following I will consider as exchange currents only the lowest order meson exchange currents (MEC) and isobar configurations (IC) of non-relativistic order. Higher order contributions, in particular MEC contributions to the charge density, will not be discussed, because I feel, a more consistent treatment of all relativistic effects is necessary[2].

2. Deuteron break-up near threshold

As a first topic I will consider photo- and electrodisintegration near threshold, because it is one of the most important cases, where one has very clear evidence for the presence of mesonic degrees of freedom in nuclei[3,4]. The reason, why the threshold region is so interesting, is that there the isovector M1-transition (essentially a spin-flip) from the deuteron ground state to the $^1S_0$ resonant antibound state dominates the transverse form factor. The deuteron as a loosely bound object has a rather extended spatial structure. Thus, the one-body transition form factor drops fast with increasing momentum transfer. In this situation the two-body character of EC helps to distribute the momentum transfer roughly equally between the two particles and correspondingly the EC form factor drops considerably slower and even surpasses the one-body form factor leading to the observed increase. I will illustrate this with two experimental results.

Fig. 1 shows a recent analysis of a d(e,e') experiment from Saclay[5] at a backward angle, where the transverse form factor dominates. The inelastic cross section is averaged from 0 to 3 MeV n-p final state energy E_{np}. The rapid fall-off of the theoretical cross section in fig. 1 without EC is readily seen and the failure of the conventional theory using the Reid soft-core potential to describe the experiment. Inclusion of MEC and IC increases the theoretical result considerably and leads to a quantitative agreement at lower momentum transfer. For higher q_μ^2 the qualitative behaviour of the theory is still quite satisfactory, in particular the lowering of the normal cross section above $q_\mu^2 \sim 15$ fm^{-2} is well accounted for. However, the theory is significantly too low. But one should keep in mind that only the lowest order π-current has been used, without regularization and no heavier meson currents are considered, even though the effects partially cancel. Furthermore, in this region of high momentum transfer one certainly has to study relativistic corrections to wave functions and operators. On the other hand, there appears some disagreement to earlier experiments in the overlapping region of $q_\mu^2 = 6$ to 10 fm^{-2}, where the new data ly systematically above the earlier ones.

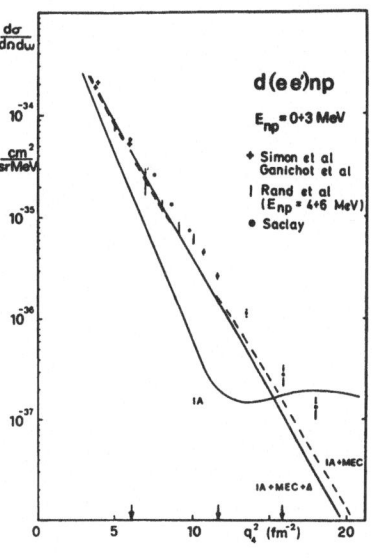

Fig. 1: d(e,e') cross section, $\theta_e = 155^0$. Theoretical curves using Reid soft-core for normal impulse approximation (N) and inclusion of MEC and IC. Arrows indicate the position of the spectra of fig. 2.

Some explicit theoretical spectra are displayed in fig. 2 to show the threshold behaviour in greater detail, which is covered up in taking the average. The $^1S_0$-resonance

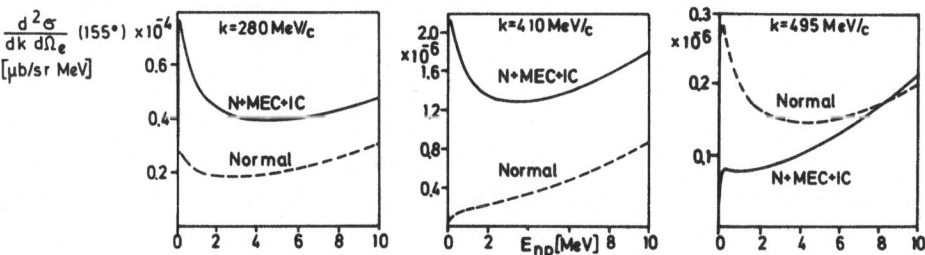

Fig. 2: Inelastic spectra for d(e,e') between threshold and E_{np} = 10 MeV for three different momentum transfers.

is clearly seen for incoming electron energies of 280 and 495 MeV. While in the former case the EC add coherently and increase the cross section, they have opposite sign in the latter case and result in a decrease. Particularly impressive is the spectrum for 410 MeV electron energy, where one is near the minimum of the normal form factor so that the $^1S_0$-resonance is not seen any more. But with EC it sticks out again and one observes a tremendous increase of the cross section.

As second example, fig. 3 shows the inelastic longitudinal and transverse form factors separated experimentally via Rosenbluth-plots[6]. It is a beautiful example how much more detailed information one can obtain from such a separation. First of all, one notices that the longitudinal form factor is very well described by the normal theory without exchange effects. This supports the old hypothesis of Siegert that in lowest order the charge density is not affected by the exchanged virtual mesons. In fact, explicit model calculations of MEC contributions to the charge density show that they are of relativistic order and, thus, should be small in this region of momentum transfer. The transverse form factor shows beautifully the $^1S_0$-resonance near break-up threshold similar to

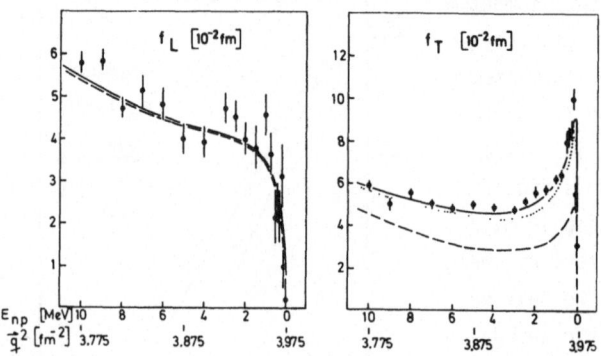

Fig. 3: Inelastic form factors f_L and f_T as a function of E_{np} around $(q^{c.m.})^2 = 4$ fm^{-2}. The dashed curves are obtained by using the Reid soft-core potential without interaction effects, while the dotted curves include MEC and the full curves both MEC and IC.

the theoretical spectra of fig. 2. The necessity of MEC and IC is readily seen and the agreement is quite impressive. Near threshold the EC add about 100 per cent. Only in the maximum the theory seems a little too low.

This brings me to the situation in photo break-up near threshold or the inverse reaction, the thermal n-p capture, which I have mentioned in the introduction. It has been noted long ago that the experimental cross section σ_c = 334 mb for thermal neutrons (v = 2200 m/s) cannot be reproduced by the conventional non-relativistic theory. A very careful analysis of Noyes[1] using the effective range theory with experimental input finds a discrepancy of 9.5 ± 1.2% with respect to σ_c^{exp} for a singlet effective range r_s = 2.7 fm (or 7.7 ± 1.5% for r_s = 2.4 fm). It is now interesting to note and I thank Dr. Tzara for drawing my attention to this fact, that all the subsequent efforts to calculate the missing interaction or exchange effects have concentrated on the relative contribution and did not worry about the total absolute value[3].

The justification for this attitude lies in the fact that the relative contribution of π-MEC is rather insensitive to the potential model and about 3% for the M1 matrix element. But it should be a little disturbing if the model one uses to calculate the correction is not able to reproduce the main part. This failure of many realistic NN-potentials comes from a wrong value for the singlet scattering length a_s on which σ_c depends very sensitively. It is illustrated in table 1, where also the contributions of IC ($\Delta\Delta$ in the deuteron, $N\Delta$ in the $^1S_0$-state) are listed. It is obvious that only the Bryan-Gersten potential having the right scattering length a_s reproduces the normal cross section of the effective range theory*. But the relative change from EC effects is rather independent from the potential model. π-MEC give between 5.5 and 6.1% increase and IC additional 2%. Regularization of the EC leads to a slight decrease, which is partially cancelled from ρ-MEC contributions. Thus, EC effects considerably close the gap between theory and experiment, but about 2% remain still missing.

potential	a_s(fm)	σ_c^N(mb)	σ_c^T(mb)	$\Delta\sigma_c^{MEC}$	$\Delta\sigma_c^T[\%]$
RSC[7]	-17.10	180.02	195.1	6.1	8.4
RSC(Λ=4fm^{-1})			194.0	5.5	7.7
dT-S[8]	-17.97	196.28	211.7	5.8	7.8
Paris[9]	-18.39	203.29	219.4	5.7	7.9
BG[10] (Λ=4)	-23.67	302.78	325.8	5.7	7.6
BG(π+ρ,Λ)			325.0	5.4	7.4
exp[1]	-23.69	302.5 ± 4.0	334.2 ± 0.5		10.4 ±1.3

Table 1: Scattering length a_s and capture cross section σ_c without (N) and with (T) exchange effects. Relative contributions with respect to σ_c^N of MEC alone are also listed. Λ indicates regularization.

In conclusion, there is clear evidence for EC effects in the M1 transition near deuteron break-up threshold and the theory is in satisfactory agreement with experiment even though some uncertainties are still existing. One may speculate that these are connected with uncertainties in the description of IC. But to answer this, we need a fully consistent description, which is still missing.

3. Exchange effects in electric transitions

What can we say about EC effects in electric transitions? We are used to answer that such effects are small thanks to the famous Siegert theorem, which expresses the dominant part of the transverse electric multipoles via current conservation by the longitudinal or charge multipoles, in conjunction with Siegert's hypothesis that the charge density remains unaffected by EC in lowest non-relativistic order. As mentioned before, this is supported by meson theories. Even though this theorem is of great value for calculations, one should not use it as an argument that EC effects are small in electric transitions. It tells us only that the remaining EC effects, which

---

* This influences also the hight of the $^1S_0$-resonance in fig. 2, where the BG potential produces a higher peak than the other realistic potentials in agreement with experiment[6].

are not implicitly taken into account, are rather small. In order to see how much EC effects are buried in Siegert's theorem, one has to compare the results obtained with and without the use of this theorem. Recently, I have discussed this question in detail[11]. Let me briefly repeat the main arguments. It is sufficient to consider only E1 since higher multipoles are much less sensitive to EC effects.

The E1 Operator is given by

$$\hat{M}^{[1]}(\hat{\mathfrak{J}}) = \frac{1\sqrt{2}}{\omega} \int d^3r \hat{\mathfrak{J}} \cdot \nabla j_1(\omega r) Y^{[1]}(\hat{r}) + \hat{M}_b^{[1]}(\hat{\mathfrak{J}}) = \hat{M}_a^{[1]}(\hat{\mathfrak{J}}) + \hat{M}_b^{[1]}(\hat{\mathfrak{J}}), \qquad (1)$$

where $\hat{\mathfrak{J}}$ is the <u>total</u> nuclear current density operator including EC. The first term $\hat{M}_a$, which is the dominant one, can be related to the charge multipole C1 with the help of current conservation

$$\hat{M}_a(\hat{\mathfrak{J}}) = -\frac{\sqrt{2}}{\omega} [\hat{H}, \hat{C}(\hat{\rho})], \qquad \text{"Siegert theorem"} \qquad (2)$$

with the charge dipole operator

$$\hat{C}^{[1]}(\hat{\rho}) = \int d^3r \hat{\rho} j_1(\omega r) Y^{[1]} \qquad (3)$$

where $\hat{\rho}$ is the <u>total</u> charge density operator. The advantage of eq. (2) is the appearance of the charge density $\hat{\rho}$ instead of the current density $\hat{\mathfrak{J}}$ in the light Siegert's hypothesis.

Now we can define two normal E1 operators:
(I) Use the Siegert theorem (eq. (2)) and define as normal contribution

$$\hat{M}_n^I := -\frac{\sqrt{2}}{\omega} [\hat{H}, \hat{C}(\hat{\rho}_{[1]})] + \hat{M}_b(\hat{\mathfrak{J}}_{[1]}) \qquad (4)$$

with the one-body charge and current density operators $\hat{\rho}_{[1]}$ and $\hat{\mathfrak{J}}_{[1]}$, respectively.
(II) Do not use the Siegert theorem and define the normal contribution by

$$\hat{M}_n^{II} := \hat{M}(\hat{\mathfrak{J}}_{[1]}) \qquad (5)$$

From a pragmatic point of view the approach (I) is very useful in conventional nuclear physics, since it allows a good description of E1 transitions including MEC contributions without knowing them explicitly. On the other hand, it masks the actual size of the MEC contributions. The difference between these two operators tells us how much EC effects are buried in approach (I). The additional exchange contributions are given by

$$\hat{M}_{ex}^I := \hat{M}_b(\hat{\mathfrak{J}}_{[2]}) \quad \text{and} \quad \hat{M}_{ex}^{II} = \hat{M}(\hat{\mathfrak{J}}_{[2]}) \qquad (6)$$

respectively, where $\hat{\mathfrak{J}}_{[2]}$ is the exchange current density operator.

In the approach (II) the full MEC corrections are given by M_{ex}^{II}. Of course, if the MEC is consistent with the two-body potential $\hat{V}$, both approaches give identical results for the sum of normal and exchange contributions. In many explicit evaluations, however, where mainly π-MEC is considered in conjunction with realistic potentials, this consistency is not fulfilled and then the first approach is more reliable for electric transitions for not too high momentum transfers as will be shown below.

Deuterium is a good case to study the two approaches, since for both the initial and the final states exact solutions are available. This is important, in particular with respect to the Siegert theorem. The Reid soft-core potential has been used and multipoles up to L = 4 are included.

The total cross section is shown in fig. 4. It is obvious that the true exchange contribution is rather large comparing the dotted with the full curve. But most of it is already included by using the Siegert theorem (dashed curve). The full and the dash-dot curves should coincide for a consistent MEC. However, in this case only π-MEC has been considered with monopole regularization of range $\Lambda = 4$ fm^{-1} and the comparison shows that the second approach overestimates the exchange contribution and, thus, demonstrates the practical advantage of the first approach using the Siegert theorem. We also show in fig. 4 the various experimental results. Below 80 MeV the agreement between theory and experiment is quite satisfactory, particulary for the most recent data from Saskatoon, Mainz and Louvain, even though recently a serious disagreement has been claimed to justify some very drastic changes in the deuteron wave functions[12]. Above 80 MeV there is definite need for more accurate data.

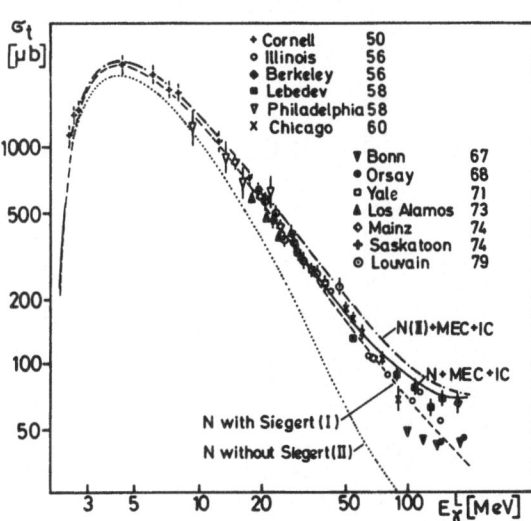

Fig. 4: Total photodisintegration cross section for deuterium with and without Siegert theorem and with exchange effects. Experimental data see ref. 13.

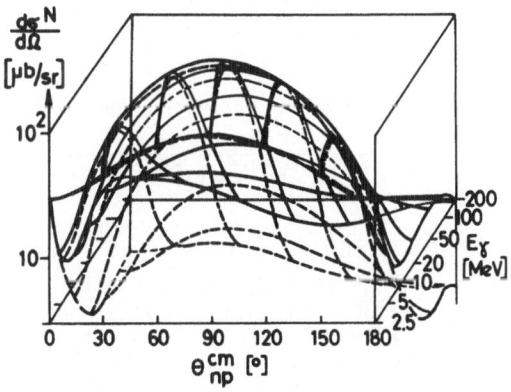

Fig. 5: Angular distribution for d(γ,p)n with (full lines) and without (dashed lines) Siegert theorem.

Fig. 5 shows the proton angular distribution up to 200 MeV with and without the Siegert theorem. Again one sees clearly the large contribution of EC as contained in the Siegert theorem, particularly at higher energies. The additional explicit EC contributions are shown in fig. 6. They become increasingly important only above 80 MeV. For two energies (80 and 120 MeV) we show in fig. 7 comparisons with experimental data.

The overall agreement is satisfactory. However, again more accurate data over a wider range of energies and angles are urgently needed.

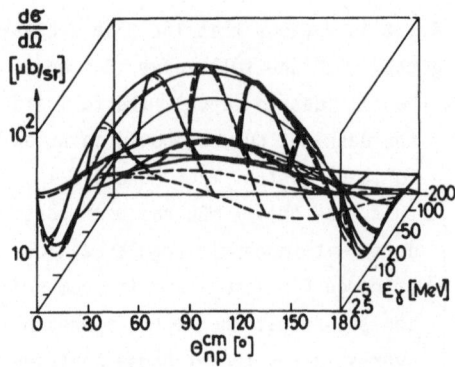

Fig. 6: Angular distributions for $d(\gamma,p)n$ with (full lines) and without (dashed lines) exchange effects.

Very sensitive to exchange corrections is the cross section for forward proton emission (0°) as is born out in fig. 8. The disagreement to the experiment[14] is probably related to the strength of the intermediate range isovector tensor force, which seems to be too strong.

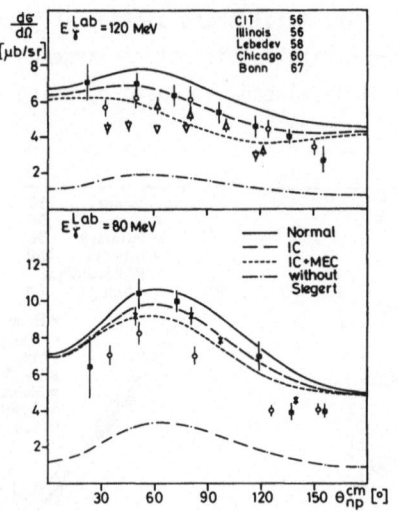

Fig. 7: Angular distributions for $d(\gamma,p)n$ with and without Siegert theorem and with separate contributions from MEC and IC. Experimental data see ref. 13.

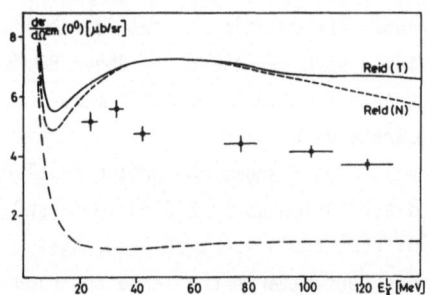

Fig. 8: Forward proton cross section in $d(\gamma,p)n$. Normal cross section with (dashed) and without (dash-dot) Siegert theorem. Full curve includes EC effects. Experimental data from ref. 14.

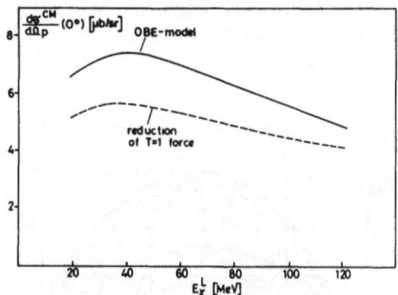

Fig. 9: Forward proton cross section for $d(\gamma,p)n$ for OBE-model and with reduction of $T = 1$ force.

To demonstrate the sensitivity on the isospin dependent force, I have taken a very simple OBE-force, which reproduces the deuteron properties. The total cross section agrees quite well with that from the Reid potential and the 0° cross section is a little higher. Then I have modified the potential by replacing $\tau_1 \cdot \tau_2$ by $(0.85 \tau_1 \cdot \tau_2 - 0.45)$ leaving the T = 0 force unchanged and weakening in the T = 1 channel the isospin dependent part by 0.4. While the total cross section is decreased by only 2 to 7% between 20 and 120 MeV the 0° cross section is lowered by about 20-25% (fig. 9). This example is used only for pedagogical purpose. But it could be that the discrepancy of the 0° cross section is not only related to the strength of the tensor force and the D-state probability but that it may at least partly come from problems of the T = 1 force in the n-p channel. Another interesting feature of the 0° cross section is the minimum around 10 MeV, which is also due to exchange effects. Unfortunately, it has not yet been explored experimentally.

Finally, Fig. 10 shows the asymmetry for disintegration with polarized photons and neutron emission under 90° in the lab system. Again one sees rather large corrections from exchange currents with increasing energy. The agreement with experimental data is satisfactory[15].

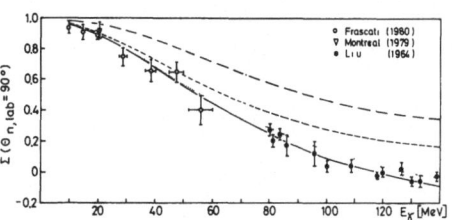

Fig. 10: Asymmetry for $d(\gamma,n)p$ with polarized photons at neutron lab angle of 90°. Notation for theoretical curves as in fig. 8. Experimental data from ref. 15.

In conclusion, there is urgent need for more accurate data on total cross section (particularly above 80 MeV), on angular distributions, especially at 0° and 180° including the minimum around 10 MeV, and on the polarization of the outgoing nucleon. The use of polarized light and deuteron oreintation seems promising.

4. Electrodisintegration

Let me now briefly come back to electrodisintegration. Recently, I have given a systematic survey on the form factors for $d(e,e')$ and structure functions for coincidence experiments $d(e,e'N)$[16]. Because of the more detailed information available from coincidence measurements the study of EC effects are very promising. One main result of this survey was that besides the threshold region, where MEC are dominant, another region between the quasi-free ridge and the photon line is of great interest with respect to the rôle of isobar configurations. Fig. 11 shows the transverse form factor along $E_{np}/MeV = 60 \ q^2/fm^{-2}$. The drastic enhancement with encreasing E_{np} is readily seen with a smooth transition into the region of real π- and Δ-production.

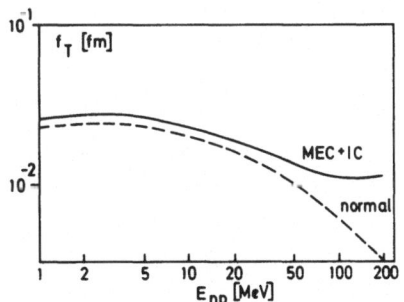

Fig. 11: Transverse form factor for $d(e,e')$ along $E_{np}/MeV=60 \ q^2/fm^{-2}$.

For a coincidence experiment the cross section is determined by four structure functions $f_{\mu\nu}(E_{np}, q^2, \theta_{np}^{cm})$, which depend also on the angle of the observed nucleon.

$$\frac{d^3\sigma}{d\Omega_{np} dk' d\Omega_{e'}} = \frac{\alpha R k'}{6\pi^2 q_\nu^4 R} \{ \rho_{00} f_{00} + \rho_{11} f_{11} + \rho_{01} \cos \phi_{np}^{cm} f_{01} + \rho_{-11} \cos 2\phi_{np}^{cm} f_{-11} \} \quad (7)$$

Here, $\rho_{\mu\nu}$ describe the virtual photon polarization. The behaviour of the structure functions along $E_{np}/\text{MeV} = 60\, q^2/\text{fm}^{-2}$ is displayed in figs. 12-15. The diagonal lon-

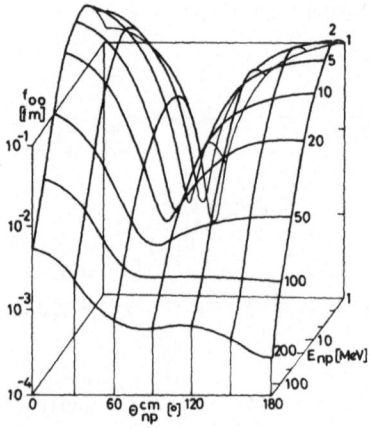

Fig. 12: Structure function f_{00} along $E_{np}/\text{MeV} = 60\, q^2/\text{fm}^{-2}$.

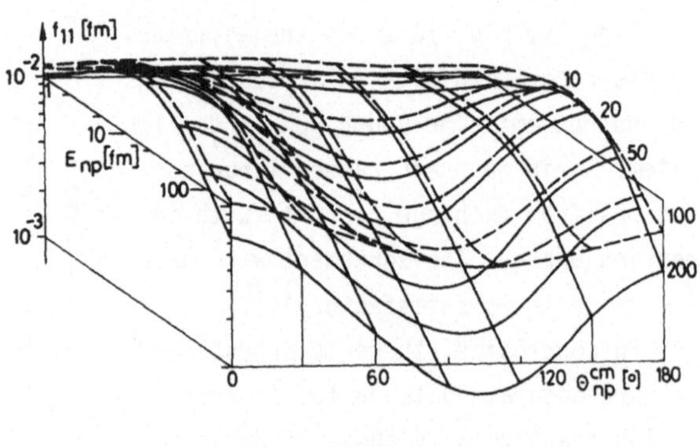

Fig. 13: f_{11} along $E_{np}/\text{MeV} = 60\, q^2/\text{fm}^{-2}$. Full curves without and dashed curves with EC effects.

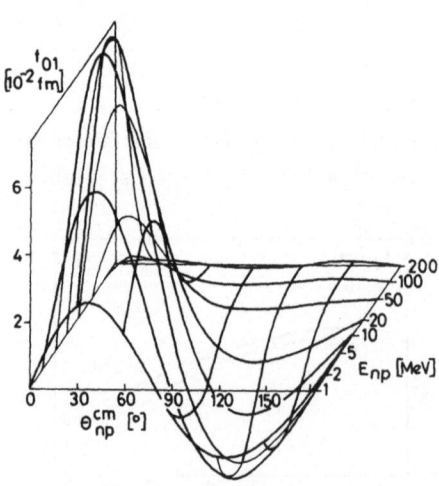

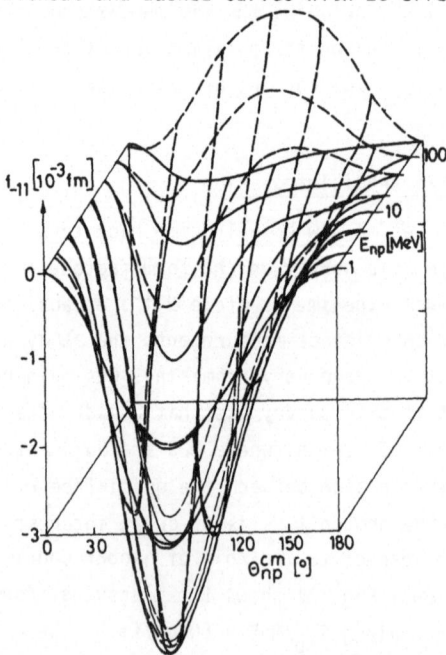

Fig. 14: f_{01} along $E_{np}/\text{MeV} = 60 q^2/\text{fm}^{-2}$.

Fig. 15: f_{-11} along $E_{np}/\text{MeV} = 60 q^2/\text{fm}^{-2}$. Notation as in fig. 13.

gitudinal f_{00} shows a peaking in the forward and backward direction at lower E_{np}, which is washed out with increasing energy transfer, and, finally, the backward peak disappears. Since the charge density is little affected by IC, there is only a small change from EC, too small to be shown separately. The diagonal transverse f_{11} starts with a rather isotropic distribution at low E_{np} and develops then a pronounced minimum around $90°$ to $120°$ with increasing E_{np}. Exchange effects are considerably. They lift up the whole surface and wash out the minimum. The interference functions f_{01} and f_{-11} are smaller in magnitude. While f_{01} (fig. 14) is little influenced by EC effects, one observes in f_{-11} (fig. 15) a dramatic change with increasing E_{np}. Even the sign changes above $E_{np} = 100$ MeV. An experimental determination of these four structure functions would therefore be very rewarding even though I am aware of the great experimental difficulties, which an out-of-plane measurement poses, as would be necessary for such a separation.

5. Exchange effects in photon scattering

As last topic I will discuss EC effects in photon scattering. The scattering amplitude T is given by a resonance term T^{res} and a so-called sea gull term or two-photon amplitude B originating from second-order terms of the electromagnetic interaction with the nucleus.

$$T_{fi} = B_{fi} + T_{fi}^{res} \tag{8}$$

The corresponding graphs are shown in fig. 16. EC effects contribute both, the resonance amplitude via exchange currents and to the sea gull term B as has been pointed out by Christillin and Rosa-Clot[17]. This is a consequence of gauge invariance. The sea gull term consists of a kinematic term B^{kin} describing the Thomson scattering from Z protons

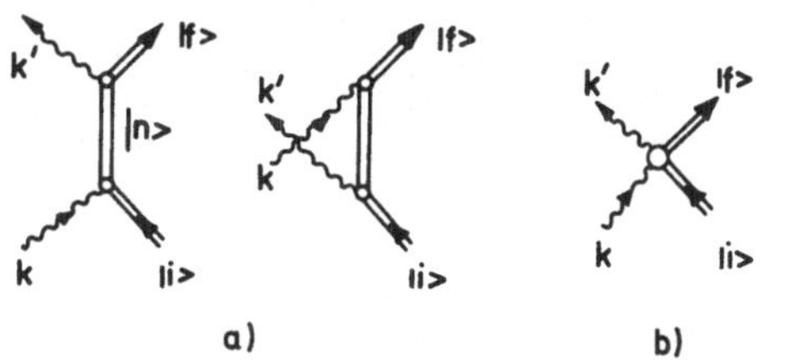

Fig. 16: Photon scattering diagrams for resonance (a) and sea gull terms (b).

$$B_{fi}^{kin} = \frac{\varepsilon' \cdot \varepsilon}{M} <f| \sum_j e_j^2 e^{i(k-k') \cdot r_j} |i>, \tag{9}$$

where k and k' are the momenta of incoming and outgoing photon with polarization ε and ε', respectively. In the presence of an isospin and/or momentum dependent potential V an additional exchange sea gull term B^{EC} is required by gauge invariance. This is easy to see in the low energy limit $k \to 0$. Then one has

$$B_{ii}^{kin}(0) = \frac{Ze^2}{M} \varepsilon' \cdot \varepsilon,$$

$$T_{ii}^{res}(0) = -\frac{NZe^2}{AM} \varepsilon' \cdot \varepsilon - <i|[\varepsilon' \cdot D,[V, \varepsilon \cdot D]]|i>, \tag{10}$$

while the limit of the total amplitude should be according to the low energy theorem

$$T_{ii}(0) = \frac{(Ze)^2}{AM} \underset{\sim}{\varepsilon}' \cdot \underset{\sim}{\varepsilon}. \tag{11}$$

Comparison gives then the condition

$$B_{ii}^{EC}(0) = <i|[\underset{\sim}{\varepsilon}' \cdot \underset{\sim}{D},[V, \underset{\sim}{\varepsilon} \cdot \underset{\sim}{D}]]|i>. \tag{12}$$

Christillin and Rosa-Clot have constructed an explicit exchange sea gull term using gauge invariance only[17]. However, this procedure is not unique, since only the longitudinal terms are fixed. But one needs the transverse parts. This is similar to the situation for EC, where current conservation is not sufficient to fix the exchange current. However, one can construct B^{EC} using an explicit model, like the OPE-model. Using the principle of minimal coupling[18], one obtains essentially four contributions to $B^{\pi-MEC}$ corresponding to the graphs in fig. 17. While the low energy limit is given by eq. (12) with V^{OPE}, the high energy limit in the forward direction is determined by the graph d) in fig. 17 corresponding to the Thomson scattering off an exchanged virtual π-meson.

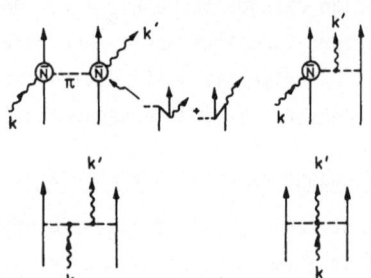

Fig. 17: Sea gull terms from pion exchange interaction.

$$B_{ii}^{\pi-MEC}(\underset{\sim}{k},\underset{\sim}{k}) \underset{k \to \infty}{\to} \frac{e^2}{m_\pi} n_\pi, \tag{13}$$

where n_π is the average number of exchanged charged pions. As a first step, M. Weyrauch[19] has recently evaluated $B_{ii}^{\pi-MEC}$ in the forward direction and compared the contribution to the elastic resonance ampltude as obtained from dispersion relations

$$T_{ii}^{res}(\omega) = T_{ii}^{res}(0) - \frac{\omega^2}{2\pi^2} P \int_0^\infty d\omega' \frac{\sigma t(\omega')}{\omega'^2-\omega^2}. \tag{14}$$

The results, normalized to zero at $E_\gamma = 0$, are shown in fig. 18. One readily sees that the energy dependence is weak and, therefore, the renormalized contribution to the total amplitude is small. Taking the high energy limit of the $B^{\pi-MEC}$, we obtain as average number of charged pion $n_\pi = 0.02$, i.e., about 1% pions compared to nucleons. We have also evaluated the approximation of ref. 17, which is shown in fig. 19

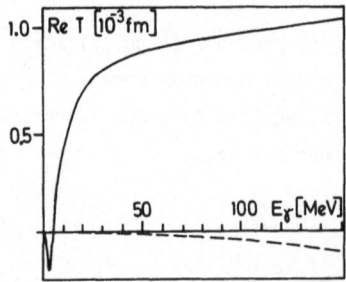

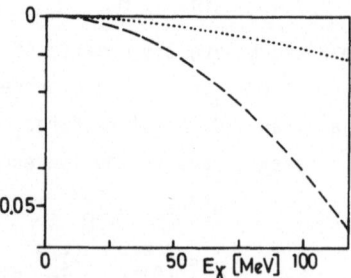

Fig. 18: Forward elastic scattering amplitude for the resonance term (full) and sea gull term (dashed). Both are normalized to zero at $E_\gamma = 0$.

Fig. 19: Comparison of $B^{\pi-MEC}$ (dashed) with approximation of ref. 17 (dotted).

in an enlarged scale. It underestimates $B^{\pi-MEC}$ considerably. At present, we are investigating finite scattering angles and, furthermore, contributions of IC to B^{EC} in order to assess the magnitude of exchange effects in photon scattering. This will be very important for the interpretation of present[20] and future γ-scattering experiments.

References:

1. H.P. Noyes, Nucl. Phys. 74 (1965) 508
2. J.L. Friar, review talk at Int. Conf. on Few-body problems, Eugene, Oregon, 1980
3. D.O. Riska, G.E. Brown, Phys. Lett. B38 (1972) 193
4. J. Hockert et al., Nucl. Phys. A217 (1973) 14
   J.A. Lock, L.L. Foldy, Ann. Phys. 93 (1975) 276
   W. Fabian, H. Arenhövel, Nucl. Phys. A258 (1976) 461
   B. Mosconi, P. Ricci, Nuovo Cim. 36A (1976) 67
5. M. Bernheim et al., preprint, Saclay, 1980
6. G.G. Simon et al., Nucl. Phys. A324 (1979) 277
7. R.V. Reid, Ann. Phys. 50 (1968) 411
8. R. de Tourreil et al., Nucl. Phys. A242 (1975) 445
9. M. Lacombe et al., Phys. Rev. C21 (1980) 861
10. R.A. Bryan, A. Gersten, Phys. Rev. D6 (1972) 341
11. H. Arenhövel, Proc. of the workshop "Intermediate energy nuclear physics with monochromatic and polarized photons", Frascati, 1980 (to be published)
12. E. Hadjimichael, D.P. Saylor, Contribution to Symposion on Perspectives in electro- and photonuclear physics, Saclay, 1980, Phys. Rev. Lett. 45 (1980) 1776
13. P.V.C. Hough, Phys. Rev. 80 (1950) 1069
    B.A. Whalin et al., Phys. Rev. 101 (1956) 377
    D.R. Dixon, K.C. Bandtel, Phys. Rev. 104 (1956) 1730
    Iu.A. Aleksandrov et al., Sov. Phys. JETP 6 (1958) 472
    A.L. Whetstone, J. Halpern, Phys. Rev. 109 (1958) 2072
    J.A. Galey, Phys. Rev. 117 (1960) 763
    R. Kose et al., Z. Phys. 202 (1967) 364
    J. Buon et al., Phys. Lett. 26B (1968) 595
    B. Weissmann, H.L. Schultz, Nucl. Phys. A174 (1971) 129
    J.E.E. Baglin et al., Nucl. Phys. A201 (1973) 593
    J. Ahrens et al., Phys. Lett. 52B (1974) 49
    D.M. Skopik et al., Phys. Rev. C9 (1974) 531
    M. Bosman et al., Phys. Lett. 82B (1980) 212
14. R. Hughes et al., Nucl. Phys. A267 (1976) 329
15. R. Caloi et al., preprint LNF-80/15(P), Frascati, 1980
    W. Del Bianco et al., Bull. Ann. Phys. Soc. 24 (1979) 648
    F.F. Liu, Phys. Rev. B138 (1965) 1443
16. H. Arenhövel, Proc. Symposion on Perspectives in electro- and photonuclear physics, Saclay, 1980 (to be published)
17. P. Christillin, M. Rosa-Clot, Nuovo Cim. 28A (1975) 29, and A43 (1978) 172
18. H. Arenhövel, Z. Phys. A297 (1980) 129
19. M. Weyrauch, diploma thesis, Mainz, 1980
20. B. Ziegler, invited talk at this workshop

PION PRODUCTION OFF LIGHT NUCLEI

J.M. Laget

DPh-N/HE, CEN Saclay, 91191 Gif-sur-Yvette Cedex, France

ABSTRACT

The analysis of pion photoproduction reactions induced on deuterium is briefly summarized : the deviations from the multiple scattering series are a possible way to study the NΔ interaction. These two-body matrix elements are used to analyse the inclusive (total absorption of real or virtual photons) or one-arm (particle spectra) reactions induced on light nuclei. It is shown that the incoming (real or virtual) photon interacts mainly with the one-body and the two-body currents alone.

I do not have enough time to make a complete review of pion photo- or electroproduction reactions off light nuclei, and I will rather attempt to summarize what kind of progress have been made during the last year (since the Vancouver Conference[1])). I have selected two topics.

On the one hand the analysis of the pion photoproduction reactions induced on the two-nucleon system has reached a great degree of sophistication. However, significant deviations, from the multiple scattering series, still remain. A controversy on their nature has started at the Vancouver Conference, and I will try to show what is the outcome of the understanding of their nature : are they a signature of possible dibaryonic states, or are they the way to study the more general problem of the short and intermediate range part of the NΔ system? I will only recall the main points and refer to a recent review article[2]) where a more throughout discussion is given.

On the other hand, I will show how these two-body matrix elements, which have led to a good description of real[3]) or virtual[4]) pion photoproduction reactions on deuterium, can be used to analyse and understand the pion photo- and electro-production reactions induced on light nuclei. These reactions tell us how the pions, which have been created in the very inner part of nucleus, propagate before escaping or being definitively reabsorbed.

I. THE TWO NUCLEON SYSTEM

In the $\Delta(1236)$ region, the final state interaction effects play an important role and, although more experimental data are needed, the main mechanisms are understood. The exchange current corrections to the $\gamma N \rightarrow N\pi$ amplitude are directly linked to the two pion photoproduction reactions at one nucleon. In deuterium, the pion multiple scattering series converges quickly, provided its first few terms are correctly computed (Fermi motion effects!) : this is an expansion in terms of the dominant nearby singularities. The beam qualities (duty cycle, energy and intensity) of the Saclay Linac have made it possible to choose the relevant parts of the three body phase space, where each mechanism dominates the cross section, and to study it in detail. A good example is given in Fig. 1, where the ratio of the experimental yield of the $D(\gamma,pp)\pi^-$ reaction to the quasi-free yield, which is strongly reduced by the choice of the high values of the momenta of the emitted nucleons, is plotted[5]). The (logarithmic) singularity of the (on-shell) pion nucleon scattering diagram is responsible for the violent variation of this angular distribution, whereas the meson exchange current corrections to the $\gamma N\Delta$ vertex[6]) have to be considered to reproduce the large angle (or large energy) data. I note, incidentally, that the maximum available energy of the present generation of electron linacs is not high enough to map completely the kinematical domain, where these meson exchange current corrections are dominant. Obviously, a 1 GeV, 100 % duty cycle machine would be the best tool to complete these kinds of experiments.

The long range and the intermediate (or shorter) range part of the NΔ interaction have also been disentangled. The long range part reduces to the pion multiple scat-

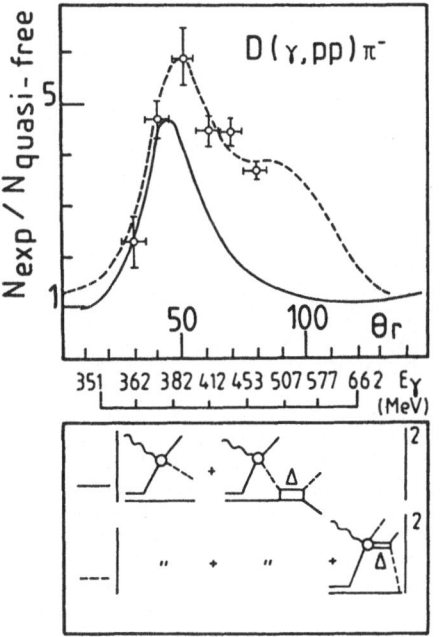

Fig. 1 - The angular distribution of the nucleon emitted in the $D(\gamma,pp)\pi^-$ reaction with a constant momentum $p_r = 400$ MeV/c when $Q = 1200$ MeV and $\omega = 90°$. The ratio between the experimental yield[5]) and the yield which should have been obtained if only one nucleon were active is plotted. The full line curve includes the pion-nucleon rescattering contribution, and the dotted curve also includes the meson exchange current corrections to the single pion photoproduction amplitude.

tering and is rather a "Δ regeneration mechanism". The intermediate part looks like the interaction between two stable particles and can be treated with very similar methods provided that the Δ width is properly taken into account. It might be responsible for the deviations from the multiple scattering series which have been observed in the study of the $\gamma D \rightarrow \pi NN$ reactions. A good example is given in Fig. 2 which shows selected angular distributions of the proton emitted with a constant momentum around $p_2 = 150$ MeV/c, when the invariant mass Q of the pair built with the pion and the other nucleon is kept constant (the quasi-free contribution is constant). The relative deviations from the quasi-free mechanism are plotted and are well accounted for by the multiple scattering of the Δ constituent (nucleon or pion).

The excess of the cross section which appears near the NΔ threshold (Fig. 2b) is well accounted for when the NΔ *interaction* is also taken into account. Using the K-matrix representation of the NN↔NΔ coupled channels, I have determined this NΔ interaction by fitting the real part of the $^1D_2$(NN) and $^3F_3$(NN) phase-shifts [ref.[7,8])] which exhibit the largest inelasticity. It is remarkable that I am able to reproduce, at the same time, the excess of the $\gamma D \rightarrow pp\pi^-$ reaction cross section near the NΔ threshold (Fig. 2b), and the resonant behaviour

Fig. 2 - The effects of the NΔ interaction on the $D(\gamma,p\pi^-)$ reaction yield (dot-dashed line curves) for selected kinematical conditions. The full line curves represent the multiple scattering contribution. See ref.[2]) for details.

of the $^1D_2$(NN) phase shift (Fig. 3) and the $^3F_3$(NN) phase shift (see ref.[2])), with *constant K-matrix elements* which are very close

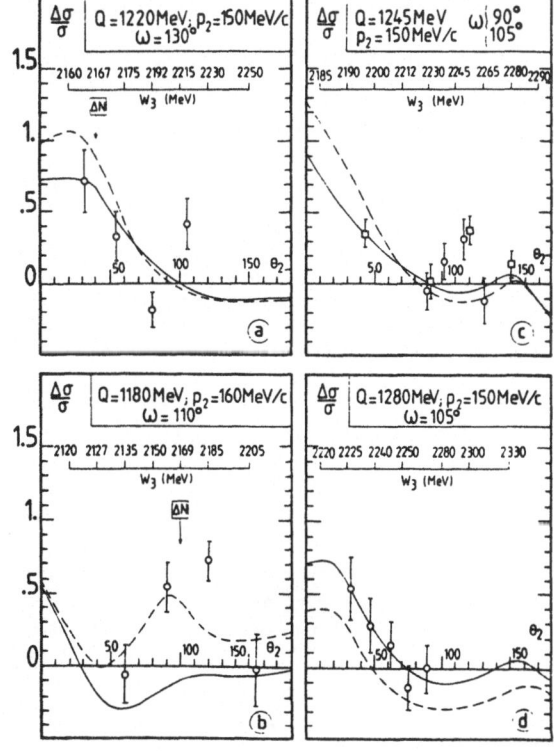

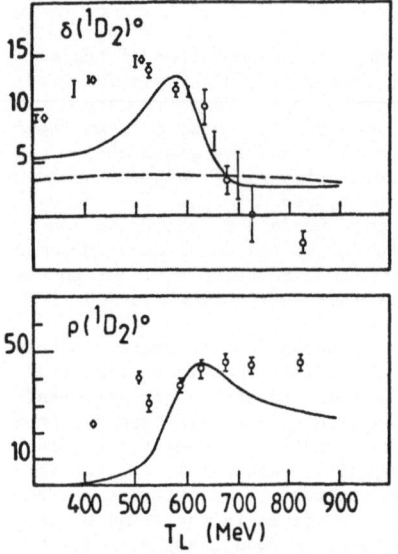

to the two baryon OBE amplitudes : the opening of the NΔ channel induces strong variations with energy in the coupled channels (threshold effects). It is also remarkable that I was not able to reproduce the excess of the $\gamma D \to \pi^- pp$ reaction by using a resonant K-matrix parametrization of the coupled NN$\leftrightarrow$NΔ channels : I refer the reader to ref.[2] for a more detailed discussion.

Fig. 3 - *The constant K-matrix fit to the $^1D_2$(NN) phase shift. The real δ and the imaginary parts ρ are defined by $2iq_1 T_{11} = - [\cos\rho \exp(2i\delta)-1]$. The broken line curve is obtained when no coupling between the two channels is allowed.*

Although I am not yet able to reproduce the excess of the cross section which appears (Fig. 2c) near W_3 = 2250 MeV, I think that it is very likely that the effects of the coupling between the NN and NΔ channels are responsible for the deviations from the multiple scattering series, which have been seen in the $\gamma D \to pp\pi^-$ reaction cross section. However, before making any definite statement, we must wait for the full solution of this coupled channel problem, starting with the OBE driving terms and including all the partial waves (I have only taken into account the coupling of the NΔ channel to the $^1D_2$ and $^3F_3$ (NN)waves).

These deviations from the multiple scattering series, have also been considered as a hint for possible underlying dibaryonic resonances. The problem is still open. Since G. Tamas discusses this topic elsewhere in this issue[9]), and since I have also considered it in ref.[2]), let me put the emphasis on the two following points.

On the one hand, the NΔ system is an example of the more general baryon-baryon system, the interaction of which is schematically depicted in Fig. 4. Since, in general, these baryons are not stable, they couple to the continuum, and their constituents can propagate freely between them : this is the *regeneration mechanism* which accounts for the *very long range part* of their interaction. The *intermediate range* and *long range* part is presumably driven by the OBE amplitude, and should be treated as a coupled channel problem. The *short range* part is still unknown and its description may require, in a way or another, to consider the quark degrees of freedom of the baryons. Before being able to say something sensible, about this short range part, from the analysis of the experiments, I really think that we must understand very well the intermediate range part, and decide to what extent it reproduces the violent structure which has been already seen. This is a challenge for theorists.

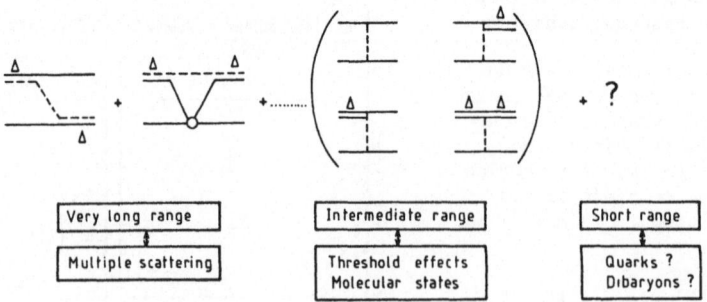

Fig. 4 - *The schematic decomposition of the NΔ amplitude into its very long range, its intermediate and its short range part.*

On the other hand, the $N\Delta \to N\Delta$ elastic scattering amplitude is as important as the $NN \to NN$ elastic or the $NN \to N\Delta$ inelastic scattering amplitudes. The analysis of the $\gamma D \to pp\pi^-$ reaction, which has been extensively studied at Saclay, is the best way to reach it. However the present generation of electron linacs does not allow us to improve significantly the quality of the data. A significant increase of the duty cycle would be a major improvement, which would make these coincidence experiments more easy. A considerable experimental effort must be undertaken, in order to map out the full three-body phase space, and to be able to perform a partial wave analysis : it is the only way to know which partial wave deviates significantly from the multiple scattering expansion.

The use of virtual photon instead of real photon would also allow us to complement the study of the $D(\gamma,p\pi^-)$ reaction by the study of the $D(e,e'p\pi^-)$ reaction, and to measure the electromagnetic form factors associated with each dominant mechanism, and more particularly to probe the short range part of the $N\Delta$ interaction : I refer the reader to refs.[10,11] where the emphasis is put on this point.

Nevertheless, I recall that the one-body and two-body matrix elements in this multiple scattering series are well under control. Let us see now how to use them as building blocks to compute the cross section of the reactions induced by a photon on heavier nuclei.

II. LIGHT NUCLEI

II.1. One-arm experiments

Fig. 5 shows schematically what may happen to a pion which is created inside a nucleus.

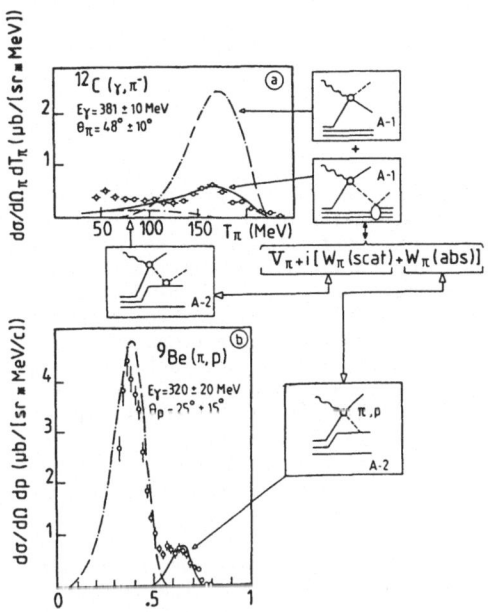

Fig. 5 - *The spectra of the pions emitted in the $^{12}C(\gamma,\pi^-)$ reaction[13], and the proton emitted in the $^9Be(\gamma,p)$ reaction [ref.[14]]. The meaning of the curves is explained in the insets, as well as the connection between the different parts of the spectra and the pion optical potential.*

When it escapes the nucleus and is emitted at a given angle (part a), its momentum spectrum is very different from the spectrum of the pions emitted on a quasi-free nucleon. Since the pion energy is close to the $\Delta(1236)$ resonance energy, it interacts strongly with the A-1 nucleons : when the pion distorted wave function (which is computed, according to ref.[12], in the semi-classical approximation) is used instead of a plane-wave, the quasi-free contribution is reduced by a factor four and comes close to the experimental data[13]). This is the most economical way to couple the elastic pion scattering channel to the channels which break up the A-1 residual nucleus : the quasi elastic pion scattering on the A-1 nucleons (which leads to the optical potential $V_\pi + i\, W_\pi(\text{scatt.})$) and the pion absorption by the A-1 nucleons (which leads to the potential $i\, W_\pi(\text{abs.})$). Of course these pions, which have disappeared from the pion elastic scattering channel, appear elsewhere in the phase-space. The inelastically scattered pions have lost energy and fill in the low energy part of the measured pion spectrum which exhibits a significant excess of cross section. I have computed the corresponding diagrams using the two-body matrix elements of the $\gamma D \to pp\pi^-$ reaction cross section[3]) : I have only changed the two-nucleon wave function and used an harmonic oscillator wave-function, which reproduces the single particle properties of ^{12}C (binding energies, radii and spectral functions). The Fermi motion and the binding energy of the active two-nucleon pair have also been taken into account. This calculation is very similar to the analysis of the $^3He(\gamma,\pi^\pm)$ reactions which I have reported elsewhere (see Fig. 47 in ref.[2])). I have also taken into account of the rescattering of the pion on the A-2 nucleons. It reduces the two-nucleon cross section by only a factor two (instead of four at the quasi-free peak) because a 100 MeV pion suffers less scatterings than a 200 MeV pion. Again those scattered pions lose energy and, as above, should fill in the lower energy part of the pion spectra. Although I have not yet computed these diagrams which involve three active nucleons, I think that the pion spectra can be understood as the incoherent sum of mechanisms which involve few nucleons in the nucleus : after the second scattering the pion loses enough energy to escape the nucleus without suffering any interaction (no more than 20 % of the pions are lost below T_π = 50 MeV).

The pions which have been absorbed in the nucleus do not escape but are responsible for the peak which appear in Fig. 5b, at the high energy part of the spectrum of the protons emitted at a given angle[13,14]). Its maximum corresponds to the two-body photodisintegration of a proton-neutron pair at rest inside the nucleus, and its width in due to the Fermi motion of the center of mass of the pair. It follows the two-body kinematics of the $\gamma + (np) \to pn$ reaction, when the proton angle and the incoming photon energy are varied.

The peak which appears at lower energies corresponds to the recoil proton associated with the quasi-free pions.

In fact the effective phase space for creating a pion at a nucleon is increased, since, besides real pions, virtual pions, which remain inside the nucleus, can also be created. This is really a meson exchange contribution, and I have used an improved version of the old "quasi-deuteron" model[15]) :

$$\sigma = L \frac{NZ}{A} \sigma_D^{ex} \qquad (1)$$

which tells us that the cross section of the $A(\gamma,pn)B$ reaction is proportional to the number NZ of neutron-proton pairs, and to the cross section of the $\gamma D \to pn$ reaction. It is assumed that the shape of the two-nucleon wave function is the same as the real deuteron wave function, and the scaling coefficient L/A is nothing but the ratio of their norms : the density of a nucleus is higher than the density of the deuterium. Contrary to the classical version of the model, I use only the (π+ρ) meson exchange part of the deuteron photodisintegration cross section which I have computed in ref.[4]), and which is shown in Fig. 6. The value L=10 of the Levinger factor has been obtained by fitting, below the pion threshold the total photoabsorption cross section on Li and Be isotopes (see Fig. 11) measured at Mainz[16]). As P. Carlos will show you[17]), the same model gives also a good accounting for the total photoabsorption cross section on heavy nuclei recently measured at Saclay. If the experimental values of the deuteron photodisintegration cross section had been used, instead of σ_D^{ex}, the fit would have been catastrophic between E_γ = 50 and 140 MeV.

Fig. 7 shows the excitation function of the area under the quasi-deuterons peak which appears, in Fig. 5b, at the high energy part of the spectra of the protons emitted at θ_p = 25° in the $^9Be(\gamma,p)$ reactions. The curve is obtained assuming that the Levinger factor remains constant between E_γ = 50 MeV and 400 MeV at the value L=10. These model underestimates the cross section below E_γ = 300 MeV and overestimates it above (I note incidentally that all the curves which are shown in ref.[10]) have been computed with L=10). However when the energy of the incoming photon increases the part of the wave function, which the quasi-deuton matrix element is sensitive to, is not the same. I refer the reader to ref.[4]), especially Fig. 11, where

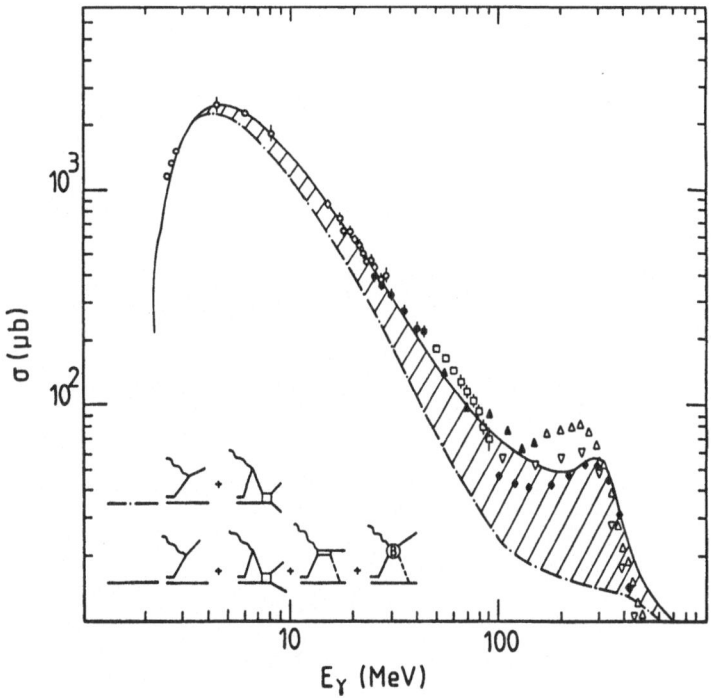

Fig. 6 - The two-body photodisintegration of the deuteron (see ref.[4]). The meson exchange current contribution σ_D^{ex} corresponds to the hatched area.

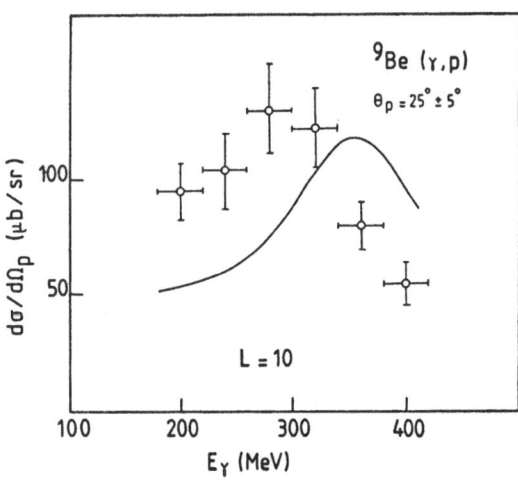

Fig. 7 - The area integrated under the quasi-deuton peak observed at $\theta_p = 25° \pm 5°$ in the $^9Be(\gamma,p)$ reaction[14]. The curve is the meson exchange contribution when L=10.

this point is discussed at length. Since the relative wave function of the two-nucleons inside the nucleus is not exactly the same as the deuteron wavefunction (it is expected to have the same shape only at short distances), the factor L should vary with energy. To what extent is still an open question. Of course the best way to answer is to compute the cross section of the $A(\gamma,pn)$ reaction cross section, starting from a correlated two-nucleon wave-function, which, unfortunately, is not well know. Presumably the $^3He(\gamma,pn)$ or $^4He(\gamma,pn)$ reactions would be the best place to check this two-nucleon mechanism : an exact calculation can be performed, since the wave functions are well known, even at short distances. However a measurement of these reaction cross sections is badly needed.

Nevertheless, the existing (γ,p) measurements can be used to determine the experimental values of the factor L. In Fig. 8, I have plotted the ratio of the area under the quasi-deuteron peak, which has been measured in the study of the $^9Be(\gamma,p)$

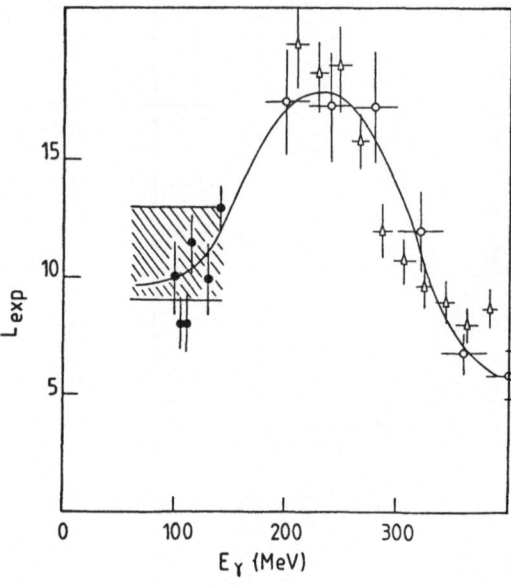

Fig. 8 - *The experimental value of the Levinger factor deduced from total photoabsorption measurement performed at Mainz (solid circles[16]) and Saclay (hatched area[17]) and from the analysis of the $^9Be(\gamma,p)$ reaction (open circles[14]) and the $^{12}C(\gamma,p)$ reactions (open triangles[13]). The curve is a mean value of L_{exp} which is used to compute the quasi-deuton cross section in Figs. 11 and 12.*

reaction[14]) and the $^{12}C(\gamma,p)$ reaction [ref.[13]], and the quantity NZ/A σ_D^{ex}. It is remarkable that these two experiments, which have been performed by two independent groups, using different methods (tagged photons[13]) and photon difference method[14]), lead to the same energy dependence of L_{exp}. I have also plotted the values of L_{exp} which come out the analysis of the total absorption cross section on light nuclei[16]) and heavy nuclei[17]) below the pion threshold.

The curve is a mean value which I will use to compute the integrated cross section of the $A(\gamma,pn)$ reaction cross section in Figs. 11 and 12.

II.2. Integrated cross section

Fig. 9 shows the integrated cross section of the $^{12}C(\gamma,\pi^+)$ reaction measured at Bonn[13]). It exhibits a different shape than the free nucleon cross section. This is due to two trivial effects. On the one hand, the Fermi motion of the nucleon, which absorbs the incoming photon, reduces the height and broadenes the free nucleon cross section. On the other hand, the binding of the target nucleon shifts the cross section towards higher energies : besides the emitted pion and proton energies, the incoming photon has to provide the target nucleon with its binding energy ($\sim$ 17.5 MeV in the p-shell and $\sim$ 38.5 MeV in the s-shell) to make it free. The losses of the pions flux, due to true absorption (which I have computed, in a semi classical way[12]), assuming that a part of the pions emitted at a nucleon are reabsorbed by a correlated pair among the A-1 nucleons, and retaining only

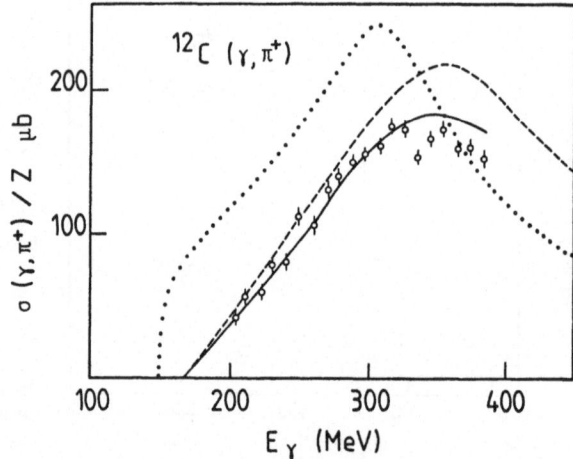

Fig. 9 - *The integrated cross section of the $^{12}C(\gamma,\pi^+)$ reaction as measured at Bonn[13]. Dotted line curve : free nucleon cross section. Broken line curve : quasi-free pion production (including Fermi motion and binding effects). The full line curve includes also the correction for "true absorption" of the pion (see text).*

the true absorptive part i W_π(abs.) of the optical potential) reduces the quasi-free cross section by less than 15 % at the Δ(1236) peak, but does not affect very much the cross section at lower energies.

This result may appear surprising, since the outgoing pion is known to suffer strong final state interactions (Fig. 5a). The explanation is given in Fig. 10. When the optical wave-functions of the pion and the proton, emitted in the quasi-free mechanism, are used instead of plane-waves, the quasi-free contribution is strongly reduced : this DWIA treatment is fully described in ref.[12]. However, I am dealing with an inclusive cross section and I have to add the contribution of the mechanisms which contribute to the losses of the flux of the particles which are created at one nucleon : the inelastic scatterings of the proton or the pion which break up the A-1 nucleons. As above (Fig. 5a), I have computed their contribution using the two-body matrix elements which are given in ref.[3]), an harmonic oscillator wave function and counting the number of active nucleon-nucleon pairs in ^{12}C. The sum of the contribution of these two channels is very close to the experimental data and to the quasi-free contribution, when it is corrected for true pion absorption effects (Fig. 9).

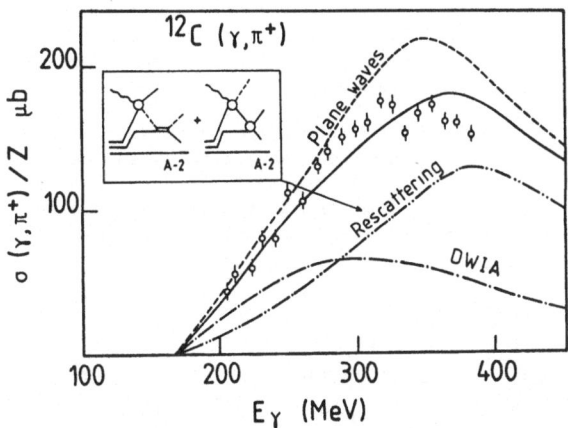

Fig. 10 - The contribution to the $^{12}C(\gamma,\pi^+)$ integrated cross section of the one body (dot dashed line curve) and the two body mechanisms (double dot dashed line curve). Their sum is the full line curve. The quasi-free cross section is also shown (broken line curve).

The lesson is that the incoming photon sees only the one nucleon current, and does not know what happens to the pions which escape the nucleus. This is a consequence of unitarity. In other words, the pion or the proton propagate nearly on shell (see ref.[3])) far away from the target nucleon and escape the interaction volume of the incoming photon, before suffering a scattering.

The quasi-free pion photoproduction contribution alone does not reproduce the total photo-absorption cross section on ^{9}Be [ref.[16]], which is depicted in Fig. 11, but the meson exchange current contribution (computed in the quasi-deuton approximation) accounts very well for the excess of the measured cross section. Contrary to the results I have reported in ref.[10]), I have corrected for true pion absorption the pion quasi-free photoproduction contribution and I have used the experimental value* (Fig. 8) of the Levinger factor instead of the constant value L=10.

---

*In the analysis of the (γ,p) reaction presented in Fig. 8, I have not taken into account of the proton rescattering effects. At T_p = 200 MeV, only 20 % to 30 % of the protons are lost at the quasi-deuton peak, and the experimental values of the Levinger factor L_{exp} might be larger by 20 to 30 %. This leads to an improvement of the fit of the total photoabsorption measurement.

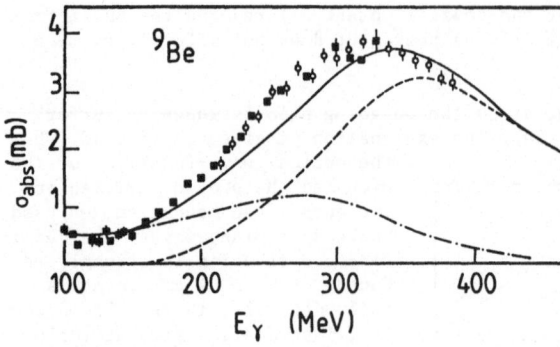

Fig. 11 - The total photoabsorption cross section for ^{9}Be as reported in ref.^{16}) (full squares) and in ref.^{13}) (open circles). Broken line curve : quasi-free pion production (including Fermi motion and binding effects). Dot-dashed line curve : meson exchange current contribution.

Here the pion is virtual and must be reabsorbed by another nucleon within the interaction volume of the incoming photon. Besides the one nucleon current the photon is also sensitive of the two-nucleon current.

This model gives also a good accounting for the "deep inelastic" electron scattering cross section (Fig. 12), which is nothing but the total absorption cross section for a virtual photon. In addition to the quasi-free pion production and the exchange current contributions, the quasi-elastic electron scattering mechanism should also be considered. The agreement is very good and it is not necessary (within the theoretical and experimental uncertainties) to introduce possible exotic pion-nucleus state in order to reproduce the data : if there were a collective state, it should represent less than 10 % of the measured cross section.

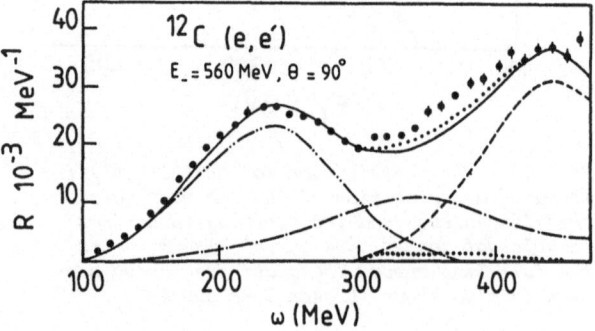

Fig. 12 - The spectrum of the electron inelastically scattered at $\theta = 90°$ by ^{12}C when the incoming electron energy is $E = 560$ MeV. The energy loss ω is plotted on abscissa. The double dot-dashed line curve corresponds to the electron quasi-elastic scattering. The broken line curve corresponds to the quasi-free pion production corrected for true absorption of the pion. The dot dashed line curve corresponds the meson exchange contribution. The dotted line curves show the influence of the longitudinal response function. The data come from ref.^{18}).

III. CONCLUSION

Contrary to pion induced reactions, photon induced reactions allow a clean separation between the entrance and the outgoing channels.

The inclusive cross sections tell us how the photon in absorbed and I have tried to show that, very likely, it sees only the one-nucleon and two-nucleon current inside the nucleus.

The exclusive cross sections tell us how the photon energy is distributed, and allow the study of the history of a pion created in the *very inner region* of the nucleus.

Two questions are open? On the one hand, it would be interesting to know how many times the pion scatters before escaping the nucleus (presumably a few). On the second hand, the energy variation of the Levinger factor L_{exp}, in our phenomenological description of the meson exchange current corrections, should be understood :

is it due only to differences between the shape of the correlated two-nucleon wave function and the deuteron wave function, or is it due to a mechanism which involves degrees of freedom beyond the OBE mechanisms?

The most important mechanims of the photon induced reactions on the two-nucleon system are well under control, and can be safely used as building blocks to analyse the reactions induced on light nuclei. However one question is still open. Are the deviations from the multiple scattering series, which have been reported in the study of the $\gamma D \rightarrow \pi pp$ reaction, a way to reach the intermediate range part of the $N\Delta$ interaction, or a signature of a more genuine short range effect, which, in a way or another, would require to consider also the quark degrees of freedom of the nucleon?

All these questions have to do with the short range part of the two body current inside a nucleus, and I believe that the study of the exclusive reactions, induced by virtual and real photons on few-body targets and light nuclei, will help us to improve our understanding of this important part of the two-baryon interaction.

REFERENCE

[1] J.M. Laget, Nucl. Phys. A335 (1980) 267.
[2] J.M. Laget, Phys. Rep., in press.
[3] J.M. Laget, Nucl. Phys. A296 (1978) 388.
[4] J.M. Laget, Nucl. Phys. A312 (1978) 265.
[5] P.E. Argan et al., Phys. Rev. Lett. 41 (1978) 86.
[6] J.M. Laget, Phys. Rev. Lett. 41 (1978) 89.
[7] R. Bryan, Proc. 8th Intern. Conf. on few-body systems and nuclear forces, Graz, 1978, eds. H. Zingl et al., vol.II (Springer Verlag, Berlin, 1978).
[8] I.P. Auer, Nucl. Phys. A335 (1980) 193.
[9] G. Tamas, This conference.
[10] J.M. Laget, From real to virtual photons, symposium on perspectives in electro and photonuclear physics, Saclay, 1980, eds. A. Gérard et al., Nucl. Phys., in press.
[11] J.M. Laget, The $N\Delta$ interaction, ibid.
[12] J.M. Laget, Nucl. Phys. A194 (1972) 81.
[13] B.A. Mecking, in "Nuclear physics with electromagnetic interaction", Springer Verlag (1979), Lecture Notes in Physics 108.
[14] S. Homma et al., Phys. Rev. Lett. 45 (1980) 706.
[15] J.S. Levinger, Phys. Rev. 84 (1951) 43.
[16] J. Arhens et al., Nucl. Phys. A335 (1980) 67.
[17] P. Carlos, This conference.
[18] J. Morgenstern, in ref.[10]), and private communication.

SCALING LAWS IN HIGH ENERGY ELECTRON-NUCLEAR PROCESSES

Marc CHEMTOB
DPh-T, CEN-Saclay
B.P.N°2, 91191 Gif-sur-Yvette
Cedex - France

ABSTRACT : We survey the parton model description of high momentum transfer electron scattering processes with nuclei. We discuss both nucleon and quark parton models and confront the patterns of scaling laws violations, induced by binding effects, in the former, and perturbative QCD effects, in the latter.

1. INTRODUCTION

Whether or not we have a transition from nucleons to quarks in nuclei is an issue which is closely linked to the understanding of the quark-hadron transition. As we progress in our understanding of quark binding, it seems as if the idea of the quark compositeness of nuclei becomes more and more natural. This situation is not unique to nuclear physics. One even currently hears talk of a quark substructure of the photon !

Three important qualitative elements justify optimism. These are : (i) The association of the quark-hadron transition with a sort of phase transition (either continuous or sharp) from a weak coupling, perturbative phase at short distances (0.1 fm) to a strong coupling, non-perturbative phase at large distances (1 fm) ; (ii) The insensitivity of the perturbative phase to confinement effects, so long as one deals with color singlet observables immune to the non-trivial infrared divergences (not removable by standard renormalization or resummation methods) ; (iii) The linkage of the short and large distance behavior in terms of multiparticle wave functions.

The soundest basis for the quark compositeness of hadrons is perhaps the partonic basis, which includes high-energy reaction processes, notably, lepton-induced deep inelastic and exclusive large momentum transfers processes. For nuclear systems too, it is very reasonable to expect the partonic basis, as opposed to the spectroscopic, to be the best suited one for exposing the nuclear quark compositeness. What one means by this is often qualified in words by talking of a continuity between nucleon and quark dynamics, of a transition in the time sequence of reaction processes from nucleon-meson to quark-gluon subprocesses, or also of point-like coupling and substructure of nuclei. The argument goes saying that by increasing the momentum transfer (squared) Q^2 and enough patience, one should be able to reach the quark stratum. Now, the basic question at this point is, precisely, how patient should one be, and to what extent this does not require an incredibly high sensitivity.

Few-body nuclei offer the most favourable test cases. The measurements of high-energy electron scattering processes carried out in the last few years, and the theoretical works they stimulated, justify some optimism. We shall continue here the discussion started in a recent talk by us[1]. Let us state briefly our standpoint : Nuclear processes with $Q^2 \geq 1$ (GeV/c)^2 expose new aspects of nuclear structure in relativistic regimes, for the analysis of which an indispensable tool is the light-cone, Infinite-Momentum-Frame (IMF), parton model framework. This approach informs us that there should exist an interval of transfers, typically $Q^2 \lesssim 4$(GeV/c)^2, where a nucleon constituent description retains its validity. The idea of a quark-nucleon transition in nuclei anticipates on the smooth joining at the higher Q^2 of this description to a quark-parton description.

Useful reference properties of high-energy processes, in general, are those properties based on the scaling notion. This is the notion that as Q^2 exceeds all mass scales in the problem (rest masses, binding energies), then : (i) The theory becomes equivalent to a scale-invariant theory where all dimensionful parameters are absent, (ii) The equivalence is expressed in terms of asymptotic expansions for the observables.

Our object in this talk will be : (i) To survey some of the recent progress in understanding the underlying basis of the scaling laws for nuclear systems and (ii) To discuss and confront the patterns of violations of these scaling laws based on predictions of the quark and nucleon constituents descriptions, respectively.

2. NUCLEAR STRUCTURE ON THE LIGHT-CONE

The light-cone formalism offers a useful tool which meets with our basic needs : A systematic, tractable relativistic framework in which one can switch smoothly from a nucleon to a quark constituent picture. Let us introduce very briefly some general elements. In the light-cone formalism, one works with kinematic variables defined as the space-time rotated (longitudinal and transverse) momenta $p^+ = (p^0 + p^3)$, $\vec{p}_T$ so that the energy variable is given by $p^- = (p^0 - p^3) = (A^2 + \vec{p}_T^2)/p^+$, as required by the on-shell condition $p_\mu p^\mu = p^2 = A^2$. A bound system A (momentum p, rest mass A) is described by a wave function $\psi(x_i, \vec{k}_{Ti})$, depending on the A constituents light-cone momentum variables defined as $k_i^+ \equiv x_i p^+$, $\vec{k}_{Ti}$. This description is, in principle, valid for the ground state as well as for the excited discrete and continuum states. The theoretical basis of the light-cone approach rests on two properties : (i) The identification of $\psi(x_i, \vec{k}_{Ti})$ with the positive-energy components of the multiparticle Bethe-Salpeter amplitude where all the fields light-cone time coordinates $\tau_i \equiv (x_i^0 - x_i^3)$ (i=1,...,A) are set equal. (This restriction is frame-dependent, i.e., it depends on the total system's momentum p^+. Alternatively, p^+ can

be regarded as a parameter whose value defines a class of frames. In the IMF, $p^+ \to \infty$; then $x_i^0 - x_i^3 \to x_i^0$ and the identification reduces to the equal-time restriction).
(ii) The association to the field-theory description of $\psi(x_i, \vec{k}_{Ti})$ of a time-ordered perturbation theory (the so-called TOPT∞), having the property that field fluctuation (vacuum polarization and pair creation) effects are removed away.

The discrete states wave functions $\psi(x_i, \vec{k}_{Ti})$ are normalizable functions obeying approximate equations of the form :

$$\left(A^2 - \sum_{i=1}^{A} \frac{\vec{k}_{Ti}^2 + m_i^2}{x_i} \right) \psi(x_i, \vec{k}_{Ti}) = \iint [dy][d\ell_T] V(y_i, \vec{\ell}_{Ti}; x_i, \vec{k}_{Ti}) \psi(y_i, \vec{\ell}_{Ti}) \quad , \quad (1)$$

where :

$$[dy] = \prod_i dy_i \, \delta\left(1 - \sum_i y_i\right) , \quad [d\ell_T] = \prod_i \left(\frac{d^2\vec{\ell}_{Ti}}{2(2\pi)^3}\right) 2(2\pi)^3 \, \delta^{(2)}\left(\sum_{i=1}^{A} \vec{\ell}_{Ti}\right) ,$$

and V is a connected irreducible scattering amplitude for 1+...+A → 1+...+A. Note the analogy with Schrödinger equation, following from the interpretation of $(\vec{k}_{Ti}^2 + m_i^2)/x_i$ as single-particle energies. One meets sometimes alternative notations for the wave functions and Eq.(1) with additional explicit factors x_i. Such factors, however, can always be absorbed into a redefinition of the potential V. The bound state Eq.(1) does not determine by itself the normalization condition on ψ. In fact, there is no unique criterion to normalize ψ. One can invoke a probabilistic parton-model interpretation bearing on conserved quantum numbers. Alternatively, one can use the correspondence with the Bethe-Salpeter amplitude Ψ, for which a normalization condition does exist by virtue of its definition as an ordered product of fields. In the approximation of an energy-independent kernel (no dependence of V on A^2), and assuming that the normalization factors $1/\sqrt{x_i}$ of constituents are factored out of Ψ (schematically, $\int dk^- \Psi(k) \sim \psi(x_i, \vec{k}_{Ti})/(\prod_i x_i)$), then one obtains the normalization condition $\int [dx][dk_T] |\psi(x_i, \vec{k}_{Ti})|^2 = 1$.

The basic process for probing the single-particle aspects of bound systems is deep inelastic scattering, defined by the kinematical region of the virtual photon four-momentum transfer q as $-q^2 \equiv Q^2 \to \infty$, $(q_0)_{lab} \equiv \nu \to \infty$, with fixed ratio $x_B = Q^2/2A\nu$ (Bjorken variable). Deep inelastic scattering does not resolve exactly the short space-time constituents coordinates ($x_i^\mu \to 0$), but rather the short distances relative to the light-cone, $x_i^\mu x_{i\mu} = 0$. The interest of the light-cone formalism is in validating an impulse approximation treatment (no final state interactions). The reason for this is caught from the intuitive picture that in the IMF the interaction and traversal times of the struck constituent are very short in comparison to the time it takes the system to change configurations. Thus, the information in deep inelastic lepton scattering bears on the single-constituent structure function

$$G_{a/A}(x, \vec{k}_T) = \iint [dx][dk_T] \, \delta(x_1 - x) \, \delta^{(2)}(\vec{k}_{T1} - \vec{k}_T) |\psi(x_i, k_{Ti})|^2 \quad , \quad (2)$$

so that one can write the e.m. inelastic spin-averaged structure tensor as,

$$W^A_{\mu\nu}(p,q) = \iint dx\, d^2\vec{k}_T\, G_{a/A}(x,\vec{k}_T)\, W^a_{\mu\nu}(k,q) \quad , \tag{3}$$

where $W^a_{\mu\nu}$ is the constituent a e.m. structure tensor.

Exclusive processes, such as form factors, require in constrast a more detailed knowledge of the wave functions and their phases. The high Q^2 region here is directly sensitive to the short space-time distances $x^\mu_i \to 0$. The schematic formula, relevant to the spinless case, in terms of which one calculates form factors can be written as :

$$F_A(Q^2) = \int \ldots \int [dx][dk_T][dy][d\ell_T]\, \psi(y_i,\vec{\ell}_{Ti})$$

$$T_H\,(y_i,\vec{\ell}_{Ti};x_j,\vec{k}_{Tj};\vec{q}_T)\, \psi(x_j,\vec{k}_{Tj}) \quad , \tag{4}$$

where T_H is an irreducible connected scattering amplitude for $\gamma^*(q)+(1+\ldots+A) \to (1+\ldots+A)$.

So far we have described the general formalism. We shall now discuss successively nucleon and quark parton models from the point of view of simple, semi-phenomenological descriptions.

3. NUCLEON PARTON MODELS

The familiar idea that nucleon-nucleon interactions are mediated by meson quantas exchanges offers a natural basis for progressing in the discussion of the light-cone nuclear dynamics. It sounds as a valuable, tractable program to solve for Eq.(1) with realistic kernels constructed in terms of one-and two-meson exchange terms. These kernels would clearly be defined in terms of the same coupling constants and mass parameters for which we can claim presently a reliable knowledge. Interesting approaches for the two-nucleon system are being initiated by Karmanov[2], Namyslowski et al.,[3] and possibly by other authors.

We shall nevertheless concentrate our attention here on an approach based on a phenomenological construction of wave functions, as this approach gives a useful insight and also turns out to be reasonably successful. Interestingly enough, the simplest phase space regions of the wave functions where one can use ones intuition are the two extreme regions of non-relativistic on-shell and relativistic, far off-shell configurations, respectively. The former region is defined as $x_i \simeq m_i/A$ and is governed for nuclear systems by non-relativistic dynamics. Two kinds of simple-minded arguments can be used to establish a correspondence between $\psi(x_i,\vec{k}_{Ti})$ and the rest frame wave functions $\psi_{NR}(\vec{q}_i)$. One can identify the total internal c.m. energies in the two frames, using $(\sum_i (\vec{q}_i^{\,2}+m_i^2)^{1/2})^2 = \sum_i (\vec{k}_{Ti}^{\,2}+m_i^2)/x_i$, subject to the definition $\vec{q}_{Ti} = \vec{k}_{Ti}$ and the constraints $\sum_i x_i = 1$, $\sum_i \vec{k}_{Ti} = 0$, $\sum_i q_{3i} = 0$. The correspondence is completely determined in this way only for the two-particle case. (For equal masses, $x = \frac{1}{2}(1+q_3/(\vec{q}^2+m^2)^{1/2})$, $q_3 = \frac{(1-2x)(\vec{q}_T^2+m^2)^{1/2}}{2\sqrt{x(1-x)}}$). An alternative complete correspondence in the general case is found by identifying the individual x_i in the two frames. This can be summarized as :

$$\psi(x_i, \vec{k}_{Ti}) = \psi_{NR}(\vec{q}_i) \cdot \left[x_i = \frac{(\vec{q}_i^2+m_i^2)^{1/2}+q_{3i}}{\sum_j (\vec{q}_j^2+m_j^2)^{1/2}}, \vec{k}_{Ti} = \vec{q}_{Ti} \right]. \quad (5)$$

It is possible in fact, as advocated by some authors, that such a construction remains reliable for large intervals of the x_i. On the other hand, a simple, independent picture of the far off-shell component of the wave functions can be caught by looking for the meson-exchange subprocesses linking the constituents. The basic observation is that since one deals with large energy denominators for intermediate states, the dominant subprocesses are those involving chain-like, optimal number of meson exchanges between nucleons. The core model[4] takes the extreme assumption that the singled-out active constituent a balances its momentum against a passive core formed by all other constituents $\alpha = (A-a)$. One is then led to a wave function of the schematic form $\psi_{a/A}(x,\vec{k}_T) \simeq [A^2-(\vec{k}_T^2+a^2)/x - (\vec{k}_T^2+\alpha^2)/(1-x)]^{-T(A-a)}$, where $T=1,2,...$ is a discrete power parameter characteristic of the binding interactions. In the few-nucleon correlated clusters model[5], one takes instead the less extreme assumption that the active constituent balances its momentum against only a few (one, two, ...) other constituents forming the passive core α.

The wave functions constructs suggested by the above arguments turn out to give a consistent and reasonably accurate description of a variety of high-energy processes[1]. Useful reference predictions are those obtained in the asymptotic limit $Q^2 \to \infty$. One can show that for bounded distributions in $\vec{k}_T$, the leading terms in the form factors and the threshold inelastic structure functions are both specified by the $x_i \to 1$ region. The predictions based on the above core model wave function are of the form :

$$F_A(Q^2) \to C \frac{F_a(Q^2)}{(Q^2)^{T(A-a)}} \log(Q^2/m_T^2),$$

$$\nu W_2^A(Q^2) \xrightarrow[x_B \to 1]{} C' F_a^2(Q^2) (1-x_B)^{2T(A-a)-1}. \quad (6)$$

Here m_T is a transverse momentum cut-off, $F_a(Q^2)$ the constituent a form factor and C, C' are constants specified by the wave function at short space-time distances.

4. QUARK PARTON MODELS

Nucleon and quark parton models have several formal features in common. One can transpose the above arguments with nearly no changes to the quark case. In fact, a description of the far off-shell region $x_i \to 1$ by minimally connected graphs is on a sounder basis here, owing to the asymptotic freedom property of QCD and the implication of nearly scale-free qq interactions with large transfers. The analogs of the asymptotic expansions (6) in the quark case are then the so-called dimensional scaling counting rules,

$$F_A(Q^2) \to C \left(\frac{1}{Q^2}\right)^{n-1}, \quad \nu W_2^A(Q^2, x_B) \xrightarrow[x_B \to 1]{} C'(1-x_B)^{2(n-1)-1}, \tag{7}$$

which can be obtained from Eq.(6) by setting T=3 and n=3A, n standing for the number of valence quarks in A. The QCD-inspired quark parton models incorporate however specific aspects regarding gluon radiation processes, which we briefly discuss now.

Until a few years ago, the basis for analyzing high Q^2 processes was Wilson short-distance operator product asymptotic expansions and the renormalization group equations. This method is not only a rather abstract one, but, as realized progressively, it is unduly restrictive. With the wave function approach, advocated most recently by Brodsky et al.[6,7], one can enlarge considerably the predictive power of QCD. This approach offers at least three advantages : (i) Intuitive insight, (ii) Equivalence to the operator product expansion method, and (iii) Simple linkage between the short and large distance behavior.

The implication of asymptotic freedom in QCD is that to make sense of perturbation theory expansions, one must organize improved expansions defined by resumming to all orders the repetitive set of graphs involving large log terms, which appear in the process of renormalizing the ultraviolet divergences. The mathematical tool for this task is the renormalization group. Consider the important example of the coupling constant. Here, the resummation of one-loop insertions for the qq g vertex in leading-log approximation leads to the effective, renormalized strong fine-structure constant $\alpha_s(Q^2) = 4\pi/(\beta \ln(Q^2/\Lambda^2))$, where $\beta = 11 - \frac{2}{3} N_f$ and N_f = number of relevant quark flavors at Q^2. The convergence criterion of perturbative QCD is thus determined by the mass scales Q measured in reference to the fundamental mass parameter Λ. (Standard assignments give $\Lambda \simeq 0.1 \sim 0.5$ GeV). For $\alpha_s(Q^2)$, the resummation of two-loop insertions, giving the next-to-leading log terms, results in correction terms of the form log log Q^2, which have an appreciably slower variation with Q^2. The intuitive physical implication of the slow log decrease of $\alpha_s(Q^2)$ with Q^2 is that there is no scale Q^2, whatever large it is, at which QCD tends to a permanently scale-invariant theory. This is the property of QCD from which originate the perturbative scaling violations.

The above reasoning is a general one, and can be extended to discuss perturbative scaling violations effects in wave functions. Wave function are not gauge-invariant quantities and, as has been realized in a variety of applications, a convenient class of gauges is the class of physical axial gauges, defined in terms of arbitrary four-vectors η by the constraint on the gluon field $A \cdot \eta = 0$. (The light-cone gauges are specified by $\eta^2 = 0$). Only in such gauges is it possible to reason in terms of valence quark wave functions decoupled from the sea of $\bar{q}$ and g components. The procedure leads one to define distribution amplitudes for quarks collinear (with the total system momentum) up to the resolving momentum Q^2,

$$\phi(x_i, Q^2) = \prod_i (d_F^i(Q^2))^{-1/2} \int^{k_{Ti}^2 < Q^2} [dk_T] \, \psi(x_i, \vec{k}_{Ti}) \quad . \tag{8}$$

The resummations bear on the set of gluon exchange ladders with minimal connectedness. (The explicit factors $d_F^i(Q^2)$ on the r.h.s. of Eq.(8) are the wave function normalization factors corresponding to the finite contributions obtained after renormalization of ultraviolet divergences in the quarks self-mass insertions). Each rung of the ladder brings in an ultraviolet divergent factor $\approx \int dk_T^2/k_T^2 \, \alpha_s(k_T^2)$, where the renormalized coupling constant factor results from the finite contributions remaining after renormalization of the various self-mass and vertex insertions. By assuming a strong ordering of the successive cells of the ladders ($Q^2 \gg \vec{k}_T^2 \gg \vec{j}_T^2 \ldots$, starting from the top, most off-shell cell), the individual cells contributions factorize and one can resum the repetitive log factors. The means to do the resummation can be formulated as a differential evolution equation for the distribution amplitude, reading

$$[\frac{\partial}{\partial \xi} + \frac{n}{2} C_F] \phi(x_i, Q^2) = \int [dy] \, \tilde{V}(x_i; y_i; Q^2) \, \phi(y_i, Q^2) \quad , \tag{9}$$

where $\xi = \frac{1}{\beta} \log \log \frac{Q^2}{\Lambda^2}$, $C_F = \frac{N_C^2 - 1}{2N_C} = 4/3$ (using $N_C = 3$) and $\tilde{V}$ is a calculable connected kernel. The boundary condition for Eq.(9) must be specified independently in terms of a starting, non-perturbative amplitude $\phi(x_i, Q_0^2)$ at an arbitrary, sufficiently large, reference scale Q_0^2. An interesting program, currently in progress by Brodsky et al.[8], consists in postulating simple phenomenological constructs for $\phi(x_i, Q_0^2)$ on the basis of intuitive arguments somewhat analogous to the ones discussed in the previous section.

One can however make simple predictions on a naive level about the nature of the perturbative scaling laws violations. For form factors, the minimal gluon propagator structure of T_H in eq.(4), leads one to the schematic asymptotic prediction

$$F_A(Q^2) \rightarrow (\alpha_s(Q^2)/Q^2)^{n-1} \tag{10}$$

modulo factors ($\ell n Q^2/\Lambda^2$) raised to powers calculable with the help of Eq.(9) in terms of anomalous dimension functions of short-distance expansions of multiple products of quark fields.

For the flavor non-singlet threshold structure functions, a more detailed analysis is possible, starting with the evolution equation for the constituents structure functions, which are defined by Eq.(2). One finds for the threshold inelastic structure function at large Q^2,

$$\nu W_2^A(Q^2,x_B) \xrightarrow[x_B \to 1]{} C'(1-x_B)^{2(n-1)-1}(1-x_B)^{\frac{4C_F}{\beta} \ln \left(\frac{\ln Q^2/\Lambda^2}{\ln Q_0^2/\Lambda^2} \right)} \left(\frac{\ln Q^2/\Lambda^2}{\ln Q_0^2/\Lambda^2} \right)^{3C_F/\beta}$$

$$\times \left(1 + \frac{2C_F}{\beta} \ln^2(1-x_B)/\ln(Q^2/\Lambda^2) + \ldots \right) . \qquad (11)$$

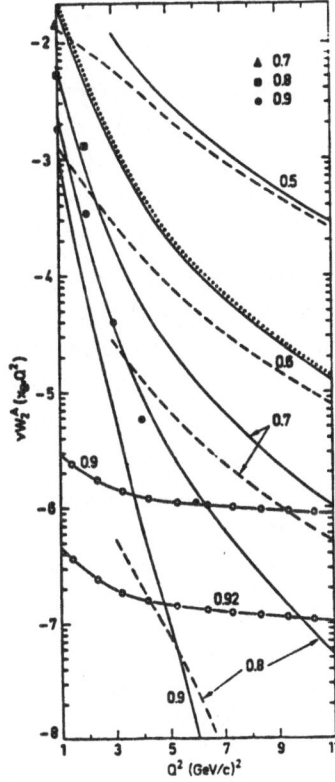 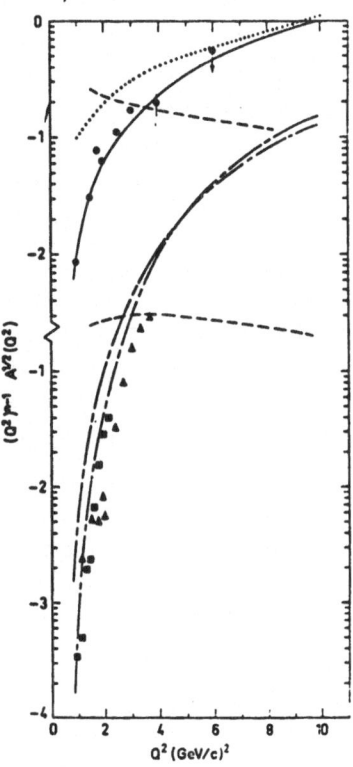

Fig.1 – Electron inelastic structure function $\nu W_2^A(Q^2,x_B)$ versus Q^2 for fixed x_B (full lines) and fixed $\xi = 2x_B / (1+(1+4A^2x_B^2/Q^2)^{1/2})$ (dashed lines). The values of x_B and ξ are as indicated. Data points for fixed x_B within ±0.02 of the central value are shown as triangles (x_B=0.7), squares (0.8) and circles (0.9). The nucleon parton model predictions are from ref.[9] (T=2,δ^2=0.8(GeV/c)^2). The predictions drawn as —·—·— are for the quark parton model, Eq.(11), using Λ=0.5GeV. The curve drawn as .... represents the square of the nucleon dipole form factor $(1+Q^2/0.71(GeV/c)^2)^{-2}$.

Fig.2 – Scalar form factors of d, ^{3}He and ^{4}He scaled by $(Q^2)^{n-1}$. Data points are shown as circles (d), triangles (^{3}He) and squares (^{4}He). The nucleon parton model predictions are from ref.[9] (T=2, δ^2=0.8(GeV/c)^2). The quark parton model predictions (dashed curves) are based on Eq.(10), using Λ=0.5GeV.

5. SCALING VIOLATIONS

In spite of the title chosen for this talk, our main emphasis in discussing some of the predictions will be on the scaling laws violations. We have no new results to report on with respect to ref.[1] and not much to add about the comparison with data. What we shall emphasize instead is an alternative suggestive representation of predictions which illustrates the patterns of scaling violations.

Let us first consider the predictions for the deuteron inelastic structure function W_2^A with the nucleon spin parton model of ref.[9]. The representation used in figure 1 allows us to visualize the variation with Q^2 at fixed scaling variable, noting that a non-constant dependence on Q^2 reflects upon the violation of Bjorken scaling. The violations are indeed very large. Clearly, this is not a surprise, since the nucleon partons are not point-like, but are instead affected form factors. Indeed, the bulk part of the non-constant Q^2 component comes from the factor $F_N^2(Q^2)$. One can appreciate on figure 1 that the removal of this factor should indeed slow down considerably the variation with Q^2 for $x_B \lesssim 0.7$, still leaving however residual Q^2-dependence for the larger x_B. In figure 1 we also illustrate the role of improved scaling variables (asymptotically equivalent to x_B) in speeding up the scaling behavior. The Nachtmann variable ξ considered here is approximately equivalent to the other currently used variable $\omega_N' = \frac{m^2}{Q^2} + \frac{1}{Ax_B}$. Improved scaling in terms of ξ is anticipated on the basis of the equality of ξ (up to small correction terms due to transverse motion) to the momentum fraction x of the struck constituent in the frame in which $\vec{q}_T = 0$. We see indeed that the use of fixed ξ gives slower Q^2 dependence than fixed x_B, but is not decisive. Thus the important conclusion here is that binding effects incorporated in wave functions are important ones, lasting to large Q^2 and increasing with x_B. On the other hand, the partial comparison in figure 1 wich data indicates that the simple model wave functions are reasonably satisfactory.

Let us now turn to the quark parton description and examine the predictions calculated on the basis of Eq.(11). The scaling violations are remarkably smooth. They would be exhanced with a larger Λ, but not excessively, owing to their log dependence. To quote definite predictions in figure 1, we have been led to determine the normalization constant C' in Eq.(11) by adjusting at the data point $Q^2 \simeq 6(\text{GeV}/c)^2$, $x_B \simeq 0.9$. (No readjustment of C' is done between the two values of x_B). Our goal of detecting a possible smooth transition with the nucleon parton predictions is clearly upset.

We arrive at similar conclusions with figure 2 by looking at the form factors in the nucleon spin parton model of ref.[9] and in the quark parton model, as given by Eq.(10), after multiplication by the dimensional scaling factor. (The latter naive QCD predictions are again normalized arbitrarily to the data points at $Q^2 \simeq 3(\text{GeV}/c)^2$). The patterns of scaling violations are again very different, with no clear continuity between the nucleon and quark descriptions. Note that the nuclear binding effects act to enhance the scaling limit, while the perturbative QCD corrections act instead to suppress it. The nucleon parton model predictions shown in figure 2 are from the spin model of ref.[9] with T=2. We note again the nice agreement with data. There is no reason for the asymptotic power law to be the same as that given by Eq.(6) for the spinless case. Indeed, as one appreciates on figure 2 (by reference to the curve Q^2 plotted close to the deuteron predictions, in order to see to what extent both curves stay parallel), there is an appreciable enhancement of the asymptotic spinless model prediction $(1/Q^2)^4$. The cause originates from both spin and binding effects. Similar properties are featured by ^{3}He and ^{4}He.

6. CONCLUSIONS

High-Q^2 processes on light nuclei offer promising perspectives for the observation of a nuclear quark substructure. What has been achieved so far is really a starting basis. Organized, systematic efforts are now needed : Experimentally, to provide us with still higher Q^2 data. Theoretically, to refine the parton models beyond the naive existing descriptions. The generally crude models which we have discussed are not sufficiently accurate to permit conclusive answers. It is now clear that binding corrections to scaling laws are significantly larger, and of entirely different nature, in comparison to the perturbative QCD scaling violation effects. Refining the treatment of meson binding effects in the nucleon parton description should be a tractable necessary task. On the other hand, the wave function approach in QCD, although it involves several difficult unsolved problems, seems more and more amenable to quantitative predictions. It would be interesting to see whether such refined quark and nucleon parton descriptions would indeed match and join at some Q^2 to one another smoothly. This question at least is not directly linked to the availability of more data.

REFERENCES

[1] M. Chemtob, Proc. Symposium on Perspectives in Electro-and Photo-Nuclear Physics, ed. A. Gérard and C. Samour, (Saclay, 1980).

[2] V.A. Karmanov, Inst. of Theor. and Exp. Physics, Preprint ITEP-8 (Moscow, 1980).

[3] P. Danielewicz and J.M. Namyslowski, Phys. Lett. 81B (1979) 110 ;
J.M. Namyslowski and H.J. Weber, Zeit. für Phys. A295 (1980) 219.

[4] R.A. Blankenbecler and I.A. Schmidt, Phys. Rev. D15 (1977) 3321.

[5] L.L. Frankfurt and M.I. Strikman, Yad. Fiz. 29 (1979) 490.

[6] S.J. Brodsky and G.P. Lepage, Proc. Summer Inst. on Part. Phys., ed.A. Moscher (Stanford, 1979).

[7] G.P. Lepage and S.J. Brodsky, Phys. Rev. D22 (1980) 2157.

[8] S.J. Brodsky, T. Huang and G.P. Lepage, Proc. XX th Int. Conf. on High Energy Physics (Madison, 1980)

[9] M. Chemtob, Nucl. Phys. A336 (1980) 299.

TOTAL PHOTONUCLEAR ABSORPTION CROSS SECTION MEASUREMENTS
BELOW THE PION PHOTOPRODUCTION THRESHOLD

P.J. Carlos

DPh-N/MF, CEN Saclay, BP 2, 91190 Gif-sur-Yvette, France

1. Introduction

Unlike the elementary nucleons which are transparent with respect to photons below k=140 MeV, the low energy photonuclear absorption spectrum of complex nuclei is dominated by the giant dipole resonance (GDR) in the region between 10 MeV and 30 MeV, depending on the nuclear mass number. This structure is to be associated with the response of the average nuclear potential to an excitation induced by a photon, and corresponds to a collective vibration mode whose characteristics have been studied over a large variety of mass numbers during the past 20 years[1]. Furthermore at high energies (2 GeV < k < 20 GeV) a large set of photon nucleus data is available[2]. Total photonuclear absorption cross sections have been measured for a large variety of nuclei (1 ≤ A ≤ 208) and interpreted in term of an effective number of nucleons A_{eff} participating in the photon nucleus interaction and defined as

$$\sigma_{TOTN}(k) = A_{eff}(k) \left[\frac{Z}{A} \sigma_{\gamma,proton}(k) + \frac{N}{A} \sigma_{\gamma,neutron}(k) \right] \quad (1)$$

where the sum into the brackets is an average over protons and neutrons for the photon-nucleon cross section. It has been found experimentally that A_{eff} is independent of k, and shows a pronounced shadowing effect $A_{eff} = A^{.91}$, which has been associated with the existence of hadronic components in the photon propagator.

In contrast, rather few total photonuclear absorption measurements were up to now available beyond the GDR region and in the region of nucleon resonances. Indeed the total photon nucleus cross section $\sigma_{TOTN}(k)$ has been measured in the low energy region (k < 140 MeV) at Mainz by B. Ziegler and his collaborators[3] for nuclear mass numbers up to A=40. These measurements have been recently extended up to 340 MeV approximately for ^{9}Be, ^{6}Li and ^{7}Li [ref.[4]]. Very recently, using the bremsstrahlung tagging facility at the 500 MeV Bonn synchrotron, J. Arends et al.[5] have determined $\sigma_{TOTN}(k)$ for a variety of nuclei from He to Pb and for 215 MeV < k < 385 MeV. Finally the Saclay group undertook to extend for 30 MeV < k < 140 MeV, and for medium and heavy nuclei[6] the method used at Livermore and at the 60 MeV Saclay linac[1] to measure $\sigma_{TOTN}(k)$ in the GDR region. This intermediate excitation energy range is of particular interest because the wavelength $\lambdabar$ of the absorbed photon is much smaller than the nuclear radius ($\lambdabar \simeq 2$ fm, when k=100 MeV), and one can thus hope to observe some nucleon-nucleon correlation effects which should show up in the total integrated cross section

$$\Sigma_o = \int_o^{m_\pi = 140 \text{ MeV}} \sigma_{TOTN}(k) dk \quad (2)$$

and in the energy dependence of $\sigma_{TOTN}(k)$, since above say $k \simeq 30$ MeV the tensor force and the tensor correlations play a predominant role in the photonuclear transition amplitudes between the ground state and the final state at energy k, through the so-called quasideuteron effect[7]. However the required precision to get an absolute value for $\sigma_{TOTN}(k)$ is difficult to access because here the $\sigma_{TOTN}(k)$ values are much smaller than the corresponding values in the resonance region, and their determination from the raw data is more or less model dependent according to which experimental method is used. It seems thus useful to recall briefly the main characteristics of the methods used in Mainz, Bonn and Saclay, and to see how they can complement each other.

2. Recall of the Mainz, Bonn and Saclay methods for $\sigma_{TOTN}(k)$ measurements

These three techniques present the great advantage of using monochromatic photons whose actual number impinging on the photonuclear target is counted for each measured point $\sigma_{TOTN}(k)$.

i) **In the Mainz method**, for a given photon energy k, which is identified in an incident bremsstrahlung spectrum by two Compton spectrometers placed in front of and behind a removable absorber of length $l(g/cm^2)$ one measures the attenuation $\frac{N(k)}{N_o(k)}$ of the photon flux. The total absorption cross section $\sigma_{ABS}(k)$ is then given as

$$\sigma_{ABS}(k) = \frac{A}{\mathcal{N}l} \ln \frac{N_o(k)}{N(k)} \tag{3}$$

where $\mathcal{N}$ is the Avogadro's number and A the atomic weight of the absorbing material. Now, $\sigma_{ABS}(k)$ is the sum of the looked for cross section $\sigma_{TOTN}(k)$ and of the total atomic absorption cross section $\sigma_{TOTA}(k)$, due essentially to Compton and pair production effects. One has then to compute a model dependent value for $\sigma_{TOTA}(k)$ whose uncertainty $\Delta\sigma_{TOTA}(k)$, mostly due to the uncertain Coulomb correction term, increases with A. Some typical values for $\sigma_{TOTN}(k)$ [ref.3,6], $\sigma_{TOTA}(k)$ [ref.8], $\sigma_{TOTN}(k)/\sigma_{TOTA}(k)$ and $\Delta\sigma_{TOTA}(k)/\sigma_{TOTA}(k)$ [ref.9] are given in table 1 for k=80 MeV and for O(H_2O), Ca, Sn and Pb respectively.

Clearly for medium and heavy nuclei ($Z \gtrsim 20$) this attenuation method becomes unsuitable since the uncertainty $\Delta\sigma_{TOTA}(k)$ becomes larger than the $\sigma_{TOTN}(k)$ to be measured. However this method has been successfully used for light nuclei and the experimentally obtained results were of central importance in connection with sum rule evaluations[10] since for the first time the experimentally determined integrated cross sections

$$\Sigma_o = \int_o^{m_\pi = 140 \text{ MeV}} \sigma_{TOTN}(k) dk = 0.06 \frac{NZ}{A} (1+K) \text{ (MeV.b)} \tag{4}$$

were found to correspond to an enhancement factor K over the classical dipole sum, $\sigma_o = 0.06 \frac{NZ}{A}$, of the order of unity.

Table 1

Typical values for $\sigma_{TOTN}(k)$ and $\sigma_{TOTA}(k)$ at k=80 MeV

	O(H$_2$O)	Ca	Sn	Pb
σ_{TOTN} (mb)	1.5	3	8	14
σ_{TOTA} (mb)	509	2380	12850	30860
$\dfrac{\sigma_{TOTN}}{\sigma_{TOTA}}$	2.9×10^{-3}	1.3×10^{-3}	6×10^{-4}	4.5×10^{-4}
$\dfrac{\Delta\sigma_{TOTA}}{\sigma_{TOTA}}$	$< 10^{-3}$	$< 2 \times 10^{-3}$	$\sim 5 \times 10^{-3}$	$\sim 10^{-2}$

ii) <u>In the Bonn method</u>, a tagged bremsstrahlung photon beam, 215 MeV < k < 385 MeV, is used to measure separately and simultaneously the double differential cross section for the photoemission of protons and charged pion photoproduction. The detecting thresholds are 60 MeV for the protons and 40 MeV for the charged pion respectively. The measured double differential cross section is then integrated to get the total cross section for the emission of an observable charged particle. Taking the measured charged particle multiplicity into account, the proton and the pion data are then combined to obtain a lower limit $\sigma_{TOTN}^<(k)$ for the total hadronic cross section $\sigma_{TOTN}(k)$. To get the complete $\sigma_{TOTN}(k)$ values, one has to correct the experimental $\sigma_{TOTN}^<(k)$ results for the loss of events due to unobservable low energetic charged particle events and completely neutral ($\pi°$,n) channel events. The ratio r between $\sigma_{TOTN}^<(k)$ and $\sigma_{TOTN}(k)$ is obtained from an intranuclear cascade model[11], and it turns out that in this energy domain this ratio r is close to unity only for light nuclei.

Although this excitation energy range is out of the scope of this talk, it is interesting to compare, when they are available, the final data obtained at Mainz and at Bonn by using two completely different methods. This is shown in Fig. 1 for the Be case where an excellent agreement between both sets of data can be seen. However for medium and heavy nuclei the computed r values are rather low, typically of the order of 0.25 for Sn and Pb and for k=215 MeV. Moreover for k < 140 MeV one should expect a r value still lower due to the closure of the photopion channels and to an increased hindrance for proton emission due to Coulomb barrier effects. Therefore although the Bonn method, like the Mainz method, is well adapted to give reliable $\sigma_{TOTN}(k)$ values in the case of light nuclei, the model dependent uncertainties ($\Delta\sigma_{TOTA}(k)$ for the Mainz method, and Δr for the Bonn method) make any extension of these methods questionable for heavier nuclei, particularly in the 30 MeV < k < 140 MeV energy range where $\sigma_{TOTN}(k)$ takes its lowest values. Clearly one needs a different experimental approach for large A nuclei which must be less model dependent.

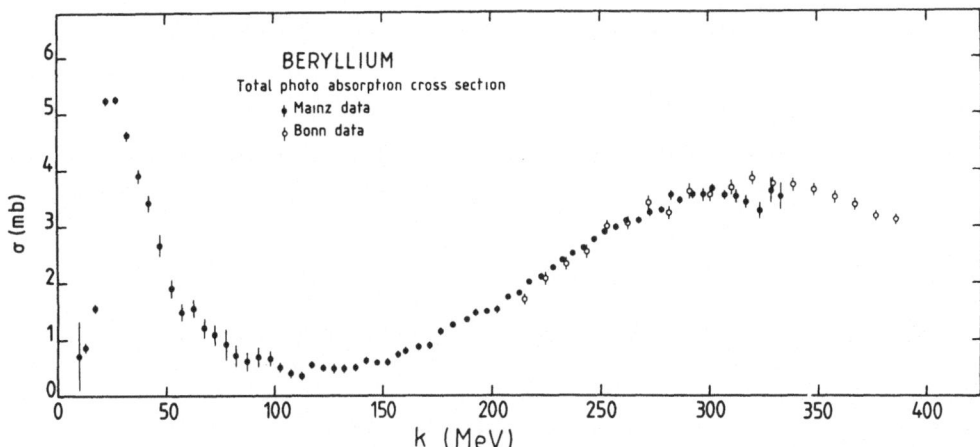

Fig. 1 - Total photoabsorption cross section $\sigma = \sigma_{TOTN}(k)$ for ^9Be. Experimental data are from references 3-5.

iii) In the Saclay method[12-13] one combines the use of a variable energy quasi monochromatic photon beam (20 MeV < k < 135 MeV), produced at the 600 MeV Saclay linac by in-flight annihilation of monoenergetic positrons, and the detection of the emitted photo neutrons in a very efficient 4π neutron detector whose detecting efficiency is $\varepsilon \simeq 0.65$. The photoneutron events are classified according to the number of emitted neutrons xn independent of whether some of these events may have included the emission of charged particles. Then for a given "monochromatic photon" energy k and for a given neutron multiplicity x one can measure separately and simultaneously the following partial photoneutron cross section (x = 1,2,3...).

$$\sigma_{xn}(k) = \sigma(\gamma,xn) + \sigma(\gamma,xnp) + \sigma(\gamma,xn2p) + \ldots \qquad (5)$$

The total photonuclear absorption cross section is now evaluated as

$$\sigma_{TOTN}(k) = \sigma(\gamma,p) + \sigma(\gamma,2p) + \ldots \sigma_{1n}(k) + \sum_{x=2} \sigma_{xn}(k). \qquad (6)$$

Clearly in the Saclay method one does not observe the purely charged particle channels. However available experimental data for ^{142}Nd, ^{181}Ta and ^{208}Pb [ref.[22]], together with intranuclear cascade model predictions[14] lead to an expected value of less than 5 % for the $\sigma(\gamma,p) + \sigma(\gamma,2p) + \ldots$ contribution to $\sigma_{TOTN}(k)$

$$\sigma_{TOTN}(k) - \sigma_{1n}(k) - \sum_{x=2} \sigma_{xn}(k) < 5 \times 10^{-2} \sigma_{TOTN}(k). \qquad (7)$$

Moreover in addition to statistical errors spoiling each $\sigma_{xn}(k)$ measurement, a more troublesome and systematic error arises from atomic interactions of the monochromatic part of the photon spectrum in the nuclear target, where positron electron pairs are created. The ensuing "target produced" bremsstrahlung photons, whose intensity is on the average much higher for low energy photons, can then produce spurious photoneutron events which clearly depend on the nuclear target thickness. Therefore several measurements with targets of different thicknesses t are carried out, for a few sampled k values because such measurements are time consuming.

Typical results obtained for Pb are shown in Fig. 2 where are plotted, for various "monochromatic photon" energies k, the variations with the target thickness t (mm) of the partial sums

$$\sigma^1(k) = \sigma_{1n}(k) + \sum_{x=2} \sigma_{xn}(k)$$

$$\sigma^2(k) = \sum_{x=2} \sigma_{xn}(k).$$
(8)

Now extrapolating for a given k the measured $\sigma^1(k)$ and $\sigma^2(k)$ towards t=0, one gets the corresponding actual values for these partial sums. One then observes that $\sigma^1(k) \simeq \sigma^2(k)$, within the overall error bars spoiling the $\sigma^2(k)$ data which can be taken as a reasonable measurement of $\sigma_{TOTN}(k)$. However between 30 and 50 MeV there might be a small positive $\sigma_{1n}(k)$ contribution which participates in the total integrated cross section only as

$$\int_{30 \text{ MeV}}^{m_\pi=140 \text{ MeV}} \sigma_{1n}(k) dk \le 4 \times 10^{-2} \sigma_o.$$
(9)

Exactly the same conclusions were experimentally reached for Sn, Ce and Ta nuclei[13]. The $\sigma^2(k) \simeq \sigma_{TOTN}(k)$ data for Pb are shown as an example in Fig. 3, together with previously measured $\sigma_{TOTN}(k)$ values in the GDR region[15]. One observes that below approximately 30 MeV the experimental cross section falls at a rate that might be expected from the high energy portion of the GDR. However above say 35 MeV the actual measured cross section is much higher than would be implied by a Lorentz energy dependence σ_L adjusted to fit the GDR[15].

The same features were observed for Sn, Ce, Ta and U, and will be now discussed in terms of quasideuteron model and integrated cross sections.

3. The energy dependence of $\sigma_{TOTN}(k)$ for medium and heavy nuclei

Following a suggestion proposed 30 years ago by J.S. Levinger[16] the total photonuclear absorption cross section for a nucleus $_Z^A N$ and for 50 MeV < k < 140 MeV is often expressed as

$$\sigma_{TOTN}(k) = L \frac{NZ}{A} \sigma_D(k)$$
(10)

where $\sigma_D(k)$ is the total photodisintegration cross section for the deuteron and L is the Levinger's parameter. This expression stresses the importance of correlated neutron-proton pairs in the absorption mechanism (quasi deuteron effect). From (10) one

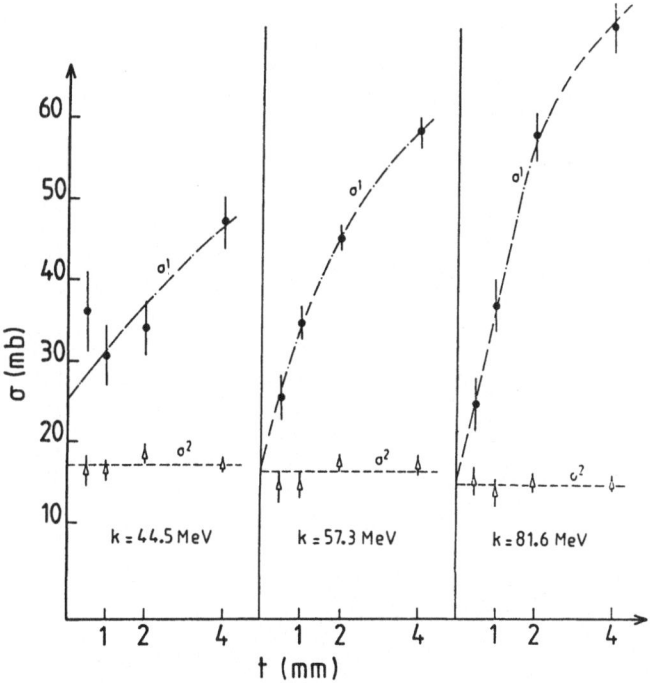

Fig. 2 - $\sigma^1(k)$ and $\sigma^2(k)$ data for different "monochromatic photon" energies k and for various thicknesses of a Pb target.

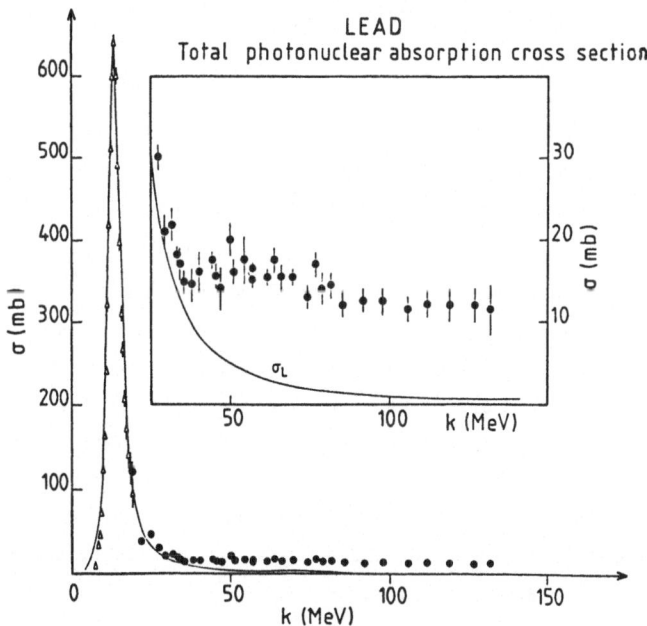

Fig. 3 - The $\sigma = \sigma_{TOTN}(k)$ data for Pb in the GDR region (△) together with the recent Saclay results (●) compared with the Lorentz curve σ_L fitting the GDR data.

can evaluate the L parameter from the corresponding experimental $\sigma_{TOTN}(k)$ and $\sigma_D(k)$ values. Such evaluated L values are shown in Fig. 4 for Ce, Ta and Pb. Similar results are obtained for Sn and U, but have been omitted for clarity sake. Clearly L is not A dependent, at least for 120 < A < 240, but it must be energy dependent.

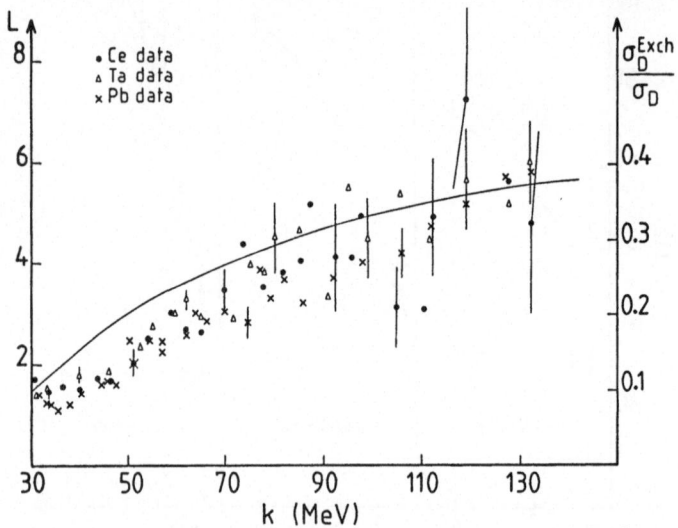

Fig. 4 - Experimental L values for Ce, Ta and Pb compared to the $\sigma_D^{Exch}(k)/\sigma_D(k)$ ratio.

It was then argued by J.S. Levinger[17] that one must introduce some "quenching effect" of the quasideuteron effect when k decreases, due to Pauli blocking effect for the neutron (or proton) emitted from the quasideuteron which absorbs the photon. Therefore introducing an exponential damping factor in (10) which now reads

$$\sigma_{TOTN}(k) = L \frac{NZ}{A} \sigma_D(k) \exp(-\frac{D}{k}) \tag{11}$$

he was able to fit fairly well the Pb data (Fig. 5) with L=8 and D=60 MeV. Although similar agreements were observed in fitting the Sn, Ce, Ta and U data[13], no theoretical evaluation of the damping factor D does exist up to now.

Furthermore J.M. Laget[18] recently proposed to associate the absorption of a 40 to 140 MeV photon by a quasideuteron only with the exchange part $\sigma_D^{Exch}(k)$ in the deuteron total photodisintegration cross section, corresponding to transition amplitudes where a virtual meson is emitted by a nucleon and then reabsorbed by the other nucleon of the deuteron. The variations with k of the ratio $\sigma_D^{Exch}(k)/\sigma_D(k)$ is represented by a solid line in Fig. 4, and show a great similarity with the evolution with k of the experimentally L values found for Ce, Ta and Pb. Therefore, for k > 40 MeV the $\sigma_{TOTN}(k)$ cross section would no longer be proportional to $\sigma_D(k)$ but rather to $\sigma_D^{Exch}(k)$. This is exactly what is observed for the $\sigma_{TOTN}(k)$ values of Sn, Ce, Ta and Pb which can be represented by a general expression

$$\sigma_{TOTN}(k) = (11 \pm 2) \frac{NZ}{A} \sigma_D^{Exch}(k). \tag{12}$$

A typical example of such a representation is given in Fig. 5 in the case of Pb.

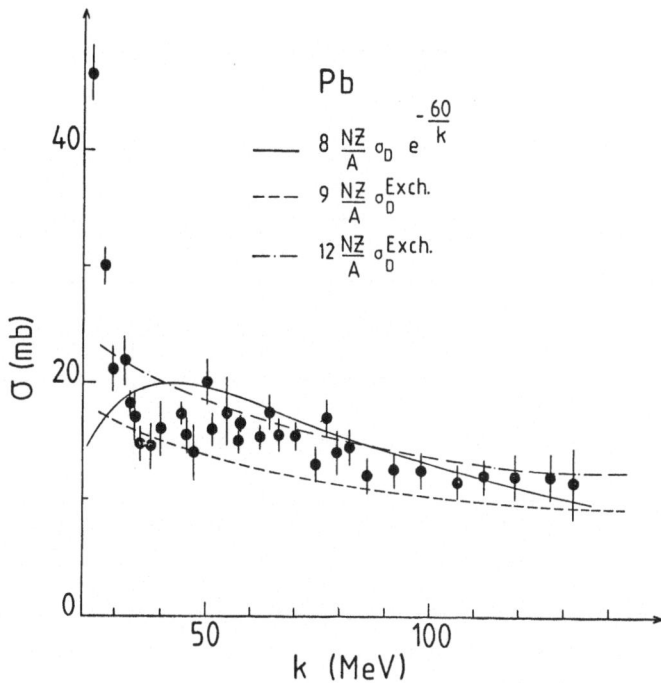

Fig. 5 - The $\sigma_{TOTN}(k)$ values for Pb compared with the modified quasideuteron model of J.S. Levinger and the J.M. Laget's predictions.

4. The total integrated photonuclear absorption cross section $\Sigma_o = \int_o^{m_\pi = 140 \text{ MeV}} \sigma_{TOTN}(k)dk$

The experimental Σ_o values for Sn, Ce, Ta, Pb and U expressed in unit of $\sigma_o = 0.06 \frac{NZ}{A}$ MeV-b are summarized in Table 2. They were obtained from previous Saclay measurements[1] up to k_o (M term) and from the present Saclay data[13] (N term) whose corresponding integrated values are also given. Furthermore to take the contribution of $\sigma_{1n}(k)$ into account, the experimentally estimated value of

$$\int_{k_o}^{m_\pi = 140 \text{ MeV}} \sigma_{1n}(k)dk \leq 4 \times 10^{-2} \sigma_o$$

has been added. Since the best knowledge of the contribution of the pure charged particle channels has been evaluated to be smaller than $5 \times 10^{-2} \sigma_{TOTN}(k)$, it is clear that the final uncertainties are largely including this eventual small and yet unmeasured contribution. It can be concluded that one can define for large A nuclei an average value for the enhancement factor K as $\underline{K = 0.76 \pm 0.10}$. Moreover there does not seem to be any A dependence for K in this range of mass numbers 120 < A < 240. One can also notice that this value K=0.76 does not disagree with an average K value fitting the Mainz data[3].

Table 2

Experimental total integrated cross section Σ_o

	Sn	Ce	Ta	Pb	U
$\sigma_o = 0.06 \frac{NZ}{A}$ (MeV-b)	1.74	2.04	2.61	2.97	3.4
k_o (MeV)	29.7	25	25	25	18
$M = \int^{k_o} \sigma_{TOTN}(k)dk$	1.15±0.09	1.04±0.07	1.11±0.09	1.17±0.08	0.88±0.05
$N = \int_{k_o}^{m_\pi=140\,MeV} \sigma^2(k)dk$	0.55±0.06	0.63±0.05	0.66±0.06	0.57±0.05	0.76±0.06
Σ_o	1.74±0.15	1.71±0.15	1.81±0.15	1.78±0.15	1.68±0.15

In the long wavelength approximation and for a non relativistic nuclear hamiltonian a sum rule for the energy integral $\Sigma_o(E1)$ of the electric dipole absorption cross section (the Bethe-Levinger sum rule)[19] has been evaluated and the corresponding enhancement factor reads

$$K_{Dipole} = \frac{mA}{e^2\hbar^2 NZ} <|[D_z,[V,D_z]]|> \qquad (13)$$

where m is the nucleon mass, D_z the dipole operator and V the two body nucleon-nucleon potential. If one is satisfied with a comparison within 10 %, which corresponds to actual experimental uncertainties, then this expression for K_{dipole} should be a good approximation for K, despite higher multipole contributions and retardation effects which are included in the experimental cross section $\sigma_{TOTN}(k)$ and despite the finite upper limit of integration in Σ_o. The large obtained values for K stimulated a series of calculations[10] of the double commutator (13) which established that two-body correlations notably those induced by the strong triplet even tensor force are responsible for the large increase in K over the Bethe-Levinger value (K ≃ 0.4).

Furthermore a quite different approach has been proposed by M. Gell-Mann et al. (GGT sum rule)[20] and modified later by W. Weise[21]. On the basis of dispersion relations, derived from the analyticity of the photon-nucleus forward scattering amplitudes, a sum rule for the total photonuclear absorption cross section including all multipoles and retardation effects, was established, which integrally combines photon-nucleus amplitudes at low ($k < m_\pi$) and high energies ($k \gtrsim m_\pi$). The underlying physical picture is that the enhancement factor K in Σ_o below pion threshold is related to physical meson photoproduction for $k \gtrsim m_\pi$. The direct coupling of a photon to exchange currents below pion threshold can then be interpreted as virtual meson photoproduction processes in the nucleus. In other words, if one takes seriously into account the exchange of mesons between nucleons in a nucleus[7], then the exchange part of the nuclear forces increases the photoabsorption below pion threshold by effectively increasing the number of charges responsible for the absorption, and the same exchange part alters the pion photoproduction cross section of Λ free particles in the isobar resonance region when they became bound together in a nucleus. This is what was actually observed in the limited energy range covered so far by the Mainz group[4] and by the Bonn group[5].

This important result for our understanding of photonuclear reactions at intermediate energies strongly suggests to extend such measurements of $\sigma_{TOTN}(k)$ towards higher energies and for a larger set of nuclei.

[1] B.L. Berman et al., Rev. Mod. Phys. 47 (1975) 713.
[2] W. Weise, Phys. Rep. 13C (1974) 53 and in Lecture notes in Physics (Springer-Verlag 1977), vol.61.
[3] J. Ahrens et al., Nucl. Phys. A251 (1975) 479.
[4] B. Ziegler, in Lecture notes in Physics (Springer Verlag) vol.86 (1978) and vol. 108 (1979).
[5] J. Arends et al., Proceedings of the Int. Conf. on nucl. phys., Berkeley (1980) to be published.
[6] A. Leprêtre et al., Phys. Lett. 79B (1978) 43 ; R. Bergère, in Lecture notes in Physics (Springer Verlag 1979) vol.108 ; P.J. Carlos, J. de Phys. 41 (1980) C3-149.
[7] G.E. Brown et al., Nucl. Phys. A338 (1980) 269.
[8] E. Storm et al., Nuclear Data Tables A (1970) 7 ; J. Ahrens, private communication.
[9] H.A. Gimm et al., NBS Technical note 968 (1978).
[10] A. Arima et al., Nucl. Phys. A205 (1973) 27 ; W.T. Weng et al., Phys. Lett. 46B (1973) 329 ; M. Gari et al., Phys. Rev. Lett. 41 (1978) 1288.
[11] T.A. Gabriel et al., Phys. Rev. 182 (1969) 1035 ; T.A. Gabriel, Phys. Rev. C13 (1976) 240.
[12] A. Veyssière et al., Nucl. Instr. Meth. 165 (1979) 417.
[13] A. Leprêtre et al., to be published.
[14] J.R. Wu et al., Phys. Rev. C16 (1977) 1812 and private communication.
[15] A. Veyssière et al., Nucl. Phys. A159 (1970) 561.
[16] J.S. Levinger, Phys. Rev . 84 (1951) 43.
[17] J.S. Levinger, Phys. Lett. 82B (1979) 181.
[18] J.M. Laget, Nucl. Phys. A312 (1978) 265 and private communication.
[19] J.S. Levinger et al., Phys. Rev. 78 (1950) 115.
[20] M. Gell-Mann et al., Phys. Rev. 95 (1954) 1612.
[21] W. Weise, Phys. Rev. Lett. 31 (1973) 773.
[22] K. Shoda et al. Phys. Rev. C4 (1971) 1842 ; T. Saito et al., Phys. Rev. C16 (1977) 958 ; A. Suzuki et al., Nucl. Phys. A257 (1976) 477 ; H. Dahmen et al., Nucl. Phys. A164 (1971) 140.

QUASI-DEUTERON EFFECTS AT INTERMEDIATE ENERGIES

Berthold Schoch
Institut für Kernphysik, Universität Mainz
6500 Mainz, W-Germany

1. Introduction

Nuclear reactions are the principal source of information regarding nuclear systems. Thereby studies with photons and electrons posess the advantage that it is relatively easy to calculate the effects of these particles on the charges and motion of the charges within the nucleus. Nevertheless before it is possible to extract nuclear structure information out of e.g.knock-out reactions the following questions have to be adressed: When will a single particle description suffice to describe the interaction between a nucleus and the projectile. When will the interaction with two and more particles be important. In this report the absorption mechanism of photons ($E_\gamma \geq 50$ MeV) and pions will be investigated. Despite the different interaction force similar kinematic conditions are imposed on the absorption process. Energy and momentum conservation forbid the photon and the pion to be absorbed by one free nucleon (see fig. 1). Although this argument does not hold

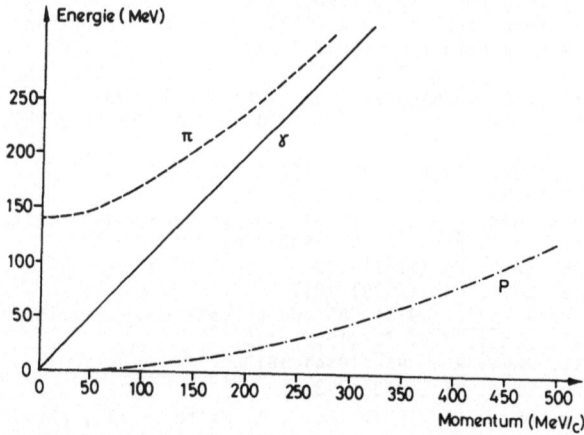

Fig.1: Momentum versus energy for γ and π as projectiles and the nucleon as a product in a nuclear reaction

rigorously for bound nucleons, it is still approximately true in the following sense: A nucleon bound in the nucleus has a definite binding energy E_s and a probability distribution of momenta $|\phi(p)|^2$ where $\phi(p)$ is the Fourier transform of the nucleons wave function. Because of the mismatch of momentum and energy transferred the nucleon must have had the additional momentum (see fig. 1) before it had absorbed the pion or the photon.

For photon energies $E_\gamma \geq 50$ MeV and all pion energies shown in fig. 1, the additional momentum is larger than the Fermi-momentum. The probability $|\phi(p)|^2$ to find such high momenta is very small and qualitatively insufficient to describe the actually observed absorption. Absorption by a pair of nucleons seems a more probable process: The two nucleons roughly split the energy and momentum of the probe among each other. This model is often called the quasi-deuteron one.

2. Absorption followed by two nucleon emission

The total absorption cross section [1] on ^{9}Be (fig.2) serves as an example of

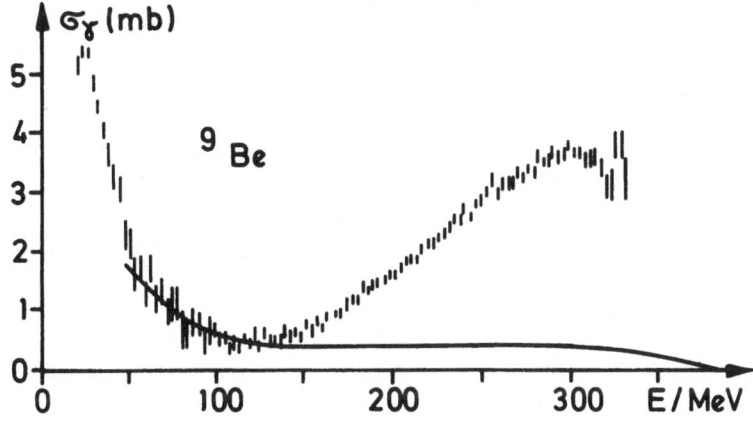

Fig.2: Total absorption cross section on ^{9}Be

the total cross section for photons on nuclei. The prominent peaks seen are the excitation of the giant dipole resonance (GDR) at the low energy side and the pion production dominated by the excitation of the Δ-resonance at the high energy side. On the one hand a collective excitation of the nucleus is connected with the absorption of the photon in the GDR-region, on the other hand the production of the pions takes place on quasifree nucleons in the nucleus. The quasi-deuteron absorption seems to be the main absorption mechanism in the energy region in between. The photodisintegration cross section represents only a small part of the total cross section (see fig.2) in the pion production region. Just for this energy region proposed Levinger [2] 30 years ago a model in which the cross section σ of the photodisintegration of a complex nucleus is related to the cross section σ_D of the photodisintegration of the deuteron: $\sigma = (L/A) \cdot N \cdot Z \cdot \sigma_D$. Thereby L/A stands for the density of (n-p)-pairs per unit nuclear volume and N·Z the number of neutron-proton pairs.

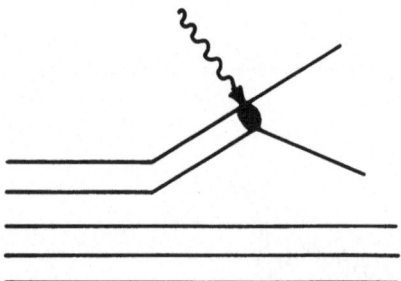 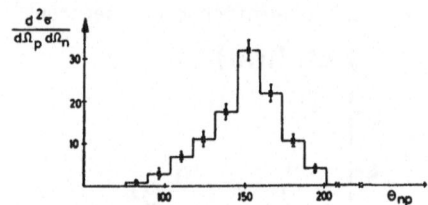

Fig.3: Quasi-deuteron mechanism

Fig.4: Opening angle of (n-p)-pairs in the reaction $^{16}O(\gamma,pn)$

This expression can be extended to differential cross sections[3] with the prediction that the angular distribution of (n-p)-pairs is like in the photodisintegration of the deuteron smeared out by the Fermi-motion of the nucleons. As fig.4 shows for the reaction $^{16}O(\gamma,np)N$ this is indeed the case [4]. The number of (n-p)-coincidences are plotted as a function of the sum of the proton and neutron angle in the energy region 215 MeV $\leq E_\gamma \leq$ 385 MeV. For the photodisintegration of the deuteron the sum of the angles is situated between 148° and 157° in this kinematics. Whereas the photodisintegration cross section close to the (3,3)-resonance is an order of magnitude smaller than the total absorption cross section (see fig. 2) the (γ,np)-reaction is believed to be the most important partial cross section below pion threshold. However, exclusive measurements of the differential cross sections for this reaction are needed to test these asumptions. A step further in this direction has been made with the reactions (π^-,nn) [5] and (π^+,pp) [6].

Fig. 5 shows a missing mass spectrum of the reaction $^9Be(\pi^-,nn)^7Li$. The neutrons which are preferentially emitted at an relative angle of 180° (see fig. 6 for the case of Li) have been analyzed with a time of flight spectrometer. With good resolution the missing mass spectrum would consist of sharp peaks at energies equal to the energies of the excited states in Li, becoming continuous at higher excitation energies. The spectrum shows maxima at about 10 MeV and then decreases slowly with increasing excitation energy. The ground state (3/2-) and the 0.48 MeV first excited state (1/2-) are weakly populated. In the α-α-n cluster picture for the ^9Be target, the transitions to these levels correspond to a removal of the weakly bound neutron

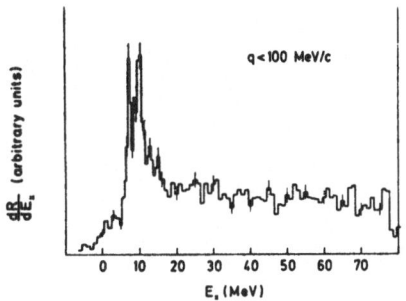
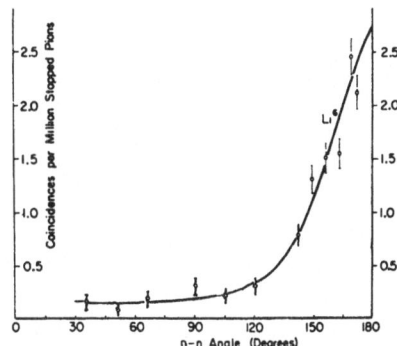

Fig.5: Missing mass spectrum of the reaction $^9Be(\pi^-,nn)$ Li

Fig.6: Opening angle of (n-n)-pairs in the reaction $Li(\pi^-,nn)$[7].

and a proton from one of the α clusters. The weak rate is therefore probably explained by the rather large distance between the proton and the neutron which absorb the pion. These results indicate the potentialities of these kind of reactions: The reactions (γ,pn) and (π,NN) can yield information concerning pair correlations in nuclei. This has been recognized for years [8] but for the most part, experimental information has been too uncertain or incomplete to give more than a general picture (Levinger model) of what is going on. With the advent of a new generation of accelerators with high duty factors and improved particle detection techniques high quality data may be available in the next future.

3. Absorption followed by one nucleon emission

The experimental situation concerning the absorption of photons and pions with the emission of one nucleon leaving the residual nucleus in a well defined state has been improved considerably in recent years. Fig. 7 shows as an example the differential cross sections for [9,10,11,12] the nucleon emission angle $\theta = 90°$ and the reactions $O^{16}(\gamma,n)O^{15}_{gs}$ and $O^{16}(\gamma,p)N^{15}_{gs}$. An exponential decrease at lower energies and a flattening is observed for the cross sections of both reactions. In the past mainly two possibilities for the reaction mechanism have been considered (fig.8). In the first graph the photon interacts with one nucleon and A-1 nucleons as spectators, in the second graph the photon interacts with a correlated (n-p)-pair like in the (γ,np)-reaction. The high momentum components of the single particle wave function can be extracted [13] in the case that the first graph dominates the reaction, where-

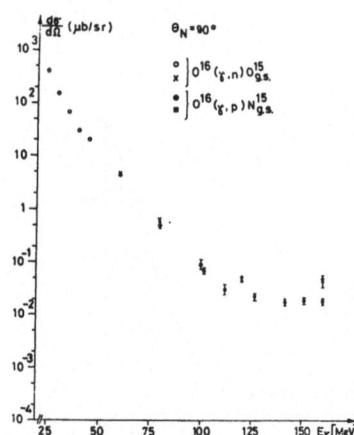

Fig.7: Exclusive (γ,p) and (γ,n)-cross sections on ^{16}O

Fig.8: Absorption of the photon on a single particle and a correlated pair

as a reaction mechanism via the second graph provides information similar to (γ,np) processes discussed above. The question to be answered before any reliable nuclear structure information can be extracted is therefore: What are the contributions of the different graphs in a wide kinematical region. Good quality data of exclusive (γ,p) as well as (γ,n)-reactions on selected nuclei can provide the information necessary. Fig. 9 shows the ratio [10] of the differential cross sections

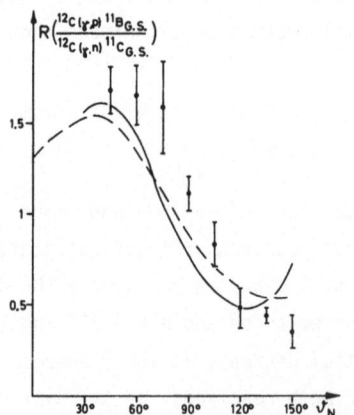
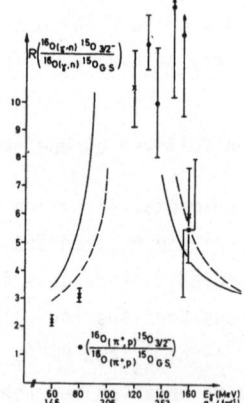

Fig.9: Ratio of the differential cross sections for the reactions $^{12}C(\gamma,p_0)^{11}B$ and $^{12}C(\gamma,n_0)^{11}C$

Fig.10: Ratios of the reactions γ,n and π^+,p [14] on ^{16}O leading to the (3/2-)-state and the ground state respectively in ^{16}O.

for the reaction $^{12}C(\gamma,p_0)^{11}B$ and $^{12}C(\gamma,n_0)^{11}C$ as a function of emission angle at E_γ = 60 MeV. The predictions of a modified quasideuteron model [15] (q.d.m.) (dashed line) and a microscopic calculation of Hebach, Wortberg and Gari [16] are shown. While the q.d.m.-calculation is solely based on graph two in fig.8, contains the calculation of H.W.G. the contribution of the two graphs whereby the second one dominates. Therefore, the qualitative agreement of the ratio is a strong indication for the dominance of the quasi-deuteron mechanism for the single nucleon emission too. Fig. 10 shows the ratios of the differential cross sections of the reactions $^{16}O(\pi^+,p_n)^{15}O_{3/2}$ and $^{16}O(\pi^+,p_n)^{15}O_{gs}$ as a function of momentum transfer. In the picture of the quasi-deuteron model nuclear structure effects should show up because a momentum transfer $\vec{q}$ (see fig. 8) has to be accepted by the residual nuclei beeing in different final states. In the framework of a single shellmodel a neutron is removed from the $p_{3/2}$-shell and $p_{1/2}$-shell respectively. Therefore, a ratio of two would be exspected due to the occupation number of these subshells. In fact for small momentum transfers the ratio starts out close to this value and reaches a maximum in the vicinity of the first minimum of the formfactor for $^{15}O_{gs}$. Because of a quadrupol contribution such a minimum is absent in the case of $^{15}O_{3/2}$-. This is again a confirmation for a quasi-deuteron like reactionmechanism for γ- and π-absorption followed by one nucleon emission. As a final example in fig. 11 the differential cross section as a function of the photon angle for the reaction $D(p,\gamma)^3He$ is shown at a proton energy of 155 MeV [17,18].

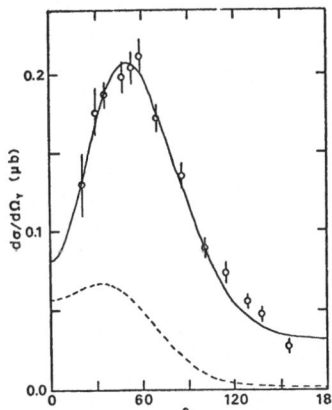

Fig.11: Measured differential cross-sections of the reaction $D(p,\gamma)^3He$ with calculations performed by Prats.

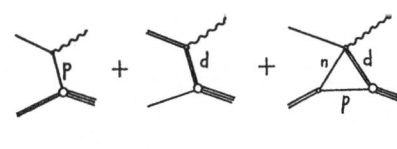

Fig.12: Graphs calculated to describe $D(p,\gamma)^3He$

Prats calculated with graphs shown in fig.12 the cross section. Thereby represents the dashed curve the result of a calculation including the two graphs on the left in fig.12. The triangular graph however which again is the quasi-deuteron contribution of the reaction (fig. 8) yields by far the most important part of the cross section.

4. Conclusions

It has been shown that photon and pion reactions in certain energy regions are dominated by the absorption on correlated (n-p)-pairs. Therefore, calculations based on quasi-deuteron models describe rather well the present experimental results. However, with more accurate data out of the reactions $(\gamma,np),(\gamma,N)$, (π,NN) and (π,N) performed so that the complete kinematics is known, the correlation function of nucleon pairs in nuclei can be studied a quantity very important for our understanding of the nuclear many body system. The dominance for absorption on pairs in reactions with a rather big mismatch in energy and momentum in the entrance channel compared with the exit channel of the reaction asks to take a close look at other reactions wether these effects show up too: in his contribution Laget has shown that with the inclusion of a quasi-deuteron contribution he can explain the enhanced cross section in (e,e')-reactions between the quasifree peak and the Δ-resonance region. Also pick up reactions, above a critical momentum transfer, are dominated by pair correlation effects [19].

To learn more about single particle wavefunctions, correlation functions and the validity of the impulse approximation, mostly used in the analyses of the data, a combined effort experimentally as well as theoretically has to be undertaken. Thereby, the use of different projectiles as probes in nuclear reactions will help to disentangle and to check the various ingredients entering in a calculation.

References

1) J.Ahrens, Nucl.Phys. A 335 (1980) 67
2) J.S.Levinger, Phys.Rev. 84 (1951) 43
3) K.Gottfried, Nucl.Phys. 5 (1958) 557
4) K.G.Hilger, Staatsexamensarbeit, Bonn 1980
5) B.Bassalleck, W.D.Klotz, F.Takeutchi, H.Ullrich and M.Furic, Phys.Rev.C 16 (1977) 1526
6) J.Favier, T.Bressani, G.Charpak, L.Massonet, W.E.Meyerhof and C.Zupancic, Nucl.Phys. A 169 (1971) 540
7) M.E.Nordberg,Jr., K.F.Kinsey and R.L.Burman, Phys.Rev. 165 (1968) 1096
8) K.Gottfried, Ann. of Phys. 21 (1963) 29

9) T.W.Phillips and R.G.Johnson, Phys.Rev. C 20 (1979) 1689
10) H.Göringer and B.Schoch, Phys.Lett. 97 B (1980) 41
11) D.J.S.Findlay and R.O.Owens, Nucl.Phys. A 279 (1977) 385
12) J.L.Matthews, W.Bertozzi, M.J.Leitch, C.A.Peridier, B.L.Roberts, C.P.Sargent, W.Turchinetz, D.J.S.Findlay and R.O.Owens, Phys.Rev. Lett. 38 (1977) 8
13) D.J.S.Findlay and R.O.Owens, Phys.Rev.Lett. 37 (1976) 674
14) D.Bachelier, J.L.Boyard, T.Hennino, J.C.Jourdin, P.Radvanyi and M.Roy-Stephan, Phys.Rev. C 15 (1977) 2139
15) B.Schoch, Phys.Rev.Lett. 41 (1978) 80
16) H.Hebach, A.Wortberg and M.Gari, Nucl.Phys.A 267 (1976) 425
17) J.P.Didelez, H.Langevin-Joliot, Z.Maric and V.Radojevic, Nucl.Phys. A 143 (1970) 602
18) F.Prats, George Washington University, Preprint 1978
19) H.J.Weber, Lecture Notes in Physics 108 (1979) 457 (Springer-Verlag, Berlin, Heidelberg, New York)

DIRECT MECHANISM IN KNOCKOUT REACTIONS WITH REAL AND VIRTUAL PHOTONS

S. Boffi
Istituto Nazionale di Fisica Nucleare, Sezione di Pavia, Pavia,
Istituto di Fisica Teorica dell'Università, Pavia, and
Istituto di Fisica "A.Righi", Università di Bologna, Bologna

Abstract: The single nucleon knockout mechanism is revisited in order to clarify unsatisfactory results concerning the single particle response of finite nuclei. The central role of the spectral density in the direct mechanism is discussed in terms of a recently derived equation for the related spectroscopic amplitudes. The results of a careful analysis of nucleon knockout reactions both with real and virtual photons are presented.

1. Introduction

Nuclear physics at intermediate energies with electrons and photons is a subject of increasing experimental effort because of the unique qualities of the electromagnetic probe. In particular, due to its capability of transferring large momenta to individual nucleons without effectively disturbing the rest of the nucleus the electromagnetic interaction is a powerful tool to study the single particle response of nuclei in knockout processes. Since long time the (e,e'p) reaction has provided a direct confirmation of the shell structure of the nucleus [1]. Recently, also the (γ,p) reaction seems to support the single particle mechanism for photon energies in the range 60 to 100 MeV [2], thus enlarging the range of explored momenta. The quality of the accumulating data requires a careful treatment of the three relevant ingredients in the theory of direct nucleon knockout reactions, i.e. i) the nuclear interaction with the external probe, ii) the response of the nucleus to this interaction, and iii) the distortion of the emitted nucleon in the final state.

In the case of incident real or virtual photons the electromagnetic interaction is well known and in the involved range of energies and momenta can be treated without important approximations by the nonrelativistic Hamiltonian of McVoy and van Hove [3], possibly extended to third or fourth order in the nucleon recoil velocity [4,5]. If final state interactions are neglected, the knockout reaction cross section is proportional to the diagonal element $S(\vec{p};E)$ in the momentum representation of the hole spectral density, which is usually interpreted as the joint probability of removing a particle with momentum $\vec{p}$ from the target nucleus leaving the residual one with the excitation energy E. Then, $S(\vec{p};E)$ gives a measure of how much the residual nucleus can be described in terms of a hole generated in the target ground state by the knockout process. Interesting exact sum rules for the energy and the nucleon number of the target are known [6] which seem to be violated by the available data [7].

In fact, the distortion of the emerging nucleon in general destroys the simple factorization in the reaction cross section between the nuclear information contained in the spectral density and the kinematic contribution of the reaction mechanism [8,9]. As a result the DWIA cross section also involves nondiagonal elements $S(\vec{p},\vec{p}';E)$ of the spectral density giving rise to difficulties in extracting the required nuclear information from the data. The distortion effects are a priori out of control and in no way can be approximated as usual by a reducing factor and an effective momentum of the initial nucleon in the factorized DWIA cross section [9,10].

In the present paper some recent work is presented aiming at gaining information on the spectral density from (e,e'p) and (γ,p) reactions. The mathematical properties of the spectral density are briefly reviewed in sect. 2; the results of a phenomenological analysis for some light nuclei are discussed in sect. 3; concluding remarks are the subject of the final section.

2. The spectral density

In the energy representation the single particle Green function in finite nuclei is defined as

$$G_{\alpha\beta}(\omega) = \langle \Psi_0 | a_\beta^+ \frac{1}{\omega - E_0 + H - i\varepsilon} a_\alpha | \Psi_0 \rangle \quad , \quad (1)$$

where a_α (a_α^+) destroys (creates) a particle in orbit α in the target ground state $|\Psi_0\rangle$, whose energy is E_0. The analytic properties of G are better exploited by inserting the complete set of eigenstates $\{|Ea\rangle\}$ of the Hamiltonian H for the system of A-1 nucleons, i.e.

$$G_{\alpha\beta}(\omega) = \oint d\omega' \frac{S_{\alpha\beta}(\omega')}{\omega - \omega' - i\varepsilon} , \qquad (2)$$

where

$$S_{\alpha\beta}(\omega) = \sum_a \oint dE \langle \Psi_0 | a_\beta^+ | Ea \rangle \langle Ea | a_\alpha | \Psi_0 \rangle \delta(E - E_0 + \omega)$$

$$= \langle \Psi_0 | a_\beta^+ \delta(H - E_0 + \omega) a_\alpha | \Psi_0 \rangle \qquad (3)$$

is the hole spectral density. The set $\{|Ea\rangle\}$ is labelled by the energy E and the index a replacing the totality of any other quantum number.

In order to determine the spectral density it is useful to consider the related Hilbert-Schmidt kernel [11,12]

$$S_{\alpha\beta}(E) = \sum_a \langle \Psi_0 | a_\beta^+ | Ea \rangle \langle Ea | a_\alpha | \Psi_0 \rangle , \qquad (4)$$

which is defined for the same set of energy values satisfying eq. (3) and is indicated here and in the following with the same symbol S as the spectral density. The kernel S(E) has a discrete spectrum for each value of E, so that it can be expanded in its eigenfunctions $\Theta(E)$ as

$$S_{\alpha\beta}(E) = \sum_a \Theta_\alpha^a(E) \lambda_a(E) \Theta_\beta^{a*}(E) . \qquad (5)$$

In eq. (5)

$$\lambda_a(E) = \sum_\alpha |\langle Ea | a_\alpha | \Psi_0 \rangle|^2 \qquad (6)$$

is the strength of the spectral density at the energy E for the channel a and gives a measure of how much the residual nucleus state $|Ea\rangle$ can be conceived in terms of a hole in the otherwise undisturbed target ground state. The total strength at the energy E is

$$\lambda(E) = \sum_a \lambda_a(E)$$
$$= \sum_\alpha S_{\alpha\alpha}(E) . \qquad (7)$$

The spectroscopic amplitudes $\Theta_\alpha^a(E)$ obey the following eigenvalue equation [12]

$$\sum_\beta \left[(E - E_0) \delta_{\alpha\beta} + T_{\alpha\beta} + M_{\alpha\beta}^{(1)}(E - E_0) \right] \Theta_\beta^a(E) = 0 , \qquad (8)$$

where T is the kinetic energy operator and $M^{(1)}$ is the hermitian part of the mass operator M appearing in the Dyson equation for the hole Green function. $M^{(1)}$ plays the role of an effective (nonlocal and energy dependent) potential whose static part is of the Hartree-Fock type. However, the energy dependent contribution is significant; without it in eq. (8) the mean removal energy would coincide with $E-E_0$.

Eq. (8) holds both for E belonging to the discrete and the continuum spectrum of the residual nucleus. In particular, for the continuum spectrum the hole spectral density has a resonance behaviour given by

$$\lambda_a(E) = \frac{1}{2\pi} \frac{\Gamma_a(E)}{\left[E - E_0 + F_a(E)\right]^2 + \Gamma_a^2(E)/4} , \qquad (9)$$

where $F_a(E)$ and $\Gamma_a(E)$ are the average values of $T+M^{(1)}$ and $M^{(2)}$ (the antihermitian part of M), respectively, i.e.

$$F_a(E) = \sum_{\alpha\beta} \theta_\alpha^{a*}(E) \left[T_{\alpha\beta} + M^{(1)}_{\alpha\beta}(E_0-E)\right] \theta_\beta^a(E) , \qquad (10)$$

$$\tfrac{1}{2}\Gamma_a(E) = \sum_{\alpha\beta} \theta_\alpha^{a*}(E) M^{(2)}_{\alpha\beta}(E_0-E) \theta_\beta^a(E) . \qquad (11)$$

From eq. (9) the necessary and sufficient condition in order that $\lambda_a(E)$ has a resonance peak in E_R is that the equation

$$F_a(E) = E_0 - E \qquad (12)$$

is satisfied by $E=E_R$. The width of this resonance is given by the positive definite quantity Γ_a.

It is possible to recognize in general that a unity upper bound exists for $\lambda_a(E)$. Only in the case where $|Ea\rangle$ is a pure hole state with respect to $|\Psi_0\rangle$, $\lambda_a(E)$ can be identified with the occupation number of the orbital in $|\Psi_0\rangle$ from which the nucleon has been removed [13].

For a more direct comparison with experiment, the momentum representation is preferable for S, i.e.

$$S(\vec{p},\vec{p}';E) = \sum_a \varphi_{Ea}(\vec{p}) \lambda_a(E) \varphi_{Ea}^*(\vec{p}') , \qquad (13)$$

where

$$\varphi_{Ea}(\vec{p}) = \sum_\alpha \theta_\alpha^a(E) u_\alpha(\vec{p}) , \qquad (14)$$

and $\{u_\alpha(\vec{p})\}$ is the selected set of orbitals. Then the total spectral strength at the energy E is

$$\lambda(E) = \int d\vec{p}\, S(\vec{p},\vec{p}';E) . \tag{15}$$

3. Phenomenological approach

The program to compute eigenfunctions and strength of the spectral density from first principles as indicated in the preceding section has not yet been undertaken. Only calculations based on the continuum shell model for the direct evaluation of the single particle Green function are available [14] with results strongly dependent on the nucleon-nucleon potential used.

In a phenomenological approach the reaction cross section can be computed by selecting different possibilities for $\varphi_{Ea}(\vec{p})$ together with different optical model potentials to simulate final state interactions. The spectral strength comes out from the comparison with experimental data as a normalization factor of the theoretical cross section.

The details of such an analysis for (e,e'p) and (γ,p) reactions can be found in refs. [15,10]. The high resolution data of the Saclay group [15] allow to distinguish removal e.g. from $p_{3/2}$ and $p_{\frac{1}{2}}$ orbit in ^{16}O. Similarly, the (γ,p_o) data [2] refer to the final nucleus in its ground state. Thus experiment is sensitive to the details of the particular hole state involved and not only to the whole charge density as in elastic electron scattering.

The main feature of (e,e'p) data on ^{16}O reported in fig. 1 is represented by an opposite asymmetric behaviour of the $p_{3/2}$ momentum distribution with respect to the $p_{\frac{1}{2}}$ one. This j-dependent asymmetry, as well as any asymmetry, is absent in the usual DWIA calculation where the reaction cross section factorizes into a product of a free electron-proton cross section and a distorted momentum distribution. In fact, even a real spin orbit term in the optical potential, which automatically destroys the factorization, is unable to produce asymmetry; only an imaginary spin orbit coupling in the distorting potential can fit the observed asymmetry, as can be seen in fig. 1. Also from fig. 1 it appears that the selected bound state wave function $\varphi_{Ea}(\vec{p})$ is not relevant over the involved range of momenta if it is able to reproduce the ground state properties. This implies that if in eq. (14) the set of natural orbitals is chosen for $u_\alpha(\vec{p})$ only one single term is important in the sum for $\varphi_{Ea}(\vec{p})$.

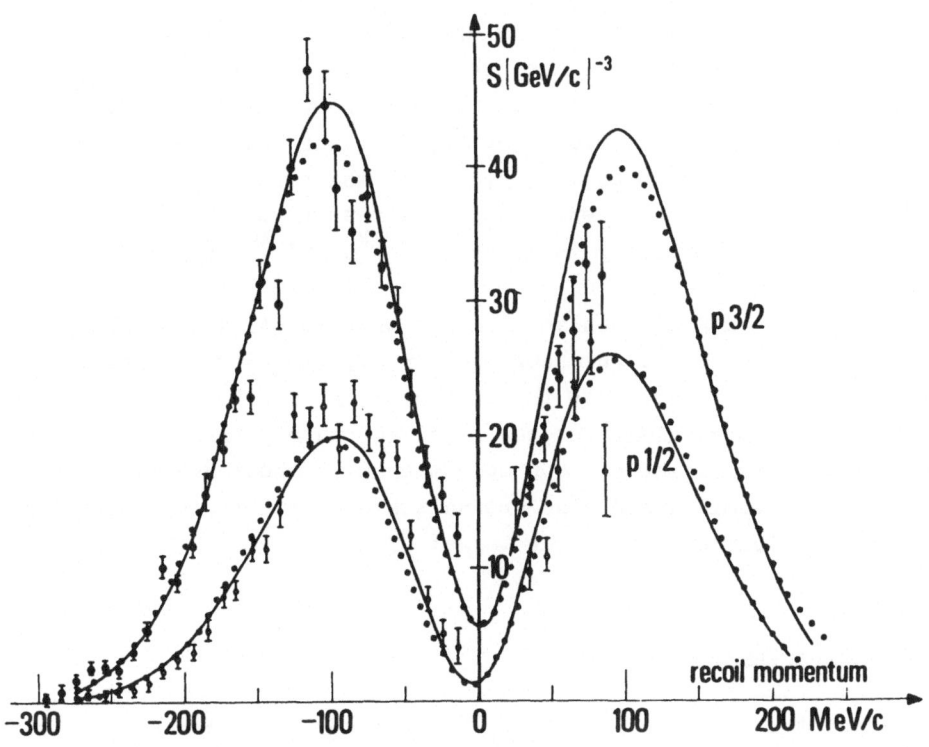

Fig. 1. Reduced cross sections versus recoil momentum of the residual nucleus for the reaction $^{16}O(e,e'p)^{15}N$ compared with experimental momentum distributions for $p_{3/2}$ and $p_{\frac{1}{2}}$ hole states. The optical potential is from ref. [16] and the bound states are from ref. [17] (full line) and ref. [18] (dotted line) (after ref. [15]).

The spectral strength used in fig. 1 is 0.59 for $p_{\frac{1}{2}}$ and 0.57 for $p_{3/2}$ [13,15].

The absorption of a real photon by the nucleus undergoing a photoreaction can be treated along the same lines as the interaction leading to (e,e'p) reactions, where a virtual photon is responsible for a direct nucleon knockout. A careful treatment of final state interactions requires a full DWIA calculation with partial wave expansion of the outgoing nucleon wave function and a scattering cross section where the nuclear contribution, i.e. the hole spectral density, and the photon-nucleon interaction, i.e. a scattering matrix containing the distorted wave of the emerging particle, are no longer factorized [10]. Also the variation of the parameters of the optical potential (in particular of the spin orbit term) affects the cross section sensitively [10].

From fig. 2 it is evident that the same combination of bound and scattering states able to reproduce (e,e'p) data also gives a satisfactory fit to (γ,p) data. In fig. 2, all the curves are obtained assuming $\lambda(E) = 0.5$, consistently with the results of the (e,e'p) analysis. A similar consistency is reached in ^{12}C and ^{40}Ca(γ,p_0) reactions [10].

In practice, a good phenomenological wave function describing the single hole bound state, when employed in eq. (13) for the spectral density, is able to reproduce the data at least for $E \leqslant 100$ MeV in a direct knockout mechanism. For the involved nuclei, this corresponds to values of the initial nucleon momentum up to about 2.5 fm^{-1}, i.e. just above the limit where also ground state correlations affect the momentum distribution inside the target [21].

4. Concluding remarks

A consistent description of (e,e'p) and (γ,p) reactions has been discussed in the frame of a direct mechanism for the knockout process. As for (e,e'p) reactions, the present day kinematic conditions at intermediate energy of the incident electron beam suggest the quasi-free scattering without any doubt; only a careful theoretical analysis is required to extract the interesting nuclear information. (γ,p) reactions imply an extension to higher initial momenta of the knocked out particle. The present analysis, limited to an incident photon energy between 60 and 100 MeV, confirms the possibility to test initial momenta up to about 2.5 fm^{-1} under the assumption of a direct mechanism. This contrasts with current work involving meson exchanges [22] and/or quasideuteron mechanism [23] whose role seems important especially when going to (γ,n) reactions. In order to clarify the situation an experiment on (γ,p) and (γ,n) reactions with polarized photons has been recently proposed [24]. In the (γ,p) reaction the direct cross section for photons polarized in the reaction plane ($\sigma_{//}$) is much larger than and with a different shape from the one for photons polarized perpendicularly to the reaction plane ($\sigma_\perp$). The lacking of convective terms in the direct mechanism produces (γ,n) cross sections which are similar and comparable with $\sigma_\perp$ of the (γ,p) reactions.

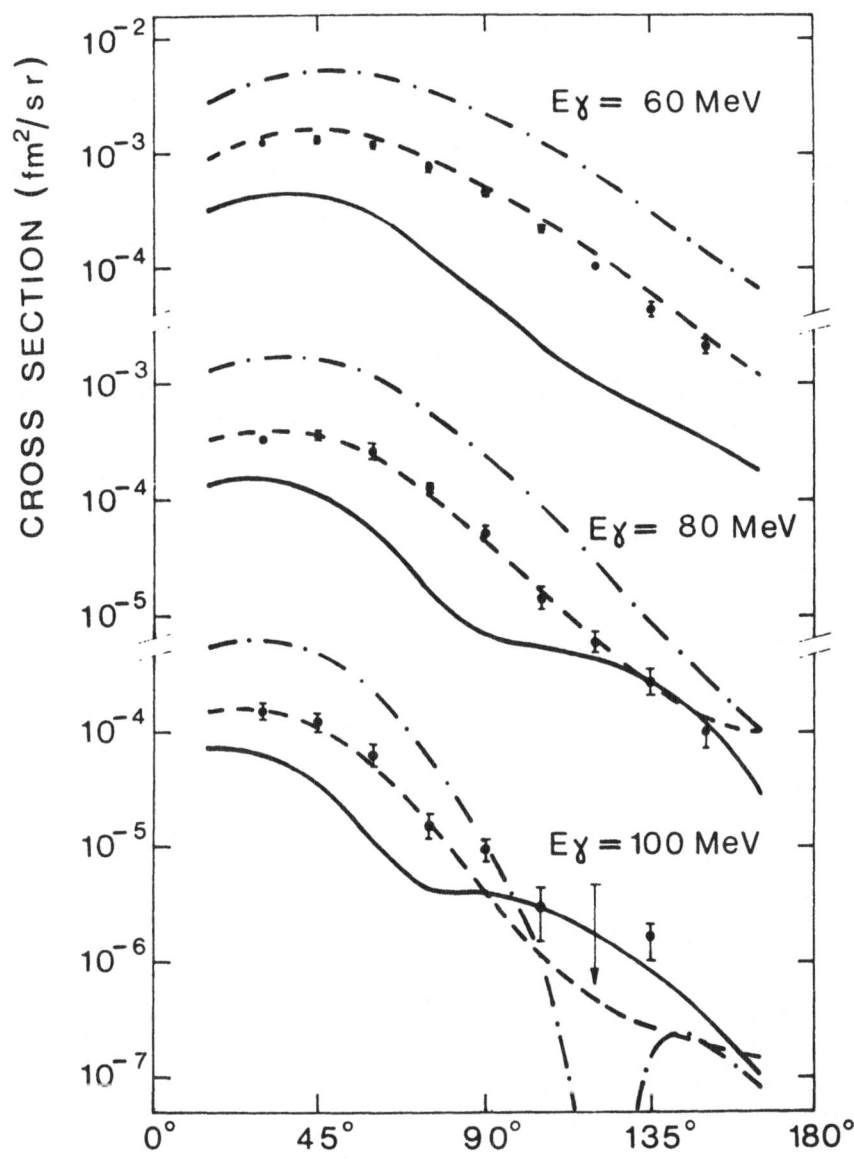

Fig. 2. Angular distribution of $^{16}O(\gamma,p_o)^{15}N$ reaction. The bound state is taken from ref. [17]; the full line is obtained with the optical potential of ref. [19], the dashed line with the one of ref. [16], and the dot-dashed one with PWIA. The experimental points are from ref. [20] (after ref. [10]).

Experimental results are desirable in order to appreciate the contribution of more complex mechanisms [22,23,25] which up to now do not seem necessary in the case of proton knockout.

In the cases where accurate data are available the spectral density has been determined with the following results: i) the spectral strength is sensibly lower than unity even for valence protons, and ii) the eigen functions of the spectral density can be approximated well by suitable natural orbitals of the target nucleus up to about 2.5 fm^{-1}.

The paper is based on work mainly performed by the author in collaboration with F.Capuzzi, C.Giusti and F.D.Pacati of the University of Pavia, whose valuable contribution is gratefully acknowledged.

References

[1] U.Amaldi Jr. et al., Phys. Lett. 25B, 24 (1967)
[2] J.L.Matthews in Lecture Notes in Physics, vol. 108 (Berlin,1979) 139
[3] K.W.McVoy and L.VanHove, Phys. Rev. 125, 1034 (1962)
[4] S.Boffi, C.Giusti and F.D.Pacati, Nucl. Phys. A336, 416 (1980); C.Giusti and F.D.Pacati, Nucl. Phys. A336, 427 (1980)
[5] C.Giusti and F.D.Pacati, Nuovo Cim. Lett. 26, 622 (1979)
[6] S.Boffi, Nuovo Cim. Lett. 1, 931 (1971); D.S.Koltun, Phys. Rev. Lett. 28, 182 (1972)
[7] J.Mougey et al., Nucl. Phys. A262, 461 (1976)
[8] D.F.Jackson, Nucl. Phys. A257, 221 (1976)
[9] S.Boffi, C.Giusti, F.D.Pacati and S.Frullani, Nucl. Phys. A319, 461 (1979)
[10] S.Boffi, C.Giusti and F.D.Pacati, Nucl. Phys. to be published
[11] S.Boffi and F.Capuzzi, Nuovo Cim. Lett. 25, 209 (1979)
[12] S.Boffi and F.Capuzzi, Nucl. Phys. to be published
[13] S.Boffi in "From Nuclei to Particles", ed. by A.Molinari (79th Course, Varenna, 1980)
[14] W.Fritsch, R.Lipperheide and U.Wille, Nucl. Phys. A241, 79 (1975)
[15] M.Bernheim et al., to be published
[16] D.F.Jackson and I.Abdul-Jalil, J.Phys. G6, 481 (1980)
[17] L.R.B.Elton and A.Swift, Nucl. Phys. A94, 42 (1967)
[18] J.P.Blaizot and D.Gogny, Nucl. Phys. A284, 429 (1977); D.Gogny, private communication

[19] M.M.Giannini and G.Ricco, Ann. of Phys. **102**, 458 (1976); M.M. Giannini, private communication
[20] D.J.S.Findlay and R.O.Owens, Nucl. Phys. **A279**, 385 (1977)
[21] J.G.Zabolitzky and W.Ey, Phys. Lett. **76B**, 527 (1978)
[22] H.Hebach, A.Wortberg and M.Gari, Nucl. Phys. **A267**, 425 (1976); M. Gari and H.Hebach, Bochum preprint, 1980
[23] B.Schoch, Phys. Rev. Lett. **41**, 80 (1978); Habilitationschrift, Mainz 1980 and to be published
[24] S.Boffi, C.Giusti and F.D.Pacati, Phys. Lett. B, to be published
[25] M.Marangoni, P.L.Ottaviani and A.M.Saruis, RT/FI(76)10, Bologna 1976

A MODEL FOR HADRONS BASED ON THE MIT BAG MODEL

J.J. de Swart
Institute for Theoretical Physics
University of Nijmegen, The Netherlands

I. Quarks, color, gluons, and bags

At present it is well accepted that hadrons, like the nucleon, are composed of quarks. These quarks are distinguished by their different <u>flavors</u>: u(up), d(down), s(strange), etc. and by their 3 <u>colors</u>. It is the color which is responsible for the interaction between the quarks in the same way that the electric charge is responsible for the electromagnetic interaction. This is described by a theory called (quantum) <u>chromodynamics</u> (QCD) which is quite analogous to (quantum) electrodynamics (QED). The forces between colored objects (like quarks) are mediated by colored, massless vectormesons, called <u>gluons</u>. There are 8 gluons that can couple to 8 color charges $g\, F_a$ (with a = 1 to 8). Here F_a is called the <u>colorspin</u> which has 8 components. These 8 components F_a are the generators of color SU(3,C) in the irrep $\underline{C}$ to which the charge belongs. Another important quantity is the length F^2 of the colorspin. This is the quadratic Casimir operator for SU(3,C); $F^2 = \sum_a F_a^2$ with eigenvalue f_c^2. Examples are:

colorless hadron $C = \underline{1}$, $F_a \equiv 0$, $f_c^2 = 0$
quark $C = \underline{3}$, $F_a = \frac{1}{2} \lambda_a$, $f_c^2 = 4/3$
antiquark $C = \underline{3^*}$, $F_a = -\frac{1}{2} \lambda_a^*$, $f_c^2 = 4/3$
gluon $C = \underline{8}$, $f_c^2 = 3$

Because experimentally no colored particles have been seen it is generally assumed that color is confined and that all observed hadrons must be colorless. Is this confinement absolutely true, or is it only a good approximation in the energy region we are working? This is unclear at present.

In the MIT bag model [1] confinement of color is build in by assuming that the quarks and gluons are restricted to some cavity or bag. It costs an energy $B = 59$ MeV/fm^3 to make such a cavity. The energy-momentum tensor $T_{\mu\nu}$ has the form

$$T_{\mu\nu} = T_{\mu\nu}^{QCD} - B\, g_{\mu\nu} \quad \text{inside the cavity}$$
$$\equiv 0 \quad \text{outside the cavity.}$$

Here $T_{\mu\nu}^{QCD}$ is the energy-momentum tensor due to the quarks and gluons as given by QCD. The extra term B in $T_{\mu\nu}$ gives rise to the aforementioned volume energy and at the same time is furnished a pressure which tries to compress the bag. This pressure needs to be balanced from the inside by the quarks and gluons:

$$B = \frac{1}{2} n^\mu \partial_\mu \left(\sum_i \bar{\psi}_i \psi_i \right) + \frac{1}{2} \sum_a (E_a^2 - H_a^2) \qquad (1)$$

Here n^μ is the normal to the bag surface, ψ_i the quark fields (where we have to sum over all quarks i), and E_a and H_a are the color electric and color magnetic fields at the surface of the bag.

II. The spherical bag

An approximation (which is usually made) is to assume [2] that the bag is spherical with a radius R. For most cases a reasonable approximation for R is

$$R \simeq r_0 N^{1/3} \quad \text{with} \quad r_0 = 0.72 \text{ fm} \quad . \qquad (2)$$

Here N is the number of quarks and antiquarks inside the bag. For the nucleon R = 1.04 fm.

Because one uses a bag fixed in space one finds for the total momentum $\underline{P}$ of this hadron

$$<\underline{P}> = 0 \quad \text{but} \quad <\underline{P}^2> \neq 0 \quad .$$

The energy of this hadron of mass M is then given by

$$E = <\sqrt{P^2 + M^2}> = E_B + E_Q + M_m \qquad (3)$$

where E_B is the bag energy, E_Q the quark energy, and M_m the color magnetic interaction energy. The bag energy $E_B = B V - Z_1/R$ contains the volume energy B V (for a nucleon about 280 MeV) and a correction term due to the zeropoint energy of the quark fields and gluon fields in the bag. Because [3] $Z_1 \simeq 1.01 \simeq 200$ MeVfm this correction becomes for the nucleon about -190 MeV.

Inside the bag we have very light quarks $m_u \simeq m_d \simeq 0$, $m_s \simeq 280$ MeV. The Heisenberg uncertainty principle requires then for each quark an energy $E \simeq p \simeq 1/R$. The exact calculation [2] gives for the quark energy $E_Q = \sum_i \alpha_i/R$, where for nonstrange quarks $\alpha_n = 2.04 = 402.5$ MeVfm. The quark energy of the nucleon is therefore $E_Q \simeq 3 \times 390$ MeV $\simeq 1170$ MeV. The discussion of the term M_m we will postpone to section IV. The main contribution to the energy of the bag and therefore to the mass of the hadron comes from the quark energy. A rough approximation for the mass of a hadron is so $M = N \alpha_n/R$. To correct for the zeropoint motion we observe that for heavy hadrons

$$E = <\sqrt{M^2 + P^2}> \simeq M + <P^2>/2M \quad .$$

Because for each quark $p \sim 1/R$ we find for a hadron that $<P^2> \sim NC/R^2$, together with the above rough approximation for the mass M we get [3]

$$E \simeq M + Z_2/R \quad ,$$

with $Z_2 \simeq 0.83 \simeq 164$ MeVfm.

For heavier hadrons we may write therefore

$$M = M_0 + M_m \qquad (4)$$

where

$$M_0 = BV - \frac{Z_0}{R} + \sum_i \frac{\alpha_i}{R} \qquad (5)$$

with $Z_0 = Z_1 + Z_2 = 1.84 = 363$ MeVfm.

For the light pion one needs to make another approximation

$$E = \langle \sqrt{P^2 + m_\pi^2} \rangle \simeq \langle P \rangle + \langle \frac{m_\pi^2}{2P} \rangle \quad .$$

It has been shown [3] by K. Johnson et al. that one can obtain a zeromass pion if one starts with zeromass quarks. The correct pionmass requires then a nonstrange quarkmass $m_n \simeq 33$ MeV.

III. The stringlike bag

Another useful approximation that can be made for the cavity is that of a fast rotating stringlike bag [4]. The quarks reside only at the ends of this stringlike bag. At each end the colors must couple to nonzero and complementary colors C and C* such that the whole bag can be colorless.

In the rest system this string has a length ℓ and a constant cross section A_0. This string (outside the ends) is filled with color electric fields. Gauss theorem gives

$$E_a = g F_a / A_0$$

where F_a is the colorspin of one of the ends and g is QCD coupling constant ($\alpha_c = g^2/4\pi$). The cross section A_0 is determined by pressure balancing eq (1). This requires that

$$B = \frac{1}{2} \sum_a E_a^2 = 2\pi \alpha_c f_c^2 / A_0^2$$

or

$$(2A_0 B)^2 = 8\pi \alpha_c B f_c^2 \qquad (6)$$

The energy density ε_0 of this string is in the rest system

$$\varepsilon_0 = B + \frac{1}{2} \sum_a E_a^2 = 2B \quad .$$

The mass of the hadron (neglecting the quark contributions) is then

$$M = (2A_0 B) \ell \qquad (7)$$

In the case of a fast rotating string the angular velocity ω is determined by the velocity $v = \frac{1}{2} \ell \omega$ of the ends. Because these ends cannot go faster than with the light-velocity $c = 1$, we have

$$\frac{1}{2} \ell \omega \simeq 1 \qquad (8)$$

The angular momentum L of such a fast rotating string is then

$$L = (\frac{1}{12} M \ell^2) \omega = \frac{1}{6} M \ell = \frac{1}{6(2A_0 B)} M^2 = \alpha' M^2 \qquad (9)$$

A neater calculation [4] shows that the slope α' of this Regge trajectory is given by

$$\alpha' = \frac{1}{2\pi(2A_0B)} = \frac{1}{2\pi\sqrt{8\pi\alpha_c Bf_c^2}} \qquad (10)$$

We note that the slope α' of the Regge trajectory depends only on the color structure of the string and is <u>independent</u> of the number of quarks and of their flavors. This explains nicely the observed universal slope $\alpha' \sim 0.9$ GeV^{-2} for the meson and baryon trajectories. In both cases the string has the color 3-3* configuration for which $f_c^2 = 4/3$. Using the values for α_c and B as determined [2] from the masses of N, Δ, and ω one finds from (10) that $\alpha' = 0.88$ GeV^{-2} in remarkable agreement with the experimental value.

A very important point to note here is, that the mass M and the angular momentum L are totally due to the gluonfields and have for large values of L no contributions from the quarks.

The radius R_0 of a string with color 3-3* at rest is (eq. 10) given by

$$R_0 = (2\pi\sqrt{\alpha'B})^{-1} = 1.47 \text{ fm} \qquad .$$

This is really quite a fat string!

IV. The color-magnetic interaction

Up till this point we considered only contributions to the mass that were spin and isospin independent. Strangeness is only broken because the mass of the strange quark s is heavier than the mass of the nonstrange quark n. We obtain this way only multiplets with a very large degeneracy. Fine structure in these multiplets arise from the color-magnetic interaction.

Because the quarks have 8 color charges $g F_a$, they have also 8 color-magnetic moments $\underline{\mu}_a \sim g F_a \underline{\sigma}$. Between quarks in relative s-waves the color-magnetic interaction is

$$M_m = m \Delta \qquad (11)$$

where

$$\Delta = - \sum_{i>j} (\sigma F)_i \cdot (\sigma F)_j \qquad (12)$$

In the spherical bag one can calculate m and

$$m = \frac{a}{R} \simeq \frac{b}{N^{1/3}} \quad \text{with} \quad a = 76.6 \text{ MeVfm}$$
$$\text{and} \quad b = 107 \text{ MeV} \qquad .$$

The sum (12) over all quark pairs can rather easily be performed. For a state with N nonstrange quarks one obtains

$$\Delta = \frac{1}{2} f_c^2 + \frac{1}{3} N(N-6) + I(I+1) + \frac{1}{3} S(S+1) \qquad (13)$$

For n^3 states with only nonstrange quarks in 1s states of the spherical bag the Pauli principle requires that either $I = S = \frac{1}{2}$, which is the N(939), or $I = S = \frac{3}{2}$, which is the $\Delta(1232)$. Their mass difference is due to the color-magnetic interaction and is equal to $4m \simeq 300$ MeV.

In a stringlike bag with two colored clusters of quarks at each end we assume that:

i) the quarks in each cluster are in relative s-waves;

ii) the color-magnetic interaction within the clusters is $m_1\Delta_1$ and $m_2\Delta_2$, and

iii) the interaction $m_{12}\Delta_{12}$ between the quarks in different clusters may be neglected.

Then

$$M = m_1\Delta_1 + m_2\Delta_2 \tag{14}$$

This means for example that we assume that for $Q\bar{Q}$ mesons the color-magnetic interaction disappears for higher L.

V. The model

It is now possible to introduce the model that has been used by us in Nijmegen to calculate masses, spin and parities of the different hadrons.

This model is chosen in such a way that it interpolates straightforwardly between the spherical bag states (L = 0) and the high L stringlike bags. This gives the mass formula

$$M = M_L + M_m \tag{15a}$$

with

$$M_L^2 = M_0^2 + \left(\frac{1}{\alpha'}\right) L \tag{15b}$$

and

$$M_m = m_1\Delta_1 + m_2\Delta_2 \; (+ m_{12}\Delta_{12}) \tag{15c}$$

The term $m_{12}\Delta_{12}$ has to be taken fully into account for the L = 0 groundstates and will practically always be assumed to vanish for $L \geq 1$. One expects, of course, the presence of spin-orbit forces and other spin dependent forces between the quark clusters at opposite ends of the string. Looking at the experimental data it seems that these are in general small. Due to the lack of accurate experimental mass determinations for the excited states it is not possible at present to give a better description of the color-magnetic interaction between the clusters.

It is important to stress that <u>the mass formula (15) contains no free parameters</u> anymore.

When the cluster i has N_i quarks coupled to total spin S_i and isospin I_i, then the quantum numbers of the hadron are given by

$$\underline{S} = \underline{S}_1 + \underline{S}_2 \;, \quad \underline{J} = \underline{L} + \underline{S} \;,$$

$$P = (-)^{L+N(\bar{Q})} \;, \quad \underline{I} = \underline{I}_1 + \underline{I}_2 \;, \text{ etc.}$$

This leads quite often to several degenerate states.

This model has been applied by us to the $Q\bar{Q}$ mesons [5] and the Q^3 baryons [6]. This is the only place where the model can really be checked. The application to the $Q^2\bar{Q}^2$ baryonium states [5], the dibaryon resonances [7] and the $Q^4\bar{Q}$ baryons [8] need still experimental confirmation, but the preliminary results are quite promising.

VI. The $Q\bar{Q}$-mesons

The MIT bag model in the spherical approximation [2,3] gives a good description of the L = 0 pseudoscalar and vector nonets. The treatment of the pseudoscalar mesons requires some extra care. For more details one must look at refs 2 and 3.

For the orbitally excited states one predicts that for each L there are four nonets. For example for L = 1 one predict nonets with $J^{PC} = 1^{+-}$; 0^{++}, 1^{++}, and 2^{++}. When one neglects the color-magnetic interaction then one expects [5] the nonstrange mesons $n\bar{n}$ all at $M \simeq 1285$ MeV and the $s\bar{s}$ mesons all at $M \simeq 1475$ MeV. In table 1 are given the assignments.

J^{PC}		1^{+-}	0^{++}	1^{++}	2^{++}
$n\bar{n}$	I = 1	B(1231)	?	A_1(1280)	A_2(1312)
$n\bar{n}$	I = 0	H(1190)	ε'(~1300)	D(1285)	f(1270)
$s\bar{s}$	I = 0	?	ε''(~1425)	E(1431)	f'(1516)

Table 1: The L = 1 $Q\bar{Q}$ mesons.

One notes a reasonable agreement. The mesons with S = 0 are perhaps about 50 to 100 MeV lighter than the S = 1 mesons. This indicates that for the L = 1 mesons the color magnetic interaction is not yet negligible, but is a factor 5 or 10 weaker than for the L = 0 ground state.

Important to note is also that none of the scalar mesons ε(~700), δ(980), and S*(980) are assigned as $^3P_0$ $Q\bar{Q}$-states. They find a very natural explanation as $Q^2\bar{Q}^2$-states as shown by R.L. Jaffe [9,10].

VII. The Q^3-baryons

The baryons are customary classified according to $SU(6) \otimes O(3)$. The lowest positive parity states (a $J^P = \frac{1}{2}^+$ octet and $J^P = \frac{3}{2}^+$ decuplet) form a [56] 0^+ multiplet. This L = 0 multiplet is well described in the original MIT-bag model [2].

We will take a better look at the L = 2 orbital excitations. These were normally classified in a [56] $\oplus$ [70] 2^+, but a more natural classification in our model is really [6] $\otimes$ [21] 2^+. Neglecting the color magnetic interaction between the ends of

the stringlike bag the fine structure of this multiplet is determined by the fine structure of the diquark Q^2. For only nonstrange quarks one predicts then only 2 levels. The lowest one at M = 1.67 GeV is a nucleon resonance with $I = S = \frac{1}{2}$ and therefore with $J^P = \frac{3}{2}^+$ and $\frac{5}{2}^+$. The higher one at M = 1.90 GeV consists of
i) nucleon resonances with $S = \frac{1}{2}$ and $\frac{3}{2}$, therefore with $J^P = \frac{3}{2}^+, \frac{5}{2}^+$, and $J^P = \frac{1}{2}^+, \frac{3}{2}^+, \frac{5}{2}^+, \frac{7}{2}^+$, and
ii) delta resonances with $S = \frac{1}{2}$ and $\frac{3}{2}$, therefore with $J^P = \frac{3}{2}^+, \frac{5}{2}^+$, and $J^P = \frac{1}{2}^+, \frac{3}{2}^+, \frac{5}{2}^+, \frac{7}{2}^+$.

We can compare this with the resonances determined in the latest Karlsruhe-Helsinki phase shift analysis [11] as given in table 2.

M	J^P	N	Δ
1.90	$7/2^+$	F17(2005)	F37(1913)
	$5/2^+$ (2x)	F15(1882)	F35(1905)
	$3/2^+$ (2x)	?	P33(1868)
	$1/2^+$	P11(2050)	P31(1888)
1.67	$5/2^+$	F15(1684)	
	$3/2^+$	F13(1710)	

Table 2: Resonance parameters of the $L^P = 2^+$ nonstrange baryon resonances as determined by the Karlsruhe-Helsinki group [11].

One notes a good agreement. Because the mass determinations have still quite some error the data give no indication of possible spin-orbit and other color magnetic effects between the clusters at the ends.

The $L^P = 1^-$ multiplet [70] ⊕ [56] 1^- requires a more extended discussion (see refs 6 and 8) and the introduction of some more parameters.

VIII. The Q^6-dibaryon resonances

Having discussed the two cases $Q\bar{Q}$ and Q^3, which can be checked against the abundant experimental data, we come to the conclusion that the model works surprisingly well. The next step is to look at those cases where the experimental data are still very scarce and incomplete. In the remaining time I would like to discuss the predictions for the dibaryon resonances.

Because one predicts very many, rather closely space dibaryon resonances it will be very hard to really check the predictions of the model. At present the best thing perhaps is to concentrate at the lower end of the mass spectrum, because there the states are not yet so dense. Personally I think that the easiest system to check the predictions is perhaps the ΛN system. The reason is that the lowest lying ΛN resonances are predicted below the pion production threshold at E_{THR} = 2.19 GeV. In the experimentally easier NN system the lowest dibaryon resonances are predicted rather far above the pion production threshold E_{THR} = 2.01 GeV. The extra complications due to inelastic processes make the data analysis in the NN case a lot more difficult.

The lowest three predicted ΛN resonances are [7]:

i) Two $L^P = 1^-$ resonances in the color 3-3* Q^4-Q^2 configuration. One at M = 2.11 GeV with S = 0 so $J^P = 1^-$ and the other one at M = 2.15 GeV with S = 1 so $J^P = 0^-, 1^-$, and 2^-. We would like to identify these with the enhancements at M = 2.14 GeV seen by T.H. Tan [12] and also other groups and at M = 2.18 GeV seen by B.A. Shahbazian et al. [13].

ii) Next comes the lowest $L^P = 0^+$ resonance in the Q^6 configuration at M = 2.24 GeV with $J^P = 2^+$. A possible candidate for this state is the resonance seen at M = 2.256 GeV by Shahbazian et al. [13].

At present it is quite important to confirm these resonances and to try to determine their spins and parities to see if the assignments are correct.

In the NN-system the lowest predicted six quark state must appear as a resonance in the I = 0 $^1P_1$ channel at M $\simeq$ 2.11 GeV. This state has the L = 1 color 3-3*, Q^4-Q^2 configuration with I = 0 and S = 1. We expect therefore degenerate states with $J^P = 0^-$, 1^-, and 2^-. The quantum numbers I = 0, $J^P = 0^-$ and 2^- are not available in the NN system. The states with these quantum numbers, called extraneous states [14], can only decay into NNπ and are therefore possibly quite narrow. One way to produce these extraneous states is by photo- or electro-desintegration of the deuteron. Other notable predictions in the I = 0 NN-system are [7]:

L = 2 Q^4-Q^2 resonances at M = 2.33 GeV with $J^P = 1^+, 2^+, 3^+$, and a
L = 0 Q^6-resonance at M = 2.36 GeV with $J^P = 3^+$.

The odd parity I = 1 NN-resonances with lowest mass have all the color 3-3* Q^4-Q^2 configuration with L = 1. In the different partial waves they are predicted at the masses given below in GeV. Also their degeneracy is indicated.

$^3P_0$	M =		2.25,	2.34
$^3P_1$	M =	2.20,	2.25,	2.34 (4x)
$^3P_2 + {}^3F_2$	M =		2.25,	2.34 (3x)
$^3F_3$	M =			2.34 (2x)

In the even parity waves of the I = 1 NN-system resonances are predicted [15] in the L = 0 Q^6 configuration in $^1S_0$ at M = 2.24 GeV and $^1D_2$ at M = 2.36 GeV. The $J^P = 0^+$ state at M = 2.24 GeV is strongly coupled to the $^1S_0$ NN-channel. Because there are no barriers present to prevent the decay, this resonance is very probably so wide, that it will be very hard to detect. The $J^P = 2^+$ state at 2.36 GeV is strongly coupled to the $^5S_2$ NΔ-channel. This coupling shifts this state almost surely to the NΔ-threshold at 2.17 GeV.

References

1. A. Chodos et al., Phys.Rev. D $\underline{9}$ (1974) 3471
2. T. DeGrand et al., Phys.Rev. D $\underline{12}$ (1975) 2060
3. J.F. Donoghue and K. Johnson, Phys.Rev. D $\underline{21}$ (1980) 1975
4. K. Johnson and C.B. Thorn, Phys.Rev. D $\underline{13}$ (1976) 1934
5. A.T. Aerts, P.J. Mulders and J.J. de Swart, Phys.Rev. D $\underline{21}$ (1980) 1370
6. P.J. Mulders, A.T. Aerts and J.J. de Swart, Phys.Rev. D $\underline{19}$ (1979) 2635
7. P.J. Mulders, A.T. Aerts and J.J. de Swart, Phys.Rev. D $\underline{21}$ (1980) 2653
8. J.J. de Swart, P.J. Mulders, and L.J. Somers, Nijmegen preprint THEF-NYM-80.15. To be published in the Proceedings of the "Baryon 1980" Conference held in Toronto, Canada.
9. R.L. Jaffe, Phys.Rev. D $\underline{15}$ (1977) 267, 281
10. R.L. Jaffe and F.E. Low, Phys.Rev. D $\underline{19}$ (1979) 2105
11. G. Höhler et al., Handbook of Pion-Nucleon Scattering, ZAED Physics Data 12-1 (1979)
12. Tai Ho Tan, Phys.Rev.Lett. $\underline{23}$ (1969) 395
13. B.A. Shahbazian, Nukleonika $\underline{25}$ (1980) 345
14. P.J. Mulders, A.T. Aerts and J.J. de Swart, Phys.Rev.Lett. $\underline{40}$ (1978) 1543
15. R.L. Jaffe, Phys.Rev.Lett. $\underline{38}$ (1977) 195 (E 617)

THE CHIRAL BAG MODEL AND THE LITTLE BAG

Vincent VENTO
DPh-T, CEN-Saclay
B.P.N°2, 91190 Gif-sur-Yvette,
France

ABSTRACT : We review the properties of the existing solutions to the Chiral bag equations of motion and discuss how the "little bag" picture could come about in this scheme. Our analysis leads to a model which is qualitatively similar to the naive quark model with pion cloud corrections. We use this latter approach to look for pion cloud signatures in experimental data.

I. INTRODUCTION

The "little bag" is an attempt to incorporate in a predictive scheme, the knowledge on the structure of the hadrons, as seen by short wavelength probes, and the interaction among them in the long wavelength regime.

The first step towards the "little bag" is what we have called the chiral bag model[1]. The latter incorporates, not only asymptotic freedom and confinement, as in the MIT bag model[2], but also chiral symmetry, a property of quantum chromodynamics (QCD) for massless quarks. For massless up and down quarks, the procedure consists in introducing a pion field outside the confinement region to assure axial current conservation. The pion field in particular, and chiral symmetry in general are viewed as the link to join the features of the two energy regimes. This pion field represents in some way the non perturbative aspects of QCD beyond confinement. How the pionic degrees of freedom arise from this underlying theory is still an open question. Several attempts in this direction have appeared in the literature[3,4]. Our approach is though phenomenological. We construct a model with minimal number of ingredients to preserve chiral symmetry. If our pion field is a complicated vacuum excitation, or some other phenomena of an extremely rich theory, QCD, is a question that remains to be answered.

Other meson-like degrees of freedom might also be included within our scheme, but in this early attempt simplicity has always been the guiding rule.

The emphasis in the little bag philosophy is that the mesonic degrees of freedom introduced via chiral symmetry play a crucial role. Why little and not chiral ? Although chirality is the main reason for introducing our approach, chirality can be minimized by perturbing around the MIT solution. The little bag, will arise when chiral effects are maximal, which in our model will require the whole non-linear structure of the theory. As we shall see there is a connection between size and large pion effects.

We use in our calculations qualitative features of the so-called perturbative expansion, mainly because there is nothing better we can do at this stage, but we are guided in our intuition by the physics coming from the lowest energy classical spherical solution to the non-linear equations we have, the hedgehog[1].

II. STATIC SOLUTIONS TO THE BAG EQUATIONS

The mathematical formulation of the chiral bag model gives rise to a system of non-linear coupled equation of motion for the fields[1]*. In order to solve them, two procedures have been followed : (1) the perturbative expansion[5,6] ; (2) the mean field approximation[1].

The perturbative expansion emphasizes confinement and assumes chiral effects to be small in comparison. The procedure is to expand the fields in terms of the coupling constant. For the observables this turns out to be an expansion in an effective dimensionless parameter ϵ,

$$\epsilon \sim \frac{1}{f_\pi^2 R^2} \qquad (II.1)$$

where f_π is the pion decay constant and R the bag radius. The method is thus useful for "large" radii.

To first order the pion field vanishes and the quark field satisfies the MIT equations of motion. In order to preserve chirality the second order in the pion field has to be included. This gives rise to a source for linear pions and reproduces the initial little bag picture[7]. The pion source can be used within the reduction formalism to compute pion emission and absorption processes. The coupling of quarks to pions gives rise to a tensor like force that produces to next order a D-state admixture. A problem arising in this scheme is that the pressure balance equation cannot be satisfied locally, unless some kind of explicit surface deformation is assumed to this order. Most calculations proceed to avoid this difficulty by averaging over the angular dependence, which is equivalent to defining the pressure balance equation by

----
*We shall omit in what follows any discussion about the gluonic degrees of freedom, which should be included in the spirit of perturbative QCD inside the bag.

$$\frac{dE}{dR} = 0 \qquad (II.2)$$

The effect of the pion field in the spectra is relatively small, although qualitatively significant. It gives rise to contributions to the energy and to the pressure which tend to shrink the bag[6,8]. If one remains within the spirit of the large bag, and neglects the deformation of the surface, one achieves a simple predictive scheme. Magnetic moments and other quantities can thus be calculated[9]. The most crucial feature of the calculation, where chirality appears in its full glory is g_A. As shown initially by Jaffe[5], the model g_A not only satisfies the Goldberger-Treiman relation, but due to the contribution from the pions becomes bigger than the experimental value of 1.25, no matter what radius one chooses. All other calculated effects disappear in the $R \to \infty$ limit, but not this one. What appeared as a crudeness of the model might open our understanding of the structure of the hadrons.

The mean field approach has given rise to a classical spherical solution to the equations of motion[1]. Unusual quantum numbers have to be defined in order to satisfy the boundary conditions. The pion field, now a c-number field, points in the radial direction at each point, thus its name : hedgehog. Taken seriously it provides us qualitatively with the picture we are seeking, the little bag.

Let us point out some of its features. For every allowed bag constant we have two solutions at different radii. The small bag solution corresponds to a maximum of the energy functional,

$$E = E(R)$$

for fixed f_π and B, where B is the bag constant, thus unstable. The big radius solution leads to a shallow minimum, i.e., stable. The rising parabola shape for small radii of the non chiral solutions is lost. The big bag solutions are dominated by non chiral effects, MIT like solutions ; for the small bag ones, the pionic effects are dominant. Beyond the small radius extremum the system collapses*. A feature of the small bag solution which is extremely exciting, is the non-relativistic behaviour of the quark wave functions. The axial vector coupling g_A, can be calculated exactly, and as a function of bag radius is monotonically decreasing, and always bigger than the experimental value, even for very large radii. From the studied quantities, g_A is the only one that presents this feature.

---

*Notice that for fixed $B > 0$ and f_π there exist in the energy functional two extrema, a maximum and a minimum, but only one of them satisfies the Goldberger-Treiman relation with fixed pion nucleon coupling constant $\frac{f^2}{4\pi} = .081$.

Other aspects of the solution, specially in relationship with magnetic moments have been discussed elsewhere[10], and we shall omit them for brevity.

III. THE LITTLE BAG

The previous section has taught us how, the pion field affects the bag picture once it is included via chiral symmetry. The main problem we face at this stage is how to eliminate the collapse feature of the energy functional at small radii. Let us conjecture that there exists some kind of mechanism in the true solution of the field equations including quantum effects, that recovers the negative slope at small radii. Three possibilities can be envisaged : (1) only the perturbative minimum remains. (2) the maximum of our energy functional is just a signal for a second minimum beyond it[11]. (3) only this second minimum remains and the perturbative minimum is washed out[12].

If we believe in the first possibility, it is clear that the theory as defined initially by the MIT group would be close to reality, although chirality enriches it tremendously by including the effects we have mentioned before, i.e., pion self-energy, pion pressure, contribution to g_A, D-state admixture etc... . If the second alternative is the chosen one, hadrons will behave differently according to determined initial conditions. Two extreme alternatives might occur in this case. One, the energy functional for fixed B and f_π is unaltered and the hadron just jumps from one minimum to the other by changing its coupling to pions. The second one is that B and/or $f_\pi(g_A)$ change while leaving the pion coupling fixed, i.e., the functional changes. If we consider the energy as a function of all its degrees of freedom R, B and f_π the different alternatives one could envisage represent motions along different paths in the space of the parameters.

If we believe blindly in the model as it stands, appart from the stabilization mechanism, charge radii tell us that large confinement regions would be a property of free nucleons and small bags would arise in nuclear matter where the "quark soup" has not been observed at nuclear matter densities.

Finally and this is the realization of the little bag idea, the third possibility implies that the complicated bag model we started from, transforms into an almost non relativistic quark model with pion cloud corrections. Effectively this leads to a quasi-particle interpretation in which the classical solution would provide a mechanism to generate the almost free quasi-quarks endowed with a mass of $\sim \frac{1}{3} m_N$. Sphericity though, produces a large deviation of g_A from its experimental value, thus the true baryons should have a large amount of D-state admixture.

This picture we envisage when properly formulated should lead to the non-relativistic quark model[13], with some corrections due to pion cloud contributions and departures from sphericity.

IV. SOME RESULTS WITHIN NAIVE QUARK MODEL APPROACH

The previous section has been mainly one of conjectures as a result of some model calculations. But while the search for the theory that provides us with some of the answers continues, let us look from a more pragmatic prospective to signatures of some of the ideas we have advocated.

We have performed a series of calculations within the naive quark model assuming a D-state admixture in the usual quark wave function[14]. In the case of g_A the result is encouraging

$$g_A = \frac{5}{3}(1 - \frac{6}{5} p_D^N) \quad . \tag{IV.1}$$

Glashow[15] obtained this result arguing by analogy to the nuclear three body problem. Other quantities within the same approach show also the right tendency $g_{\Delta\pi N}/g_{\pi NN}$[14] and $\frac{D+F}{D-F}$[15]. The amount of D-state admixture depends strongly on the initial spherical state one starts from. We shall not make at this moment a strong point on its magnitude, just say it is large compared to similar situations in nuclear physics, i.e. 7 % for triton. We advocate in our calculation for a tensor force like interaction produced by quark pion coupling in the spirit of the chiral bag model results mentioned earlier.

Pion cloud effects in magnetic moments can also be significant as was shown[10] in a calculation for strange baryons in which deformation effects were not included.

Preliminary results in the Δ-γ N E2 transition in the same spirit[16] look promising. The fact that this calculation depends on the explicit radial wave function one chooses, makes predictions though, less universal.

In our opinion a complete reanalysis of the naive quark model results with the new ingredients will most certainly not only shed some light in our understanding of the properties of the baryons but point out the experiments that need to be performed to clarify these conjectures.

V. CONCLUSION

The results obtained with the naive quark model approach, as well as the systematic deviation of g_A from the experimental value point towards a more complicated structure of the baryons than initially suspected. Chiral bag models provide qualitative understanding for these features. The duality between large and small bags is still unresolved. Most of the static properties are almost radius independent and the one that is not, is difficult to calculate within the chiral bag approach beyond lowest relevant order in the perturbative expansion.

The collapse feature of the energy functional is a deep problem. We might just overcome it by ignoring its existence and accepting the energy functional close to its stability point but not beyond it ; or try to understand it. This might bring new physics into the picture and the success of the non relativistic model points to some of us that the direction towards maximal chirality, i.e., little bag, is worth pursuing.

The calculations within the naive quark model have also shown, that this extremely successful approach might need revision, and that experimental deviations from its predictions might open our understanding for phenomena that were omitted in its original formulation. This signals the possibility of beginning to understand the overlap between the "old" and the "new" degrees of freedom and points towards new theoretical and experimental work that needs to be done. A concrete experimental proposal would be to determine more accurate E2 matrix elements.

The work on which this notes are based has been done in collaboration with G.E. Brown, A.D.Jackson, J.H.Jun, E.M.Nyman and M.Rho. The more speculative part is a consequence of lengthy discussions with G.E.Brown and M.Rho.

REFERENCES

(1) V.Vento, M.Rho, E.Nyman, J.H.Jun and G.E.Brown, Nucl.Phys. A345 (1980) 413.

(2) A.Chodos et al., Phys.Rev. D10 (1974) 2594 ; T.De Grand et al., Phys.Rev. D12 (1975) 2060.

(3) C.G.Callan, R.F.Dashen and D.J.Cross, Phys.Rev. D19 (1979) 1826.

(4) T.Saito and K.Shigemoto, Prog.Theo.Phys. 63 (1980) 256.

(5) R.L.Jaffe, 1979 Erice Summer School "Ettore Majorana".

(6) V.Vento, Stony Brook thesis ; unpublished.

(7) G.E.Brown and M.Rho, Phys.Lett. 82B (1979) 177 ;
G.E.Brown, M.Rho and V.Vento, Phys.Lett. 84B (1979) 383.

(8) F.Myhrer, G.E.Brown and Z.Xu, to be published.

(9) J.F.Logeais, M.Rho and V.Vento, unpublished.
I.Hulthage and J.Wambach, private communication.

(10) G.E. Brown, M.Rho and V.Vento, to be published in Phys.Lett.

(11) M.Rho, Contribution to the International Conference on Nuclear Physics, Berkeley 1980.

(12) G.E.Brown, Contribution to the Symposium on Perspectives in Electro-and-Photo-Nuclear Physics, Saclay (1980).

(13) H.J.Lipkin, Phys.Rep. 8C (1973) 175.

(14) A.D.Jackson and V.Vento, to be published.

(15) S.L.Glashow, Physica 96A (1979) 27.

(16) G.E.Brown, M.Rho and V.Vento, work in progress.

* * *

THE COLOUR DEGREE OF FREEDOM AND MULTIQUARK STATES

H. Högaasen
University of Oslo, Norway

In this talk I will show some examples of how one starts to understand hadron spectroscopy from the interaction between more elementary constituents.

Since november 1974 when the ψ/J and its partner ψ' were discovered, the consensus of particle physicists is that the quark model is no longer a model. The mesons are made up of (mainly) a quark and an antiquark and the nucleons (essentially) of three quarks. Many physical states that were thought to be elementary are now known to be composite. This automatically induces a shift in attitude when one looks for fundamental interactions. We all know how the extremely varied and complicated interatomic interactions were explained from electrodynamics when the internal structure of atoms was understood and quantum mechanism was used. One can now believe that nuclear forces in turn will be understood, when we have obtained a really good understanding of interquark forces. As we shall see this will not be easy, but there is progress in that direction.

What are the physical properties of the quarks? I have always found that it is an amusing activity to see what the first natural philosophers used as words and concepts. Unhappily I have not been able to read "περὶ φύσεως" by Anaxagoras in original, but according to a partial translation[1] from 1882, the building blocks of nature are distinguished from each other by "Gestalt, Farbe und Geschmack". These two last properties are today known as colour and flavour. The quarks of definite flavour come in three colour states that are denoted by red, white and blue - a quark transforms under the fundamental representation of SU_3^{colour}. Of flavours we know five: u, d, s, c and b, - the discovery of the sixth t is supposed to be only a matter of time.

The flavour degree of freedom specify the kind of quark. With only two flavours all the building blocks of ordinary nuclear physics can be quantitatively described: The proton (neutron) is made of uud (udd) and the mesons π, ρ and ω are made up from the isodoublet (u,d) and the (anti) doublet $(\bar{d}, -\bar{u})$. There seems to be no need for other flavours to make an acceptable normal universe as theory is today. Nevertheless, the study of the "abnormal" flavours as strangeness, charm and bottom has been the road to the understanding of normal matter.

The colour degree of freedom is the same for all flavours and it is the root of the interquark forces. It is impossible today to speak about colour without mentioning quantum colourdynamics or QCD. It is a gauge theory of quark fields $q(x)$, it has a Lagrangian and therefore looks in all ways as respectful as QED. The forces between quarks are mediated by an octet of gauge field represented by the vector fields for gluons $G_\mu^a(x)$, where a runs from 1 to 8, μ from 1 to 4. The Lagrangian is

$$L(x) = -\frac{1}{4} F_{\mu\nu}^a F_a^{\mu\nu} + \bar{q}[i\gamma_\mu(\partial_\mu - ig_s G_\mu^a \frac{\lambda^a}{2}) - m]q$$

Had it not been for the colour indices a on the fields, and the presence of the eight Gell-Mann matrices λ^a, this would have been like the QED Lagrangian.

Forces between quarks are therefore propagated by the octet of massless gluons coupled to the eight quark currents $J_\mu^a = \bar{q} \gamma_\mu \frac{\lambda^a}{2} q$ as $-ig_s J_\mu^a G_\mu^a$ just as the photon A_μ couples to the electromagnetic current $j_\mu = \bar{\psi} \gamma_\mu \psi$ as $-ie\, j_\mu A_\mu$.

There is a most important difference however between QCD and QED. In QCD the field tensors $F_{\mu\nu}^a$ are defined as

$$F_{\mu\nu}^a = \partial_\mu G_\nu^a - \partial_\nu G_\mu^a - g_s f_{abc} G_\mu^b G_\nu^c$$

where f_{abc} are the structure constants of the gauge group SU_3. Because of the last term in the field tensor, interaction between gluons is as strong as between quarks. QCD is therefore basically a nonlinear theory and therefore mainly unsolvable. At very small distances however, the theory is becoming asymptotically free, the coupling constant g_s is then so small that the dynamic equations can be linearized, one can apply perturbation theory and make analogy with electrodynamics. We can speak of eight (noncommuting) colour charges and eight (noncommuting) colour-magnetic moments for the coupling of gluons to quarks. In the 1 particle exchange approximation we have then the following analogs between electromagnetism and chromodynamics:

Electric potential	$V_{ij} = \frac{Q_i Q_j}{r_{ij}}$	1
Colour electric potential	$V_{ij} = \sum_{a=1}^{8} \frac{K_i^a K_j^a}{r_{ij}}$	2
Magnetic moment	$\underline{\mu} = \frac{g}{2mc} Q \underline{S}$	3
Eight colourmagnetic moments	$\underline{\mu}^a = \frac{g_c}{2mc} K^a \underline{S}$	4

Here Q is the electric charge operator, $K^a = g_s \cdot \frac{1}{2} \lambda^a$ are the colour electric charge operators, g and g_c are the g-factors and g_s the gluon-quark coupling constant.

Unhappily it is for dimensions much smaller than the size of hadrons that one can treat QCD perturbatively. When separation distances are of the order of a sizeable fraction of one fermi, the nonlinearity of QCD makes it a theory that we cannot find solutions of. (This is also the case for the good old Navier Stokes equation in hydrodynamics!)

QCD may be right or it may be wrong, and the best one in most cases can do is to use it as an inspiration for making models.

The property of QCD that is the most important for spectroscopy is the conjecture of infrared slavery: quarks and gluons are always bound together such that only colour singlets can be free particles. We can call this the principle of confinement and it explains why there is no free quarks or gluons observed. Any model for baryons and mesons must therefore incorporate this.

Bag models[3] are popular as is evident from De Swart and Vento's talks. They impose colour confinement by using as a boundary condition that there is no colour flux out of a region in space, which then is called the bag.

I would like to mention two very nice applications of bag models, as they shed light on the origin of $q\bar{q}$ forces and short distance nucleon-nucleon forces. The first application[4] is made of two heavy quarks confined in a cavity, where they then interact via massless gluons. It then makes sense to use an adiabatic Born-Oppenheimer approximation. The bag equations are solved for fixed $q\bar{q}$ separation r, and the total energy is minimized with respect to variations in the shape of the confining surface. This then gives what one can call the potential energy of the colour singlet $q\bar{q}$ pair as a function of r, and this can then be put into a nonrelativistic Schrödinger equation. In figure 1 we see the calculated potential together with the ground state wave-functions of charmonium and bottomium. The potential is an interpolation between the Coulomb at short distances and a linear potential at great distances and this is indeed a kind of potential that has been used to give quantitative description of charmonium and bottomium.

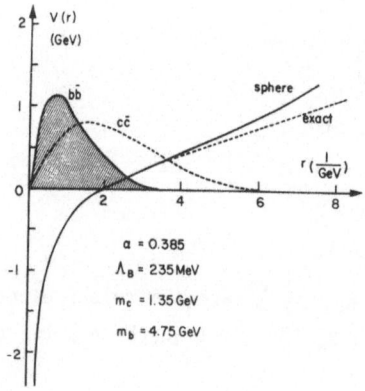

Fig. 1

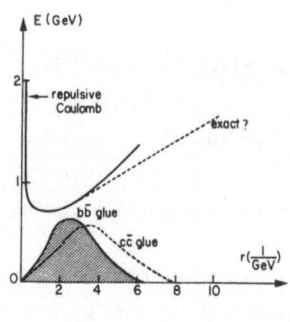

Fig. 2

In figure 2 the same is done when there is a "valence gluon" in the bag such that the $q\bar{q}$ is in a colour octet. There is then a repulsive force at very short distances, which however turn into attraction at around 0,2 fermi. It goes without saying that there is great interest to find states that contain valence gluons G. Here the calculations have been done for $q\bar{q}G$, still more spectacular would be the glueballs GG. We shall discuss "qqqG" states at the end of the talk.

A calculation which would be of great importance for nuclear physics would now be to do the same approach in a multiquark state where there are 6 quarks. Then one could compute a 3q -3q interaction potential which would be the short range nucleon nucleon potential, when the subsets of three quarks have the quantum numbers of the nucleons. To do this one should then fix the distance R between the center of mass of the two three quark states, and minimize the energy with respect to the bag shape. This is really a very hard problem and it is also an undertaking which is on more shaky theoretical grounds than the $q\bar{q}$ calculations we have just discussed. Quarks building up nucleons are light quarks, so that it is hard to justify any use of a Born-Oppenheimer approximation.

Nevertheless, de Tar has carried out this programme in a simplified form and obtained very interesting results.[5] Instead of fixing the separation distance R he introduces a variable δ which turn into R only at large distances. To describe the separation of the three quark clusters he introduces left and right orbitals by linear combinations of S and P states in the spherical bag (q_S and q_P)

$$q_L = q_S - \sqrt{\mu}\, q_P$$
$$q_R = q_S + \sqrt{\mu}\, q_P$$

The parameter μ ranges from 0 to 1 for maximal to minimal overlap between the orbi-

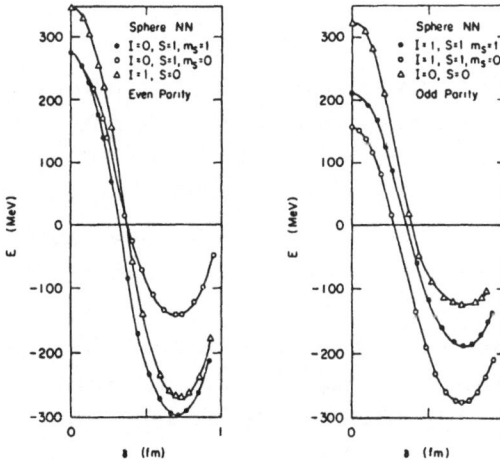

Fig. 3.

tals. δ is defined through μ as

$$\delta = \frac{2\sqrt{\mu}\;(1+\mu)}{1+\mu^2} \int q_s^\dagger(\underline{r})\; q_p(\underline{r}) Z\; d^3r$$

The resulting interaction energy as function of δ is shown in figure 3, it represents a calculation of the nucleon-nucleon interaction, may be not from first principles but nevertheless from an input which has its source in quark-quark physics. The resulting soft repulsive core and strong attractive region around $0.5 < \delta < 1$ fermi is certainly what we know that work in nuclear physics.

This can permit us to dream that one day all nuclear forces will be quantitatively understood from interquark forces. If this is the case nuclear forces will be nonlocal and there must be multibody internucleon forces in heavy nuclei.
In the MIT bag model one gets no bound states of baryons as long as one has only two flavours, and the existence of nuclei can therefore not be explained there. If one includes more flavours than two, bound states are predicted.[6-8] The one that should be least difficult to find (but still unfound) is a ΛΛ like state with a binding energy of around 20 MeV.

The last topic I want to mention is a subject I like a lot. It is the question if there are colour isomers among hadrons.

If one introduces a new degree of fredom in the description of a physical system, one always gets more states. In the history of physics there has always been a direct connection between the experimentally determined multiplicity of states and the number of degrees of freedom necessary to describe them.

The specific heat of gases shows that molecules have spatial extension: it increases when a cold gas is heated, because the rotational and vibrational degrees of freedom are unfrozen.

Closer study leads to the existence of stereo isomers: molecules with the same atoms have different physical properties because the atoms have a different spatial arrangement.

The electron spin leads to a doubling of states in the hydrogen atom. Spin of atoms was made obvious by the two silver spots in the Stern Gerlach experiment. Such examples are numerous. If Quantum Chromo Dynamics is correct, colour is as important for hadrons as electric charge is for atoms. It is therefore a problem of some importance to find evidence as direct as possible for the colour degree of freedom.

We hope that history will repeat itself and give us this proof through the complexity of hadronic spectra.

If colour isomers exist, why are they not more familiar? The reason lies in the principle of colour confinement, stating that physical systems that are not colour singlets cannot exist as free particles. By the multiplication (addition) rules for colour $3 \times \bar{3} = 1 + 8, 3 \times 3 = \bar{3} + 6, 3 \times 6 = 10 + 8$ one finds that the simplest mesons are made of $q\bar{q}$ pairs, the simplest baryons of qqq. For these states

the colour degree of freedom is completely frozen: there is one and only one way of
making a colour singlet. To unfreeze hidden (confined) colour we must heat hadronic
matter by creating at least an extra $q\bar{q}$ pair and make mesons as $qq\bar{q}\bar{q}$ systems,
baryons as $qqqq\bar{q}$. For $qq\bar{q}\bar{q}$ states we will have twice as many states if quarks have
colour than we would have with colour singlet quarks: A group $qq(\bar{q}\bar{q})$ can be either
a colour $\bar{3}(3)$ or a colour $6(\bar{6})$ states, when the total $qq\bar{q}\bar{q}$ system is a colour
singlet. Most of these states will however be belonging to the meson-meson continuum because the subgroups $q\bar{q}$ $q\bar{q}$ will, for most configurations, pair themselves in
separate colour singlets and escape from the colour confining prison as two ordinary
mesons. Such decays are called superallowed.[9]

Some multiquark states can however be prevented from dissociating in this way.
If we select configurations where an angular momentum separate coloured group of
quarks, the states can be sufficiently stable to show up as resonances.[10-12]

Some time ago I believed that colour isomers had been discovered in the meson
sector. The broad baryonium states ($3\bar{3}$) should have narrow colour isomers ($6\bar{6}$), but
the experimental evidence for these narrow states seem to fade away. It is therefore nice that in the baryon sector there are states found that have a nice interpretation as multiquark states, and that have properties that have been predicted.
They have been named mesobaryons[10] and for this there is a classification[13] that
is more than two years old and that shows signs of respectability.

The physical picture of the model is the following: the quarks inside the mesobaryonium cluster into two colour non-singlets. Inside each clusters the quarks are
in a relative s-wave, but an angular momentum L between the two clusters creates a
centrifugal barrier which prevents the quarks from recombining into unbound colour
singlets. The configurations most likely to be stable are $(qqq)-(\bar{q}\bar{q})$ where both
clusters form a colour octet (we will call these "octet-bonded" states) and
$(qq)-(qq\bar{q})$ in the colour $6-\bar{6}$ representations ("sextet-bonded" states).

The $\bar{3}-3$ configuration for the $(qq)-(qq\bar{q})$ system is highly unstable because the
colour triplet $qq\bar{q}$ can divide itself easily into a colour triplet q and a singlet $q\bar{q}$
and this singlet will not feel the confining forces.

Let us for simplicity discuss the case of the octet-bonded states: all the considerations we make will easily extend to the sextet-bonds. We use the same notation as in Ref. 13: $\theta^F(C,S)$ specifies the flavour (F), colour (C) and spin (S) quantum numbers of the three quark configuration, and $D^F(C,S)$ specifies in a similar way
the content of the $q\bar{q}$ system.

The spins $\vec{S}_1$ and $\vec{S}_2$ of θ and D combine to give a total spin $\vec{S} = \vec{S}_1 + \vec{S}_2$, which
then couples to L and one obtains a total angular momentum $\vec{J} = \vec{L} + \vec{S}$ with $J_{max} =$
$|L + S| \geq |\vec{L} + \vec{S}| \geq |L - S| = J_{min}$.

The two coloured clusters are bound together by strong colourelectric confining
forces, and in the cases where the colour flux tube ends on colour charges different
from 3 or $\bar{3}$, it cannot be broken by the creation of a $q\bar{q}$ pair. The system should

therefore have enhanced stability and should lead to resonances narrower and narrower as the angular momentum between the clusters is higher and higher. A limit for the mass of this states would be when the flux tube breaks by the creation of a pair of valence gluons.

Spin-spin forces between quarks in a relative s-state are known to be extremely important. If these were ignored, the mass of the pion and rho (or N and Δ) would be the same. The colourmagnetic forces that give rise to the spin-spin interaction are very short ranged[10]. When we compute masses the colourmagnetic interaction is therefore used only between quarks that are in relative s-waves.

It turns out that the colourmagnetic mass defect is maximal when three quarks form a colour octet, flavour singlet, with spin ½. This state $\theta^1(8.2)$ has the flavour content uds, and the lowest lying mesobaryons are made when this state is linked to a colour octet $q\bar{q}$ pair $D^9(8,3)$, which is a nonet in flavour

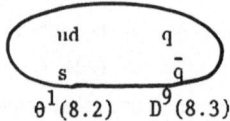

$\theta^1(8.2)$ $D^9(8.3)$

When the relative orbital angular momentum was one, it was assumed that the isotriplet

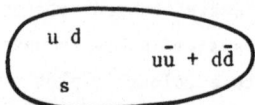

could be a state seen[14] at 2.26 MeV, and that the isosinglet

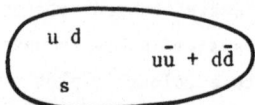

was a 2.13 GeV state[15]. Predictions were made for the other states in the nonet as well as for their modes of disintegration.

We believe that creation of $q\bar{q}$ pairs is important in the decay of multiquark baryons as it is in the disintegration of ordinary baryons that are made of three quarks. This immediately tells us that three particle decay modes are particularly important. Moreover, the dominant decay modes will be the ones containing as many s and $\bar{s}$ quarks as the disintegrating state contains. This is because it is well known that creation of $s\bar{s}$ pairs is suppressed relatively to the creation of $u\bar{u}$ and $d\bar{d}$ pairs.

Let us now look how the experimental situation has evolved since the first tentative identification of multiquark states was given. The missing five states in the nonet fall into two isospin doublets with hypercharge +1 and -1 respectively and into one isosinglet with hypercharge 0. Predicted masses were 2375 MeV for the isodoublets and 2550 MeV for the isosinglets.

The isosinglet has quark content

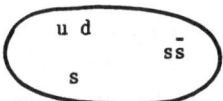

and should therefore decay into $\Sigma K \bar{K}$, $\Lambda K \bar{K}$ or $\Xi K \pi$. A 5σ signal corresponding to a mass of 2576±5 MeV and a width of 37±11 MeV has been observed by the ACNO Collaboration in the $\Sigma^- K^+ K_1^0$ channel[16].

The Y = -1   I = 1/2   state has the quark content

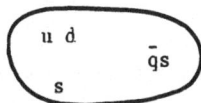

and should therefore disintegrate into $\Lambda \bar{K} \pi$, $\Sigma \bar{K} \pi$ and $\Xi \pi \pi$.

A very significant signal (~ 6σ) has been observed[17] in the $Y \bar{K} \pi$ channels corresponding to the disintegration of a Ξ^* state with mass 2373±8 MeV and total width Γ = 80±25 MeV. The authors of the article reporting this discovery remark that they do not see the two body decay modes expected and predicted if the $\Xi^*(2370)$ is a three quark state. If we accept that this state is a multiquark state the absence of two body decay modes is no wonder, and it is clear that the observed mass is what one would like. It can also be remarked that with this recent and strong evidence of a new Ξ^* the result of an earlier experiment gains in credibility. Indeed there has been reported[18] a 3σ signal of a Ξ^* state decaying into $\Xi^- \pi^+ \pi^0$ in a $K^- d$ experiment at 4.93 GeV/C. That state was assigned a mass of 2392±27 MeV and a width Γ = 75±68 MeV, and it could therefore be that this is the expected $\Xi \pi \pi$ decay mode of the $\Xi^*(2370)$.

It might therefore be possible that only one isodoublet is missing in the lowest lying SU_3 flavour multiplet of resonant multiquark baryons. This N* state with the quark content

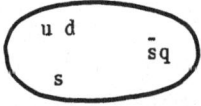

should then have a mass of around 2375 MeV and disintegrate into $YK\pi$ final states. There should in analogy with the 2,26 GeV state be a quasi two particle decay mode as YK* and it might be that this is the less difficult to find .

We think that if such a Y = +1 state is found and that its dominant decay mode leads into particles carrying strangeness it is as good a signal for a multiquark state as Y = 0 states disintegrating into three strange particles.

If it is also found at the right mass, the accumulating evidence for the existence of multiquark baryons would be overwhelming.

There has also been an experimental discovery which even if it cannot be placed inside this nonet, gives credibility to the multiquark concept. This is the $R^+(3,17)$ at 3.17 GeV with a width smaller than 20 MeV, that has been seen in two experiments at different energies[19]. As it is observed disintegrating into three strange particles ($\Sigma K \bar{K}$ + pions, $\Lambda K \bar{K}$ + pions) we would believe it to contain a hidden $s\bar{s}$ pair and have the composition qqs $s\bar{s}$. As it is narrower than the states we have discussed earlier we would believe it to carry one additional unit of angular momentum and have L = 2. It has isospin 1, and the lightest octet bonded states that contain isovectors with hidden strangeness, belong to a flavour 72 plet[13] made of $\theta^8(8,3)$ $D^9(8,4)$.

Calculating the colourmagnetic mass defects of such a state and adjusting for the mass increase due to an L = 2 excitation, the prediction for R^+ state[20] is around 3,18 GeV, a value that is again embarassingly good in view of the expected theoretical uncertainties.

Until now we have spoken only about triquarks and diquarks separated by an angular momentum barrier. The extremely unstable states we get if θ and D are in an s wave should, if our L = 1 states are correctly normalized in mass, be at a lower mass. The lightest state has quantum numbers $J^P = 1/2^-$ Y = 0   I = 0 and should have a mass of 1,44 GeV. This mass value is practically the same as has been obtained in calculations inside the MIT bag model.

Although these s-wave states cannot be expected to be seen as resonance bumps they have been looked for as poles in the P-matrix by Roisnel[21] who indeed seems to show that scattering data shows P-matrix poles where they are expected from the MIT bag model.

It follows then that our picture is not yet in disagreement with observations either for L = 0, L = 1 or L = 2.

A very interesting development would be to find states with valence gluons. The most unusual of these would be glueballs: massive states without quarks. There can also be baryons made of three quarks and one gluon. Valence gluons have an estimated effective mass of ≅ 0,7 GeV and could therefore be strongly mixed[22] with some of the mesobaryons we have been talking about. Such states would constitute a very clear proof that the gluons are strongly selfinteracting.

If colour octet bonded states are found in the baryon sector they should also be expected in the dibaryon sector. The lowest mass of the corresponding NNπ and NΛπ (NN$\bar{K}$) states are then expected at 2.45 GeV and 2.46 GeV[23].

We are somewhat embarassed by the great number of states that we predict and which have not been found; but we are pleased that the lightest candidates for multiquark baryons have quantum numbers consistent with the lightest states predicted. Because they have small production cross sections they can be found only in the decay channels and regions of phase space where nonresonant background happen to be very weak.

References:

1) Ferdinand Rosenberger,
   Die Geschichte der Physik, Braunschweig 1882 p.11.

2) H. Fritzsch, M. Gell-Mann, and H. Leutwyler,
   Phys. Lett. 47B, 365 (1973),
   S. Weinberg, Phys. Rev. Lett. 31, 494 (1973).

3) A. Chodos, R.L. Jaffe, K. Johnson, C.B. Thorn, and V.F. Weisskopf,
   Phys. Rev. D9, 2471 (1974).
   T. DeGrand, R.L. Jaffe, K. Johnson and J. Kiskis,
   Phys. Rev. D12, 2060 (1975).

4) P. Hasenfrantz, R.R. Horgan, J. Kuti, and J.M. Richard,
   Phys. Lett. 95B, p. 299 (1980).

5) De Tar, C.E. Phys. Rev. D18, p. 323 (1978)
   Phys. Rev. D19 1028, 1451 (1979)

6) R.L. Jaffe, Phys. Rev. Letters 38, 195 (1977)
   Erratum ibid p. 617.

7) A.Th.M. Aerts, P.J.G. Mulders, and J.J. de Swart,
   Phys. Rev. D17, 260 (1978).

8) H. Högaasen and P. Sorba,
   Nucl. Phys. B150, 427 (1979).

9) R.L. Jaffe, Phys. Rev. D15, 267, 268 (1977).

10) Chan Hong-Mo and H. Högaasen,
    Phys. Letters 72B, p. 121 (1977) and Nucl. Phys. B136, p. 401 (1978).
    Chan Hong-Mo et al., Phys. Letters 76B, p. 634 (1978).

11) R.L. Jaffe, Phys. Rev. D17, 1444 (1978).

12) A.Th.M. Aerts, P.J.G. Mulders and J.J. de Swart,
    Phys. Rev. D19, 2635 (1979)
    Phys. Rev. D21, 1370, 2653 (1980).

13) H. Högaasen and P. Sorba,
    Nucl. Phys. B145 119 (1978).

14) Amsterdam-CERN-Nijmegen-Oxford (ACNO) Collaboration, cited by
    R.A. Salmeron in Proc. European Conf. on Particle Physics, Budapest,
    1977 (eds. L. Jenik and I. Montway) Budapest 1977.

15) W. Lockman et al., UCLA preprint 1109 (1978).

16) ACNO Collaboration, paper submitted to
    the Tokyo Conference (1978). CERN/EP/PHYS 78-24.

17) J. Amirzadeh et al., Phys. Letters 90B, 324 (1980).

18) F.A. Dibaianca and R.J. Endorf,
    Nucl. Phys. B98, 137 (1975).

19) J. Amirzadeh et al., Phys. Letters 89B, 125 (1979).

20) M.De Crombrugghe et al., Nucl. Phys. B156, 347 (1979).

21) C. Roisnel, Phys. Rev. D20, 1646 (1979).

22) Chan Hong-Mo et al., CERN TH. 2828,
    to be published in Z. für Physik.

23) H. Högaasen and J. Wroldsen, to appear.

TOPOLOGICAL INTERPRETATION OF MULTIQUARK STATES

Basarab NICOLESCU

Division de Physique Théorique (Laboratoire associé au CNRS),
IPN Orsay and LPTPE, Université P. et M. Curie, Paris, France

Abstract : In this talk we discuss the topological selection rules which govern the physics of multiquark states in the framework of the DTU theory. These new selection rules lead us to expect that narrow multiquark hadrons are rare, are strongly coupled only to some particular channels, and appear only in some restricted mass regions.

I. INTRODUCTION : THE DUAL TOPOLOGICAL UNITARIZATION APPROACH TO HADRON PHYSICS

The Dual Topological Unitarization (DTU) approach to hadron physics represents a way of dealing with the difficult and subtle problem of confinement. It is an S-matrix "topological perturbation" theory which incorporates (and even requires) a given quark-like structure of hadrons. Its domain of validity being restricted to low- p_T physics, DTU appears as complementary to the perturbative QCD theory, which nicely describes high- p_T physics.

The DTU approach is formulated as a generalization of the description of hadron scattering via the quark duality diagrams [1] . The central idea is the recognition of a possible correspondence between the complexity of singularities of scattering amplitudes and the topological complexity of certain 2-dimensional surfaces [2]. Namely, the logical chain is the following : amplitudes are determined by their singularities ; singularities are derived from unitarity ; the content of unitarity can be expressed through graphs (Landau graphs) ; and, finally, graphs can be embedded on 2-dimensional surfaces. (In fact, 2 is the minimal number of dimensions of the manifolds on which graphs can be embedded). These 2-dimensional surfaces are therefore representations of the singularity structure of the scattering amplitudes and the unitarity products simply correspond to the connected sums of the surfaces. A simple example of quark-duality diagrams for meson-meson scattering, embedded on a bounded and oriented plane, is given in Fig.1. Notice that the orientation of the boundary in Fig.1 results from the fact that mesons are described as color-singlet quark - antiquark states. A more complicated graph in meson-meson scattering is shown in Fig.2 ; it can be shown that this graph can be embedded on a torus.

High-energy hadron phenomenology already suggests that the simplest singularities of the hadron scattering amplitudes are the leading contributions. In other words, graphs with a complex topological structure seem to be suppressed when compared with graphs with a simple topological structure. It has been therefore tempting to postulate the existence of a "topological expansion" [2], i.e. a theory in which the lowest "topological entropy" level corresponds to the leading contribution to the scattering amplitude and in which the higher "topological entropy" levels act as perturbative corrections. This approach stimulated detailed theoretical and phenomenological studies which have proved to be very successful in the case of mesons (see Ref.3 for a recent review on this subject).

However, the extension of DTU to baryons is intrinsically difficult, due essentially to the complex nature of the "topological entropy" index, which governs the topological expansion. Important questions remain to be answered. How the color is to be introduced ? Why there is confinement of quarks ? What properties of quarks are really represented by quark-duality diagrams ? If the quark lines represent the flux of energy-momentum of quarks, how this can be reconciled with the permanent confinement property ?

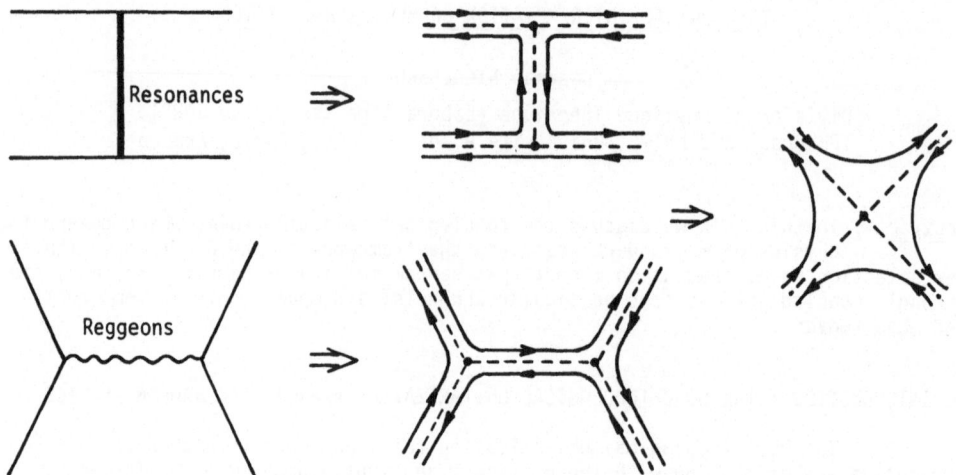

Fig.1 Example of quark-duality diagrams for meson-meson scattering. The solid oriented lines represent "quark lines" and the dashed lines - Landau arcs.

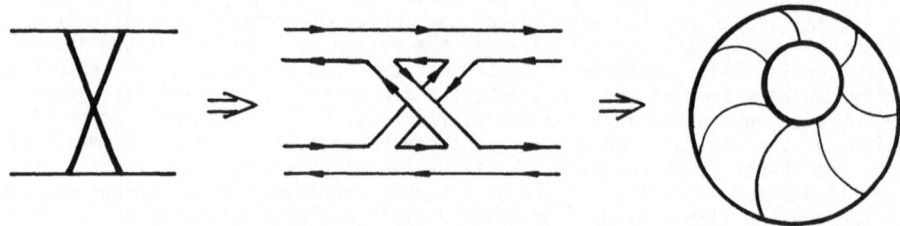

Fig.2 A graph which corresponds to exchange of crossed clusters and which can be embedded on a torus.

II. THE QUANTUM SURFACE AND THE CLASSICAL SURFACE

A major step forward was recently made by Chew [4] and subsequently by Chew and Poenaru [5], who proposed to distinguish between : a) the space-time (energy-momentum) aspects of the scattering and b) the aspects related to the internal quantum numbers (which have, in their turn, to be related to the presence of confined constituents). Namely, they propose to describe hadron interactions by a pair of surfaces, a "classical" surface and a "quantum" surface.

The classical surface has the properties previously formulated in DTU [2]. It is 2-dimensional, orientable and bounded. Moreover, it is postulated to correspond to aspects of the scattering connected to the space-time continuum.

The quantum surface, on the other hand, describes the discontinuous aspects of the scattering. It is the space of *structures*, the space of confined constituents. The quantum surface is 2-dimensional, orientable and closed. The last property is related to the conservation of quantum numbers. One can understand intuitively this relation, by taking the simplest example of a closed surface - a sphere. If the sphere represents an amplitude, different regions of the sphere are therefore associated with particles and their quantum numbers. The property of complete contractibility of the sphere describe the conservation of quantum numbers.

The classical and the quantum surfaces are not disjoint : the boundary of the classical surface is given by its intersection with the quantum surface, this boundary being precisely the familiar Harari-Rosner duality graphs [5].

Unitarity products again correspond to a connected sum of surfaces. It is interesting to note that the only closed surface which can reproduce itself is the sphere (see Fig.3). It is therefore suggested that the nonlinearity of the problem appears only at the level of the sphere.

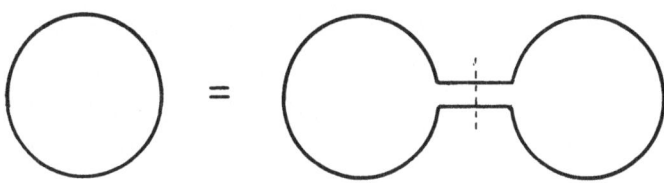

Fig.3 Connected sum of spheres leading to a sphere.

A proper mathematical definition of the overall topological entropy index $\tau_{Q,C}$ such that, under any connected sum operation, $\tau_{Q,C}$ either increases or stays stationary has been found [5]. The proof involves the triangulation of the quantum surface. The triangle - the 2-dimensional simplex - naturally appears as the "basic" object of the present construction.

III. A POSSIBLE DESCRIPTION OF THE QUANTUM SURFACE

The quantum surface was first introduced in Ref.4. A tentative definition of ordinary and multiquark hadrons, quarks and gluons, based upon this quantum surface has been made in Ref.6. Finally, a complete topological expansion theory, involving both the quantum and classical surfaces has been formulated [5]. For the purpose of this talk we will follow here mainly the simple description of the quantum surface given in Ref.6.

The hadrons have to be defined at the lowest entropy level, i.e. they correspond to regions of a sphere. Namely, in order to stay in the lowest entropy level when one considers hadrons as intermediate states in unitarity products, the particles have to be associated with "discs". A "disc" (Fig.4(a))- a region of the sphere whose perimeter touches itself only once- is associated with an "elementary" hadron, while a "non-disc" (Figs.4(b) and 4(c))is associated with a "composite" (molecular) hadron (e.g. the deuteron). The perimeter of a disc gives the "identity" of a hadron, being a representation of flavor indices, while the interior of a disc has to be related to color indices.

It is important to note the close relationship between the notion of "disc" and the notion of "multiparticle channel resonance". Namely, the topological contraction of any collection of discs belonging to a given sector (or "channel") of the "spherical" (or "ordered") Hilbert space leads *uniquely* to a given particle disc. In other words, the particles belonging to the respective channel "resonate".

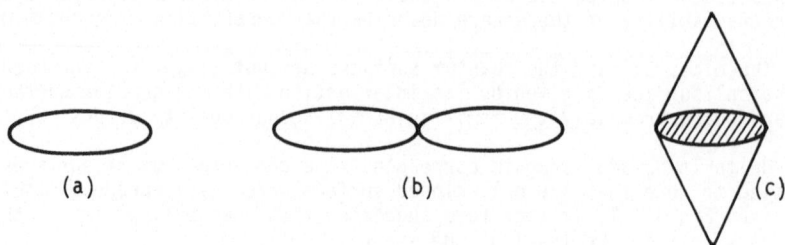

Fig.4 Example of a disc ((a)) and of non-discs ((b) and (c)).

Here one can find the very root of the physical mechanism controlling the convergence of the topological expansion.

The "building block" of hadrons is shown in Fig.5. It consists of a "core" triangle and three "peripheral" triangles. The opposite orientations of the core and peripheral triangles are needed in order to insure the overall orientation of the quantum surface.

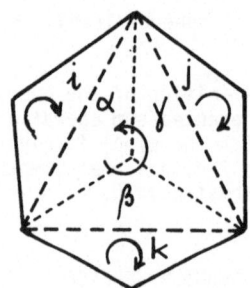

Fig.5 The basic "building block" of hadrons, consisting of three flavons and three chromons.

The core triangle is in its turn subdivided in three triangles, each of them being associated with a different value of a "topological color" index $C = \{\alpha,\beta,\gamma\}$. It is important to note that the "topological color" index C has three and only three possible values, this property being a *consequence* of the fact that the number of dimensions of the quantum surface is two.

The peripheral triangles, which build the boundary of the disc shown in Fig.5 are representations of the "topological flavor" index $F = \{i,j,k,...\}$. It can be shown that a proper topological characterization of F implies a *limited* number of flavors - eight [5] or even six [7].

A "topological quark" will be therefore represented by the configuration shown in Fig.6(a) and a "topological gluon" by the configuration shown in Fig.6 (b). Strictly speaking, a quark is not a "constituent" by itself, being built from other two topological structures : a "topological flavon" F (Fig.6(c)) and a "topological chromon" C (Fig.6(d)). One finds here an interesting analogy with the description of quarks in the recent "Quantum Structuredynamics" (QSD) grand-unified theory of Greenberg and Sucher [8].

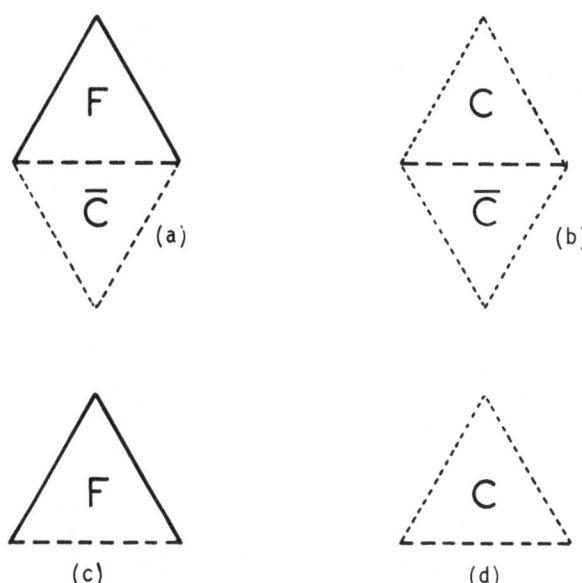

Fig.6 Topological quark (a), topological gluon (b), topological flavon (c) and topological chromon (d).

Notice that none of the structures shown in Fig.6 are "discs" - they have incomplete (or even no) boundaries. Therefore they cannot appear as physical states. The topological constituents of hadrons are, by construction, confined. In this sense, hadrons appear as "elementary".

By gluing together m basic discs and n anti-basic discs (i.e. of appropriate opposite orientation) one obtains an infinite sequence of possible hadrons. The basic disc of Fig.5 is by itself representing an ordinary q^3 baryon (B_3). For m = 1 and n = 1 one obtains two possible configurations : an ordinary $q\bar{q}$ meson (M_2 - Fig.7(a)), which is in fact a flavon-antiflavon state, and a $q^2\bar{q}^2$ baryonium (M_4 - Fig.7(b)). For m = 2 and n = 1 one obtains a $q^4\bar{q}$ multiquark baryon (B_5 - Fig.7(c)), for m = 3 and n = 1 - a q^6 dibaryon (D_6 - Fig.7(d)), etc.

A topological definition of "color-singlet" state can be given [6]. Note the existence of the "triality" property i.e., that the number of quarks minus the number of antiquarks is 0 (mod 3). Also, one obtains a nice connection between baryon number and color : the baryon number is the difference between the number of anticlockwise and clockwise oriented core triangles.

IV. TOPOLOGICAL SELECTION RULES FOR MULTIQUARK STATES ORIGINATING FROM THE QUANTUM SURFACE

Any multiquark state appearing on the quantum surface (see the examples of Fig.7) is characterized by the fact that any contiguous colors of a chromon-antichromon pair are different (i.e. these pairs do not annihilate). This property is the source of "spherical" selection rules.

Namely, one can look for the different spherical decay channels of the multiquark states by decomposing the corresponding discs into all possible collections of discs obtained by creation of flavon-antiflavon pairs. As is easily seen

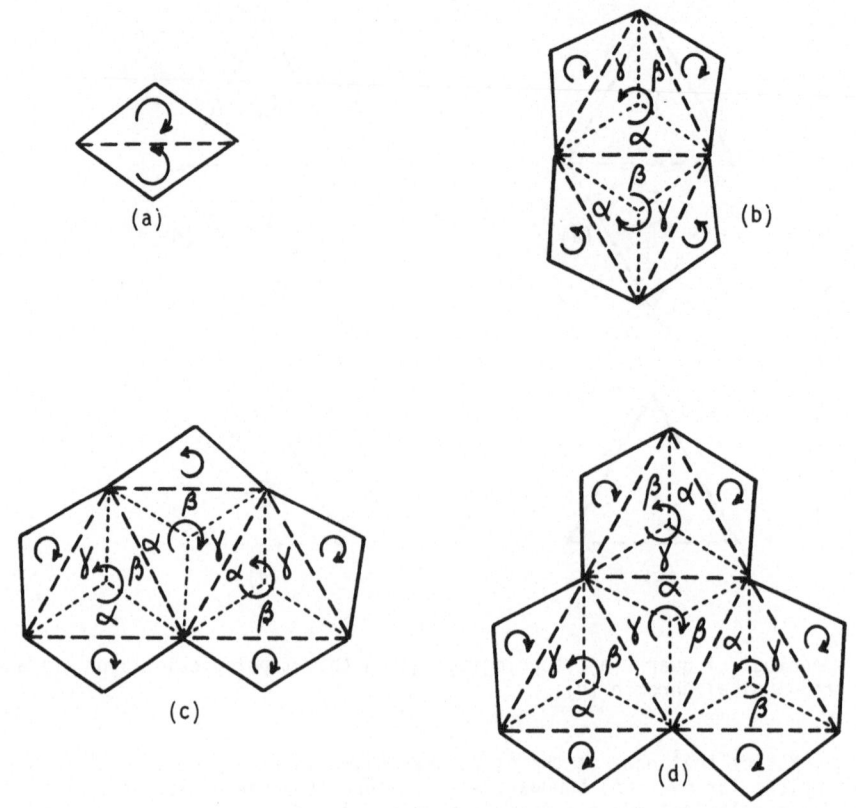

Fig.7 Ordinary meson (a), baryonium (b), $q^4\bar{q}$ baryon (c) and dibaryon (d).

from Fig.7 a baryonium state can communicate with a baryon-antibaryon pair
($M_4 \to B_3\bar{B}_3$), but not with a meson channel ($M_4 \not\to M_2M_2$). A $q^4\bar{q}$ baryon can decay, e.g.,
into baryon-baryon-antibaryon channel ($B_5 \to B_3B_3\bar{B}_3$) but not into a ordinary meson-
ordinary baryon channel ($B_5 \not\to M_2B_3$). A q^6 dibaryon decays into a 3 baryons-
1 antibaryon channel ($D_6 \to B_3B_3B_3\bar{B}_3$) but does not communicate on the sphere with a
baryon-baryon channel ($D_6 \not\to B_3B_3$), i.e. a strangeness 0 dibaryon will be weakly
coupled with, for example, the proton-proton channel.

Of course, these selection rules (which are a generalization of the
familiar OZI rule) are broken in higher orders of the topological expansion.
However, there are multiple theoretical and phenomenological indications for a
rapid convergence of the topological expansion. Therefore, it is possible that these
selection rules for multiquark states (particles and reggeons) are not too
unrealistic. In fact, by studying the t-channel "exotic" baryonium exchange
($I = 2$ or $3/2$ or $S = -2$) reactions one has already a phenomenological indication
of the validity of these selection rules [9]. Also, the spherical selection rules
lead to a topological structure of hadron cross-sections in nice agreement with
experimental data [10]. Some of the selection rules discussed here can be
obtained in the "geometrical" model of Ref. 11. Our way of approaching multiquark physics is also similar to the spirit in which a recent QCD- inspired generalization of DTU has been done [12].

One can conclude that the present study of the quantum surface indicates that the stability of multiquark hadrons is connected with a general mechanism : the non-contractibility of certain triangulation patterns. One obtains certain topological selection rules which can explain some paradoxical aspects of the physics of multiquark states.

Multiquark hadrons must generally have, in our approach, normal widths. Their density is expected to be large, and it follows that there is only a small chance of detecting them in a clear way. The only unambiguous possibility would be to detect narrow multiquark states, which may appear near or below certain "topological stability thresholds" (e.g. $B_3\bar{B}_3$ for baryonium, B_3M_4 for $q^4\bar{q}$ baryons, B_3B_5 for dibaryons). Therefore, sharp states, if they exist at all, will be those of lowest mass and lowest angular momentum. They are, in any case, *rare* events. This conclusion is reinforced by the study of the classical surface.

V. TOPOLOGICAL SELECTION RULES FOR MULTIQUARK STATES ORIGINATING FROM THE CLASSICAL SURFACE

The classical surface is rigorously speaking, a bounded 3-sheeted 2-dimensional surface, the three sheets being united by a "junction line" [5]. However, for the purpose of the simplified presentation of this talk, it will be sufficient to use its projection onto a plane. I will therefore continue to use the familiar Harari-Rosner diagrams, but with a different meaning attached to the "quark lines", which represent the boundary of the classical surface. Namely, it was shown by Stapp [13] that the "quark lines" can be consistently associated with the ± 1/2 spin indices of quarks (in fact, of the flavons). The quark lines do not describe the flow of the energy-momentum of the quarks, and this corresponds to the physical fact that quarks cannot be defined as asymptotic states in space-time (they are confined). The space-time aspects of hadron collisions are described by Landau graphs (see Fig.1 for an example), each Landau arc being associated with the energy-momentum four vector of a particle. The Landau graphs are embedded on the classical surface and there are definite rules for their contraction to a single vertex [5].

The general constraints at the lowest entropy can be expressed as follows : 1) any singularity of the scattering amplitude corresponds to a multivertex Landau graph that can be contracted to a single vertex without internal loops ; 2) any crossing of topological lines on the classical surface is forbidden.

It can be seen that these constraints lead to severe restrictions on the physics of multiquark states. I will take as an illustration the baryonium case.

The non-communication at the lowest entropy level between baryonium and ordinary mesons is clearly seen on the classical surface, due to the impossibility of avoiding crossings between the different topological lines (Fig.8). I also give, in Fig.9, two examples of possible suppressions in certain processes where recently it was claimed that there is no evidence for the baryonium production [14]. Note that even the baryonium exchange processes are suppressed (Fig.10).

In fact, the constraints originating from the classical surface are severe even for the spectrum of multiquark states. It can be shown [5,7] that a consistent theory can be very well built by considering only ordinary mesons, ordinary baryons and baryonium states. This indication throws some doubt on the possibility of detecting narrow multiquark hadrons other than baryonium (however, there can still be some broad multiquark states, e.g. dibaryons, but they will correspond to "molecular" hadrons).

One cannot really know how good the above-discussed "topological suppressions" are before doing detailed dynamical calculations. However, the study of the nature of singularities associated with the higher entropy amplitudes

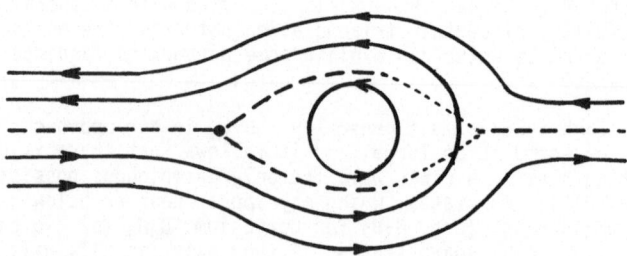

Fig.8 The baryonium-meson coupling via a baryon-antibaryon loop on the classical surface. The dashed lines indicate Landau arcs. The dotted lines indicate symbolically the necessity of crossing of the different topological lines.

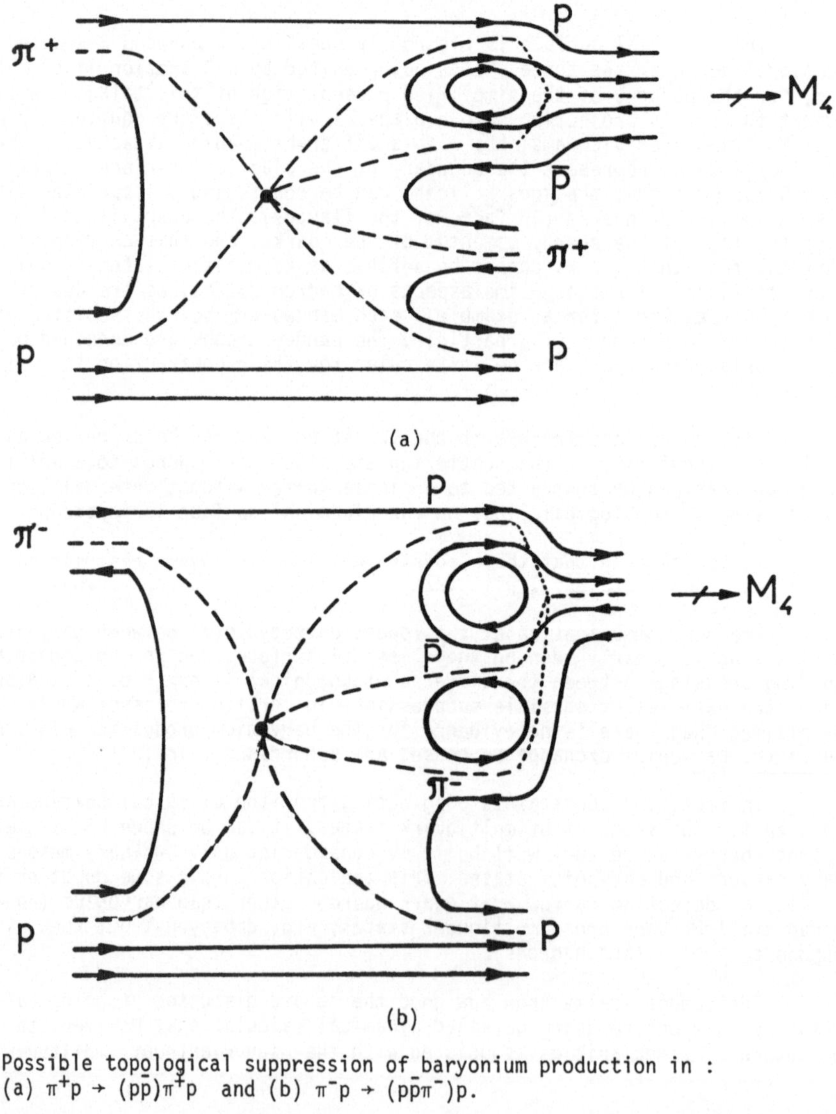

Fig.9 Possible topological suppression of baryonium production in :
(a) $\pi^+ p \to (p\bar{p})\pi^+ p$ and (b) $\pi^- p \to (p\bar{p}\pi^-)p$.

indicates the possibility of their decreasing dynamical importance [7]. Therefore one can conclude, from the study of the classical surface, that one has a chance to detect multiquark hadrons only in very particular reactions, many channels being topologically suppressed. The most probable candidates for narrow multiquark hadrons are baryonium states under the $B_3\bar{B}_3$ threshold.

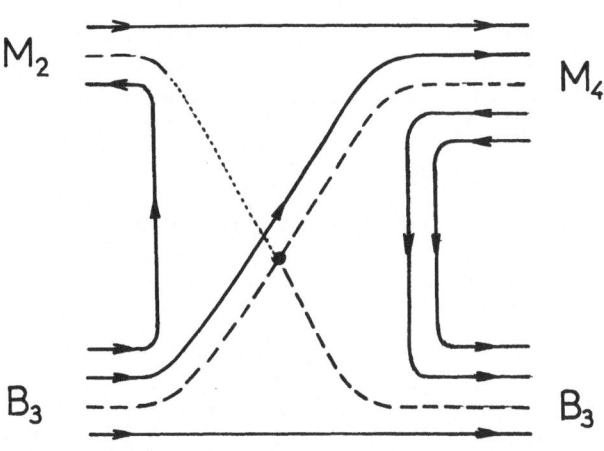

Fig.10 Possible topological suppression of baryonium production in processes involving baryonium exchange.

VI. COMPUTATION OF HADRON MASSES FROM THE TOPOLOGICAL EXPANSION

We have calculated the hadron masses in the DTU framework by imposing duality on an infinite sum of ladder graphs generated from "spherical unitarity" [15]. By making a certain simple dynamical approximation, we derived an explicit generic Regge-trajectory formula for any given process. By requiring simultaneous consistency for entire sets of processes, we were able to calculate the masses of ordinary and multiquark hadrons associated with the u and d quarks. The only arbitrary parameter is the mass of the ρ, which merely serves to set the mass scale.

The technical details and the results of our computation can be found in Ref.15. Here I give just one result, concerning the beginning of the mass spectrum for each family of hadrons (Fig.11), a result which can be of interest for the search for narrow multiquark states. In particular, it is seen from Fig.11 that the baryonium spectrum begins at a rather low mass (1.33 GeV). In fact, we predict the existence of four baryonium states under the $N\bar{N}$ threshold [15]. The lowest-mass member of the dibaryon family is at 2.18 GeV.

It is interesting to note that our treatment of the simultaneous consistency for sets of processes leads us to separate them in "bootstrap cycles". The first "cycle" involving multiquark hadrons stopped at the baryonium level. One thus finds again an indication that it is not necessary to define elementary hadrons beyond ordinary mesons, ordinary baryons and baryonium.

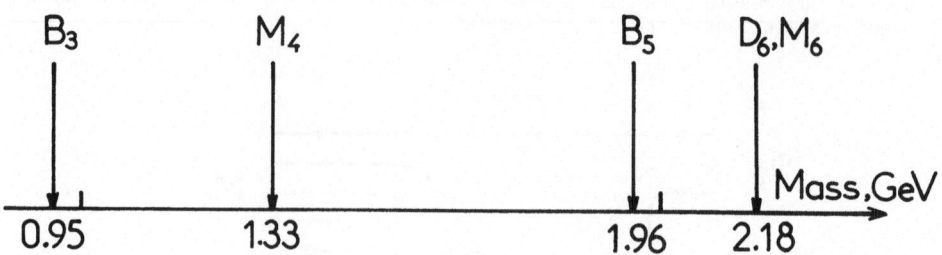

Fig.11 The beginning of the predicted mass spectrum for baryons and multiquark hadrons (Ref.15).

VII. CONCLUSIONS

The topological selection rules can explain some paradoxical aspects of the physics of multiquark states (e.g. their peculiar systematics in decays and their experimental elusiveness). We have to conclude that narrow multiquark hadrons are rare events. This is a distinctive feature of the DTU approach, when compared with most of the bag and color chemistry models. Another distinctive feature of the DTU theory is the prediction of low mass and low spin narrow baryonium states, under the baryon-antibaryon threshold.

By using the already developed argument of quark-lepton analogy in DTU [5] it will be interesting to search for the consequences of the corresponding topological selection rules in a unified description of weak, electromagnetic and strong interactions.

ACKNOWLEDGMENTS

It is a pleasure to thank Professors Anna Maria Saruis and M. Rho for the kind invitation to give this talk. I am also grateful to Professors G.F. Chew and V. Poenaru for many useful discussions and correspondences.

REFERENCES

[1] H. Harari, Phys. Rev. Lett. 22, 562 (1969) ;
J. Rosner, Phys. Rev. Lett. 22, 689 (1969).

[2] G. Veneziano, Nucl. Phys. B74, 365 (1974) ; Phys. Lett. 52B, 220 (1974).

[3] G.F. Chew and C. Rosenzweig, Phys. Reports 41, 263 (1978).

[4] G.F. Chew, Nucl. Phys. B151, 237 (1979) ; Phys. Lett. 82B, 439 (1979).

[5] G.F. Chew and V. Poenaru, Phys. Rev. Lett. 45, 229 (1980) ; Lawrence Berkeley Laboratory preprints LBL-9768 (1979) and LBL-11433 (1980).

[6] G.F. Chew, B. Nicolescu, J. Uschersohn and R. Vinh Mau, CERN preprint TH-2635 (1979) ; B. Nicolescu, Invited talk, in "Quarks, Gluons and Jets", Proceedings of the XIVth Rencontre de Moriond, Les Arcs, France, 1979, edited by J. Tranh Thanh Van, p.63-72.

[7] G.F. Chew, private communication.

[8] O.W. Greenberg and J. Sucher, University of Maryland preprint 81-026 (1980).

[9] B. Nicolescu, Nucl. Phys. B134, 495 (1978).

[10] P. Gauron, B. Nicolescu and S. Ouvry, Orsay preprint IPNO/TH 80-58 (1980).

[11] E. Elbaz and J. Meyer, Nuovo Cimento 57A, 271 (1980).

[12] G.C. Rossi and G. Veneziano, Nucl. Phys. B123, 507 (1976) ; see also L. Montanet, G.C. Rossi and G. Veneziano, Phys. Reports 63, 149 (1980).

[13] H.P. Stapp, Lawrence Berkeley Laboratory preprints LBL-10774 and LBL-11770 (1980).

[14] See the recent review of the experimental situation of baryonium production by D. Perrin, Proceedings of the 5th European Symposium on Nucleon-Antinucleon Interactions, Bressanone (Italy), 23-28 June 1980, CLEUP (1980), p. 545-574.

[15] L.A.P. Balázs and B. Nicolescu, Zeitschrift für Physik C6, 269 (1980) ; see also L.A.P. Balázs and B. Nicolescu, "Multiquark Hadrons in Topological Bootstrap", Proceedings of the 5th European Symposium on Nucleon-Antinucleon Interactions, Bressanone (Italy), 23-28 June 1980, CLEUP (1980), p. 271-283.

SEARCH FOR DIBARYONIC RESONANCES OF SMALL MASS
(Q_{DB} < 2.3 GeV)

G. Tamas

DPh-N/HE, CEN Saclay, BP 2, 91190 Gif-sur-Yvette, France

Abstract : A possible dibaryonic resonance at 2.23 GeV appears in the photo excitation of the deuterium, but not in the pion induced reaction. This result is discussed.

Is there any dibaryon resonances without strangeness or charm? And if such resonances exist, are they a two nucleon system exchanging mesons or a six-quark cluster? Several approaches may be done. The one presented here uses a non baryonic unpolarized probe (photon or meson) to excite the deuterium, also unpolarized, and look to various cross sections as a function of energy. The hope is to find a narrow resonance behaviour (compared to the Δ width). The experimental goal is then to communicate several times one hundred MeV to the two-baryon system. We can notice that for instance, an energy transfer of 300 MeV leaves the two-nucleon channel with a relative momentum of 550 MeV/c. This means that: i) many final state channels are open; ii) many partial waves may be involved; iii) a large physical background coming from the quasi-free processes will be present. So, the experimental problem is to find good conditions to get rid of these complications. In a very exclusive experiment, the rejection of this background may be excellent, but the result will be very sensitive to interference effects. On the contrary, an inclusive experiment will include a large background with less effective interferences.

THE QUASI FREE MECHANISM

It will be discussed for photons, but the conclusions are similar for pions. The quasi free mechanism in the photon absorption is very dominant above the pion production threshold, and so for the reaction:

$$\gamma + D \rightarrow \pi + \mathcal{N} + \mathcal{N}.$$

The photon interacts with one nucleon as if it was free, the second one being spectator (fig. 1). The parameters describing this process are the invariant mass $Q_{N\pi}$ and the momentum of the spectator nucleon p_R. When $Q_{N\pi}$ and $|p_R|$ are set constant the quasi-free contribution remains constant. Its order of magnitude is strongly dependent on $|p_R|$ and decreases rapidly with it. We can note that the remaining free parameter θ_R, the angle between the recoiling nucleon and the photon is related to the photon energy E_γ. Thus keeping $Q_{N\pi}$ and $|p_R|$ constant, but changing θ_R, so the photon energy E_γ, must leave the quasi free cross section invariant (except for a kinematical factor).

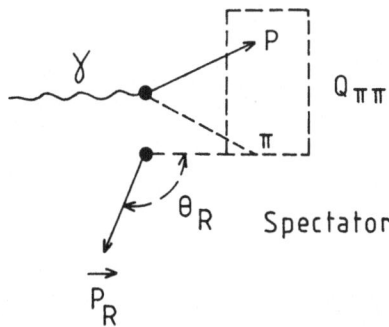

Fig. 1. Schematic representation of the quasi free process.

The quasi free process, found to be dominant [1], has to compete with more complicated ones as the pion and nucleon rescattering now well understood theoretically [2] as experimentally [3]. All this part of the photon-deuterium interaction is now under control.

THE EXCLUSIVE EXPERIMENT [4]

Using the bremsstrahlung photon beam of the Saclay Linac, the reaction

$$\gamma + D \to \pi^- + p + p$$

was investigated by detecting the emitted pion in coincidence with one of the two protons. The momenta of the particles were measured so as to completely determine the kinematics of the reaction. The advantage of this method lies in the possibility of reducing the contribution of the dominant quasi-free photoproduction by considering different momenta p_R of the recoiling nucleon. The pion was detected in a magnetic spectrometer and the proton either in a second magnetic spectrometer or in a range telescope. We kept constant, as explained before, p_R, $Q_{N\pi}$ and the angle ω of the pion relative to the incident photon in the pion nucleon rest frame. We varied only the angle θ_R of the recoiling proton and hence the photon energy E_γ. We have plotted on the figs. 2 and 3 the measured counting rate divided by the prediction of the first order quasi-free model (impulse approximation) $f(\theta_R)$. For a small value of p_R (p_R = 50 MeV/c, $Q_{N\pi}$ = 1276 MeV, ω = 90°, fig. 2) the quasi-free approximation ($f(\theta_R)$ = 1) is excellent. But for p_R = 150 MeV/c and $Q_{N\pi}$ = 1246 MeV (fig. 3) it fails completely to reproduce the data as well as the second order calculation taking in account the π and p rescattering (dashed curve) although it has successfully accounted for previous experimental results [4]. The two loops diagram added to the other one gives a better agreement, but does not explain the bump observed around 400 MeV. At the present time this discrepancy is very difficult to explain in term of contributions from rescattering processes, such as the Δ-N scattering [6]. This rapid variation of the cross section around 400 MeV, with an observed width of the order of 40 MeV, put forward the possible existence of a two-baryon resonance.

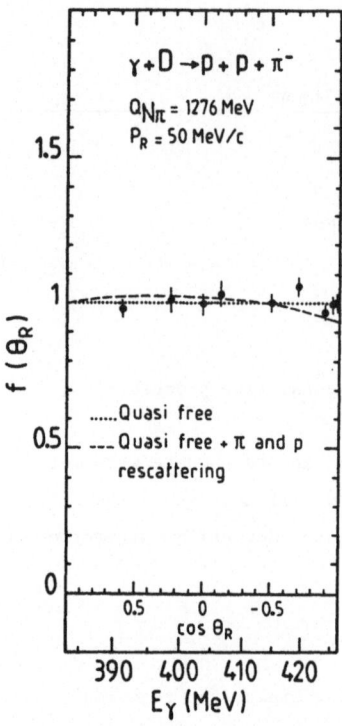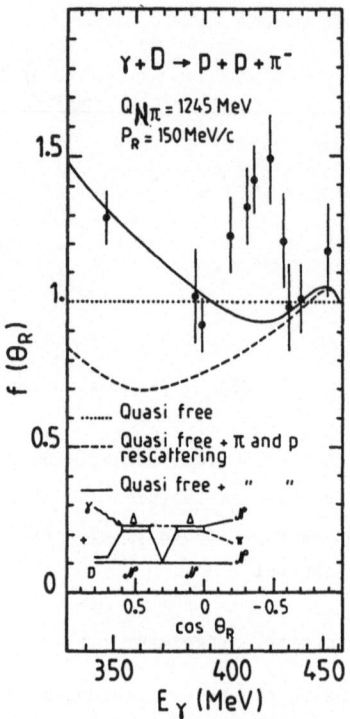

Fig. 2. Ratio of the measured counting rate to the first order quasi-free calculation $f(\theta_R)$ as a function of the recoil angle θ_R (and the corresponding photon energy E_γ) for the reaction $D(\gamma, p\pi^-)p$. The recoil momentum p_R is small (50 MeV/c). The dotted curve is a calculation taking in account also the rescattering term.

Fig. 3. Same as fig. 2 but for p_R larger (150 MeV/c). The solid curve represents a calculation taking in account the two loops diagram.

THE INCLUSIVE EXPERIMENT [5]

A more inclusive experiment, less sensitive to possible interference effects had to be designed, and consisted in studying the

$$\gamma + D \rightarrow p + x$$

with photon energies around 400 MeV. When the angle θ and the momentum p of the proton are determined, the cross section as a function of E_γ is expected to behave as shown schematically in fig. 4a : i) a peak corresponding to the photodisintegration (PD) centered at an energy $E_{np}(\theta,p)$; ii) then, above the pion production threshold E_{th}, the contribution from the standard pion photoproduction including the quasi-free production (minimized if the momentum and angle of the proton are large), as well as pion and proton rescatterings (QFR); iii) above this usual π-production background, a bump at E_{DB} corresponding to the presence of a possible dibaryonic resonance (DB) decaying with the emission of a pion. With the use of a bremsstrahlung beam, the observed counting rate is the convolution of the continuous bremsstrahlung spectrum with the above cross section. This yield, as shown in fig. 4b, is a function of the photon spectrum end-point energy E_-. A step in this yield around E_{DB} would be the signature for the dibaryonic resonance. As the effect is expected to be small (a few percent of PD), the experiment was designed so that it could provide measurements with a relative accuracy better than 1 %. Protons were detected with both magnetic spectrometers. One of them was fixed at $\theta = 60°$ and $p = 577$ MeV/c and was used to permenently check the stability of the liquid deuterium target thickness (0.47 g/cm^2). Its focal plane was split into four bins of $\Delta p/p = 0.03$. The second one was used for measurements done at three angles: 90°, 105°, 120° and for two analyzed momentum bins: 365-381 MeV/c and 381-397 MeV/c. The data were taken by alternating high and low electron energy measurements. With all these precautions the total systematic uncertainty for a given bin was smaller than the statistical one which was about 0.5 %.

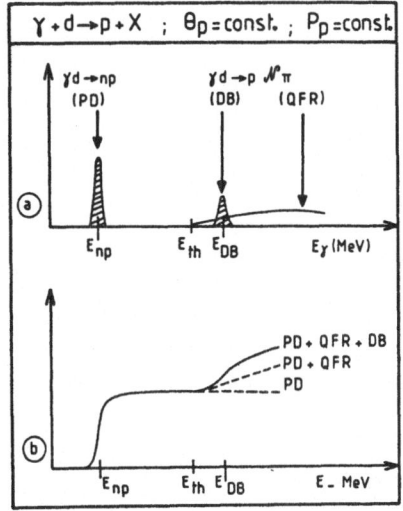

Fig. 4. Counting rate of the D(γ,p)X reaction for a fixed momentum of the detected proton versus the proton energy: a) for a monochromatic photon beam; b) for a bremsstralung as a function of the end point energy.

Since in the two-body photodisintegration reaction the energy E_γ is only determined by θ and p, $E_\gamma = E_{np}(\theta,p)$, the normalized counting rate must be a constant $C(E_{np})$ in the interval $E_{np} < E_- < E_{th}$. From the 10 different kinematical conditions, a total of 67 points fulfilled this last condition (figs. 5 and 6).

These results indicate that the bremsstrahlung spectrum was correctly determined and confirm the good stability of the set-up. So we can confidently calculate above the pion production threshold the contribution of the photodisintegration process to the yield $Y(E_-)$. We can compare its absolute value with the yield Y_π of the QFR model [2] using a Reid wave function. A fit was performed with one normalization parameter,

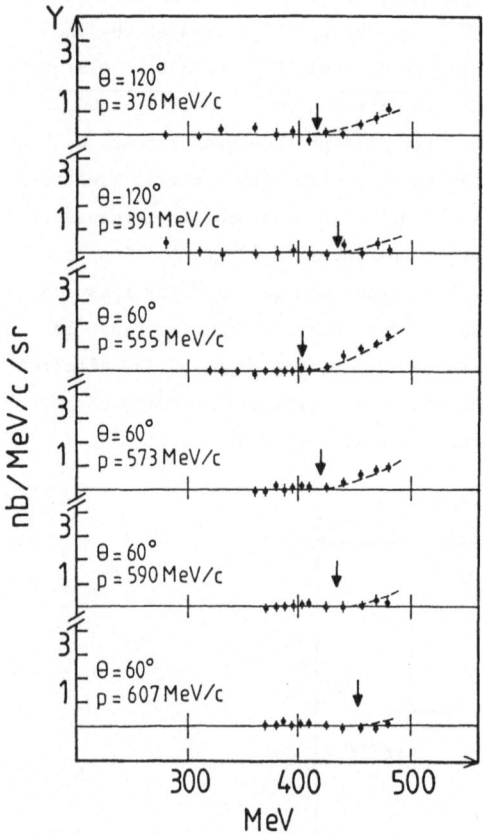

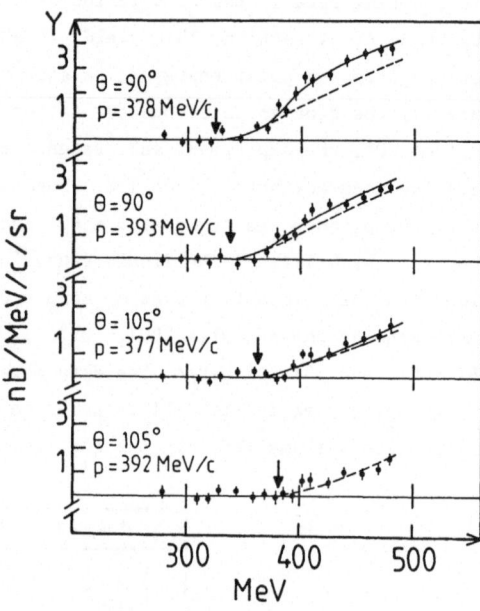

Fig. 5. Measured yield of the $D(\gamma,p)X$ reaction as a function of the bremsstrahlung end point energy E_- after subtraction of the $D(\gamma,p)n$ contribution. The arrows indicate the threshold pion energy E_{th} for the various kinematical conditions when E_{th} is larger than 410 MeV. The dotted curve represents the calculated [3] non resonant contribution only.

Fig. 6. Same as fig. 5 but with $E_{th} < 410$ MeV. The solid curve includes a resonance at 390 MeV with a width of 40 MeV.

a, for the points with $E_- > E_{th}$, keeping only the cases where E_{th} was larger than 410 MeV (i.e. $\theta = 60°$ and $\theta = 120°$) in order to get rid of the anomaly described earlier (fig. 5). The angular distribution and the shape of the curves (fig. 3) are well reproduced by the model in these cases. Nevertheless the QFR model is unable to account for the rapid variation of the cross section in the 370-410 MeV region (fig. 6). The conclusion is that the shape is no longer correct when the threshold energy E_{th} is lower than 410 MeV. We therefore introduce a term to include the influence of the excitation of a resonance. The data suggest E_{DB} = 390 MeV (Q_{DB} = 2.23 GeV) and a width of the order of 40 MeV. We can thus extract the integrated cross section b of the possible resonance from the data above 40 MeV. b is given in table 1.

PION INDUCED EXPERIMENT [6]

Therefore it was tempting to study the reaction

$$\pi^\pm + D \rightarrow p + X$$

in the same spirit as in the photon case. The kinematics is very similar in both cases

Table 1

The energy integrated cross section over the resonance for several center of mass momenta p_{CM} of the detected proton corresponding to the various experimental conditions of the $D(\gamma,p)X$ inclusive experiment.

θ_{lab} (deg.)	p_{lab} (MeV/c)	$p_{C.M.}$ (MeV/c)	b MeV. b/sr.MeV/c
90	378	417	0.389 ± 0.039
90	393	431	0.147 ± 0.036
105	377	466	0.111 ± 0.033
105	392	482	0.024 ± 0.033

and the quasi-free process can be calculated by the same way. As the pion beam is monochromatic, the situation is the one described by fig. 4a. The experiment was performed at SIN and the proton detected in the SUS1 spectrometer. The results for 372.5 and 302.5 MeV/c are presented in fig. 7 (a and b). The solid curves are the QFR model predictions done by J.M. Laget for Reid and HM2 wave functions. The arrows show the expected energy of a 2.23 GeV dibaryon resonance. The small bump appearing in this region moves with the proton momentum and is not therefore a resonance signature. So the pion induced reaction gives a very different result than the photon one. The comparison is given in table 2 and it turned out that the ratio of the pion excitation of the dibaryonic resonance to the photon one is less than 10 times smaller than the corresponding ratio for the quasi-free process.

Table 2

Photon and pion induced reaction cross section for $\gamma + D \rightarrow p + X$ (first coulumn) $\pi^+ + D \rightarrow p + X$ (second column) on the two first lines. The first line corresponds to the quasi-free and rescattering contribution. The second line corresponds to the possible dibaryon candidate. The third column gives the ratio of pion to photon cross sections. The kinematical conditions are: p_p = 370 MeV, θ_p = 90°, $Q_{NN\pi}$ = 2.23 GeV (E_γ = 390 MeV, T_π = 246 MeV). The last line gives the comparison of total cross section on the proton $\gamma+p$ and π^++p at the same projectile energy.

		γ	π	"π"/"γ"
$\genfrac{}{}{0pt}{}{\gamma}{\pi^+} + D \rightarrow p + X$ $\frac{d\sigma}{d\Omega dp}\left(\frac{nb}{sr.MeV/c}\right)$	QF + resc.	10	2000	200
	DB candidate	10	≤200	≤20
$\sigma_{tot}(\mu b)$ $\genfrac{}{}{0pt}{}{\gamma}{\pi^+} + p$		400	10^5	250

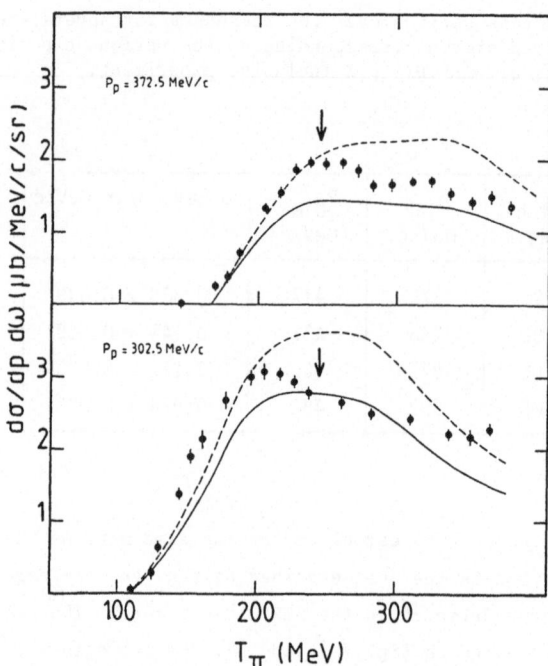

Fig. 7. The differential cross section of the $D(\pi^+p)X$ reaction as a function of the pion kinetic energy at $\theta_p = 90°$. The arrows indicate the pion energy corresponding to the total mass 2.23 GeV. The theoretical curves [2] are made with the Reid deuterium wave function (dashed line) and with the HM2 wave function (solid line). The results are shown for two proton momenta: a) p_p = 372.5 MeV/c; b) p_p = 302.5 MeV/c.

OTHER EXPERIMENTAL INDICATIONS

Let us just recall some recent experimental results.

(i) With photons, two anomalies were found in the ^{4}He$(\gamma,p\pi)$ experiment [7] consistent with dibaryonic masses of 2.16 and 2.24 GeV (this last one very close to the deuterium result).

(ii) In the π^+D backward scattering (180°) [8], done at SIN, a structure seems to appear in the excitation function around Q_{DB} = 2.23 GeV non predicted by the existing models.

(iii) The result of the $\pi^\pm-\vec{d}$ scattering measurement will be presented by E. Boschitz at this workshop.

(iv) pp polarized projectile and target experiments were analyzed by Yokosawa [9] and the results are given in table 3. We have to notice that the states around 2.22 GeV seem to be a different object than the one observed in the photon experiment as the measured widths are not the same.

So many results give some indication of possible dibaryon states. But the existing data are not a consistent set and need an effort of clarification. For instance it seems that some effects appear around 2.23 GeV in polarized pp scattering ($\Gamma \sim 100$

Table 3

Analysis of the polarized pp scattering experiments [9]
Candidates of the dinucleon resonance

(i) I=1 isospin state	$B_1^2(2.14)$	$B_1^2(2.18)$	$B_1^2(2.22)$	$B_1^2(2.43)$	$B_1^2(2.43)$
Mass, GeV	2.14-2.17	2.18-2.20	2.20-2.25	2.43-2.50	2.43-2.50
Width, MeV	50-100	100-200	100-200	~150	~150
Quantum state	$^1D_2$	triplet P?	$^3F_3$	probably $^1G_4$	triplet R_{jj}?
(ii) I=0 isospin state	$B_0^2(2.14)$	$B_0^2(2.22)$	$B_0^2(2.43)$		
Mass, GeV	2.14-2.17	2.20-2.26	2.40-2.50		
Width, MeV	50-100	100-200			
Quantum state	triplet?	$^1F_3$	triplet?		

to 200 MeV), in $\pi^+\vec{d}$ backward scattering, in π^+d backward scattering, in $D(\gamma,p\pi)$, $D(\gamma,p)X$ and $^4He(\gamma,p\pi)$ ($\Gamma \sim 40$ MeV) but not in $D(\pi^\pm,p)X$. If all these results correspond to the same object, one has to explain why the measured widths are so different and why no signal appears in the $D(\pi^\pm p)X$ experiment. On the other hand, even if the pp results and the photon results correspond to different states, we have to understand the difference between the photon and the pion experiment presented here. Several explanations can be given; i) if the excited state has an isospin T=0 it can be reached from deuterium by a photon and not by a pion. ii) From the elementary particle point of view, the pion structure ($q\bar{q}$) is very different from the photon, and they can act very differently with a 6 quark system.

It clearly appears that the dibaryon case remains widely open and needs a great amount of theoretical and experimental works. Several experiments on deuterium are in progress using the electromagnetic probe (at Bonn with the 2 GeV bremsstrahlung beam and the 500 MeV tagged beam and at Saclay with the positron annihilation in flight monochromatic beam). But more complete information will be obtained by the new generation electron machines (high duty cycle, higher energies): they will allow to reach high excitation energy (200-700 MeV) for large values of the momentum transfer, so to measure fundamental quantities as the radius of the possible six-quark bag.

REFERENCES

[1] P.E. Argan et al., Nucl. Phys. A296 (1978) 373.
[2] J.M. Laget, Nucl. Phys. A296 (1978) 388.
[3] P.E. Argan et al., Phys. Rev. Lett. 41 (1978) 86.
[4] P.E. Argan et al., submitted to publication (1980).
[5] J.M. Laget, Nucl. Phys. A335 (1980) 267.
[6] SIN-Saclay-Grenoble collaboration, to be published.

[7] P.E. Argan et al., La Toussuire, Feb. 1977.

G. Tamas, Inter. School of intermediate energy nuclear physics, Ariccia, June 1979.

[8] R. Frascaria et al., Phys. Lett. 91B (1980) 345.

[9] A. Yokosawa, Physics Reports 64 (1980) 47.

EXPERIMENTAL SEARCH FOR DIBARYON RESONANCES

E. Boschitz

Kernforschungszentrum Karlsruhe, Institut für Kernphysik
Universität Karlsruhe, Institut für Experimentelle Kernphysik
7500 Karlsruhe, Federal Republic of Germany

In this talk I want to limit myself to the search for dibaryon resonances with pions on deuterons, experiments which have recently been performed at the Swiss Institute for Nuclear Research. Before entering the discussion of the different experiments let me say a few words of introduction for those not familiar with the subject of dibaryon resonances.

The observation of a striking energy dependence in pure spin total p-p cross sections[1] have considerably revived the interest in the two nucleon system since phase shift[2,3] and dispersion relation[4] analyses of these data indicated the existence of 'dibaryon resonances' which may be some of the multiquark states predicted on the basis of the bag model. Since these first results, more experimental evidence was produced from spin observables in NN scattering, pd scattering and deuteron photo-disintegration[5]. However, the subject has remained controversial. First, the early data on $\Delta\sigma_L$ were questioned since they disagreed with pp phase shift analyses in which OBE models of the inelastic channels were used.[6] Recently $\Delta\sigma_L$ was remeasured up to 600 MeV by the Geneva group at SIN[7] and up to 516 MeV by the Basque group at TRIUMF.[8] While the preliminary data from the first group agree with the early Argonne results, the data from the second group disagree to some extent. The more important question, however, concerns the interpretation of these structures in the cross sections since strong inelastic thresholds may produce effects which are phenomenologically difficult to distinguish from genuine resonances.[8]

One way to decide if these 'structures' correspond to real resonances is to look for signals in other two nucleon channels. In fact, since the 'dibaryon resonances' in the pp channel have large inelasticities (~80 %) there must be a considerable coupling into the πd and πnp channels.

The πd system has been a research topic at SIN for many years. Pedroni et al.[9] produced very accurate $\pi^{\pm}$d total cross section data which indicated charge symmetry violation of a few percent. Gabathuler et al.[10]

have measured the differential cross section at intermediate energies with great precision to test recent Faddeev calculation[11,12] which have reached a high level of sophistication. A remarkable feature of these cross section data is the deep minimum in the angular distribution at 256 MeV incident pion energy which so far has not been reproduced by even the best Faddeev calculations. Therefore, it is suspected that some essential ingredient in the πd reaction dynamics is still missing. Kanai et al.[13] have proposed that the deficiencies in reproducing the large angle cross section data above the (3,3) resonance can be removed if effects of dibaryon resonances are included. In the differential cross sections, however, the dibaryon signals are appreciable only in the backward hemisphere where the use of the Glauber model in the calculation is questionable.

During the past year experiments with pions on deuterons have continued, this time, however, specifically directed towards the search for dibaryon signals. The following experiments have been performed:

1) Study of the reaction $\pi^{\pm}d \to R_1 \to \pi NN$, by detecting one proton at fixed angle and momentum, for various incident pion energies[14] (Saclay - Grenoble - SIN collaboration)

2) Search for an isospin 2 dibaryon resonance in the reaction
$\pi^+ d \to \pi^- + R_2$
$\quad\quad\quad\hookrightarrow \pi^+ pp$
by studying the momentum spectrum of negative pions at fixed incident pion energy and fixed production angle.
(Saclay - Grenoble - SIN collaboration)

3) Measurement of the $\pi^+ d$ excitation function at $180°$.[15]
(Orsay - Grenoble - Neuchâtel - Univ.of South Carolina - collaboration)

4) Measurement of elastic $\pi^{\pm}\vec{d}$ scattering[16]
(Karlsruhe - SIN - Erlangen - UBC - Grenoble collaboration)

5) Measurement of the reaction $\pi^+\vec{d} \to pp$[17]
(Karlsruhe - SIN - Erlangen - collaboration)

In discussing the results of these experiments I shall skip the first two since Dr. Tamas has already reported on them. I shall briefly comment on 3) then concentrate on 4) and 5).

To 3)
The $\pi^+ d$ elastic scattering excitation function for $\Theta_\pi = 180°$ was measured between 130 and 280 MeV kinetic energy.[15] The authors claim some evidence for a structure in the vicinity of 250 MeV which may be

due to a dibaryon resonance formation. However, the structure is not very pronounced, and there are questions about the π^+d data in the energy region of interest since the π^+p cross sections (determined with the same experimental set-up) at energies above 220 MeV are in strong disagreement with recent π^+p data[18] and most extensive phase shift analyses and partial wave dispersion relations[19]. A 180° π^+p cross section as small as observed by the authors would necessitate high partial waves in π^+p scattering which whould show up in other angular regions of the differential cross sections. In addition to the experimental uncertainties there are theoretical difficulties. Grein et al.[20] have pointed out that the πd amplitudes at backward angles are sufficiently small to allow for noticable interference effects with resonant dibaryon amplitudes, but the very smallness of the background amplitudes makes them quantitatively unreliable. Genuine pion absorption effects which are important at backward angles are not yet understood. Therefore, Grein et al. conclude that any claims for dibaryon resonances deduced form backward hemisphere differential cross sections alone seem to be premature.

To 4)

A more promising way to detect the interference between the resonant dibaryon and background amplitudes may be the measurement of certain spin observables in elastic πd scattering. Kubodera et al.[21] have quantitatively estimated the effects of the three most likely candidates of dibaryon resonances, the $^1D_2$(2.14-2.17 GeV), $^3F_3$(2.20-2.26 GeV) and $^1G_4$(2.43-2.50 GeV) p-p resonances on the vector and tensor polarizations in πd scattering assuming a partial width $\Gamma_{\pi d}$ of 10 MeV. This width was deduced as an upper limit from an analysis of forward hemisphere differential cross sections. A model calculation[20] indicated that the true channel coupling is somewhat less. While only relatively small effects were observed for the differential cross sections, strong dibaryon signals were predicted for the various polarization observables. In the experiment which I am going to describe in detail the vector analysing power defined as

$$iT_{11} = \frac{\sqrt{3}}{2} \frac{1}{P} \left[\frac{\sigma\uparrow - \sigma\downarrow}{\sigma\uparrow + \sigma\downarrow}\right]$$

$\uparrow \equiv$ direction of $\vec{k}\times\vec{k}'$
$\vec{k} \equiv$ incident pion momentum
$\vec{k}' \equiv$ scattered pion momentum

was determined by measuring the differential cross sections $\sigma\uparrow$ and $\sigma\downarrow$ of elastic $\pi^+\vec{d}$ scattering for the two spin states of a vector polarized deuteron target of polarization P. In such an experiment the kinematics of elastic πd scattering must be accurately established in order to separate the background produced by pions interacting with the other nuclei in the target. For this reason the experiment was performed at the high resolution pion beam and spectrometer facility at the Swiss Institute of Nuclear Research (S.I.N.). The details of this facility are described in ref.22).

The incident pion beam was defined geometrically by scintillation counters which limited the target spot size to 10×10 mm. The counters were aligned to ±0.2 mm to ensure that only the central region of the polarized target was illuminated.

The polarized deuteron target (18×18×5 mm in size) consisted of 95% perdeuterated n-butanol (98% atomic purity) and 5% deuterium oxide, doped with about 1% porphyrexide by weight. The deuterons were polarized dynamically in a magnetic field of 2.5 T at a temperature of ~0.5 K. The deuteron polarization was measured by NMR methods using a signal averager. The polarization value was determined by calibration with respect to the thermal equilibrium signal at 0.46 K and checked against the asymmetry of the enhanced polarized deuteron resonance signal. The target polarization was around (16±1.6)%.

The scattered pions were detected by the pion spectrometer within an acceptance angle of 9 deg. Multiwire chambers in front of the spectrometer allowed a dividing into smaller angular bins. At forward angles a prominent πd-peak is observed in the spectra on top of a background from inelastic processes from other nuclei in the polarized target. At larger scattering angles where the background becomes appreciable the recoil deuteron was detected by a scintillation counter telescope in coincidence with the pion. This greatly reduced the background under the πd peak.

In order to reliably extract the number of counts in the elastic πd peaks from the background, the peaks were systematically fitted by two component Gaussian line shapes, the background by a linear or quadratic form. The justification for applying a smooth structureless background under the elastic πd peak was demonstrated by replacing the deuterated butanol by normal butanol, and also by comparing background runs with and without liquid ^{3}He in the target cell. The consistency of various fitting procedures was tested. The final uncertainty in the data include the statistical errors of the πd peak, the statistical uncertainty of the background subtraction, the uncertainties from the least square fits and the errors in measuring the target polarization.

The reproducibility of data taking was established by several checks.

The results obtained for Θ_{cm} = 70°, 90°, 110°, 125° and 140° at T_π = 143 MeV and for Θ_{cm} = 55°, 70°, 85°, 100°, 115°, 130°, and 145° at T_π = 256 MeV are shown in Figure 1.

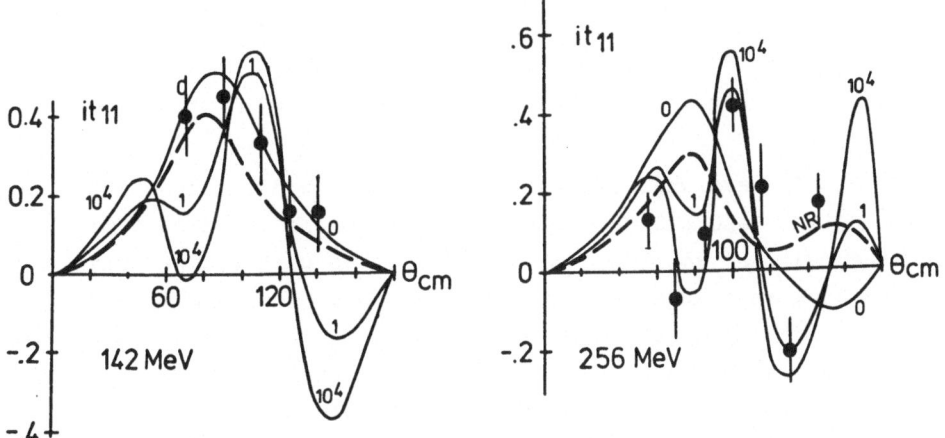

Figure 1: Comparison of the π^+d elastic scattering data with predictions from Kubadera et al.[21]

The curves shown in this figure are the predictions of Kubodera et al.[21] The dashed curves are the result of Faddeev calculations in which no dibaryon resonances are included. The solid curves are produced by including the $^1D_2$ (2140) resonance at T_π = 155 MeV, the $^3F_3$ (2220) at T_π = 245 MeV and the $^1G_4$ or $^1S_0$ (2430) at T_π = 415 MeV. The dibaryon of given J couples to two pion angular momentum states in the πd channel: ℓ_π = J±1. The various solid curves are labeled by the mixing parameter ε (the value ε = 10^4 corresponds to a pure ℓ_π = J+1 coupling and ε = 0 corresponds to ℓ_π = J-1). For a given resonance the ε parameter is a characteristic constant. At the lower energy the results are compatible with the Faddeev calculation, but also with a mixing parameter ε = 0. However, at the higher energy none of the conventional theories[11,12] agrees even qualitatively with the structure of the data. The oscillatory behaviour is surprisingly well reproduced by the predictions of ref.21 in which effects of dibaryon resonances are explicitely included (the experiment favours ℓ_π = 4). We consider this as a strong indication for the presence of at least one dibaryon resonance in the πd channel with a strength compatible with the parameters of ref.21. Independent of the details of the model the very observation of strong oscillations is a direct indication of a strong contribution from a higher partial wave interfering with the background.

To 5)

With the results obtained in $\pi^+\vec{d}$ elastic scattering obviously it is interesting to look for dibaryon signals in the reaction $\pi^+\vec{d} \to pp$. However, the situation is complicated by the fact that inspite of a considerable theoretical effort in the past years this reaction is not as well understood as the πd elastic scattering and, therefore, it is difficult to obtain a reliable background amplitude. For the most recent summary on this reaction see ref.23).

Kamo et al.[24] have analysed the reaction $pp \to \pi^+ d$ in the energy range between 391 MeV and 810 MeV. Their calculation is based on a model in which the reaction amplitude consists of three parts, a neutron exchange Born amplitude, a ΔN intermediate state contribution and a dibaryon resonance contribution. The addition of a separate dibaryon contribution was motivated by the failure of the authors to reproduce the polarization asymmetry at 591 MeV with only the ΔN admixture. However, the coupled channel method of Niskanen reproduces the asymmetry well enough, also at this energy, to eliminate the need for extra free parameters. The predictions of Niskanen and Kamo et al. for the deuteron polarization observables is shown in Figure 2.

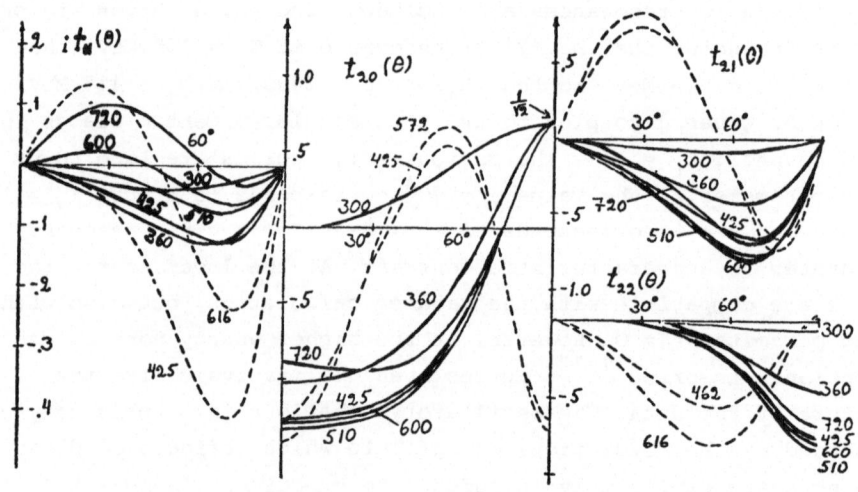

Figure 2: The vector and tensor polarizations t_{kq} of the final state deuteron in the $pp \to \pi^+ d$ reaction. Solid curves are predictions from Niskanen,[23] dashed curves calculations by Kamo et al.[24]

A fortnight ago we have finished our first measurements of the vector analysing power iT_{11} of the $\pi^+ \vec{d} \to pp$ reaction. The principle of the measurement was the same as for the $\pi^+ \vec{d}$ elastic scattering experiment. The experimental set-up was quite conventional. The two outgoing protons were detected in coincidence, by scintillation counter telescopes. The energy of both protons was measured by large NaI detectors (25 cm $\emptyset$ × 25 cm). Pulseheight discrimination in ΔE counters and the time-of-flight information established a clean signature for the reaction. Multiwire chambers in front of the NaI detectors allowed angular binning and gave additional constraints due to the coplanarity of the reaction. With this detection technique the smooth background due to $(\pi^+, 2p)$-reactions on ^3He, ^{12}C, ^{14}O in the polarized target could be reliably subtracted. The incident pion beam was carefully monitored.

We have measured iT_{11} for 8 angles between $\Theta_{cm} = 25°$ and $85°$ at $T_\pi = 256$ MeV and for 4 angles at $T_\pi = 295$ MeV. In addition the energy dependence was investigated from 180 to 320 MeV at $\Theta_{cm} = 55°$.

Since the experimental data are presently being analysed I cannot present final results (I may be able in the proceedings). At present, I can only give the trend of our on-line results. At all energies and angles the vector analysing power is positive. It is small at 180 MeV and increases monotonically to about 30% at 320 MeV. The angular distribution of iT_{11} at $T_\pi = 256$ MeV has a maximum value of about 20% at $\Theta_{cm} = 55°$.

Dr. Niskanen is presently extending his calculations of iT_{11} to our energy region. He is also studying the effects of explicite inclusion of dibaryon resonances. A comparison with his predictions may give us an idea about the most sensitive region to look for a dibaryon signal.

Let me conclude with an outlook:
Next spring we shall study the energy dependence of the $\pi^+ \vec{d}$ elastic scattering and the $\pi^+ \vec{d} \to pp$ reaction in more detail using a polarized target of larger deuteron vector polarization. We are further investigating the possibilities of a tensor polarized target. Also, a kinematically complete study of the deuteron break-up with pions is proposed by a SIN group, and the $pp \to \pi^+ d$ reaction is being studied extensively by the Karlsruhe - Neuchâtel - SIN collaboration and the Geneva group, using polarized proton beam and target.

References

1) I.P. Auer et al. Phys.Rev.Lett.$\underline{41}$ (1978) 354 and ref. therein

2) H. Hidaka et al., Phys.Lett.$\underline{70B}$, 479 (1977)

3) N. Hoshizaki, Prog.of Theor.Phys.$\underline{60}$, 1796 (1978) and $\underline{61}$, 129 (1979) see also earlier work by R.A. Arndt, Phys.Rev.$\underline{165}$, $\underline{1834}$ (1968)

4) W. Grein and P. Kroll, Nucl.Phys.$\underline{B137}$, 173 (1978)
P. Kroll, University of Wuppertal preprint WOB 78-13 (1978)

5) For a summary see: A. Yokosawa, Phys.Reports Vol.$\underline{64}$, No.2, 1980 and references therein

6) D.V. Bugg in AIP Conf.Proc.51, 362 (High Energy Physics with Polarized Beams and Polarized Targets, Argonne, 1978), ed. G.H. Thomas, Amer.Inst.Phys., New York, 1979

7) F. Lehar, Proceedings of the Int.Symp. on High Energy Physics with Polarized Beams and Polarized Targets, Lausanne, Switzerland, 1980

8) D.V. Bugg, ibid

9) E. Pedroni et al., Nucl.Phys.$\underline{A300}$ (1978) 321

10) K. Gabathuler et al. SIN preprint PR80-011, June 1980

11) N. Giraud et al., Phys.Rev.$\underline{C21}$, 1959 (1980)

12) A.S. Rinat et al., Nucl.Phys.$\underline{A329}$, 285 (1979)

13) K. Kanai et al., Prog.Theor.Phys.$\underline{62}$, 153 (1979) and Proceedings of the 2nd Meeting on Exotic Resonances, Hiroshima, Japan, 1980, HUPD-8010 (1980) 39

14) J. Arvieux et al., SIN Newsletter No.13, 1980

15) R. Frascaria et al., Phys.Lett.$\underline{91B}$ (1980) 345

16) J. Bolger et al., Proceedings of the Int.Symp. on High Energy Physics with Polarized Beams and Polarized Targets, Lausanne, Switzerland,1980

17) G. Smith et al., SIN Proposal R-79-07.1

18) D. Fitzgerald et al., Int.Conf.on Baryon Resonances, Toronto, July 1980

19) R. Koch and E. Pietarinen, Nucl.Phys.$\underline{A336}$ (1980), 331

20) W. Grein, K. Kubodera, and M.P. Locher, SIN Preprint PR-80-012, July 1980

21) K. Kubodera et al., J.Phys.$\underline{G6}$, L1 (1980)

22) J.P. Albanese et al., Nucl.Inst.and Meth.$\underline{158}$, 363 (1979)

23) J.A. Niskanen, Proceedings of the 5th Int.Symp.on Polarization Phenomena in Nuclear Physics, Santa Fe, New Mexico, August 1980

24) H. Kamo and W. Wateri, Prog.Theor.Phys.62 (1979) 1035 and Proceedings of the 2nd Meeting on Exotic Resonances, Hiroshima, Japan, 1980, HUPD-8010 (1980) 31

PRESENT STATUS OF (ee'p) EXPERIMENTS

S. Turck-Chièze

DPh-N/HE, CEN Saclay, 91191 Gif-sur-Yvette Cedex, France

I. INTRODUCTION

I would like to precise that I do not intend to discuss of all ee'p coincidence experiments presently accessible with the new high duty cycle accelerators but I wish to focus this discussion to those based on the quasi-elastic process[1] where a lot of data are now available. At that point I must emphasize the very poor state of corresponding (ee'n) experiments due to experimental difficulties.

This (ee'p) field has been developed in the last twenty years in Stanford[2], Orsay[3], Kharkov[4] for deuterium experiments and in Frascati[5], Tokyo[6], Saclay for intermediate nuclei with great enthusiasm. Technical difficulties arising from the limited possibility of the accelerators (low duty cycle) and the experimental set up (momentum resolution and rejection of accidental events) have limited nowadays the exploration of this field to two or three laboratories.

I am obliged, for time reason to limit the presentation to intermediate nuclei where new data and calculations are available this year. Nevertheless I want to mention that the light nuclei are presently widely studied in Saclay. In this case the effects of the reaction mechanism and the final state interaction are less important and better understood. Data have been already obtained on ^{3}He(ee'p) in the range of 100 MeV separation energy and 310 MeV/c initial proton momentum, the break-up in two and three bodies is clearly visible and the analysis is now in progress. The momentum distribution of D(ee'p)n cross section has been also measured in Saclay up to 330 MeV/c and will be extended up to 500 MeV/c in a near future to reach the region where the s-state is at a minimum and the cross section is mainly due to the d-state. Concerning the medium nuclei : in the first part, I shall summarize what we have learned from the first series of experiments about the interpretation based on the shell model ; in the second part, I shall discuss the recent data on ^{12}C and ^{16}O (ee'p) experiments performed in Saclay and the advances in their theoretical interpretation.

II. INFORMATION DEDUCED FROM PREVIOUS EXPERIMENTS

The first generation of quantitative experiments coming mostly from Tokyo and Saclay, with different experimental conditions, has been analysed in the framework of the shell model to yield the separation energy (E) spectrum, the distribution of initial proton momentum (p) and occupation numbers (Nα).* In the distorted wave Born approximation, the cross section was factorized as follows :

$$\sigma_{coinc} = K \sigma_{ep} S^D(E,p) \qquad (1)$$

$$= K \sigma_{ep} \sum_\alpha |\phi^D(p)|^2 P_\alpha(E) \qquad (2)$$

---

*We recall the energies, momenta relationships :

$$E = e - e' - T_{p'} - T_{A-1} \qquad \vec{p} = -\vec{P}_{A-1} = -(\vec{e} - \vec{e}' - \vec{p}')$$

where $\vec{e}$ and $\vec{p}$ are the initial electron and proton momenta and $\vec{e}'$, $\vec{p}'$, $\vec{P}_{A-1}$ the outgoing electron, proton and recoil nucleus momenta.

where K is a kinematical factor, σ_{ep} a relativistic ep cross section in which the off shell effects are taken into account by the mean of an effective mass of the initial proton. Equation (2) shows that the spectral function S^D is directly connected to the hole state wave functions.

The analysis of all the data leads to some important and general remarks :

- The shape of the momentum distributions, the position and the signature of the first hole state peaks are in quantitative agreement with the model. On the other hand, at high separation energies the width of the hole states increases so rapidly that the 1s state is not even visible in the ^{28}Si energy spectrum. The extraction of total occupation probability :

$$N = \sum_\alpha \int S_\alpha^D (E,p) \, p^2 dp$$

always leads to number smaller than 1, even in deuterium : ^{2}D : 0.85, ^{12}C : 0.58, ^{28}Si : 0.69, ^{40}Ca : 0.81, ^{58}Ni : 0.94. The energy weight Koltun sum rule estimation[7] which connects the total energy per proton E_Z/Z to the mean values of kinetic and separation energies by the relation :

$$E_Z/Z = 1/2 \left[<T> - <E> - \frac{M}{M_{A-1}} <T> \right]$$

exhibits an experimental discrepancy of about 2 or 3 MeV, in the range of about 60 MeV separation energy (residual nucleus excitation energy) and 300 MeV/c initial proton momentum. This suggests some displacement of the strength towards higher energy and momentum range.

The disagreement observed for the total occupation number and the sum rule raises some questions about the validity of the shell model description and the influence of the correlations and asks for improved theoretical calculation.

III. PRESENT STATUS OF (ee'p) EXPERIMENTS ON INTERMEDIATE NUCLEI

III.1. The reaction mechanism of the (ee'p) process

Much efforts have been developed from the theoretical point of view, to leave the naive picture of the shell model and treat the distortion of the outgoing proton with less ambiguity in the imaginary part of the optical potential.

A complete non factorized coincidence cross section calculation has been developed by Boffi et al.[8] They succeed in two main improvements : an exact non relativistic calculation of the off shell effects which are important at high momenta and high energies ; the introduction of a proper treatment of the spin orbit interaction. This formalism has the advantage to use for the bound state the best available HF calculations (in good agreement with high momentum electron scattering) and for the distortion of the outgoing proton the "sophisticated" optical potential of D. Jackson[9] in which the imaginary part is adjusted on the polarization proton data. The same momentum distribution ($\ell=1$) of the ^{12}C(ee'p)^{11}B experiment has been studied in four kinematical conditions to investigate the influence of the final state interaction. Among them two will be also considered for the two subshells ($\ell=1, j=+1/2, +3/2$) in the ^{16}O(ee'p)^{15}N case to estimate the influence of spin orbit interaction. These calculations show that the appropriate treatment of the off shell effects and the FSI can create a variation of the absolute value of the spectral function. These effects can be calculated and are at most of 20 % for different kinematical conditions. This theoretical prediction is in qualitative agreement with recent Saclay data (except for ^{12}C "parallel" data* where the χ^2 is always greater than 30)^{10}). As an example, Fig. 1 shows the momentum distribution of $p_{1/2}$ and $p_{3/2}$ in ^{16}O (ee'p) experiments. The observed experimental asymmetry of opposite sign for the two distributions, could only be explained by introducing a spin orbit term in the imaginary part of the optical potential as was done by D. Jackson. The two bound states [Gogny (GO)^{11}) (HF calculation) and Elton-Swift (ES) (phenomenological potential)] selected in the figure have little influence on the shape of the momentum distribution up to 300 MeV/c. Nevertheless, the asymmetry is small in most of the cases and of the order of the experimental precision. A closer comparison requires the improvement of the experimental accuracy of the present data.

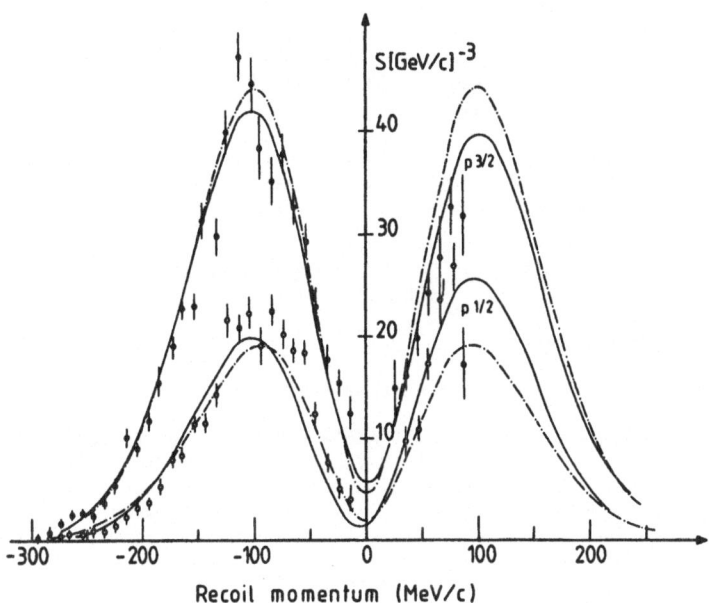

Fig. 1 - The same ^{16}O(ee'p)^{15}N momentum distribution for $p_{1/2}$ and $p_{3/2}$ components, in the two perpendicular kinematical conditions. The Saclay measurement is compared with a calculation using 2 sets of potentials for the bound state and the optical potential. GO+JA (full line) and ES+GK (-·-).

---

*Both kinematical conditions keep $|\vec{p}'|$ constant. In the parallel kinematics the outgoing proton ($\vec{p}'$) is ejected in the direction of the transfer momentum ($\vec{q}$), if $|\vec{p}'| > |\vec{q}|$, the outgoing proton momentum is antiparallel to the recoil nucleus momentum ($\vec{p}_B$) if not the two final momenta are parallel. In the perpendicular kinematics, $|\vec{p}'| = |\vec{q}|$ = cte, $\vec{p}_B$ is nearly $\perp \vec{q}$, two coplanar conditions are considered with $\vec{p}'$ on both sides of the transfer momentum $\vec{q}$.

III.2. The accuracy of the measurement

I would like to emphasize this experimental point. In addition to statistical errors and to the subtraction of the background of accidental events we must consider the influence in this kind of coincidence experiments of the cross section reduction ; we measure in fact a six differential cross section $\sigma(|\vec{e}'|,\theta_e,\phi_e|\vec{p}'|,\theta_p,\phi_p)$ depending on the outgoing electron and proton, and reduce it to a twice differential cross section, depending on the separation energy E and the initial proton momentum $|\vec{p}|$. By the way we mix different kinematical conditions compatible with the experimental set up which lead to the same (E ± ΔE, p ± Δp) bin but correspond to different electron proton cross section. In fact the experimental measurement is compared to a calculation relative to a mean kinematical condition and we reduce the variation due to the electron proton cross section by dividing each event by its corresponding cross section. This procedure introduces the factorization of the coincidence cross section in a bin, and leads to about 7 % incertitude. We must keep in mind this point in the interpretation of coincidence measurement when using large solid angle spectrometers. Does it mean that we need a calculation of the cross section reproducing the phase space of the experimental set up.

For this reason we display here our results in term of spectral function (rather than cross section) and compare them to the equivalent distorted theoretical spectral function deduced from the cross section by the same procedure :

$$K \; |C_\alpha(p',p)|^2 \int_{\Delta E} P_{\alpha\alpha}(E) dE \Big/ K \; \sigma_{ep} \text{ mean.}$$

III.3. $^{16}O(ee'p)^{15}N$ experiment

III.3.1. *What we learn from the outer shells?*

An extended measurement on this light nucleus has been performed at Saclay in the range of separation energy from 0 to 100 MeV and 0 to 300 MeV/c proton momentum. The main contributions to the p-shell (Fig. 2) are clearly peaked at 12.5 MeV ($p_{1/2}$) and 18.5 MeV ($p_{3/2}$), some smaller components $p_{3/2}$ appear also between 20 and 25 MeV, as were predicted by (d,^{3}He) experiment[12]). It is a reasonable assumption to consider these p-shape contributions as single hole state and so the corresponding experimental momentum distributions have been compared to Boffi theoretical calculations for different sets of bound state and optical potentials. The results are summarized in table 1 for selected potentials.

Bound state- optical potential	$P_{1/2}$ 10-15 MeV			$P_{3/2}$ 15-20 MeV			$P_{3/2}$ 20-25 MeV		
	χ^2	η	N	χ^2	η	N	χ^2	η	N
Elton-Swift. Plane waves	BAD	0.98	-	BAD	0.99	-	BAD	0.99	-
Elton-Swift-Glassgold Kellog	2.7	0.66	0.45	2.7	0.71	0.53	1.7	0.69	0.13
Elton-Swift-Jackson	2.4	0.50	0.59	3.2	0.66	0.57	1.5	0.67	0.13
Gogny-Jackson	4.4	0.51	0.59	3.1	0.65	0.57	1.7	0.66	0.13

The shape of the momentum distribution is well reproduced by all the distorted calculations as we can see by the χ^2 for different models and energy regions. Fig. 3a presents the case of $1p_{1/2}$ for ES-GK potentials. The absorption factor

$$\eta = \int_0^{p_{max}} S_D \, d\vec{p} \Big/ \int_0^\infty S_{PW} \, d\vec{p},$$

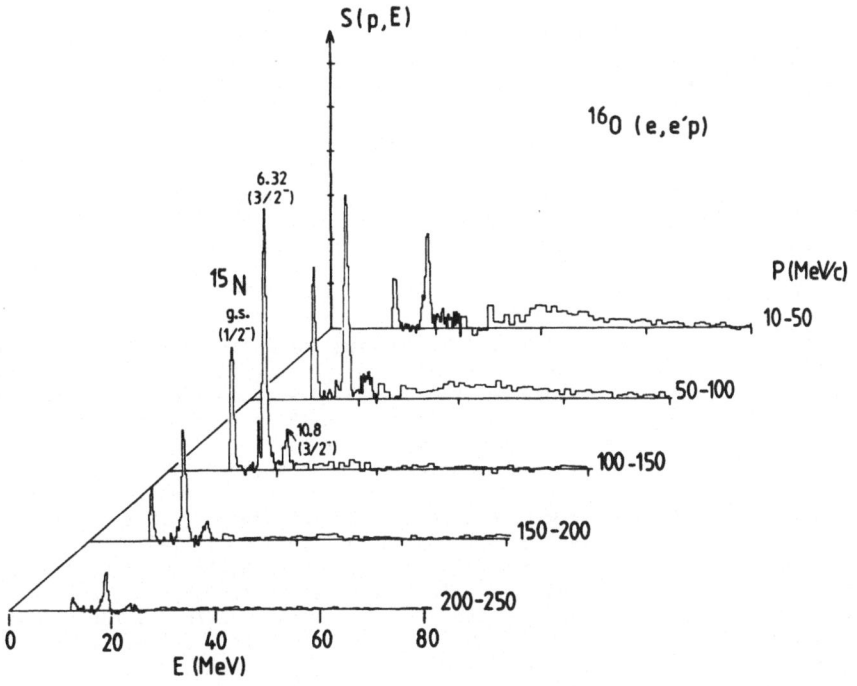

Fig. 2 - Separation energy spectrum for different ranges of recoil nucleus momentum. The inner shell which spreads about 40 MeV is maximum for p=0 MeV/c, the p contributions corresponding to E < 25 MeV are maximum around 100 MeV/c [Saclay experiment].

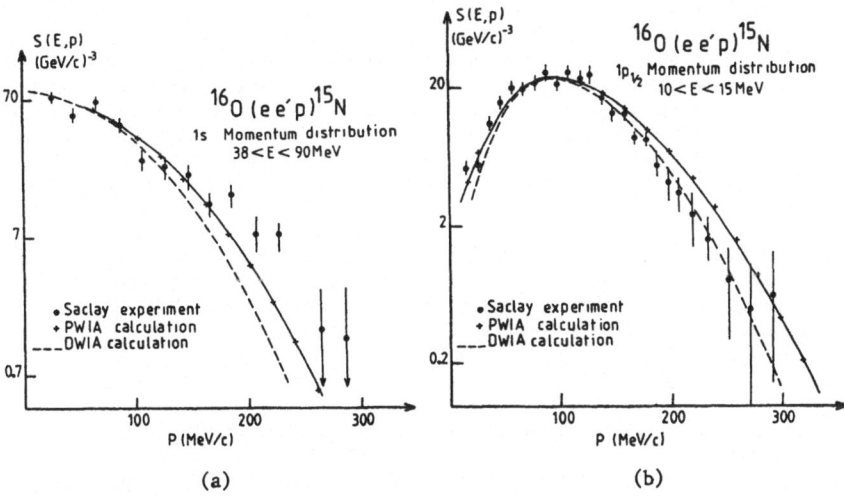

Fig. 3 - Momentum distribution for the ground state of the recoil nucleus ($p_{1/2}$) and for a large range of excited energy (38-90 MeV). The Saclay experimental distributions are compared with the same PWIA and DWIA calculation. In Fig. 3a the agreement with the distorted calculation is very good ; on Fig. 3b the data seem dominated by a s shape distribution but very high components are observed for p > 200 MeV/c.

is equal to 0.98 for the plane wave calculation, so we could be confident that the results would not be influenced by momenta larger than 300 MeV/c. The introduction of D. Jackson potential increases the absorption of about 25 % for the first peak. The spectroscopic strength, which is up to now theoretically uncertain appears here as a normalization factor between experiment and calculation[8]

$$N_\alpha = \int S^{exp} \, d\vec{p}/(2j\alpha+1 \, \eta_\alpha).$$

This number can depend on the quality of the optical potential (25 % effect between GK and DJ) but is nearly independent of the bound state chosen. Nevertheless N_α is definitly smaller than 1 for the two p contributions ($N_{1/2}\,p \simeq 0.59$, $N_{3/2}\,p \simeq 0.7$).

III.3.2. *Large separation energies*

Above 25 MeV (Fig. 2), the large bump spreading out up to 80-90 MeV is expected to be mostly contribution of s-state. Nevertheless the experimental shape is not yet theoretically understood. The range covered by the measurement shows how important must be the rescattering or correlation effects : in the momentum distribution, the high components at high momenta are very important (Fig. 3b) even if for p < 200 MeV/c the s shape appears to be in good agreement with the data ; the spectroscopic number, in the hypothesis of pure 1s state, increases with the range of energy up to values greater than 1, together with a deterioration of the χ^2 ; 12 % of the total strength is found between 60 and 100 MeV and the sum rule is filled with an excess of + 1.93 MeV. So no serious interpretation of the inner shells could be done up to now with a sample formalism based on single hole state. Moreover a correction for experimental rescattering process must be done for p > 200 MeV/c and large separation energy. An estimation made by G. Capitani leads to contributions of the order of 50 % for E > 60 MeV ; this correction will be done in a near future.

IV. CONCLUSION

The new measurements of the quasi free process (ee'p) on ^{12}C and ^{16}O, combined with the recent development of theoretical formalism have demonstrated that the reaction mechanism and the off shell effects are now well understood for outer shells (in the precision of about 10 %) where the single hole state seems to be a reasonable assumption. It appears that the corresponding absolute spectroscopic strength is definitively smaller than 1. A reliable calculation of these quantities will be precious for the interpretation of these quasi free experiments ; some estimations[13] of the target ground state correlations show that the $1p_{1/2}$ is a little less occupied than the $1p_{3/2}$ in ^{16}O but the occupation numbers are both around 0.9. Concerning the inner shells, data are now available on a very large range, and ^{16}O is one of the best candidate to test the correction of rescattering process and the calculations with correlations.

The understanding of the inner shells is very important because we have noticed that the theoretical spectral function of the s-state is very dependent of the bound state model in opposition with the corresponding p-state spectral function (as shown in table 1) where the variation reaches only 20 % at 280 MeV/c.

REFERENCE

This work is due to the collaboration of : P. Barreau, M. Bernheim, J. Morgenstern, D. Tarnowski, S. Turck-Chièze : DPh-N/HE, CEN Saclay ; J. Mougey : Institut Laue Langevin, Grenoble ; D. Royer : DEMT/SEEN, CEN Saclay ; G.P. Capitani, E. De Sanctis, INFN Frascati ; S. Frullani : INFN Sanità, Roma ; G.J. Wagner : Max Planck Institut, Heidelberg ; E. Jans, IKO, Amsterdam.

[1] J. Mougey et al., Nucl. Phys. A262 (1976) 461.
[2] M. Croissiaux et al., Phys. Rev. 127 (1962) 613.
[3] P. Bounin et al., Ann. de Phys. 10 (1965) 475.
[4] Y.P. Antoufiev et al., Sov. J. Nucl. Phys. 22 (1976) 121.
[5] U. Amaldi et al., Phys. Lett. 25B (1967) 24.

[6] K. Nakamura et al., Nucl. Phys. A268 (1976) 381.
[7] D.S. Koltun, Phys. Lett. 28 (1972) 182.
[8] S. Boffi et al., Nucl. Phys. A336 (1980) 437 and references there in.
   S. Boffi, communication of this conference.
[9] D.F. Jackson and I. Abdul-Jalil, J. Phys. G6 (1980) 481.
[10] M. Bernheim et al., Nucl. Phys. to be published.
[11] J. Decharge and D. Gogny, Phys. Rev. C21 (1980) 1568.
[12] V. Bechtold et al., Phys. Lett. 72B (1977) 169.
[13] W. Fritsch et al., Nucl. Phys. A241 (1975) 79.

FUTURE (e,e'p) EXPERIMENTS AT IKO

C. de Vries, T. de Forest Jr., C.W. de Jager, E. Jans,
J.H. Koch, L. Lapikás, R. Maas, H. de Vries,
P.K.A. de Witt Huberts

Instituut voor Kernphysisch Onderzoek,
Amsterdam

I INTRODUCTION

While inclusive electron scattering can be used to map out the general nuclear response surface, the exclusive (e,e'p) reaction probes more specific aspects of nuclear structure (ref. 1). These aspects range from single particle components in the nuclear ground state over decay modes of giant resonances to the structure of states high in the continuum.

In the past, the main work in this field (ref. 2, 3, 4) has been carried out at Saclay (ref. 5) with a 1-2% duty factor beam. The new electron accelerator at IKO will have a duty factor as high as 10% and in general a much improved figure of merit, due to a series of technological advances. Using IKO as an example, we show what experiments are possible with future medium energy accelerators.

The next section gives a brief survey of the figures of merit of the IKO facility. That these figures are indeed realistic is shown in the Status Report in the Appendix. Section III gives four examples of (e,e'p) experiments planned at IKO to demonstrate the lines of research being pursued.

II FIGURES OF MERIT FOR COINCIDENCE EXPERIMENTS

Although the importance of coincidence experiments has long been recognized, actual experimental progress in this field has always been hampered by the limitations set by the equipment parameters. Continuous improvements in beam duty factor, opening angles of the spectrometers, coincidence resolving time, etc. have shown to be crucial to obtain reasonable coincidence counting rates while suppressing the accidental rates to a tolerable level. A thorough discussion of the experimental conditions leading to estimates of count rates for different types of experiments can be found in ref. 6 and 7.
Here we shall elaborate in some detail upon the figures of merit which are important for planning coincidence experiments at IKO. Several other laboratories in Europe, U.S.A. and Japan plan electron scattering facilities (ref. 8) that will further improve the IKO parameters (2.5% at 500 MeV and 10% at 250 MeV). The IKO apparatus has been designed with emphasis on improving the beam duty factor to the 2.5-10% level, increasing the quality of the spectrometers (solid angle, intrinsic resolution) and obtaining a small coincidence resolving time (see Appendix A). Another important advance is that the missing energy resolution can be made as low as 150 keV. This can be achieved by adapting the dispersion matching mode to coincidence experiments. Details of this novel mode of operation are given in Appendix B.
Table 1 lists for particularly interesting kinematics the crucial parameters which govern the possibilities for the experiments considered here. The numbers are given separately for discrete and continuum final states. For the latter, accurate missing energy resolution is not crucial, while the size of the momentum bite is important. Combining the parameters of Table 1 one obtains two quality factors: Q_1, which determines the coincidence count rate R, and Q_2, which determines the real to accidental ratio, R/A.

Table 1 Quality factors for coincidence experiments (e,e'p) at IKO

	discrete final state	continuum final state
E_o^{max} (MeV)	500	500
df^{max} (%)	2.5	2.5
t_c (ns)	0.8	0.8
$\Delta\Omega_e$ (msr)	5.6	5.6
Ω_p (msr)	17	17
$\Delta k_e/k_e$ (%)	± 2.5	± 5.0
$\Delta k_p/k_p$ (%)	± 5.0	± 5.0
δE_m (MeV)	0.15	1
Q_1 (msr^2)	0.48	0.95
Q_2 (%/ns/MeV)	21	3.1

Kinematics chosen: k_e = 500 MeV/c, $k_{e'}$ = 400 MeV/c, k_p = 400 MeV/c

E_o^{max} = maximum available primary energy

df^{max} = duty factor at E_o^{max}

t_c = coincidence trigger resolving time

$\Delta\Omega_{e,p}$ = maximum solid angle for electron (QDD), proton (QDQ) spectrometer

$\Delta k_{e,p}/k_{e,p}$ = usable momentum acceptance for electron, proton spectrometer, leading to a phase space range ΔE_m = 20 MeV for discrete final states, and to ΔE_m = unlimited for continuum final states

δE_m = missing energy resolution in DM mode, resulting from intrinsic resolution of BHS, QDD and QDQ; contribution from target not included. For continuum final states the value 1 MeV is chosen arbitrarily.

Q_1 = $\Delta\Omega_e \Delta\Omega_p \Delta k_e \Delta k_p/k_e/k_p$, quality factor that determines countrate R

Q_2 = $df/t_c/\delta E_m$, quality factor that determines real to accidental ratio R/A

III EXAMPLES OF FUTURE (e,e'p) EXPERIMENTS

The recent work in (e,e'p), especially at Saclay (ref.5), has already yielded some beautiful results. However, it has also become evident that - at present - such experiments are time consuming. Any systematic research or more precise measurements can only be carried out if the experimental conditions are significantly improved. To show what can be achieved under more advanced conditions, we discuss below four different experiments, using the above figures of merit (Table 1). Each experiment focusses on a different aspect of the nuclear dynamics, illustrating the wide scope of (e,e'p) research.

1° $^3He\ (e,e'p)\ ^2H$

It is clearly of interest to extend the study of nuclear single particle aspects towards higher momenta. Few body targets, such as ^3He, allow comparison with the most sophisticated theoretical predictions.

A liquid ^3He target, able to handle about 1-2 µA of beam current is under construction. Its rectangular, flat shape allows an optimal geometrical choice of target angle relative to outgoing proton and electron directions (see ref. 9). Together with the contributions from straggling in the cell windows and in the heat shields, this yields a contribution of the whole target assembly to the missing energy resolution of < 200 keV.

The quality factors Q_1 and Q_2 given in Table 1 allow to extend the range of momentum distribution of the two-body break-up channel beyond that obtained by the recent Saclay experiments. This is illustrated in fig. 1. Two different theoretical predictions (ref. 10 and ref. 11) are shown. The experimental points are taken from ref. 12. New measurements (ref. 13) extend the experimental momentum distribution covered at Saclay to $P_r \simeq 310$ MeV/c.

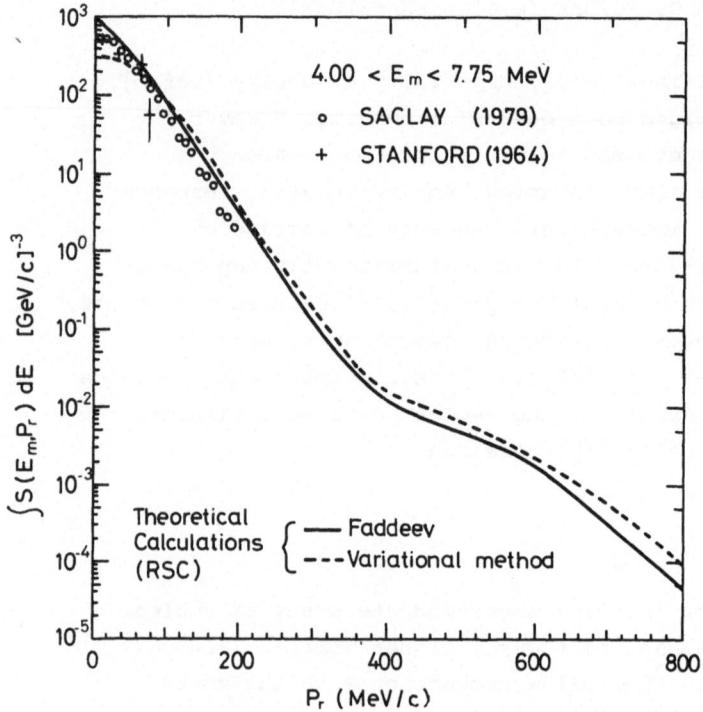

Fig. 1 Theoretical predictions and experimental results for the momentum distribution in the two-body break-up channel (^{3}He(e,e'p)^{2}H).
Solid line: Faddeev calculation (ref. 10)
Dashed line: Variational calculation (ref. 11)

Extrapolating from the Saclay experiments with the IKO quality factors and the theoretical predictions, we arrive at a coincidence count rate R = 1/MeV/hr at P_r = 500 MeV/c and R/A = 6. To reach even higher deuteron recoil momenta, with a still tolerable R/A = 1, one would have to increase the beam current. However, this is not possible with this liquid helium target. For targets which are not hampered by heat dissipation such a trade-off is indeed feasible and hence considerable reduction in beam time can be obtained. For the three-body break-up channel, where missing energy resolution is unimportant, we expect real count rates similar to those of the Saclay experiment, but a ratio R/A that is a factor 8 larger.

2°   *Proton knockout from valence shells:* $^{12}C(e,e'p)^{11}B$

With the (e,e'p) reaction, it is possible to determine the momentum distribution, energies and widths of the target hole states. This type of experiment profits from the improved energy resolution.

In the reaction indicated several final states will be populated (figure 2). The coincidence count rate can be deduced from the quality factors in Table 1 and the numbers given in ref. 9. With an effective target thickness of 80 mg/cm^2 and with 30 µA beam current one obtains $R = 3\,s^{-1}$ with a real to accidental ratio $R/A \simeq 10$. As a result, parts of the spectral function $S(E_m, P_r)$ that are a factor 25 smaller than observed before can be measured. Therefore, the largest measurable recoil momentum for a $1p_{3/2}$ proton knockout leading to the ^{11}B ground state becomes $P_r \simeq 380$ MeV/c. Another important result of the improved missing mass resolution is the possibility to separate hitherto unresolved final states. Interesting examples are the "pick-up forbidden" states at 4.44 MeV and 6.74 MeV (see fig. 2). The $5/2^-$ state at 4.44 MeV may correspond to knockout from the normally unoccupied $1f_{5/2}$- orbital, being a signature of ground state correlations in ^{12}C. This state has been populated in the (d,^{3}He) reaction (ref. 14), but a detailed analysis is complicated by interfering two step mechanisms. A much cleaner investigation of this state should be possible with the (e,e'p) reaction where the two step mechanism is known to be small.

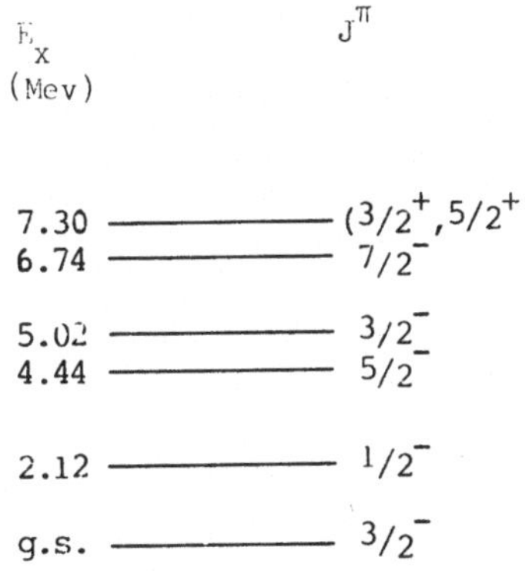

E_x (MeV)	J^π
7.30	$(3/2^+, 5/2^+)$
6.74	$7/2^-$
5.02	$3/2^-$
4.44	$5/2^-$
2.12	$1/2^-$
g.s.	$3/2^-$

Fig. 2 Level scheme of ^{11}B

3° *Proton knockout in de "dip" region*

Any theoretical interpretation of the observed inclusive (e,e') cross section makes assumptions about the final nuclear reaction channels. A serious constraint for the theoretical interpretation is therefore provided by actually measuring the reactive content, i.e. by looking at (e,e'p) and (e,e'π). This is of particular importance in the dip region between the pion production region, where a long standing disagreement exists between experiment (ref. 15) and theory (ref. 16).
The following estimates, taken from ref. 6, show the feasibility of such experiments at IKO:

^{12}C (e,e'p) in the dip region:

Current	: 15 µA
Duty cycle	: 2.5%
Target	: (∼ 150 mg/cm^2)
Kinematics	: k_e = 500 MeV/c, k_e' = 300 MeV/c, θ_e = 60°, k_p = 600 MeV/c
Estimated rates	
Real coincidences:	R ≃ 0.18/sec
Real/accidental	: R/A ≃ 0.5

4° *Giant resonance study: ^{40}Ca (e,e'p) ^{39}K*

Giant resonances have been studied extensively in the past with real and virtual photons in reactions such as (γ,p), (γ,n) and (e,e'). However, the location and the multipole assignment of the giant resonances is often difficult to extract from such data alone. A coincidence measurement, such as (e,e'p) to a specific final state, helps to determine the multipole character through the angular distribution of the emitted proton.

As a typical example we choose ^{40}Ca. Figure 3 shows the expected energies (ref. 17) of protons emitted from isovector giant electric resonances leading to final states in ^{39}K.

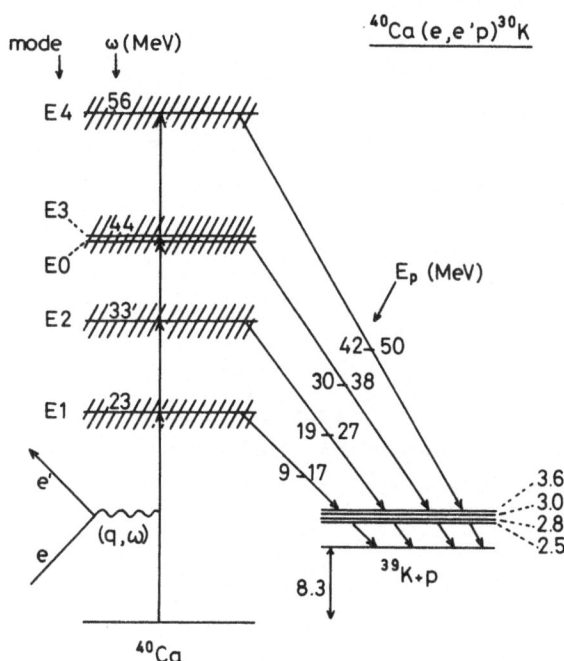

Fig. 3 Giant resonance level scheme: ^{40}Ca and its decays to low lying states in ^{39}K. The excitation energies of the multipole resonances are taken from ref. 17.

The IKO equipment is especially suited to study the highly excited isovector resonances, because a) the intrinsic coincidence resolving time (0.8 ns) of the detectors is then hardly affected by uncertainties in reconstruction of the path length difference in the proton spectrometer ($E_p > 20$ MeV: $\Delta t_c < 1$ ns) and b) the good missing mass resolution which still allows to separate the final states

in ^{39}K. Using the estimates of refs. 6 and 18, such an
experiment seems feasible with a 10% d.f. beam. With a
10 mg/cm^2 target and a current of 10 μA coincidence
count rates for each giant resonance level are expected
to be in the order of 10-100/hr (depending on initial
energy and the electron spectrometer angle) with R/A
larger than 10.

Two possibilities are interesting: 1) to determine the
population of different final states, 2) to distinguish
monopole and quadrupole giant resonances through the study
of their angular decay patterns.

Other coincidence measurements

Electro-induced coincidence experiments - other than (e,e'p) -
considered are:

A(e,pπ$^-$)Ae': This probes the γ+n → p+π$^-$ reaction inside the
nucleus. Count rate estimates (taken from ref. 6)
for most favorable kinematics: ^{12}C (e,pπ$^-$) ^{11}C,e':
R = 2 counts/sec., at R/A = 40.

^{3}He(e,e'T)π$^+$: For light nuclei one can measure the electron in
coincidence with the recoiling nucleus.
Count rates taken from ref. 6:
R = 50/hr for 17 MeV outgoing tritons and R < 0.5/hr
for 66 MeV. No limitation regarding R/A exists
for this kinematically complete experiment.
Tritons with the energies given above can be
handled by the IKO QDQ focal plane detector system
(see Appendix A).

General A(e,e'x) experiments (x = d,α,τ,π):
Such experiments will also be possible in future,
but they will have to wait until enough experience
with the coincidence technique has been obtained.

Appendix A

STATUS REPORT ON THE AMSTERDAM ELECTRON SCATTERING FACILITY

In June 1980, first test results were obtained with the 500 MeV beam handling system (including the dispersion matching mode of operation), the high resolution QDD spectrometer and its focal plane detector system.

A.1 *Accelerator (MEA)*

Table 2 shows the beam conditions during these test runs, the design values for MEA and the best conditions achieved earlier in specific accelerator test runs.

Table 2

Beam parameters of MEA

	Design values	Best values obtained	Values used during the test runs
energy (MeV)	500	120	90 and 115
peak current (mA)	20	20	1-2
repetition rate (Hz)	2000	200	50
burst length (μs)	50	30	4
energy spectrum (%)	<0.3	0.1	$\sim$0.5

A.2 *Beam handling system (AFBU)*

This system consists of two parts, each deflecting the beam over 90°. The first part (with two 45° bending magnets) is an achromatic system containing the energy-defining slit. The second part (again two 45° bending magnets) can be tuned

in normal mode (NM) or in dispersion matching mode (DM).
The dispersion at the target can be tuned from 5-13 cm/%
whereas the intrinsic resolution is < 1.0×10^{-4}.
Specific details have been published before (ref. 19 and 20).
Although more refined tuning has still to be done the system
functioned very close to the design specifications which
have been set for high resolution experiments.
Gratifying is also that tuning is rather easy and that the
whole system (which is computer controlled) showed stable
operation.

A.3   *500 MeV electron scattering facility (EMIN)*

We report here on the performance of the QDD spectrometer and
its associated focal plane detection equipment. The tests
have been done with center line equipment, such as a toroid-
type absolute current monitor upstream of the target, a 45 cm
diameter scattering chamber provided with a sliding foil for
each of the two spectrometers (QDD and QDQ) and a remotely
controlled target ladder. Two large quadrupoles downstream of
the target refocus the multiply scattered electron beam onto
the 100 kW beam dump, located directly behind a 2 m thick
concrete wall 15 m from the target.
Fig. 4 shows the two-spectrometer arrangement, Table 3 gives
its parameters.
A more detailed description of the high resolution QDD spectrometer (for single arm experiments) and the large solid angle
QDQ spectrometer (for coincidence experiments) has been given
elsewhere (ref. 20).

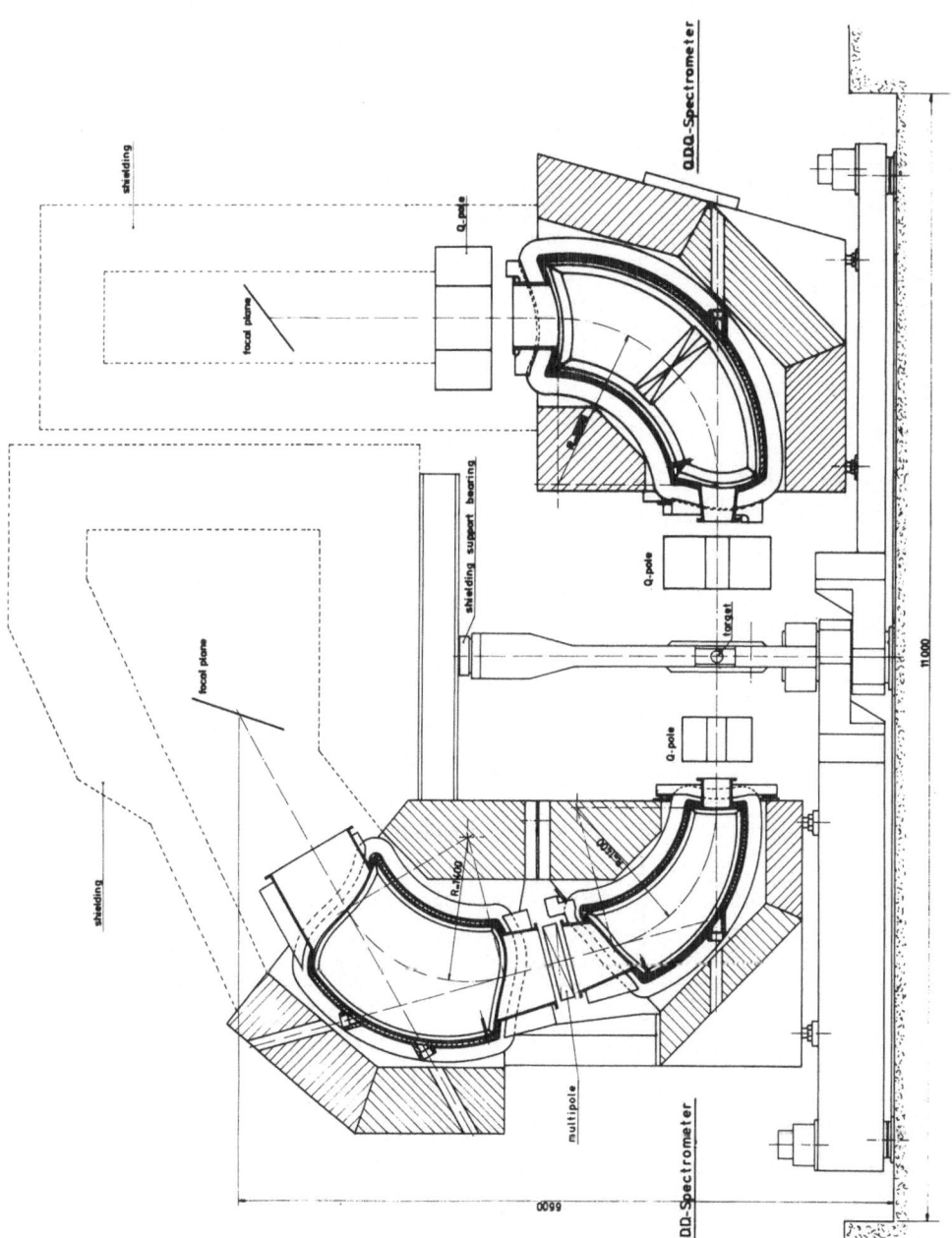

Fig. 4 Layout double spectrometer arrangement

Table 3

Main Spectrometer Parameters

	QDD	QDQ
General		
radius of curvature	140 cm	160 cm
maximum momentum	600 MeV/c	750 MeV/c
corresponding field	1.43 Tesla	1.56 Tesla
Dipole magnets		
Deflecting angle	2 x 75°	90°
Gap	7 cm	12.8 cm
Focal plane		
angle between focal plane and ref. trajectory	41°	50°
momentum acceptance	± 5%	± 5%
solid angle	5.6 msr (80x80 mrad)	17 msr (140x140 mrad)
focussing conditions	$<x\|\Theta>=0$ (point-to-point)	$<x\|\Theta>=0$ (point-to-point)
	$<y\|y>=0$ (parallel-to-point)	$<y\|\phi>=0$ (point-to-point)
dispersion	$<x\|\delta> = 6.78$ cm/%	$<x\|\delta> = 7.28$ cm/%
magnification	$<x\|x>= 0.60$ cm/cm	$<x\|x>= -1.17$ cm/cm
		$<y\|y>= 2.49$ cm/cm
momentum resolution	$<1\times10^{-4}$ for $\Delta p/p=\pm1$% $<3\times10^{-4}$ for $\Delta p/p=\pm5$%	$<1\times10^{-3}$ for $\Delta p/p=\pm5$%

Notes:

Both spectrometers have a flat focal surface.
The dipoles have seventh order polynomial entrance and exit profiles.
The quadrupoles contain sextupole up to dodecapole components.
The QDD has a cross-over in the non-dispersive direction ($<y|\phi>=0$) between the two dipoles.

The QDQ is presently being tested. Considering the good
results with the QDD spectrometer, no problems are anticipated
with the QDQ spectrometer, manufactured with comparable
precision as the QDD by Bruker Physik, Karlsruhe. Hence the
design values of both can safely be used for planning
future experiments.

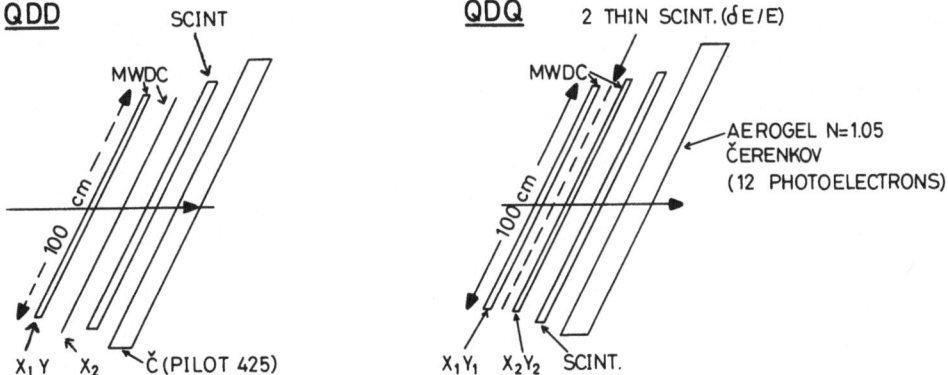

Fig. 5   Detector telescopes in QDD and QDQ

The layout of the wire chambers and backup trigger counters
is shown in fig. 5. The chambers are of the multiwire
proportional type with the additional feature of drifttime
interpolation. Some properties of these counters are given
in Table 4.

The wire chambers cover the full momentum width ($\Delta p/p = 10\%$) of the focal plane. The chamber gas used is the standard 50/50 isobutane - argon mixture.

The trigger of the QDD detector consists basically of a scintillator in coincidence with a Cerenkov counter.

From the three wire chamber hits a full reconstruction of the particle trajectory can be performed with resolutions given in the table. One event can be processed in 150 ns and stored in a fast buffer with a 128-event capacity.

In between beam bursts the buffered events are transmitted via a memory (1.5 µs per cycle) to a PDP-11/34 processor. Initial tests have shown stable functioning of the wire chambers.

Table 4

Summary of the main properties of the wire chamber

number of sense wires	280
wire length (X_1, X_2)	110 mm
wire pitch	4 mm
anode-cathode gap	4 mm (QDD)
	5.6 mm (QDQ)
coarse channel width	2 mm
fine channel width	0.25 mm (2.5×10^{-5})
corresponding angular resolution (FWHM): Θ	7 mrad
ϕ	7 mrad

The QDQ focussing conditions (see Table 3) require four wire planes (also with drifttime interpolation) for full definition of the particle trajectory. A scintillator (S), if needed in coincidence with a Cerenkov detector, provides the trigger for the wire chamber read-out electronics. The Cerenkov detector consists of aerogel material, with an index of refraction n = 1.05, to be used for pion detection purposes.

The QDQ system can be optionally equipped with a stack of
two thin scintillators for heavy charged particle discrimination.
The inter-spectrometer coincidence is established from the
scintillator signals. Our tests indicate that the photomultipliers contribute 500 ps to the coincidence resolving
time t_c. The precision of the flight path reconstruction
implies δt_c = 250 ps (200) for the QDD (QDQ) arm and $\beta = 1$
particles.
Hence the expected overall coincidence time resolution is
smaller than 1 ns.

A.4   *Test results*

Fig. 6 indicates how most of the first test results have been
obtained. With the so-called "magic" target, containing
strips of C, Al and Nb, we have obtained straight-forward
and simultaneous results on the proper functioning of the
dispersion matching mode of the beam switch yard, overall
system resolution, energy calibration, dispersion and location
of the focal plane.

A typical spectrum is also shown in fig. 6. From the
figure it is apparent that a resolution which is independent
of the $\Delta p/p$ of the primary beam, directly indicates proper
match between the beam switch yard and the QDD spectrometer
optics. The observed peak widths were 2.5×10^{-4} at 90 MeV
and 2.0×10^{-4} at 115 MeV for small solid angle. After
unfolding effects such as straggling, kinematic broadening
etc., we can conclude from the present results that the
intrinsic resolution is in the order of 1×10^{-4}.

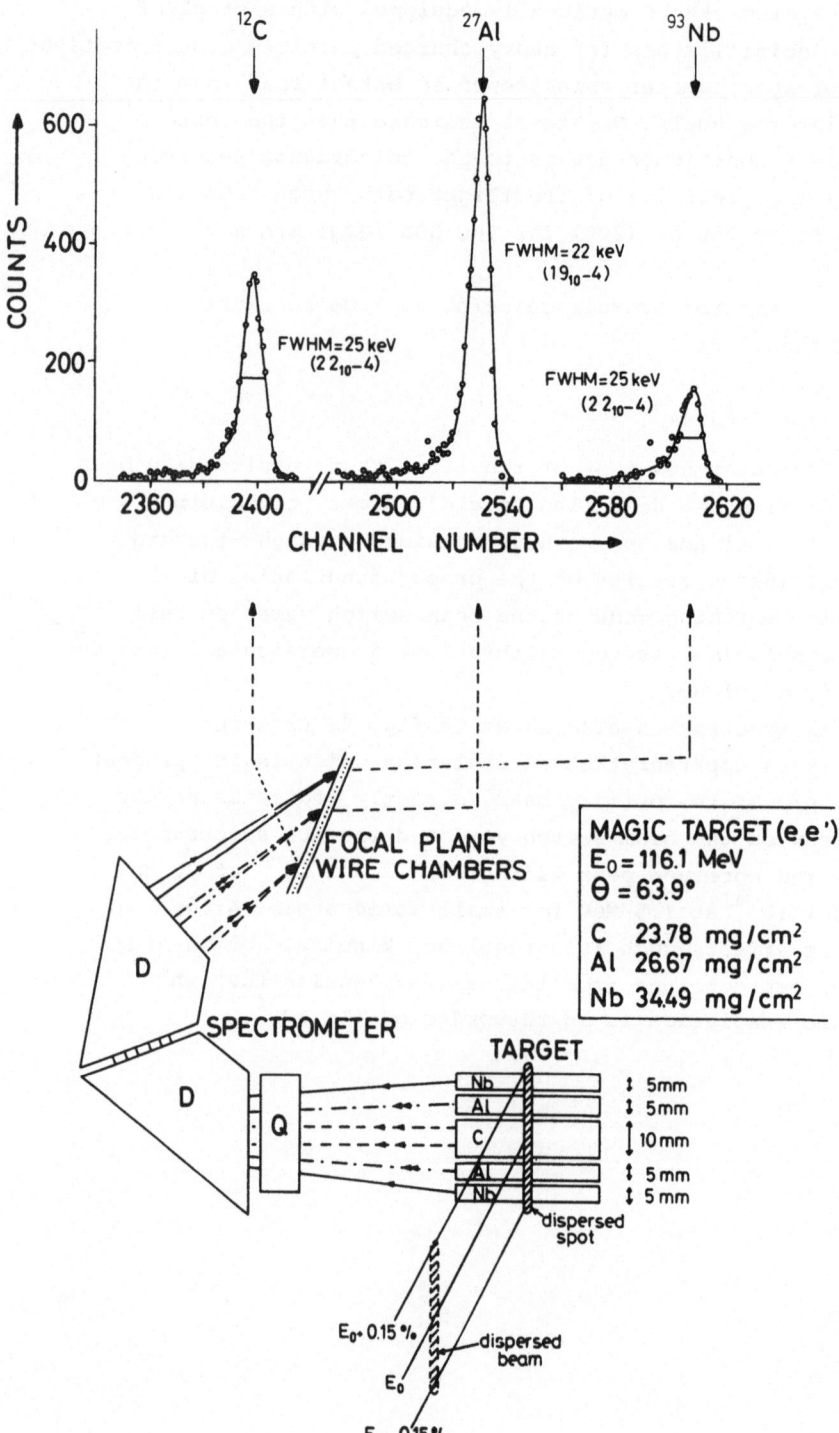

Fig. 6: The "magic target" which has been used during the summer 1980 tests and the elastic scattering spectra from ^{12}C, ^{27}Al and ^{93}Nb.

Appendix B

The missing energy E_m, a relevant quantity in coincidence experiments, is defined as

$$E_m = p_0 - p_1 - p_2^2/2M$$

where p_0, p_1 and p_2 are the momentum of the incoming electron, scattered electron and the (non-relativistic) knockout particle of mass M respectively. We neglect the kinetic energy of the recoiling rest nucleus.

It can be shown quite generally (see ref. 21) that a non-zero spread Δp_0 in primary beam energy p_0 does not affect the missing energy resolution, ΔE_m, if one adapts the dispersion matching (DM) mode of operation for coincidence experiments. In this mode the primary beam dispersion, D_0, at the target has to be chosen such that

$$D_0 = p_0/(p_1/D_1 + p_2^2/M/D_2)$$

where D_1 and D_2 are the known dispersions $<x|\delta>/<x|x>$ of QDD and QDQ spectrometer, respectively. Inserting the IKO numbers (see Table 3) and typical values for p_0, p_1, p_2, we find D_0 in the range 6-8 cm/%, which is easily amenable to our beam handling system (see A.2). Hence, the known advantages of the DM mode for single arm experiments - overall energy resolution independent of Δp_0 for high target currents - also apply to coincidence experiments.

Apart from target thickness effects, ΔE_m can therefore be improved considerably ($\Delta E_m < 150$ keV) compared to the normal mode of operation (point focus), where Δp_0 contributes already e.g. 400 keV at $p_0 = 400$ MeV and $\Delta p_0/p_0 = 0.1\%$ primary beam slit setting.

References

1. G. Jacob and Maris, Rev. Mod. Phys. 38(1966)121, Rev. Mod. Phys. 45(1973)6
2. A. Johansson, Phys. Rev. 136(1964)1030
3. V. Amaldi et. al., Phys. Lett. B25(1967)24
4. K. Nakamura et. al., Phys. Rev. Lett. 33(1974)853 and Nucl. Phys. A268(1976)381
5. J. Mougey, Proc. Int. Conf. on Nuclear Physics with Electromagnetic Interactions, Mainz 1979
6. J.S. O'Connell, B. Schoch, "Intermediate Energy Electromagnetic Interactions with Nuclei", MIT Workshop, June 1977
7. B. Schoch, J.S. O'Connell, "Coincidence Measurements at IKO", IKO Internal Report, October 1979
8. Proceedings Charlottesville Conf. on "Future possibilities for Electron Accelerators", January 1979
9. J. Mougey, Thesis, Université de Paris-sud, 1976
10. A. Dieperink et. al., Phys. Lett. 63B(1976)261
11. C. Ciofi degli Atti, E. Pace (private communication)
12. E. Jans et. al., Proc. Int. Conf. on Nuclear Physics, Berkeley, 1980, in the present figure radiative corrections have been applied
13. E. Jans (private communication)
14. F. Hinterberger et. al., Nucl. Phys. A106(1968)161
15. P.D. Zimmerman et. al., Phys. Lett. 80B(1978)45
16. T.W. Donnelly et. al., Phys. Lett. 76B(1978)393
17. A. de Shalit and H. Feshbach, Theoretical Nuclear Physics, Vol. I, New York, 1974
18. HEPL PHY76-80168 "Proposal to the Nat. Sc. Found., Stanford 1978
19. J. Bergström, R. Maas, Internal IKO Report (1976)
20. C. de Vries, Proc. Int. Conf. on "Nuclear Physics with Electromagnetic Interactions", Mainz 1979 and Internal IKO Report: "Status Report on the two-spectrometer (QDD and QDQ) arrangement in the EMIN hall, March 1980
21. L. Lapikás and P.K.A. de Witt Huberts, Proc. Symposion on "Perspectives in Electro- and Photo Nuclear Physics, Saclay 1980

SACLAY ACTIVITIES IN ELECTRO-AND PHOTONUCLEAR PHYSICS AT INTERMEDIATE ENERGIES
AND FUTURE PROSPECTS

Claude SCHUHL

Chef du Service de Physique Nucléaire à Haute Energie
Laboratoire de l'Orme des Merisiers
F 91191 Gif-sur-Yvette Cedex

The validity of the decision to build Saclay's 600 MeV electron linear accelerator (A.L.S.) is substantiated by ten years of experimental work and is shown by the scientific relevance of the experimental results obtained with its primary and secondary beams, on various experimental areas. This is primarily due to the unique qualities of the electromagnetic probe for the detailed study of nuclear systems : the electromagnetic interaction is described by one of the best grounded theories in physics, namely quantum electrodynamics, whose predictions have been verified to accuracies of order 10^{-9} and electromagnetic interaction is carried by a lepton having no strong interaction with the target nuclei and capable of transferring large momenta to the target's elementary constituents.

The A.L.S. delivers an electron beam of 600 MeV maximum energy, with 1 to 2 % duty cycle and a mean intensity of 300 µA, to six experimental areas :

- the primary electron beam in a hall (HE1) for high resolution spectrometry ($\Delta E/E = 10^{-4}$),

- three photon beams either monochromatic or from bremsstrahlung in the energy range from 20 to 600 MeV (HE3 and BE halls),

- two secondary $\pi^{\pm}$ meson beams with energies between 15 and 100 MeV and a particle flux of 10^6/sec. These channels are tunable to muon beams whose pulsed structure is particularly remarkable.

For example, experiments were performed for electron scattering up to 604 MeV with an intensity of 20 µA, for photonuclear reactions with 40 nA average current of 20 MeV to 450 MeV in $\Delta p/p \simeq 1$ % and for pions and muons channel with a 135 kW power beam (320 µA × 420 MeV).

I. SHORT REVIEW OF ACTIVITIES AT A.L.S. (Accélérateur linéaire de Saclay).

Physicits of our laboratory will describe in details some of the experiments performed at A.L.S., so I shall give a squeezed review of the different domains of activities. Many of them were obtained in cooperation with other laboratories : Amsterdam, Bâle, Bologna, Bordeaux, Cambridge, Clermont-Ferrand, Frascati, Grenoble, Louvain, Lyon, Mainz, Orsay, Roma, Torino, Urbana,... . I apologize for giving no reference to the different works.

A. EXPERIMENTS PERFORMED WITH THE ELECTRON BEAM

1. Electron scattering

1.a. *Elastic scattering of electrons*

Elastic scattering of electrons at large momentum transfers is the most powerful and precise way for studying certain properties of the ground state of the nucleus. From the variation of the cross section as a function of momentum transfer, one determines the charge and magnetization distributions. Besides the qualities of the accelerator (high energy and large intensity), the special properties of the "900" spectrometer - absence of background due to the good shielding and excellent resolution - better than 10^{-4} - makes it the ideal tool for measuring very weak cross sections (less than 10^{-37} cm^2 sr^{-1}), as well as for separating the energies of the final states. It is a privilege to be able to do measurement up to momentum transfers of 4 or 5 fm^{-1}.

- *Charge elastic scattering* was studied on various nuclei : ^{40}Ca, ^{48}Ca, ^{58}Ni, ^{116}Sn, ^{124}Sn and ^{208}Pb.

- *Magnetic elastic scattering* has allowed the first measurements of a valence neutron and proton orbital to be made. It was performed with the "900" spectrometer at 155° of the incident beam. This method can only be applied in a few particular cases, and such orbitals have been studied in five $f_{7/2}$ and $g_{9/2}$ nuclei, namely ^{49}Ti, ^{51}V, ^{59}Co, ^{87}Sr, ^{93}Nb. In all these cases, the radii predicted by Hartree-Fock calculations using a density-dependent interaction are too large by 2 to 4 %, the effect being more pronounced for neutron orbitals. The introduction of mesonic exchange currents would seem to reduce part of the disagreement.

- *The magnetic form factor of* ^{3}He has been measured up to a momentum transfer of 5 fm^{-1}, corresponding to an energy of 604 MeV, the present maximum available. The data indicate a diffraction minimum around q^2 = 18 fm^{-2} and the beginning of the next maximum.

1.b. *Electron inelastic scattering*

- *The magnetic structure of the deuteron* has been studied at momentum transfers ranging from 2.5 to 4.2 fm^{-1}. The inelastic cross section measured in the continuum region, which corresponds to electrodisintegration of the deuteron close to threshold, is not reproduced by the impulse approximation model. The data can only be reproduced with the introduction of mesonic currents. This experiment is one of the simplest and unambiguous confirmation of mesonic currents.

- *On heavy nuclei* one can only analyse the detailed structure of the transition densities if one measures at high transfers. Thus it was possible to analyse the rotational band 0^+, 2^+, 4^+, 6^+ of samarium isotopes. The first excited states 3^- of ^{208}Pb was studied. We are considering the study of platinum and osmium isotopes.

1.c. *Deep inelastic electron scattering*

In a deep inelastic experiment on ^{12}C, the excitation function has been measured for a region of excitation energies between 30 and 300 MeV, including the quasi-elastic peak and the beginning of the (3.3) resonance. Longitudinal and transverse components of the cross section have been separated, demonstrating the dominance of the transverse part, especially beyond the quasi-elastic peak. In this region the cross section is sensitive to meson exchange currents and to the propagation of the Δ in the nucleus.

2. (e,e'X) reactions

2.a. *(e,e'p) reactions*

The one per cent duty cycle allows us the study of (e,e'p) reactions by detecting in coincidence the scattered electron and the outgoing proton in the focal planes of two magnetic spectrometers : the "600" and the "900". The range of missing energies extends up to 100 MeV and that of analysed momenta up to 300 MeV/c, the kinematics is the coplanar one.

- *The study of the reaction d(e,e'p)* has been carried out at 500 MeV for two values of the transferred momentum : 1.75 and 2.25 fm^{-1}. The impulse distribution has been measured up to 340 MeV/c. This has permitted an analysis of the wave function of deuterium and especially of the nucleon-nucleon medium range interaction.

- *In the case of the 2s-1d shell nuclei,* the 1s and 1p states are well separated in energy and clearly identifiable by the shape of their momentum distributions for the ^{12}C and ^{16}O nuclei. The situation is more confused for the heavier nuclei ^{28}Si, ^{40}Ca and ^{58}Ni where the number of shells is larger and the deepest lying states overlapping. Among other things, one should notice the very large spreading of the 1s state. Clearly there is a problem with the shell concept for so deeply lying nucleons.

- *A measurement of the spectral function of ^{3}He* has been performed by means of the reaction ^{3}He (e,e'p). The initial proton momentum distribution was sampled up to 310 MeV/c. The missing energy varies from 25 MeV to 95 MeV. The energy resolution of 1.2 MeV allows to separate unambigously the two body disintegration in which the deuteron is the residual nucleus and the three body break up.

2.b. *Other (e,e'x) reactions*

- Explorating experiments done on ^{6}Li and ^{9}Be have shown the possibility of measuring quasi-free scattering on nucleon clusters in nuclei.

- The pion form factor and thus its radius was determined from the measurement of the π^+ electroproduction cross section on hydrogen. The analysis of the results yields a pion radius of 0.74 ± 0.12 fm.

B. EXPERIMENTS PERFORMED WITH THE PHOTON BEAMS

1. Photonuclear reactions below 140 MeV

During the last decade, systematic studies of the total photonuclear absorption cross section were made at the AL60 of Saclay ($E_\gamma <$ 35 MeV), then in the low energy hall of the A.L.S. (B.E.) between 25 and 140 MeV. These experiments have all been performed using monochromatic photons obtained by annihilation in flight of positrons.

Since 1974, the aim of the studies undertaken at the A.L.S. is essentially to complete the total photonuclear absorption measurements in the poorly studied region between 35 and 140 MeV. In this energy region, where the photon wavelength varies from 5 to 1.5 fm, one hopes to observe the nuclear correlation effects in nuclei. These effects show up first in the photonuclear cross section integrated up to 140 MeV and which was found equal to (1 + K) times the Thomas-Reiche-Kuhn sum rule (60 $\frac{NZ}{A}$ MeV mb) with K = 0.8 ± 0.15. In addition the study of neutronic multiplicities in the various (γ, xn) channels has yielded a certain number of preliminary results on the competition between the two de-excitation mechanisms after photon absorption with energy less than 140 MeV : the fast mechanism of the precompound nucleus and the slow evaporation mechanism of the compound nucleus.

2. Photonuclear reactions at high energies

2.a. *Photoproduction in the Δ resonance region*

At intermediate energies, the medium range part of the elementary N-N interaction begins to be felt. Therefore the coupling to one or many pions channels : NNπ, etc. must also be taken into account. In the region of energy available to the A.L.S. and up to the GeV region, the NNπ interaction is dominated by the two-body Δ-N interaction. The electromagnetic probe has special advantages for creating a Δ in the nucleus in the neighbourhood of a nucleon. Indeed the creation mechanism is well understood and the Δ-$\vec{\text{N}}$ interaction in the final state can be studied.

A systematic study of this phenomenon has been made with bremsstrahlung and monochromatic photons on all the photodisintegration reactions of ^{3}He and ^{4}He and for pion photoproduction on ^{2}H, ^{3}He and ^{4}He.

$\gamma d \to pp\ \pi^-$: a significant example

The wave function of deuterium is well known in the momentum range considered. Therefore the effort was focused on the systematic study of the Δ-N interaction. The duty cycle of the accelerator has permitted the detection of a pion and a proton in coincidence. It was first verified that 90 % of the cross section was due to the pion photoproduction on a quasi free nucleon. A reduction of several orders of magnitude of this contribution was obtained by transferring large momenta owing to the large energies available.

The importance of the rescattering mechanism has thus been demonstrated. This implies that, an important part of the Δ-N interaction is associated with the exchange of a real pion. It is therefore a long range interaction and can be reduced to a series of two-body scattering.

The existence of dibaryonic states ?

However, important deviations appear relative to the previous model for the same reaction in different kinematics. The deviation can be due either to the medium range part of the interaction ($r \lesssim .5$ fm), i.e. one or two pion exchange or to the short range part of to dibaryonic states which would explain the observed rapid variations as a function of the photon energy, also visible in a more inclusive $\gamma(d,p)X$ reaction. The existence of such states would impose severe constraints on the baryon-baryon interaction. It should be noticed that they are predicted by the quark bag models and other kinds of models. Experiment designed to confirm the existence and properties of such states located at 2.16 GeV, 2.24 GeV and 2.32 GeV were performed.

2.b. *Photoproduction experiments at threshold*

- The photoproduction of very low energy pions is a privileged field to study the "soft pions" physics governed by the P.C.A.C. principle (partially conserved axial current), analogous to the C.V.C. principle (conserved vector current). The π^+ photoproduction, initiated with the collaboration of the University of Louvain-la-Neuve, gave a precise measurement of the axial form factor of some light nuclei : d, ^{3}He, ^{6}Li, ^{12}C, ^{14}N.

- The π° photoproduction has been measured near threshold on three light nuclei : d, ^{3}He, ^{4}He, relatively to photoproduction on the proton but interpreted from a different point of view.

C. EXPERIMENTS PERFORMED WITH THE PION AND MUON BEAMS

I want to mention the most important experiments performed with the pion and muon beams :

- elastic scattering of $\pi^\pm$ on H, D ^{4}He between 20 and 100 MeV.

- Exchange $\pi^- p \to \pi^\circ n$ and $\pi^- p \to n\gamma$.

- Study of the mechanism of reactions involving pion absorption : $^{12}C(\pi^-,p)^{11}B$, $^{16}O(\pi^+,p)^{15}O$.

- Nuclear fragmentation by π^- at rest or in flight on Si, Ni, Fe.

- μ^+ lifetime measurement with a high accuracy : $\tau_{\mu^+} = 2.197187 \pm 0.000117$ μs.

- μ^- capture in liquid hydrogen given $\Lambda_c = 460 \pm 20$ s^{-1}.

- Muon capture rates in ^{10}B, ^{12}C and ^{14}N.

II. FUTURE PROSPECTS

A. MEDIUM RANGE FUTURE

The improvements already underway are : a monochromatic tagged photon channel (HE3B), an improvement in intensity and resolution due to the "Stradivarius" transport system, an increase in maximum energy to 720 MeV, and the installation of new equipment. The latter includes a large solid angle spectrometer in the HE3A hall, and new computation facilities for the experiments in the HE1 hall. These allow us to extend the various experiments previously mentioned. Thus, for the next five years our laboratory will maintain its position at the forefront of nuclear physics.

1. Elastic and inelastic electron scattering experiments

The "Stradivarius" optics will permit systematic use of an energy resolution less than 10^{-4} and a tenfold increase of the intensity by using the spectrometer as an energy-loss system. A high resolution detector ($\Delta E/E \simeq 2.5 \times 10^{-5}$) covering the entire focal plane $\Delta p/p = 10$ %) and the installation of a new data handling on-line computer, will permit peaks of very weak magnitude to be extracted. These facilities will together be able to provide data at large momentum transfer ($q > 3$ fm^{-1}) required to determine transition densities.

2. The (e,e'p) experiments

For the (e,e'p) coincidence experiments the reaction mechanism study undertaken for ^{12}C will be extended to ^{4}He. Experiments on light nuclei (A < 4), to be complemented by (e,e') scattering measurements on ^{3}He and eventually on ^{3}H where magnetization and charge density distributions will be performed and constitute a global and accurate test for wave function calculations of few nucleon systems and exchange currents distributions.

3. Deep inelastic electron scattering

The energy increase of the machine up to 720 MeV will provide a larger choice of kinematical conditions for coincidence experiments. It will also allow deep inelastic (e,e') experiments to cover a larger part of the (q, ω) plane and to scan the Δ region. In order to better understand the reaction mechanisms, the experiments must be focused on light nuclei (A < 4), with separation of the transverse and longitudinal components. They will be further extended to medium and heavy nuclei.

4. Photonuclear experiments

Very exclusive reactions of the type (γ,pπ), done in coincidence, provide means to understand the details of the reaction mechanism if a more complete set of data is available. This will become possible when the equipment of the HE3A area will be complemented with a large solid angle and large $\Delta p/p$ spectrometer ($\Omega \simeq 25$ msr ; $\Delta p/p \simeq 20$ %). It will result in an increasing of the counting rates by an order of magnitude. The monochromatic beam in HE3A equipped with such a spectrometer will allow a systematic study of more inclusive reactions such as $(\gamma,\pi^{\pm})$ and (γ,p) for which only partial informations are available, and to further extend all these experiments to different nuclei.

As early as 1980, the HE3B equipment where tagged annihilation photons are used should allow the systematic measurement of photon total absorption cross sections between 150 and 500 MeV, in very good experimental conditions. The data will provide global constraints on photonuclear reaction mechanisms. For the heavy nuclei (A > 100), the analysis of photoneutronic de-excitation channels will proceed at the same time using HE3A's monochromatic beam.

Photon total absorption experiments under 140 MeV (in the BE hall) will be extended to a larger range of nuclei in order to learn more on the variation of the absorption of a photon on a quasi-deuteron as a function of the mass number. Experiments in which a proton is detected will be undertaken using the "Neptune" spectrometer, particularly to obtain the proton angular distributions in the photodisintegration of the deuteron between 80 and 120 MeV ; thus will be revealed the multipolarities involved. In light nuclei, studies of the partial channels γp_0, γp_1 and γp_2 will be undertaken in order to investigate the reaction mechanisms.

5. Experiments with pion and muon beams

The advent of meson factories has reduced the international competitivity of the pion and muon channels. Nevertheless, the mastering of the pion beam monitoring down to a 1.5 % error in the low energy region (< 100 MeV) and the time structure of the muon beams, provide the opportunity to run experiments particularly suited to these beams.

B. LONG RANGE FUTURE

1. Higher energy and duty cycle

The long range perspective of nuclear physics are those of a fully evolved science, where the dynamics of the nucleus are tackled not only in terms of nucleons but also in terms of mesons and, more recently, in terms of quarks and gluons.

With its present features, the A.L.S. has already helped to lift the veil covering this physics. A more suitable tool is essential to uncover more of its unknown aspects. We will describe the limitations of the A.L.S. for the type of experiments presently done and sketch the possibilities offered by a larger duty cycle and a higher energy.

1.a. *Why a high duty cycle ?*

All high intensity electron accelerators prior to the A.L.S. had small duty cycle ($\simeq 10^{-3}$). With a duty cycle of 1 to 2 %, the A.L.S. made a considerable progress in this regard. However, instantaneous intensities remain too high. To avoid excessive noise pile-up in the detectors, those consist of well protected and closed spectrometers with small solid angles. Therefore, detection efficiency is relatively low. Moreover, concidence experiments are limited by the counting rate of accidental events. For both reasons, a 100 % duty cycle would be a considerable improvement.

1.b. *Why increasing the energy ?*

Besides a large four momentum transfer, a general argument in favour of an accelerator of higher energy is the possibility offered to have photons exchanged in a wider range of polarization. It permits to cover for various excitations of the nucleon the (q^2, ε) plane where q^2 is the squared momentum transfer and ε the longitudinal polarization of the virtual photon.

2. Domains open to such a machine

What would be the possibilities opened by a machine having 2 GeV maximum energy, duty cycle close to 100 % and 100 µA beam intensity.

2.a. *Measurement of nuclear form factor*

α) Electron scattering

Electron scattering experiments performed at the A.L.S. have shown that nuclear densities are determined with good accuracy only when momentum transfers of the order of 4 fm^{-1} are reached. In order to obtain a spatial resolution R of the order of .15 fm higher transfers must be reached the maximum value of which being about 1.5/R, that is 10 fm^{-1}.

β) Pion electroproduction at threshold

Variation of the axial nuclear form factor as a function of four-momentum transfer could be determined through the reaction (e,e'$\pi^{\pm}$). On the other hand, the matter form factor could be deduced from the reaction (e,e'$\pi°$) in the case of light even-even nuclei and, to a lesser extent, in the ^{2}H and ^{3}He nuclei with detection of the recoil nucleus. Such experiments are not feasible at the A.L.S. because of its too small duty cycle.

2.b. *Measurements of impulse distributions*

(e,e'N) reaction are the best to measure the impulse distributions of a bound nucleon in the nucleus. In such experiments, where the scattered electron and the outgoing nucleon are detected in coincidence, a high duty cycle is again of foremost importance. A factor from 10 to 100 on the lowest cross sections presently measurable would permit reaching the region where short range interactions begin to play an important role.

Another field of research should be pointed out, namely reactions of the type (e,e'2N), a direct method for measuring nucleon-nucleon correlations, first step towards multiple coincidence experiments. Here again we are faced with the foremost importance of high duty cycle.

2.c. Δ-nucleon interaction and nuclear excited states (dibaryons ?)

The interest for a wide scanning of the (q^2,ε) plane appears clearly if one tries to extend to virtual photons the experiments performed at the A.L.S. with real photons. This could be made through the (e,e'p), (e,e'π) and (e,e'pπ) exclusive reactions. Possible separation of the transverse and longitudinal components will allow the study of the Δ-nucleon interaction and that of the form factor of highly excited states.

2.d. Meson electro- or photoproduction on the proton

Pion electroproduction on the nucleon would allow an accurate study of the operator describing this elementary reaction. The measurement of baryonic resonance form factors could also be refined.

High quality monochromatic photon beams would allow studying heavy meson photoproduction, either strange (K) or non-strange (ρ,ω), from which the as yet badly known corresponding form factors could be extracted.

3. Conclusion.

The interest of a 2 GeV accelerator with a large duty cycle seems already evident if one wants to pursue a detailed study of mesonic degrees of freedom and partly unveil the quark and gluon degrees of freedom. This machine would therefore permit nuclear physics to make an important advance towards the understanding of nuclear structure at very short distances and hadronic physics in general. However, it is possible that the harvest of experimental results and the progress of theory in the coming years will require the need of a higher energy machine.

Whithout any doubt all the european community of nuclear physicists would be greatly stimulated by the prospect of recovering an "avant-garde" position at the core of the physics of the infinitesimally small.

COINCIDENCE MEASUREMENTS
WITH HIGH ENERGY ELECTRONS

J.S. O'Connell
National Bureau of Standards
Washington, DC 20234/USA

I. INTRODUCTION

During the last year an ongoing workshop on "Future Directions of Electromagnetic Nuclear Physics" was held in which a large fraction of the photonuclear community of the United States participated. The purpose of this project was to study what new measurements with photons and electrons ought to be made to advance our knowledge of nuclear structure and reactions. A large amount of the effort went towards exploring coincidence measurements. More than twenty proposals were generated in which the experimental feasibility and theoretical justification of specific electromagnetic reactions were examined. The goal of the workshop is to have the needs of nuclear physics dictate electron accelerator parameters rather than the other way around. A document describing the results of the workshop should be available early in 1981 (St81).

I want to discuss topics that fall under the heading of coincidence measurements with high energy transfer ω or high momentum transfer q or both. High ω measurements lead us into the energy domain where meson creation dominates the photon absorption process and where meson and isobar propagation and decay in the nuclear medium can be studied. High q measurements lead us into the small distance domain where one hopes to study features of the nucleon-nucleon interaction that are obscure at longer wave lengths.

A somewhat fanciful metaphor to illustrate this concept is to consider a nuclear microscope at different magnifications. With a 4 Fermi diameter field-of-view one sees noninteracting nucleons moving in circular orbits (Fig. 1a). Increasing the magnification by a power of two (1b) shows the finite size of the nucleons and their occasional collisions with exchange of a pion. At the next step down (1c) with 1 Fermi aperture the transformation of the nucleon into a new kind of particle can be seen. At 1/2 Fermi field (1d) heavier mesons are seen keeping the nucleons apart. Finally at the highest magnification used to date (1e) one begins to glimpse the constituents of the nucleons themselves, the quarks and gluons.

"Seeing" inside the nucleus with high energy photons and electrons is a complicated business especially at high ω when the nucleus emits a number of decay products. This situation is where coincidence measurements are of value in providing product identification and correlations to help sort out what was happening in the microscopic volume in which the photon was absorbed.

There are two primary coincidence arrangements (Oc77) involving electrons: tagged bremsstrahlung (Fig. 2a) and direct electroproduction (2b). In both cases signals

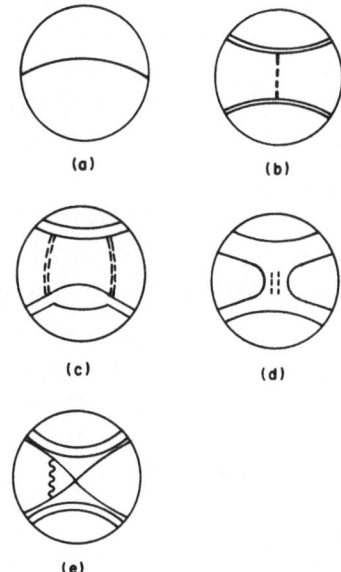

Fig. 1. Fields-of-view in a nuclear microscope. The diameters decrease about a factor of two in each step from a to e; the first field is about 4 fm in diameter. (a) nucleon bound in a single-particle potential, (b) long-range one-pion-exchange, (c) medium-range attraction through delta creation by two pion exchange, (d) short-range repulsion by vector meson exchange, (e) very short-range anticorrelation through gluon initiated quark exchange.

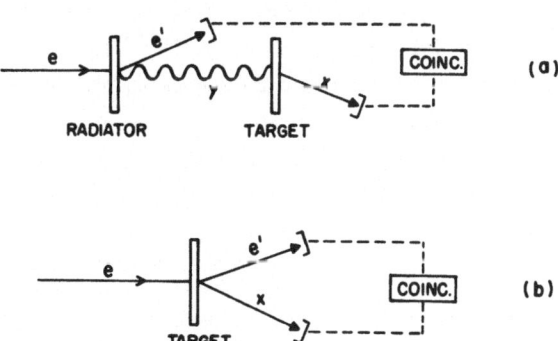

Fig. 2. Schematic of coincidence arrangements for (a) tagged bremsstrahlung (real photons), (b) direct electroproduction (tagged virtual photons).

from an electron detector and a product detector are required to be in time coincidence to establish that a reaction was induced by a real or virtual photon with energy $\omega = E_e - E_{e'}$. The practical difference in the two arrangements is that in (2a) the e' detector accepts nearly all electrons of energy $E_{e'}$ and counts at a high rate while the product detector counts at a low rate. In (2b) the e' detector accepts only a small fraction of the electrons that have caused nuclear excitation while the product detector counts at a high rate because many other virtual photons of energy ω (but different q) were absorbed by the nucleus.

Both methods have their advantages and disadvantages. The primary advantage of the direct electroproduction (e,e'x) is that the momentum transfer may be varied (at constant ω) to measure the transition form factor of a specific final state. The feasibility of either method for a specific reaction depends on a favorable ratio of true coincidence to accidental rate $N_{acc} = \frac{\tau}{d.f.} N_e \cdot N_x$ where τ is the circuit resolving time (typically 10^{-8}) and d.f. is the electron beam duty factor (ratio of average to instantaneous e beam current). Pulsed linacs have d.f. $\simeq 10^{-2}$ while continuous wave linacs and storage rings have d.f. $\simeq 1$.

II. EXAMPLES OF (e,e'x) MEASUREMENTS

x = nucleon

Other talks in this workshop cover the progress and plans for (e,e'p) measurements at Saclay and Amsterdam. I think it is not too strong a claim to say that the results of past quasi-free measurements have provided the most unambigious information available on single-particle ground state densities and orbital binding energies. The success of these first generation electromagnetic coincidence measurement demonstrates the promise and the power of the method.

It would be interesting to know the corresponding structure function for the neutron orbitals. The near equality of proton and neutron radii and shape distribution functions suggests that for $N > Z$ nuclei the hole energies and momentum distribution functions should reflect the higher neutron densities. For example one would naively predict the Fermi momentum of the neutrons to be $(N/Z)^{1/3}$ that of the protons. The (e,e'n) knockout experiment can be performed by measuring the neutron with a time-of-flight spectrometer. A duty factor of 0.2 (50 ns on, 200 ns off) could be used for neutrons in the range 30 to 100 MeV (Sc79). Coincidence counting rates comparable to past (e,e'p) measurements could be approached because of the larger allowable target thickness and the increased d.f.

x = 2 nucleons

The subject of two-nucleon correlations in the nuclear ground state has been extensively discussed in the theoretical literature. However few (if any) measurements can be unambiguously related to the ground-state pair-distribution function in relative coordinate $\vec{r} = \vec{r}_1 - \vec{r}_2$ or momentum $\vec{p} = 1/2(\vec{p}_1 - \vec{p}_2)$ space. A less ambiguous function more closely related to two-nucleon knockout reactions is the two-nucleon transition amplitude defined as the overlap of the initial and final nuclear wave functions. There is evidence that the two-nucleon transition amplitude influences less inclusive reactions.

A recent calculation (Ho80) on quasi-free electron scattering concludes that about half of the (e,e') reaction is found to correspond to multinucleon removal. Indeed a summation of single particle (e,e'N) cross sections does not give the inclusive (e,e') cross section. The one-nucleon momentum distribution as inferred from (γ,p_0) measurements (Ma79) exhibit a tail above proton momentum 400 MeV/c that is not accounted for by mean-field calculations. The theoretical conjecture (Bo80) is that these momentum components are generated during violent, short-range NN collisions. These might include, for example, the transition of one (or both) nucleons to a delta. If this is the case two nucleons are temporarily outside the Fermi sea and will be liberated together if an electromagnetic quantum places one of them on the mass shell.

The experimental feasibility of a kinematically complete measurement depends on a theoretical estimate of the magnitude of the differential cross section, the availability of suitable detectors and electron beam, competing background reactions, and ratio of accidental to true coincidence rates.

As an example the $^{12}C(e,e'2p)$ counting rates have been estimated (Li81) for a geometry in which two high-energy protons with low relative momentum are ejected at $90°$ to the incident e beam. The measurement is sensitive to initial two-nucleon relative momentum components inside the nucleus near $p = 1/2q\cos\theta/2$ (see Fig. 3). The estimated cross section becomes too small to measure ($<10^{-40}$ cm^2/MeV^2-sr^2 assuming both protons are captured by one spectrometer) for initial relative momenta greater than $\sim$ 750 MeV/c. The measurement is limited at low relative momentum ($\sim$ 200 MeV/c) by the loss of angular correlation between protons as they multiple scatter on exiting the nucleus. The two-nucleon transition amplitude can therefore be determined with a spatial resolution of approximately 0.3 fm.

The complimentary measurement with real photons is (γ,pn) for which some information is already known for $A \leqslant 16$. There is strong evidence that the quasi-deuteron process is an important doorway for the ejection of a single high-energy nucleon. The reabsorption by the nucleus of the second nucleon requires a second high-momentum transfer interaction. Thus the ratio of one- to two-nucleon ejection following photon absorption is a measure of the mean free path (averaged over the nuclear volume) of high energy nucleons in nuclear matter.

<u>x = pion</u>

The measurement of the $(\gamma,\pi^{\pm})$ and (γ,p) cross section differential in particle energy and angle with tagged bremsstrahlung at Bonn (Ar80,Ro80) is filling an important void in our knowledge of the total photon absorption cross section in nuclei in the delta energy region. The data (corrected by a Monte Carlo calculation for missed neutrals) show a reduction in peak height and broadening of the delta resonance in nuclei when compared with that expected from A free nucleons. This trend is accounted for by the isobar-hole model as due to medium modifications of the free delta width by

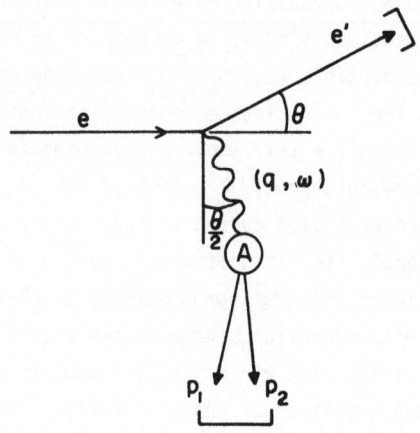

Fig. 3. The A(e,e'2p) reaction to probe interacting nucleon pairs in the nucleus.

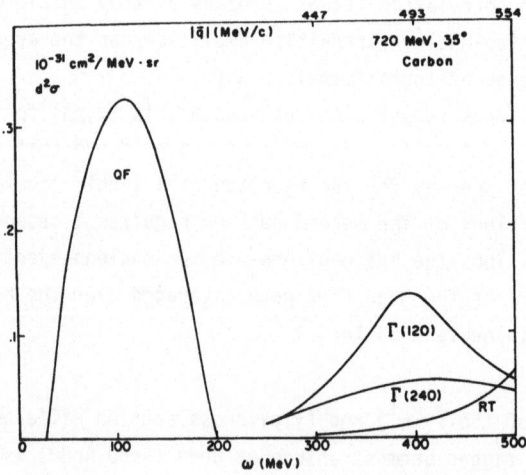

Fig. 4. Electron scattering cross sections for excitation of the quasi-free peak, the delta peak, and the elastic radiative tail.

Pauli blocking, Fermi motion, binding, non pionic decay $N\Delta \to NN$, Δ-hole residual interaction, and coupling to multiparticle-hole states.

The total absorption cross section in the delta region has been measured for a few light nuclei by the Mainz group with a Compton spectrometer as the photon detector using the attenuation method. This method can be extended to heavier nuclei when high duty factor bremsstrahlung beams become available which will permit the use of a e^+e^- pair spectrometer as the photon detector. The data acquisition time will be reduced by more than two orders of magnitude using the pair coincidence technique (Ah79).

An alternative method of measuring the total photon absorption cross section in the delta region that does not require coincidence is (e,e') at forward angles. Although this method has been tried in the past (Vl76, Gl79) the incident electron energies and scattering angles (1.2 GeV at 14^0) and (2 GeV at 15^0) were large enough that the quasi-free peak was excited and its high energy tail interfered with the extraction of the delta peak shape. An estimate (Oc81) of cross sections at lower momentum transfer (750 MeV at 20^0) shows one can suppress quasi-free scattering while still keeping the elastic radiative tail within tolerable limits (Fig. 4).

The technique of forward inelastic electron scattering is readily extended to coincidence measurements when high duty factor electron beams are available. Table 1 shows the predicted $(e,e'\pi^+)$ rates for carbon. This technique has a number of advantages over tagged bremsstrahlung: (1) coincidence counting rates are one to two orders of magnitude higher, (2) the energy resolution capability is higher because of the small spot size on target, and (3) the virtual photon is linearly polarized (Pe74). It may be possible to produce an aligned delta inside the nucleus by this method.

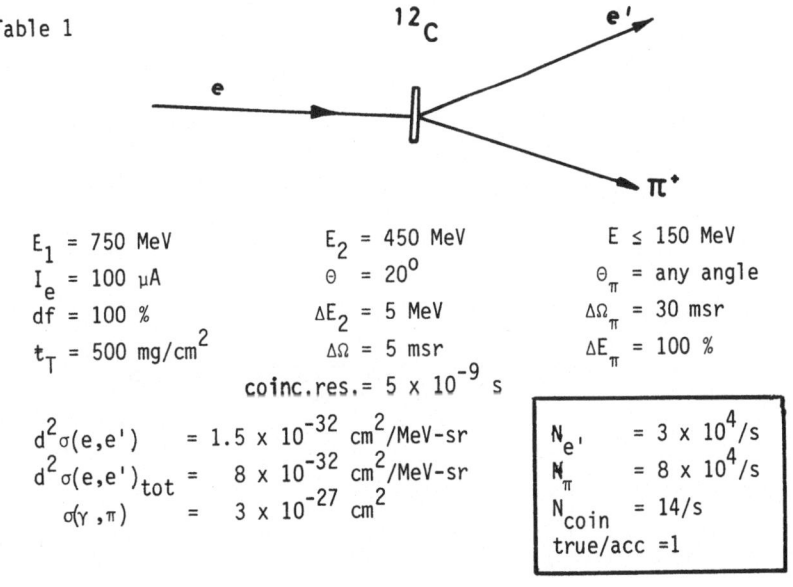

Table 1

E_1 = 750 MeV
I_e = 100 μA
df = 100 %
t_T = 500 mg/cm^2

E_2 = 450 MeV
θ = 20^0
ΔE_2 = 5 MeV
$\Delta\Omega$ = 5 msr
coinc.res.= 5 x 10^{-9} s

$E_\pi \leq$ 150 MeV
θ_π = any angle
$\Delta\Omega_\pi$ = 30 msr
ΔE_π = 100 %

$d^2\sigma(e,e')$ = 1.5 x 10^{-32} cm^2/MeV-sr
$d^2\sigma(e,e')_{tot}$ = 8 x 10^{-32} cm^2/MeV-sr
$\sigma(\gamma,\pi)$ = 3 x 10^{-27} cm^2

$N_{e'}$ = 3 x 10^4/s
N_π = 8 x 10^4/s
N_{coin} = 14/s
true/acc =1

x = nucleon plus pion

Photoproduction of deltas with real photons (γ,pπ⁻) in nuclei has been measured at the Bonn synchrotron using tagged bremsstrahlung (Ar80, Ro80). Coincidence measurements at Saclay (La79) on light nuclei have been analyzed to extract the NΔ interaction.

The electroproduction of deltas (e,e'Nπ) requires a triple coincidence but estimated event rates show the measurements are feasible (Re81). Even (e,e'ΔΔ) is possible (50 events per hour with a 2 GeV electron beam (Pe79)). The knockout of a delta from the nucleus by a virtual photon can arise either by inducing the transition of a nucleon to a Δ or by striking a preexisting Δ. The former process is mainly transverse while the latter process will have a large longitudinal component. One of the first questions to be asked of this measurement is how does the delta transition form factor differ from its free space q dependence.

x = gamma ray

The elastic scattering cross section of real photons in the delta region has not been measured for complex nuclei. The forward scattering amplitude can be predicted from the dispersion relation and optical theorem. Figure 5 show the amplitudes in the delta region are much larger than in the giant resonance region (Le80). Tagged bremsstrahlung can be used for these measurements (Ha81) but there are advantages to

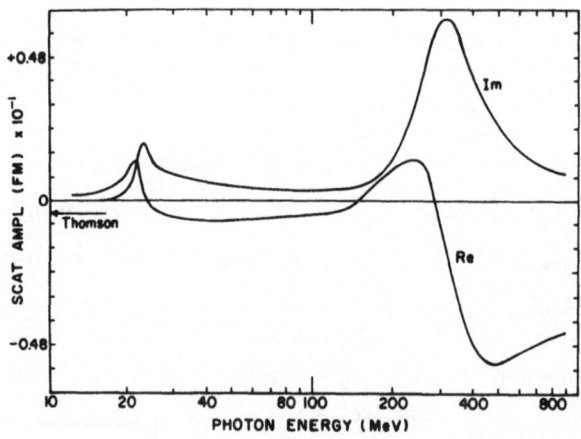

Fig. 5. Real and imaginary scattering amplitudes for elastic photon scattering from carbon at zero degrees (Le80).

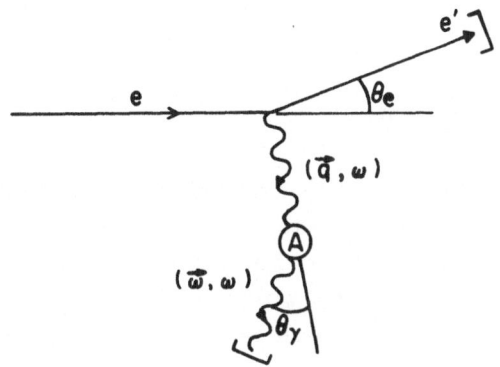

Fig. 6. Elastic photon scattering measured by the A(e,e'γ) reaction.

forward-angle electron scattering in coincidence with a photon detector. Figure 6 shows the geometry. The direction of the virtual photon $\vec{q}$ is near perpendicular to the incident electron beam. Thus 'forward photon scattering' is in an experimentally accessible direction. Wide-angle bremsstrahlung will form a background to this process but it may be tolerable at photon angles away from the initial and scattered electron directions. The photon scattering data serves as a check on the total absorption measurement and its angular distribution helps identify the multipolarities of the transitions (Ge76).

Few-nucleon Targets

In the previous examples the target nucleus was taken to be A>>4 and the results of the measurements assumed to be characteristic of nuclear matter or reflecting average nucleon properties. A number of specific (e,e'x) measurements have been suggested for the targets H, D, ^{3}H, ^{3}He, ^{4}He whose focus is either on elementary particle features or on the free two-nucleon interaction.

A. The neutron electric form factor is largely unknown, but is needed to give the body wave function of the A≤4 nuclei. Two methods have been suggested (Ar81a) to determine this function. (1) Measure G_{En} by scattering electrons from deuterium and performing a Rosenbluth separation. The neutron and electron would be observed in coincidence D(e,e'n)P to reduce theoretical uncertainties. (2) Measure G_{En} by scattering longitudinally polarized electrons from deuterium and measuring the polarization of the recoiling neutron with a polarimeter. The asymmetry, which is expected to be about 4%, is directly proportional to the electric form factor.

B. A measurement of the tensor polarization T_{20} in D(e,eD) elastic scattering (Ho81) in the momentum transfer range $4 < q < 6 fm^{-1}$ would give new information on the short range and tensor part of the deuteron wave function through measurements of the monopole-quadrupole interference form factor. T_{20} changes sign at the first diffraction minimum of the monopole form factor.

C. A new generation of experiments is being planned to measure the parity violating interference between the weak and electromagnetic interactions in the scattering of polarized electrons from hydrogen (So81). The goal of these experiments is to measure independently each of the four electron-nucleus electro-weak coupling constants with a precision an order of magnitude better than the SLAC experiment. The experiment needs beam energies from 600 to 2000 MeV with high duty factor to simplify the polarized electron source and lower the instantaneous rate of detected events to the level where individual pulses can be counted.

D. New data on elastic and inelastic form factors of the $A \leq 4$ nuclei at high momentum transfer have stimulated discussions about how to describe the transition from the non-relativistic nucleon-meson description of nuclei to a relativistic quark-gluon description. Data on two, three, and four nucleon targets can be more readily related to the NN potential than measurements on heavier targets because the wave functions are derived from Schrödinger and Faddeev equations. Also of interest is the fact that within this small range of A one finds matter densities that make large excursions around that of normal nuclear matter.

The magnetic form factors of $A = 2, 3$ need to be known to higher values of momentum transfer to understand the spin-flip coupling to quarks. The evidence from the charge form factor of deuterium indicates that the quark degrees of freedom become important at $q \simeq 2$ GeV/c. Coincidence measurements at high momentum transfer are necessary to separate the elastic scattering from the much larger inelastic yield (Ar81b).

The breakup reactions on $A \leq 4$ can also be measured out to 2 GeV/c (Ch81). The approach to scaling and the agreement with quark counting rules will be two of the theoretical themes explored by these reactions.

III. CONCLUSIONS

The potential for A(e,e'x) measurements is beginning to be critically examined. Nuclear disintegration by electrons and photons shares many of the problems of interpretation found in reactions induced by hadronic projectiles. These theoretical difficulties should be viewed more as a challenge than as an obstacle. The strength of nuclear physics is the diversity of its probes and targets. In the past progress in finding the simplicities that nature has hidden in the potentially chaotic nuclear many-body system often came from the systematics of the data rather than one individual measurement. Nuclear physics differs from elementary particle physics in this respect.

The theme of this workshop has been the diversity of topics that are studied with real and virtual photons. Coincidence experiments in all energy ranges will give new insights on nuclear structure and reaction mechanisms. This new measurement tool assures us that real and virtual photons will continue to play a fundamental role in the evolution of our understanding of the atomic nucleus.

REFERENCES

(Ah79) J. Ahrends, Proceedings of the Vancouver Conference "High Energy Physics and Nuclear Structure" 1979, North-Holland Pub. Co.

(Ar80) J. Arends et al. "Experimental Study of the Photo emission of Protons off ^{12}C Using Tagged Photons in the Energy Range 200 - 385 MeV", Bonn preprint May 1980.

(Ar81a) R.G. Arnold and F. Gross, "Measurements of Nucleon Electric Form Factors" in *Future Directions* (St81).

(Ar81b) G.R. Arnold and P.E. Bosted, "Electric Form Factors of D, ^{3}He, and ^{3}H at Large Momentum Transfer" in *Future Directions* (St81).

(Bo80) O. Bohigas and S. Stringari, Phys. Letts. 95B, 9 (1980).

(Ch81) B.T. Chertok, "Measurements of Coincidence Electroproduction From Light Nuclei in the Range 0.7 to 2 GeV/c in *Future Directions* (St81).

(Ge76) H. Genzel et al. Z. Physik A279, 399 (1976).

(Gl79) U. Glawe et al., Phys. Lett. 89B, 44 (1979).

(Ha81) E. Hayward, "Photon Scattering Experiments" in *Future Directions* (St81).

(Ho80) Y. Horikawa, F. Lenz, and N.C. Mukhopadhyay, Phys. Rev. C22, 1680 (1980).

(Ho81) R.J. Holt, "Tension Polarization in Electron-Deuteron Elastic Scattering" in *Future Directions* (St81).

(La79) J.M. Laget, in Proceedings of the Vancouver Conference "High Energy Physics and Nuclear Structure" 1979 p. 207, North-Holland Pub. Co.

(Le80) R. Leicht, Mainz, private communication.

(Li81) J.W. Lightbody, "A(e,e'2-nucleon)A-2 Reaction Studies" in *Future Directions* (St81).

(Ma79) J.L. Matthews, Proceedings of the 1979 Mainz Conference on "Nuclear Physics with Electromagnetic Interactions", Lecture Notes in Physics 108, p. 369, Springer-Verlag.

(Oc77) J.S. O'Connell and B. Schoch, "Electromagnetic Coincidence Experiments" in Proceedings of the MIT June Workshop in Intermediate Energy Electromagnetic Interactions with Nuclei 1977.

(Oc81) J. O'Connell, J. Blomqvist, and B. Schoch, "Measurement of Photon Cross Sections in the Delta Region by Forward Angle Electron Scattering" in *Future Directions* (St81).

(Pe74) M. Perl, "High Energy Hadron Physics" (1974) p. 465, John Wiley and Sons.

(Pe79) S. Penner, Proceedings of the 1979 Mainz Conference on "Nuclear Physics with Electromagnetic Interactions", Lecture Notes in Physics 108, p. 99, Springer-Verlag.

(Ro80) H. Rost, "Experimentelle Untersuchung totaler Wirkungsquerschnitte für Photoreaktionen an verschiedenen Kernen im Bereich der Δ(1232)-Resonanz", Bonn preprint 1980.

(Re81) R.P. Redwine and H.E. Jackson, "Proposal for $(\gamma,\pi N)$ and $(e,e'\pi N)$ Experiments at a 1 - 2 GeV High Duty Factor Electron Accelerator" in *Future Directions* (St81).

(Sc79) B. Schoch, Proceedings of "The International School of Intermediate Nuclear Physics", Arriccia (Rome) 1979.

(So81) P.A. Souder et al., "Measurements of Parity Violation in the Scattering of Polarized Electrons from Protons" in *Future Directions* (St81).

(St81) Proceedings of the workshop: *Future Directions in Electromagnetic Nuclear Physics*, Edited by P. Stoler to be published.

(Vl76) V.G. Vlasenko et al.; Yad. Fiz. 23, 504 (1976); Sov. J. Nucl. Phys. 23, 265 (1976).

POLARIZATION EXPERIMENTS

K.H. Althoff

Physikalisches Institut

Universität Bonn, 5300 Bonn, West Germany

Polarization experiments have been a powerful tool in elementary particle physics for more than 20 years. The first information about a certain reaction is obtained by measuring the differential cross section $d\sigma/d\Omega$ where one averages over all spin states. If one looks for smaller effects one can not afford to neglect all the information that can be obtained if definite spin states of the incoming and outgoing particles and target were used.

To demonstrate the significance of these polarization experiments I have picked out as an example the <u>photoexcitation of the nucleon</u> $\gamma + N \rightarrow N + \pi$ and the <u>photodisintegration</u> of the deuteron $\gamma + d \rightarrow p + n$. Only measurements of the recoil nucleon polarization and the target asymmetry are covered in this report. Photon asymmetry measurements will be discussed in the next talk by Matone.

A short description of polarized electron sources and the significance of some polarization experiments with electrons will be given at the end of the report.

Nucleon Resonances

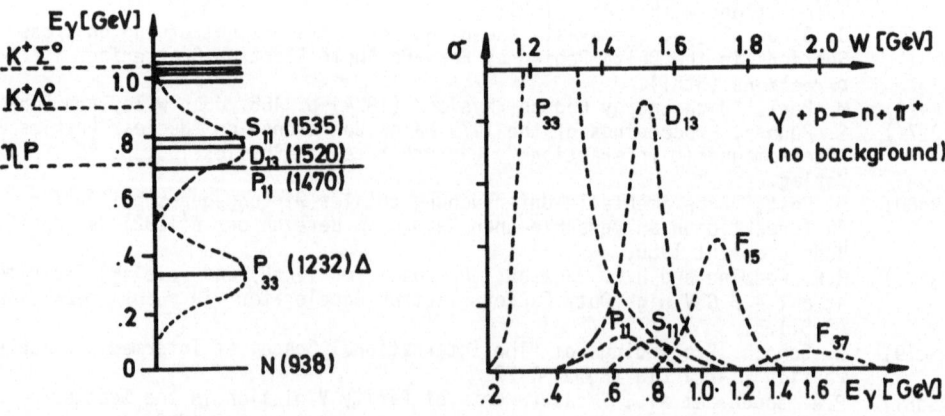

Fig. 1   Spectrum of nucleon resonances and a qualitive picture of the relative strength for the first excited states.

The comparison of the experimentally obtained photo-couplings to the nucleon resonances with quark-model calculations is still an interesting question. Exact measurements especially of the smaller amplitudes can serve as a test of refined theories.

As indicated in Fig. 1 the main difficulty in determining the amplitudes comes from the widths of the resonances compared to their separation. Also the opening of new channels with increasing photon energy E_γ contributes to the complexity of the problem.

The excitation of the resonances can be made by real- or virtual photons.

Real photons

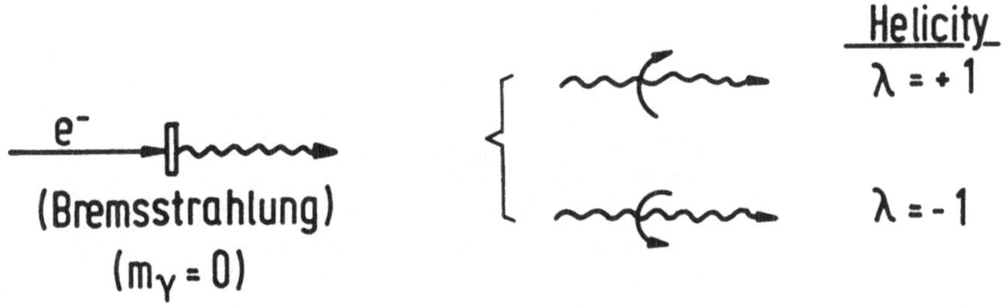

Helicity
λ = +1
e⁻
(Bremsstrahlung)
(m_γ = 0)
λ = -1

Virtual photons

e⁻
π
N
q
e⁻
N
(m²_γ = q²)
λ = +1
λ = 0
λ = -1

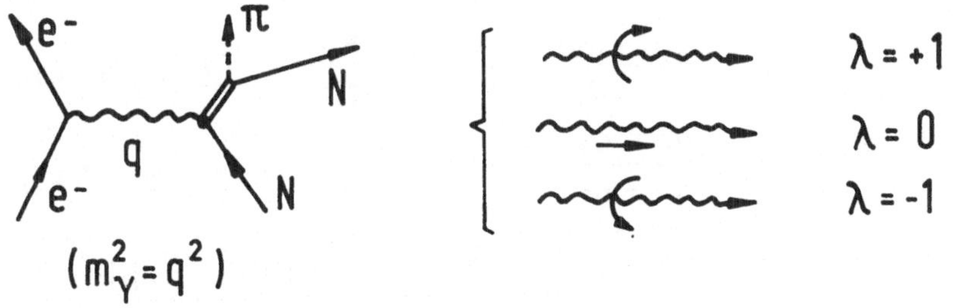

For the description of the different observables we use the helicity formalism in the (πN)-center of mass system.

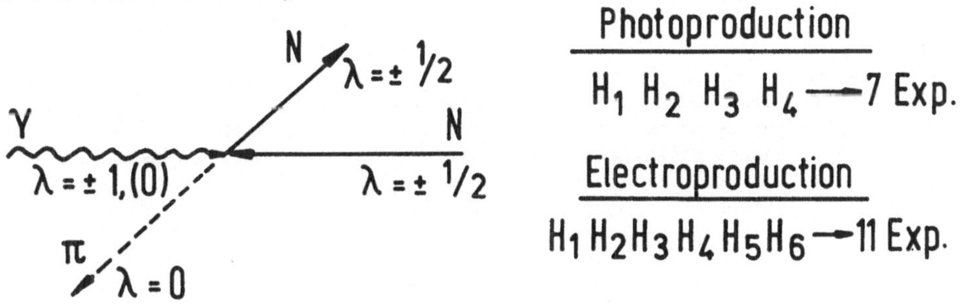

N
λ = ± 1/2
γ
N
λ = ± 1,(0)
λ = ± 1/2
π
λ = 0
Photoproduction
H1 H2 H3 H4 → 7 Exp.
Electroproduction
H1 H2 H3 H4 H5 H6 → 11 Exp.

This description has the adventage that the observables can be written as simple algebraic expressions. On the other hand these amplitudes have mixed angular momenta and parity. Therefore we will also use the multipole description in few examples. In Fig. 2 the different observables are defined and expressed by the helicity amplitudes. The observables are: The differential cross section $d\sigma/d\Omega$, the recoil nucleon polarization P, the photon asymmetry Σ and the target asymmetry T. A is the analyzing power of the second scattering target, P_γ the polarization of the photons and P_y the target polarization. K is the ratio of free to bound target nucleons.

For a complete set one needs double polarization experiments. Polarized photons <u>with</u> a polarized target for instance have first been used in Daresbury[1]. An important feature of this method is the fact that the third observable can be calculated using the other two.

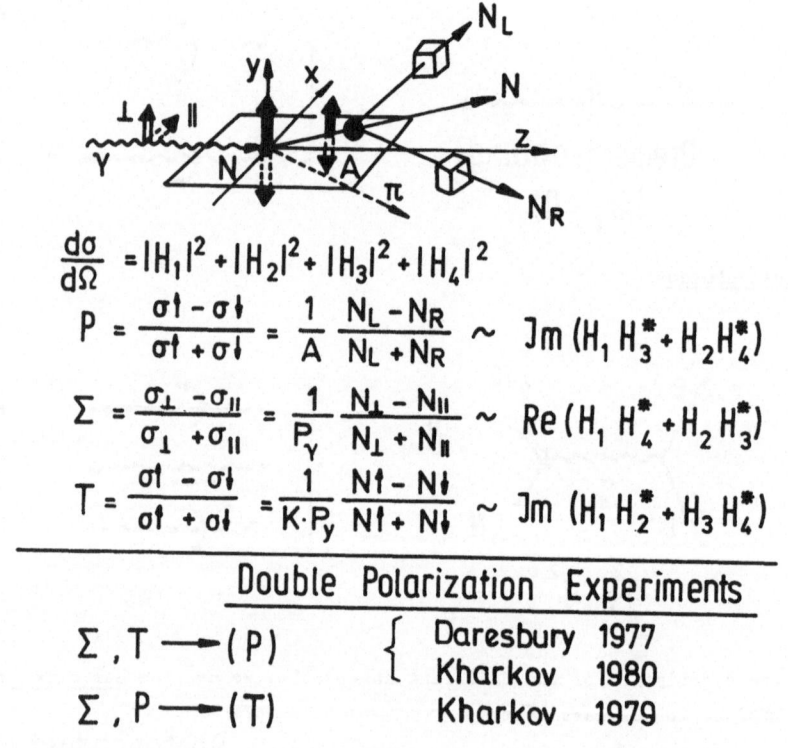

Fig. 2   Definition of the physical observables. P = Recoil nucleon polarization, Σ = Photon asymmetry, T = Target asymmetry, P_γ = Photon polarization, P_y = target polarization, K = ratio of free to bound nucleons.

Recoil Nucleon Polarization

The polarization of the recoil particle is determined by measuring the left-right-asymmetry of the scattered particle. The analyzing power A depends on the particle energy and the scattering angle θ. Three different target-types have been used so far. Liquid helium is a good analyzer at lower energies and carbon and liquid hydrogen at high energies. The most important disadvantage of this method is the low efficiency. Only half a percent of the produced particles can be used for the analysis. A low counting rate and large statistical errors are typical. Another uncertainty is the influence of inelastic levels which mostly can not be separated. In spite of these difficulties this method was quite successful before polarized photons and targets were introduced in high energy photon physics.

In Fig. 3 the elastic scattering cross section and the analyzing power for protons from carbon is plotted. This shows that only a small angle and energy range is useful for a certain scattering target[2].

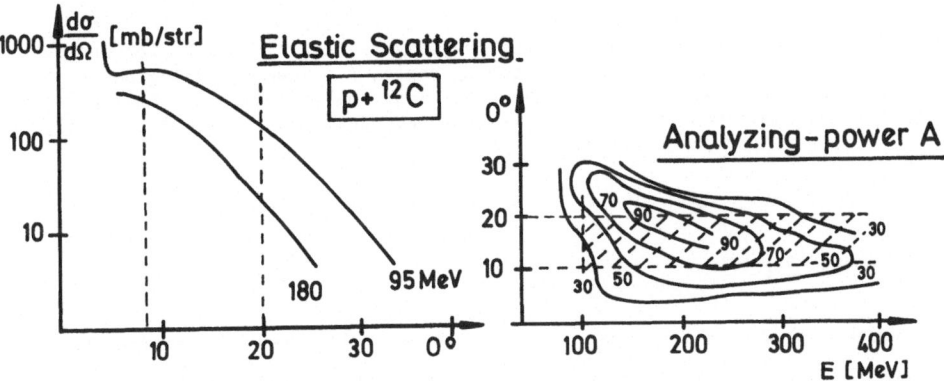

Fig. 3 Elastic differential cross section $d\sigma/d\Omega$ for protons on carbon and analyzing power A. Similar pictures can be shown for the other scattering targets at different particle energies.

In table 1 some properties of the three targets are shown for protons and neutrons. The measurement of the recoil neutron polarization is even more difficult since the counting rate is further reduced by the low efficiency of the neutron counters.

Protons

	E [MeV]	$\bar{\Theta}$	$\bar{A}$	Eff.	Inel. levels [MeV]
^{4}He	4-60	60°	0.5	0.5%	23; ^{4}He(pd)^{3}He
^{12}C	80-300	15°	0.7	1%	4,3; 7,6; 9,6; ....
H$_2$	>200	15°	0.4	0.5%	>400

Neutrons

	E [MeV]	$\bar{\Theta}$	$\bar{A}$	Eff.	Inel. levels [MeV]
^{4}He	5-40	60°	0.4	0.5%	22; ^{4}He(n,d)T
H$_2$	30-150	30°	0.3	0.5%	>400

Table 1. Some properties of the scattering targets ^{4}He, ^{12}C and H$_2$ for protons and neutrons. $\bar{\Theta}$ = mean scattering angle, $\bar{A}$ = mean analyzing power. If the energy resolution is not sufficient the influence of the inelastic contributions have to be taken into account. A decrease of the analyzing power is the consequence.

The Δ-Resonance-Region

At low photon energies it is more convenient to use multipoles to describe the amplitudes. Since the electromagnetic interaction conserves parity the P$_{33}$-state can be excited by E2 and M1 transitions.

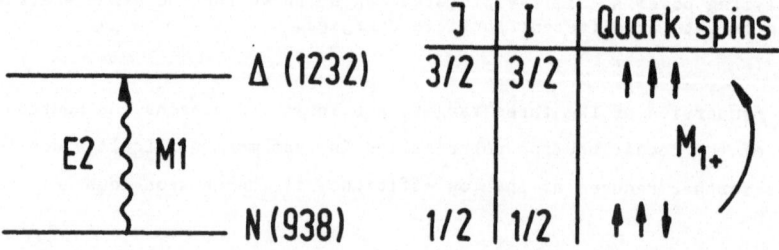

According to the quark model N and Δ belong to the same multiplett, all quarks are in a 1s-state. The transition is induced through a magnetic dipole radiation via a spin flip of one quark. In our language this is the M_{1+} multipole amplitude leading to the final πN-state with orbital angular momentum $l = 1$ and the nucleon spin $(+)1/2$ adding up to $J = 3/2$.

Since this description often causes confusion the final states excited by the E1, E2, M1 and M2 transitions via the matrixelements $E_{1\pm}$ and $M_{1\pm}$ are listed in table 2.

Multipole	Matrixelement	(J^P)	(πN-System)
E1	E_{0+}	$1/2^-$	$S_{1/2}$
E1	E_{2-}	$3/2^-$	$D_{3/2}$
M1	M_{1-}	$1/2^+$	$P_{1/2}$
M1	M_{1+}	$3/2^+$	$P_{3/2}$ (Δ)
E2	E_{1+}	$3/2^+$	$P_{3/2}$ (Δ)
E2	E_{3-}	$5/2^+$	$F_{5/2}$
M2	M_{2-}	$3/2^-$	$D_{3/2}$
M2	M_{2+}	$5/2^-$	$D_{5/2}$

Table 2. Resonances excited by EL and ML transitions via the matrixelements $E_{1\pm}$ and $M_{1\pm}$. l is the angular momentum of the πN-system. The nucleon spin is parallel (+) or antiparallel (-) to the angular momentum l.

As discussed before M1 and E2 transitions can excite the Δ-resonance via the multipole amplitudes M_{1+} and E_{1+}. The naive quark model alows only the M_{1+}. The experimental data however indicate a E_{1+} contribution of the order of 10%. An exact determination of this multipole might give some information about the size of the quark bag since the quadrupole transition is sensitive to the charge distribution inside the bag. Only polarization experiments are able to determine the small amplitudes. An evaluation however is still difficult because of non-resonant contributions and influences of higher resonances. Further complication is caused by the isospin structure of the photon which carries isospin $I = 0$ and $I = 1$ leading to isoscalar and isovector parts of the amplitudes ($E_{1\pm}(o)$, $E_{1\pm}(1/2)$, $E_{1\pm}(3/2)$,.). Several multipole analyses have been performed[3,4,5,6,7,8]. There are still large discrepancies for the small amplitudes especially near the resonance energy.

Let us compare the two reactions $\gamma + p \rightarrow \pi^0 + n$ and $\gamma + p \rightarrow \pi^+ + n$.

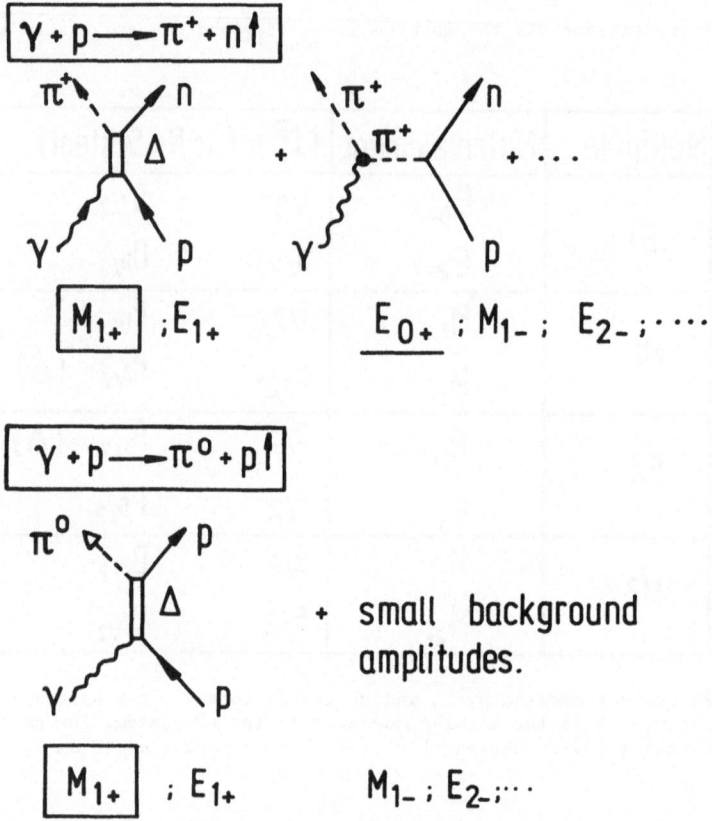

In the case of π^+-production the electric dipole amplitude E_{0+} gives an important contribution since the photon can couple directly to the charge of the pion. This is not the case for π^0-production. The small background amplitudes can only be determined in polarization experiments. This will be shown in a simple example. The recoil nucleon polarization can be written (neglecting E_{2-} and M_{2-} contributions):

$$P \sim \sin\Theta (a + b \cos\Theta + \cdots)$$

$$a \sim \text{Im}\, [E_{0+}^* (M_{1+} + 2M_{1-} + 3E_{1+})]$$
$$b \sim \text{Im}\, [M_{1-}^* (M_{1+} + 3E_{1+})]$$

At $\theta = 90°$ the polarization P is proportional to a. Since E_{0+} is small for $\pi°p$ and large for π^+n the polarization at $\theta = 90°$ should be quite different. In Fig. 4 this is demonstrated.

The b-term contains the interference between M_{1+} with M_{1-} and E_{1+}. The angular distribution should go as $\sin\theta \cdot \cos\theta = \sin 2\theta$. This can be clearly seen for the $\pi°p$-reaction[9]. Information about M_{1-} and E_{1+} can such be obtained.

The point at $\theta = 90$ for the π^+n-reaction was the first experiment where the polarization of the recoil <u>neutron</u> was measured[10]. The other data were obtained using polarized photons and a polarized target[11].

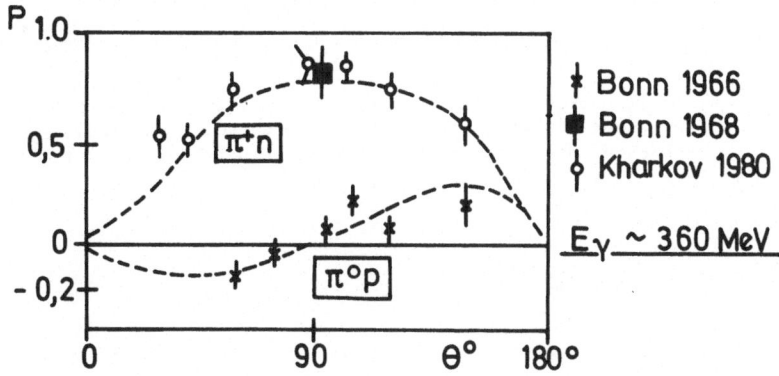

Fig. 4   Recoil nucleon polarization for E_γ around 360 MeV. The data ✚ and ✳ are recoil measurements. The data ◊ were obtained with polarized photons and a polarized target. Reactions: $\gamma + p \rightarrow \pi° + p\uparrow$ and $\gamma + p \rightarrow \pi^+ + n\uparrow$. ^{4}He and H_2 were used as analyzer.

Photodisintegration $\gamma + d \rightarrow p + n$

As another example for a recoil proton polarization measurement consider the reaction $\gamma + d \rightarrow p\uparrow + n$. The large polarization for this reaction between 300 and 700 MeV can not be easily explained by conventional models. A Tokyo group suggested a dibaryon contribution[14]. Whether this is conclusive or not, these measurements initiated a great activity hunting for dibaryons which are suggested by the bag-model.

In Fig. 5 some old data from Bonn[12] and Stanford[13] at $\theta = 60°$ and the data from Tokyo[14] and Kharkov[15] at $\theta = 90°$ are shown.

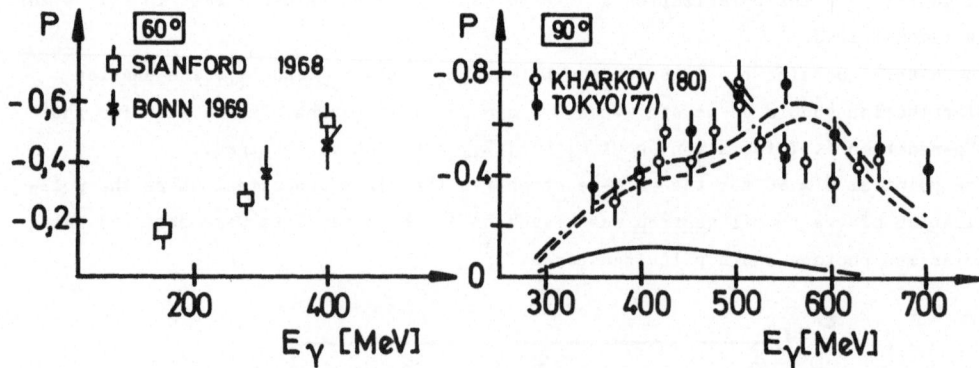

Fig. 5 Recoil proton polarization for the reaction $\gamma + d \rightarrow \vec{p} + n$. All groups have used ^{12}C as analyzer. — No dibaryons included. —·— and ... with dibaryon contribution. (Right figure is taken from Ref.[16]).

Polarized Nucleon Targets

Because of the pioneering work at CERN[17] on polarized targets and on materials with good radiation resistance measurements also with real and virtual photons could be performed with great success. Some remarks about the target technology might be of interest. To achieve a high nucleon polarization one needs low temperatures and a high magnetic field.

$$N_2 = N_1 \cdot e^{-\frac{\mu_p \cdot B}{kT}}$$

$$E_{Mag} = \mu \cdot B \quad ; \quad \mu_e = 660 \, \mu_p$$

$$P = \frac{N_1 - N_2}{N_1 + N_2} = \tanh \frac{\mu_p B}{kT}$$

$$\left. \begin{array}{l} B = 25 \, KG \\ T = 1°K \end{array} \right\} \quad \begin{array}{l} p \sim 0.3\% \text{ for protons} \\ p \sim 92\% \text{ for electrons} \end{array}$$

But if we take for instance a magnetic field of 25 KG and a temperature of 1°K we get only a polarization of about 0.3% for the protons, for the electrons however 92%. By applying a suitable RF-field, about 70 GHz in this case, one can dynamically increase the population of one level. This dynamic polarization works because the relaxation time for the nucleon spin orientation is much larger than for the electron spin.

A target material often used is butanol:

$$H - \underset{\underset{H}{|}}{\overset{\overset{H}{|}}{C}} - \underset{\underset{H}{|}}{\overset{\overset{H}{|}}{C}} - \underset{\underset{H}{|}}{\overset{\overset{H}{|}}{C}} - \underset{\underset{H}{|}}{\overset{\overset{H}{|}}{C}} - OH$$

With T = 0.5 K and B = 25 KG one reaches about P~80% for the polarization of the free hydrogen nucleons. If one replaces the H by deuterium one can get a polarized deuteron target with about P~25%. The technology is developing in the following directions: lower temperatures (^{3}He/^{4}He-dilution cryostat), higher magnetic fields (super conductivity) and new material with more free, polarizable nucleons (NH_3 for instance)[18].

The Second Resonance Region

Looking at Fig. 1 one can see a group of three resonances around a mass of about 1500 MeV, corresponding to E_γ of about 700 MeV.

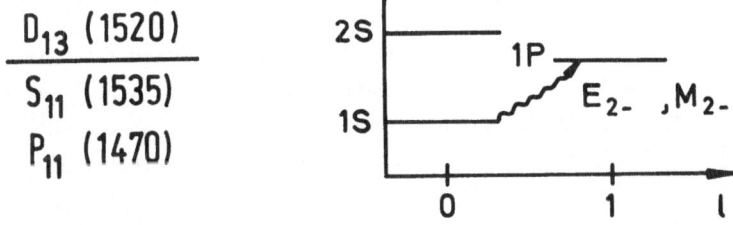

The dominating resonance D_{13} is excited by E_{2-} and M_{2-}-amplitudes. (according to the quark model with one quark excited to a 1p state.)

If we neglect for a moment the other two resonances and consider the reaction
$\gamma + p \to \pi^+ + n$, we have to take the amplitudes E_{2-}, M_{2-} and the E_{0+} (charged pion production!) into account.

For the target asymmetry we get the expression:

$$T \sim \frac{1}{d\sigma/d\Omega} \sin\theta \, (a + b\cos\theta + ...)$$

$a \sim$ interference M_{1-} and $(E_{2-} + M_{2-})$ (neglected)

$b \sim \text{Im } E_{0+} (E_{2-} + M_{2-})$

so we are left with a very simple expression

$$T \sim \frac{1}{d\sigma/d\Omega} \cdot b \cdot \sin 2\theta$$

If the cross section $d\sigma/d\Omega$ were constant with θ, the angular distribution of T would look like a $\sin 2\theta$-curve. (solid line).

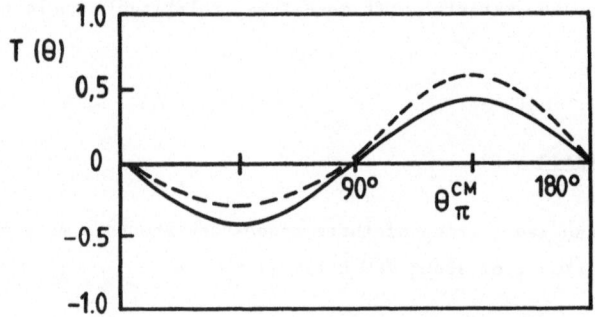

Including the angular dependence of $d\sigma/d\Omega$ we get the dotted line. In Fig. 6 the data are shown together with some theoretical predictions. A remarkable similarity with our $\sin 2\theta$-curve based on very simple assumptions can be seen[19].
A closer look however shows that the a-term can not be neglected completely and additional amplitudes contribute. This can be explored only in a detailed analysis. But the simple version of the analysis already shows how sensitive polarization parameters like T are to small contributions of different amplitudes.

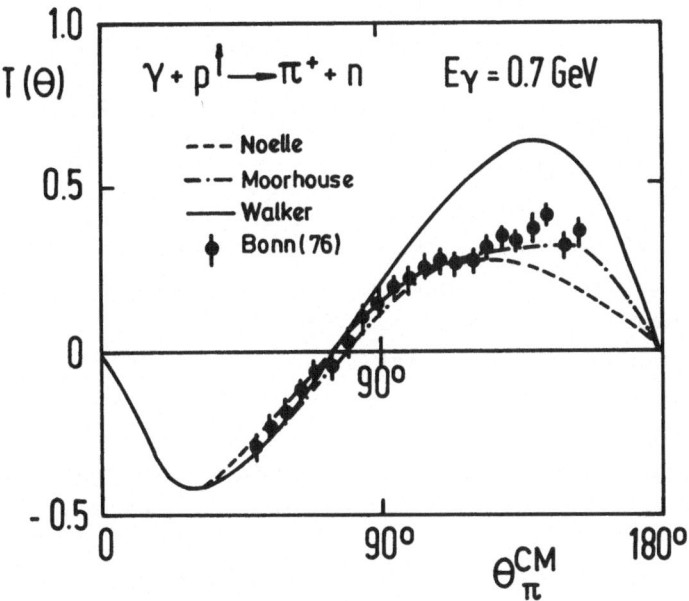

Fig. 6   Target asymmetry for the reaction $\gamma + \vec{p} \to \pi^+ + n$ at $E_\gamma = 700$ MeV.

More information about the radiative decay matrix elements and a comparison with experimental data and quark model calculations are presented in the paper of Kajikawa at the conference in Toronto 1980 (Ref. 16). The overall agreement especially between a quark model with spin orbit coupling included and experimental data is fairly good. Astonishing however, is the bad agreement for the photo coupling of the most prominent resonance, the Δ-resonance. There is no explanation for this discrepancy up to now.

Why Neutron Targets?

As already mentioned before the photon is an isospin mixture.

$$\omega^0 \quad I = 0 \quad (A^S)$$

$$\rho^+ \, \rho^0 \, \rho^- \quad I = 1 \quad (A^V)$$

It can for instance couple directly to the isoscalar ω^0 or the isovector $\rho^{+,0,-}$, corresponding to an isoscalar amplitude A^s and an isovector amplitude A^v which can have $A^{1/2}$ and $A^{3/2}$ contributions in our case.

Pion production from protons and neutrons can be written in the form:

$$\gamma + p = \pi^+ + n \sim A^s + A^v$$

$$\gamma + n = \pi^- + p \sim A^s - A^v$$

To seperate A^s and A^v contributions proton <u>and</u> neutron targets have to be used. Since practically no free neutron target exists one uses deuterium. This introduces some additional problems:
1) Fermi-motion between the nucleons
2) Final-state interaction
3) relatively low polarization

An additional problem arises from the Nuclear-Magnetic-Resonance signal (NMR) of the deuteron which has a complicated shape due to the electric quadrupole interaction.

The spin of the deuteron is one since proton and neutron spins are parallel. About 95% are in the S-state (no orbital angular momentum). If the quadrupole interaction did not exist we would have 3 equally spaced levels and one single symmetrical NMR-signal. The quadrupole interaction shifts the levels depending on the angle between the magnetic field B and the electrical field gradient dE/dr of the electron shell. Since we do not have a single crystal the two lines are smeared out.

Finally there is the 5% contribution of the D-state giving rise to an opposite orientation of the neutron and an additional correction.

In Fig. 7 the angular distribution of the target asymmetry T(θ) for the reaction
γ + n↑ → π⁻ + p at a photon energy of 700 MeV has been plotted[20]. Compared to the reaction γ + p↑ → π⁺ + n (Fig. 6) a quite different behavior of the experimental data
is obvious. The theoretical predictions available at that time could not explain the
data. New analyses based on these data and on newer data from other laboratories
have improved considerably. This clearly shows the significance of polarization
experiments also for neutron targets.

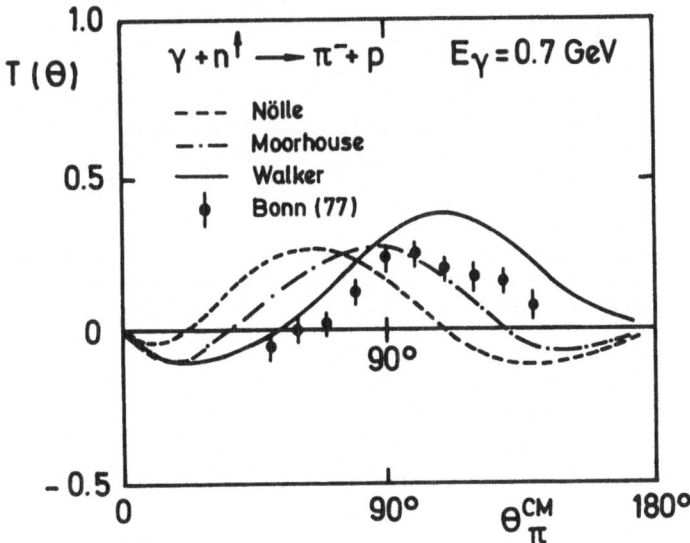

Fig. 7   Target asymmetry for the reaction γ + n↑ → π⁻ + p at 700 MeV.

Target Asymmetry for Photo-Disintegration of Deuterium

The large polarization of the recoil proton in the reaction γ + d → p↑ + n as discussed before (Fig. 5) and the possible explanation by dibaryon contributions led us
to measure the target asymmetry T for polarized deuterons. Also here we expected a
large T value around $E_\gamma \sim 500$ MeV.
As seen in Fig. 8 at θ = 130° the data are consistant with T = 0[21)22)]. Measurements
of the Tokyo group[22] at different angles also showed that one does not need contributions of dibaryon resonances to explain the data. On the other hand a reliable
theory for the deuteron disintegration is not available up to now.

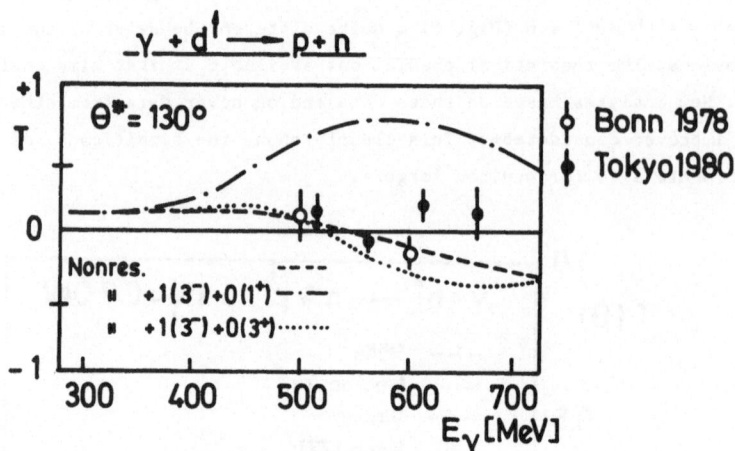

Fig. 8 Target asymmetry for the reaction $\gamma + \vec{d} \rightarrow p + n$. The curves are theoretical predictions with and without dibaryon resonances. The Bonn data were taken at $\theta = 135°$.

Polarization Experiments with Electrons

High intensity polarized electron sources for accelerators are available now, which made a new generation of experiments possible.

The first high intensity sources have been developed in Bonn[23] and Yale[24]. Two methods have been used:

1) Polarized atomic beams and unpolarized flash light.
2) Unpolarized atomic beams and polarized laser light. (Fano-effect)

A few years ago a very intense source has been developed at SLAC [25] where a Ga As-cristall and a polarized laser has been used.

In tab. 3 some information concerning the sources is given.

e^--Source	Ionisation	e^-/Puls	P	Laboratory
Polar. Atomic Beam (Li, Rb)	Flash Light	$10^8(10^9)$	60%(85%)	Bonn 69, Yale 72(79)
Atomic Beam Cs, Rb	polar. Laser	10^9	80%	Bonn 74
Semi-Conductor Ga As	polar. Laser	10^{11}	40%	Slac 76

table 3 Types of polarized electron sources

Very important high energy polarized electron-hadron scattering experiments were performed to test invariance principles and to measure the internal spin structure of the nucleon[26]. Polarized electrons have been scattered from unpolarized nucleons, but also double polarization experiments with polarized targets were made.

References

1. P.J. Bussey et.al, Nucl. Phys. B154 (1979) 205, 492
2. P. Lütter, Thesis, Bonn University PIB 1-91 (1970)
3. P. Nölle et.al., Nucl. Phys. B26 (1971) 461
4. W. Pfeil and D. Schwela, Nucl. Phys. B45 (1972) 379
5. F. Berends and D. Weaver, Nucl. Phys. B30 (1971) 575
6. P. Feller, et.al., Nucl. Phys. B104 (1976) 219
7. V.B. Ganenko et.al., Sov.J.Nucl. Phys. 24 (1976) 594
8. A.W. Smith and N. Zagury, Preprint University Rio de Janeiro, 16 (1979)
9. K.H. Althoff et.al., Z. Phys. 194 (1966) 144
10. K.H. Althoff et.al., Phys. Lett. 26B (1968) 640
11. V.A. Get'man et.al., Conf. on High Energy Phys., Madison (1980)
12. R. Kose et.al., Z. Phys. 220 (1969) 305
13. F.F. Liu et.al., Phys. Rev. 165 (1968) 1478
14. T. Kamae et.al., Phys. Rev. Lett. 38 (1977) 468
    H. Ikeda et.al., Phys. Rev. Lett 42 (1979) 1321
15. A.S. Bratashevskij et.al., Symp. Lepton and Photon Inter. at High Energies, Batavia (1979)
16. R. Kajikawa, Preprint Nagaya DPNU-31-80 (1980)
17. S. Mango et.al., Nucl. Instr. 72 (1970) 45
    P. Roubeau, Cryogenics 6 (1966) 207
18. U. Härtel et.al., Symp. on Pol. Beams and Targets, Lausanne (1980)
19. K.H. Althoff et.al., Phys. Lett. 63B (1976) 107
20. K.H. Althoff et.al., Nucl. Phys. B116 (1976) 253
21. G. Glasmachers, Diplom Thesis, Bonn IR-79-22 (1979)
    K.H. Althoff et.al., Internal Report BONN, IR-80-33 (1980)
22. N. Araji et.al., Contribution to the Conference on "Baryons", Toronto (1980)
23. G. Baum and U. Koch, Nucl. Instr. and Meth. 71 (1969) 71
24. V.W. Hughes et.al., Phys. Rev. A5 (1972) 195
25. C.K. Sinclair et.al., Symp. on Pol. Beams and Targets, Argonne (1976) 424
26. V.W. Hughes, Symp. on Pol. Beams and Targets, Argonne (1978) 171

EXPERIMENTS WITH MONOCHROMATIC AND POLARIZED PHOTON BEAMS

L. Federici, G. Giordano, G. Matone, P. Picozza, R. Caloi, L. Casano,
M.P. de Pascale, M. Mattioli, E. Poldi, C. Schaerf, P. Pelfer, D. Prosperi,
S. Frullani and B. Girolami

presented by
G. Matone
INFN - Laboratori Nazionali di Frascati, Frascati, Italy

Introduction.

A great deal of interest has born during the last few years about the use of monochromatic photon beams with high degree of polarization and low background.

One of the main motivations for this interest was undoubtedly the successful results the Frascati /1/ and SLAC /2/ laboratories obtained by making laser photons to collide with high energy electrons. The photon beam at SLAC was obtained with a ruby laser (1.78 eV) and the 20 GeV electron linear accelerator providing photons up to several GeV, whereas the Ladon beam at Frascati operates with an Argon Ion Laser (2.41 eV) and the Adone storage ring in the energy region between 5 MeV and 80 MeV /1/.

By limiting myself to consider this second case, I can summarize here the results obtained so far in the following table

Energy (MeV)	Intensity γ/s	Resolution %	Polarization
5	2×10^4	~ 1	~ 1
80	2×10^5	~ 8	~ 1

letting the reader refer to the published papers of the LADON group for any complementary details on this subject (see refs. quoted in ref./1/). These numbers are now planned to be improved both in intensity and monochromaticity by modifying the laser arrangement on the machine /3/. In any case they appear extremely encouraging to initiate good experimental research in the photonuclear reactions studies. This beam came into operation in 1979 and the first experimental results on deuterium photodisintegration by polarized photons are now available /4/.

The use of storage rings is clearly favourite with respect to Linacs where the duty-cycle is in general very poor. Moreover the new generation of storage rings, completely dedicated to synchrotron radiation studies, are now planned to be built both in Europe and in USA and the general technical requirement advanced for them is to have very good emittance together with a very high stored current /5/. In order to extend the wavelength region available from a synchrotron radiation source, special components will be foreseen such as superconducting "wigglers" and "undulators". In this second case, the objective of optimum source brightness leads to the criterium to keep the horizontal and vertical angular divergency down to $\sim 10^{-5}$ rad /5/. This fulfils the needs for a good synchrotron radiation beam but at the same time optimizes the conditions for having also a good backscattered photon beam. These two merging interests lead to the consideration of having a nuclear facility like that installed on different machines /6/. Quantitative predictions have been treated with some details by the LADON group. Here different lasers and different methods have been discussed and the results can be summarized as follows:

Machine	E_e (GeV)	Laser (Power)	Beam spot sizes and divergencies				I_e (mA)	Photon energy (MeV)	I_γ (γ/sec)	$\Delta E/E$ (%)
			horizontal		vertical					
			(mm)	(mrad)	(mm)	(mrad)				
NSLS	2.5	Ar (100 W)	0.29	1×10^{-2}	0.045	6×10^{-3}	500	~ 210	10^7	1
ALFA 3	3.5	Ar (100 W)	0.28	2.5×10^{-2}	0.04	1.5×10^{-2}	280	~ 400	$\sim 2 \times 10^6$	1
ESRF	5.0	Ar (100 W)	0.58	2×10^{-2}	0.09	1.2×10^{-2}	500	~ 780	$\sim 4 \times 10^6$	1.5
LEP	50	CO_2 (1 KW)	2.5	5×10^{-3}(?)	0.25	2.5×10^{-3}(?)	10	~ 100 (16°) ~ 4.000 (160°)	2×10^5	22
LEP	50	YAGx4 (3MW)	2.5	5×10^{-3}(?)	0.25	2.5×10^{-3}(?)	10	~ 10.000 (32°) ~ 40.000 (160°)	100/burst at 20 b/sec	12

The case of LEP must be considered separately. The beam emittance in the case of LEP is not as good as one would have. Nevertheless, as a pure exercise, it has been included in the list with an hypothetical angular divergency 4-times better than what is the designed number /7/. Should this be considered feasible, the LEP case would become one of the most exciting facilities since it could provide photon beams tunable from ~ 100 MeV up to ~ 40 GeV, with reasonable monochromaticity and good polarization.

Over the landscape of these different possibilities other considerations must be done. The expenses for a Laser installation on a storage ring are modest with respect to the total investement for the machine itself, and moreover this activity can be conceived to run in parasitic mode without affecting the life of the synchrotron radiation community. Thus there are serious reasons to think that several of these possibilities will become really available in the years to come. If we include LEP in this list, we could imagine a situation for the next decade, where polarized photon beams from few MeV up to several GeV will be operating at the same time. This means that the scientific problems we could investigate will range from the low up to the very high energy region where the typical problematic of the high energy physics will be heavily involved.

A common feature of all these possibilities will be that the intensity will never be greater than $(10^7 - 10^8)$ γ/s at best. This necessarily will require apparatus with very big solid angle, that mainly means (4π) detectors.

At this point the correct way to proceed this discussion would be to display the main topics where the polarization is very useful in the understanding of the physics implied. For obvious reasons, this attempt can not be systematic and thus I will limit myself to the proposition of few significative examples where the importance of the problems and the role of the polarization can easily be appreciated. The exposition will scan the entire energy interval and is very far from being exhaustive. The topics discussed are a selection among others and the choice that has been made is only fruit of my personal feeling on what is going to be an exciting future development in intermediate and high energy nuclear physics.

1. - The low energy region ($E_\gamma \leq 100$ MeV)

1.1. - The dynamic collective model

The first flash I would like to give is on the old problem of the validity of the Dynamic collective model in the description of the nuclear giant resonances /8/. Undoubtedly this is one of the most interesting topic where with a monochromatic and polarized photon beam one can really settle the problem of the coupling between the giant dipole resonance and the surface degrees of freedom. This coupling is the direct signature for the DCM whose qualitative success spread quite far and wide since the pioneer work of Fuller and Hayward /9/ on deformed nuclei.

The underlying idea of the model is to describe the photon scattering on nuclei as given by the sum of the contributions coming from the expansion of the electromagnetic field according to the total angular momentum transfer j and the angular momentum transfers L and L' at the two vertices of the Fig. 1 /10/, where

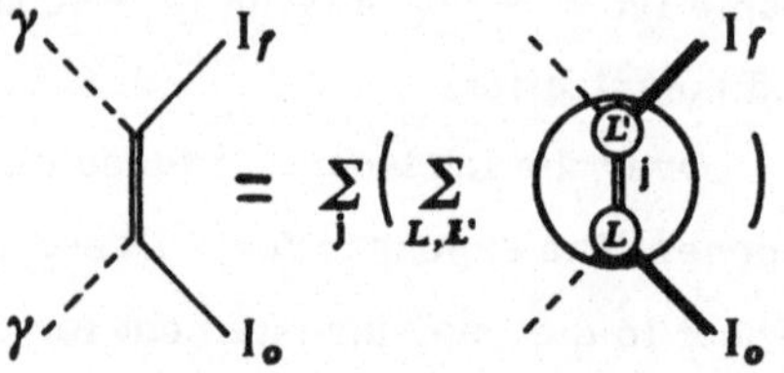

FIG. 1 - Decomposition of the scattering amplitude according to angular momentum transfer.

$$|I_f - I_o| \leq j \leq I_f + I_o, \quad |L - L'| \leq j \leq L + L'.$$

In the region of the GDR where pure dipole radiation contributes, $L = L' = 1$ and $j = 0, 1, 2$. Consequently, the differential cross section for an unoriented scattering target, can be written as

$$\frac{d\sigma}{d\Omega} = \sum_{j=0}^{2} \frac{|A_j|^2}{2j+1} g_j(\theta) , \qquad (1)$$

where $g_j(\theta)$ are known angular distribution functions depending only upon the relative orientation of the polarization vectors of the incident and scattered photons. The scattering amplitude A_j are directly related to the nuclear polarizabilities (0-scalar, 1-vector, 2-tensor) and are the quantities which contain the information about the nuclear structure /10/. The scalar polarizability describes the isotropic part of the scattering amplitude, the vector polarizability measures the "optical activity" of the nucleus while the tensor polarizability measures the optical anisotropy of the nucleus /10/. These are quantities accessible to photon scattering experiments and thus serve as a common meeting ground with theory. For example, in the framework of the usual hydrodynamic model a spherical nucleus has three degenerate GDR-states. Therefore, vector and tensor polarizabilities vanish since the nucleus is optically isotropic. No inelastic scattering will occur. But if a coupling is conceived between the volume GDR-oscillations and the quadrupole surface vibrations, as suggested in the DCM, the nucleus, due to the instantaneous deformation, becomes optically anisotropic and inelastic scattering into the 2^+-vibrational states will happen.

This inelastic tensor component of the scattering amplitude has been calculated in detail and, for a spherical vibrational nucleus typically as much as 30% of scattering at backward angles is predicted to the first excited 2^+ state /11/. But recent experiments performed with the bremsstrahlung beam obtained with MUSL-2 at the Illinois University seem to contradict quite remarkably this prediction /12/. No more than 15% of the elastic scattering has been found to go to the first excited state in ^{60}Ni, as shown in Fig. 2. And even worse is the case of the heavy deformed ^{166}Er where the inelastic transitions have been found drastically smaller (by a factor $\simeq$ 3-5) than the prediction of the DCM /13/ (see Fig. 3).

In conclusion something seems to be wrong either in the experiments or in theory. A definite way out could be the knowledge of the polarization of the incoming photon beam. In fact, in that case, expression (1) specializes in the two following /14/:

$$\frac{d\sigma^\perp}{d\Omega} = \frac{1}{3}|A_0|^2 + \frac{7}{30}|A_2|^2 , \quad \frac{d\sigma^\parallel}{d\Omega} = \frac{1}{3}|A_0|^2 \cos^2\theta + \frac{1}{5}|A_2|^2 (1 + \frac{1}{6}\cos^2\theta) . \qquad (2)$$

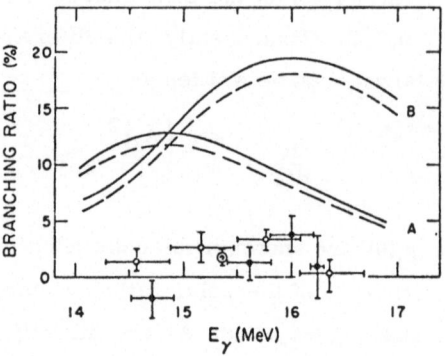

FIG. 3 - The ratio of inelastic to elastic scattering in ^{166}Er as given in ref. /13/.

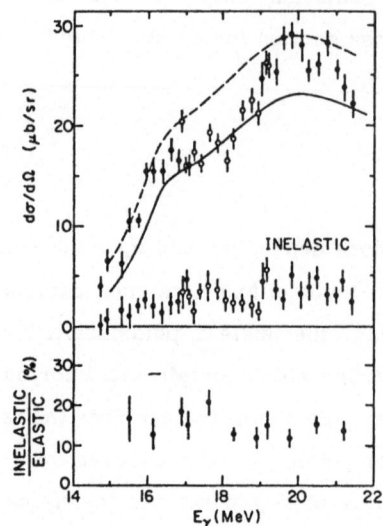

FIG. 2 - The ratio of inelastic to elastic scattering in ^{60}Ni as given in ref. /12/.

These say, that the nucleus has no coherent (j = 0) scattering along the polarization vector at $\theta = \pi/2$, so that a measurement of photon scattering in this direction is a direct measure of the incoherent (j = 2) scattering. And this measurement does not depend on the energy resolution of the detectors which could be a serious difficulty when the vibrational levels are so close to the ground state that is not easy to separate the two contributions in the scattering, by γ-ray spectrometry /14/.

1.2. - Proton polarizabilities

Another point where the knowledge of the photon polarization is determinant, is the old question of the measurement of the proton electric and magnetic polarizabilities α and β. These two quantities are fundamental structure parameters which, together with charge and magnetic moment, fully control the behaviour of the proton system in a static or slowly varying electromagnetic field. Therefore they can be obtained by Compton scattering experiments at low energy, where the differential cross section can be expressed in terms of the expansion /15/

$$\left(\frac{d\sigma}{d\Omega}\right)_p = \left(\frac{d\sigma}{d\Omega}\right)_o - \frac{e^2}{4\pi M_p} \omega^2 \left\{\alpha(1+\cos^2\theta) + 2\beta\cos\theta\left[1 - \frac{3\omega}{M_p}(1-\cos\theta)\right]\right\} + O(\omega^4) \quad (3)$$

ω and M_p being the photon energy and the proton mass respectively. Moreover, $(d\sigma/d\Omega)_o$ is the cross section for the proton thought as structureless and the second term is a structure correction depending on the above mentioned polarizabilities.

The use of monochromatic and polarized photons represents a substantial improvement in the determination of these two quantities. First of all the monochromaticity removes all the usual difficulties one has with bremsstrahlung beams. Moreover, the polarization allows to make the linear combination of the parallel and perpendicular cross section /16/

$$f_1(\theta, \alpha) = \frac{1}{2}\left[\frac{d\sigma^\perp}{d\Omega} - \frac{d\sigma^\parallel}{d\Omega}\right], \qquad f_2(\theta, \beta) = \frac{1}{2}\left[\frac{d\sigma^\perp}{d\Omega}\cos^2\theta - \frac{d\sigma^\parallel}{d\Omega}\right],$$

which depend only upon α and β respectively. It is immediate to see how sensitive this method could be in the determination of α and β separately.

According to the present understanding, these two quantities can be related to the structure functions usually defined in the deep inelastic scattering and the following sum rules can be deduced /17/ :

$$\alpha + \beta = \lim_{q^2 \to 0} \frac{1}{2\pi^2} \int_{\nu_{th}}^{\infty} \frac{\sigma_T(q^2, \nu')}{\nu'^2} d\nu', \tag{4}$$

$$\alpha = \frac{(\lambda e)^2}{16\pi M_p^3} + \lim_{q^2 \to 0} \frac{1}{2\pi^2} \int_{\nu_{th}}^{\infty} \sigma_T(q^2, \nu') \frac{R(q^2, \nu')}{q^2} d\nu'. \tag{5}$$

Besides the usual definitions, the other quantities are defined as follows:

$$\sigma_T(q^2, \nu') = \text{total photoabsorption cross section}, \qquad R(q^2, \nu) = \frac{\sigma_L(q^2, \nu)}{\sigma_T(q^2, \nu)}.$$

While the first is the very well known Damashek and Gilman sum rule yielding the result /18/

$$\alpha + \beta = (14.2 \pm 0.03) \times 10^{-4} \text{ fm}^3, \tag{6}$$

the second one requires R to vanish asymptotically more rapidly than ν^{-1} and fournishes an independent evaluation of α expressed through the experimental determination of $R(q^2, \nu)$ /17/. The relationship between Compton amplitude and deep inelastic scattering can be understood just looking at the topological structure of the electron-proton inelastic cross section (see Fig. 4). The dashed bottom part of the figure ($W^{\mu\nu}$) in the limit of $q^2 \to 0$ describes the Compton scattering of real photons /19/.

According to the quark-parton model, in the scaling region the photon interacts with a quasi-free quark and the structure functions are independent of

q^2 (see Fig. 5). In particular if the quarks have spin 1/2, $R = 0$. Within QCD this picture is modified by the presence of gluons which lead to logarithmic q^2

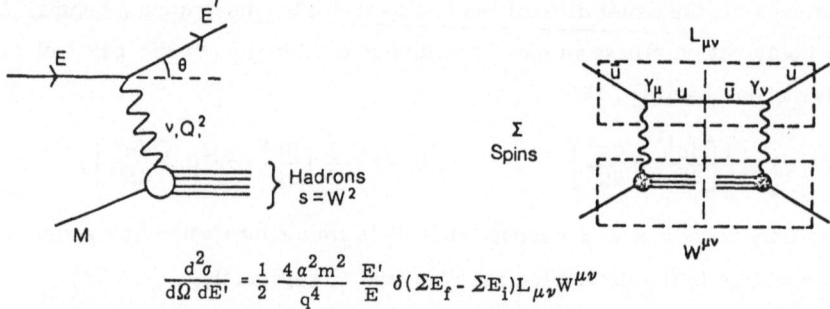

FIG. 4 - Tensor structure of electron-proton inelastic cross section with obvious significance of the symbols.

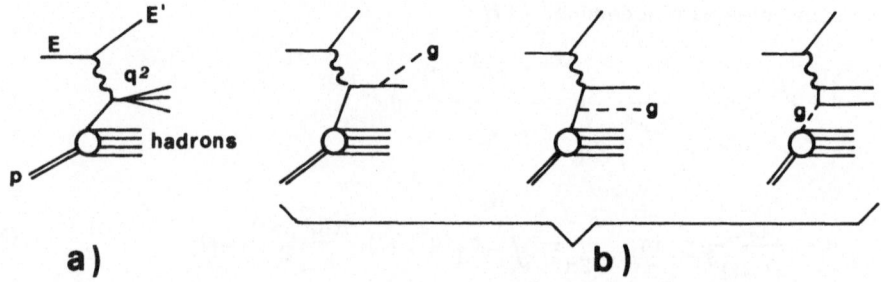

FIG. 5 - a) Parton model picture of deep inelastic scattering;
b) Leading order QCD diagrams involving gluons.

dependence of the structure functions and a well defined behaviour (non-zero) for R (see Fig. 5) /20/.

Thus an unconventional way to look at the second term in the right hand side of eq. (5) could possibly be related to the scaling violation in the deep inelastic scattering. Therefore numerical evaluations within QCD-theory would be highly desiderable.

Taking into account the old experimental determination of R obtained at SLAC and a scaling behaviour in the asymptotic region, a numerical evaluation of this sum rule has been attempted in ref. /17/ yielding the result

$$\alpha = (9.3 \pm 2.0) \times 10^{-4} \text{ fm}^3 \qquad (7)$$

which could now be improved with the new data from the muon experiments at CERN and Fermilab /20/. The present experimental results are the following:

	$\alpha \times 10^4$ (fm^3)	$\beta \times 10^4$ (fm^3)
V. J. Gold'anski et al. (1960) /21, 22/	10 ± 5	4 ± 5
P. Baranov et al. (1974) /23/	10.7 ± 1.1	-0.7 ± 1.6

A fit of these experimental data between 50 MeV and 110 MeV has been made with eq. (3) and the constraint (6) and according to this, the present experimental evidence gives for the proton the following determinations /17/

$$\alpha = (12.4 \pm 0.6) \times 10^{-4} \text{ fm}^3, \qquad \beta = (1.8 \pm 0.9) \times 10^{-4} \text{ fm}^3. \qquad (8)$$

It has been already argued how this result has to be considered as a surprising conclusion because one knows that all photoabsorption process on nucleons are dominated by the $\Delta(1236)$ resonance whose excitation is of magnetic nature /22/. Moreover this situation is complicated by the presence of the π^0-meson pole contribution /24/ which appears in order $O(\omega^4)$ and higher. This term is quite significant at low energy: at 100 MeV and $\theta = 150^\circ$ it accounts for $\sim 10\%$. In conclusion a new and more precise experimental determination of α and β is seriously needed.

2. - Medium energy region (100 MeV $\lesssim E_\gamma \lesssim$ 1 GeV)

The physics phenomenology in this energy range, has been recently discussed in a dedicated Workshop held at Frascati /25/ under the joint sponsorship of the ESF (European Science Foundation) and INFN (Italian Institute of Nuclear Physics). Being impossible to summarize here the work done on that occasion, I will limit myself to report few ideas that strongly rely to the knowledge of the incoming photon polarization.

It has been known for a long time that the meson exchange currents (MEC) and virtual excitation of the internal nucleon degrees of freedom, must be taken into account in any nuclear electromagnetic and hadronic process. In particular when the energy of the photon is sufficient enough to produce real pions in the region of nuclear isobar excitations, i.e. like Δ's, the dynamics of the pion in nuclear medium must be taken into account through the propagation of the Δ and the subsequent Δ-hole interaction. These models have been developed by different groups and given various generic names (isobar-doorway, isobar-hole, collective N^* resonances, giant (3,3) resonances) but they contain essentially the same ingredients /26/.

When a bound nucleon is excited for example, into the (3,3) resonance in analogy with the Brown-Bolsterli schematic model, a Δ-particle N-hole pair is being created. This Δ decays rapidly ($\tau \sim 10^{-23}$ s) into a nucleon and a pion, where the pion can either leave the nucleus or be absorbed by another bound nucleon, thereby creating a new particle-hole pair. The probability for the resonant reabsorption of the pion is very large inside a complex nucleus; its mean free path is much smaller than the average nucleon-nucleon distance /27/

$$\lambda = 1/\varrho\sigma \ll d.$$

Therefore this resonant reabsorption mechanism is expected to be dominant, at least in large nuclei: as it gives rise to a strong ph-ph coupling it has to be discussed in the frame of a RPA-calculation. As a result of such a calculation we expect a coherent superposition of such ph-configurations; that means a collective excitation of the whole nucleus.

This collective mode of excitation corresponds to a resonant pion current which floats through the nucleus and which is characterized by precise set of quantum numbers: it is needless to say that this argument can be qualitatively extended for all baryonic resonances, and in this sense one has to regard the nucleus to consist of baryons rather than of nucleons /27/.

The investigation of these spectra with photons can be particularly interesting when they are polarized. In fact, in this way it should be possible to systematically disintangle the multipole excitation strength distribution, in a way very similar to that used in the low energy region (GDR). Moreover, since the photons are strongly coupled to vector mesons, special mechanism can be investigated, like ϱ propagation in nuclei. The ϱ-meson can either split into two pions giving rise to a double ($\Delta \bar{N}$) configurations ($\Delta\Delta\bar{N}\bar{N}$); or it can excite bound nucleons into the $N^*(1520)$ and create subsequently coherent nuclear N^* excitations. These two modes are mixed and simple and double pion-photoproduction could give information on these two different mechanisms /28/.

To conclude with, if this picture will be shown to be correct, then the next fashinating question to be asked is whether or not this multibaryon concept of the nucleus, as a generalization of the conventional multinucleon model, can be further extended toward a general description of the nucleus as a multiquark system. In this sense Fig. 6, taken from ref. /29/, clarifies substantially the meaning of this concept. At present time very little is known about coherent multiquark excitations and nothing can be really said about

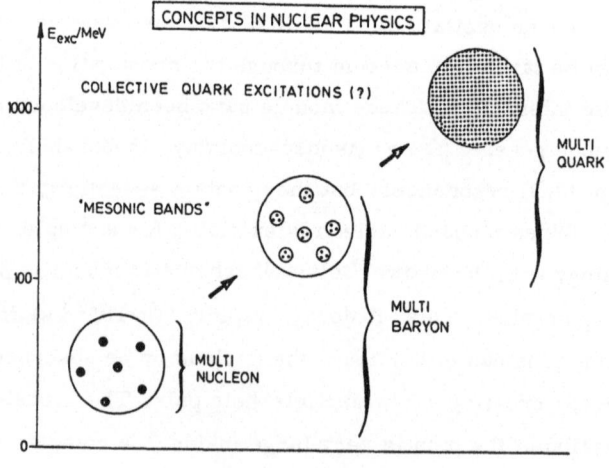

FIG. 6

the applicability of such a quark model. Nevertheless this big "affair" of nucleus as a multiquark system will be one of the most challenging matters for any of us and for the years to come.

3. - High energy region (20 GeV ≲ E_γ ≲ 40 GeV)

I will conclude my discussion giving few remarks on the photon physics in the high energy region where a lot of "high energy physics" is still quite a topical subject. With regards to this, it's worthwhile reminding the proposal to move the "SLAC hybrid facility" into the 20 GeV backscattered laser beams at SLAC /30/ or to look through the list of the experiments presently running at CERN (WA57; NA1; NA14; WA4 in ref./31/ and Fermilab. Electroproduction processes in the limit of $q^2 \to 0$, usually interpreted as the scattering of an almost real photon, have been extensively discussed also on the occasion of the ECFA-study of an ep facility for Europe /32/.

Nowadays, the main interest of the high energy people for the photon physics lies in the unique property of the photon to have both a hadronic (meson like) and a point-like component. We know that for low transverse momentum photon-hadron physics, the photon can be thought of as a superposition of vector mesons. On the other hand for high transfer momentum, the photon interaction with hadronic matter is predominantly mediated by the bare $q\bar{q}$ component in the photon wave function (when seen in a rapid moving frame).

In a similar way to hadron-hadron scattering, the production at high momentum transfer can reveal the jet structure associated with hard scattering at quark and gluon level. The basic hypothesis of these models are that quark jets arise from a direct hard collision between the quark in the photon and one in the nucleon as shown in Fig. 7 /33/.

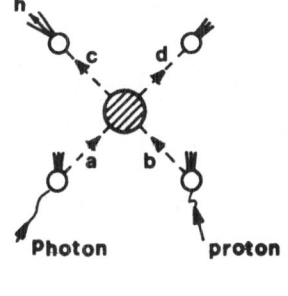

FIG. 7

The large p_T hadron production is assumed to occur as a result of elastic scattering of quarks $a+b \to c+d$, followed by the fragmentation of c into observed hadron h. In the gluon jet model based on QCD by Fritzsch and Minkowski /34/ the final product of such a photon-quark scattering will be instead a quark and a gluon manifesting themselves still by hadronic jets. These two different dynamical pictures give the idea of the problematic involved in this kind of physics. But this phenomenology can be even more attractive if one thinks that if the photon is polarized, so will be the constituent quarks.

In this sense polarized photon beams can provide polarized quark beams. A very convenient source of **polarized quarks turns out to be circularly polarized** photon beams. Suaya and Townsend /31/ have recently shown that fragmentation of polarized quarks into hadrons could yield an independent test of those QCD predictions that rely strongly on the intrinsic angular momentum of quarks and gluons. In their model, the process which should be identified is shown in Fig. 8.

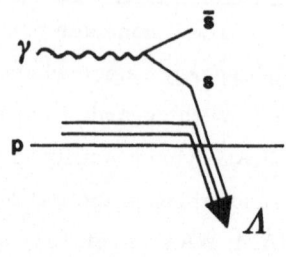

FIG. 8

Fast forward Λ's at high P_T are produced when an s quark from circularly polarized photon fragmentation combines with a -ud- pair from the proton. The fast s quark spin is aligned with the photon helicity direction. Owing to the vector nature of QCD, the s quark helicity in the interaction is preserved and the Λ acquires the original s quark helicity. Suaya and Townsend /35/ have shown that averaging over quark polarization between $x=0.2$ and $x=1$ (Feynman variable) gives $P = +0.3$. Similar argument holds for the Σ production where the result is $P_\Sigma = -0.1$.

As it has been discussed previously, the possibility to use the LEP machine as a source for a backscattered photon beam of very high energy, appears to be very problematic because, even accepting to work with a wide-band photon beam, still remains the problem of the bad duty-factor ($\sim 10^{-5}$) which imposes severe limitations. A practical possibility could be to use the bubble-chamber technique as in the SLAC arrangement /30/. The bubble-chamber with its (4π) geometry, excellent detection ability for both high multiplicity events and strange particle decays close to the vertex is the ideal detector, particularly when backed up with good downstream track measurements, particle identification and γ-detection systems.

According to the SLAC estimate /30/ for the counting rate, using 75 cm fiducial volume of hydrogen, 15 expansion/sec and 50 γ/pulse, one obtains

$$\text{events}/\mu\text{b}/\text{day} = \varepsilon_T \times 2.7 \times 10^{-6} N_\gamma/\text{day} \simeq 90 \text{ events}/\mu\text{b}$$

where:

$N_\gamma/\text{day} = 50\,(\gamma/\text{p}) \times 15 \text{ pps} \times 8.6 \times 10^4 \text{ s/day} \simeq 6.5 \times 10^7\,\gamma/\text{day}, \quad \varepsilon_T \simeq 50\%$,

and this number is of a considerable interest not only for the above mentioned possibility but also for other lines of research.

This reference goes mainly to the high mass vector production and the so called "search for charm". In particular (e^+e^-) experiments have not been successful in clearly identifying charmed baryon pairs which suggests that these states are not strongly produced diffractively. Photoproduction could furnish good chances for this recognition.

References

/1/ - L. Federici et al., Nuovo Cimento 59B, 247 (1980).
/2/ - J. Ballam et al., Phys. Rev. Letters 23, 498 (1969); 24, 955 (1970); 24, 960 (1970); 24, 1367 (1970); 25, 1223 (1970); Phys. Rev. D1, 94 (1970); D5, 545 (1972); D5, 1603 (1972); D7, 3150 (1972); D8, 1277 (1973); Phys. Letters 41B, 635 (1972); Nuclear Phys. B29, 349 (1971); B76, 375 (1974); B122, 383 (1977).
/3/ - R. Caloi et al., Frascati Reprint LNF-79/42 (1979); LNF-79/6(Int.).
/4/ - R. Caloi et al., Frascati Preprint LNF-80/15, submitted to Phys. Rev. Letters.
/5/ - European Synchrotron Radiation Facility, ed. by Y. Farge and P. J. Duke (European Science Foundation, 1979).
/6/ - G. Matone, The Monochromatic and Polarized Photon Beam with the ESRF Machine, in the Proceedings of the Workshop on Intermediate Energy Nuclear Physics with Monochromatic and Polarized Photons, Frascati, July 1980; ed. by G. Matone and S. Stipcich (LNF, 1980).
/7/ - A. Hoffmann, CERN Report ISR-TH/AH/PS (1980).
/8/ - M. Danos and W. Greiner, Phys. Rev. B4, 284 (1964); H. Arenhövel, G. Gueuss and V. Rezwani, Phys. Letters 39B, 249 (1971).
/9/ - E. G. Fuller and E. Hayward, Phys. Rev. Letters 1, 465 (1958).
/10/ - H. Arenhövel, Photon Scattering by Nuclei: Theory and Experiment, in 'Photonuclear Reactions and Application', Asilomar Conference (1973), pag. 449.
/11/ - H. Arenhövel and H. J. Weber, Nuclear Phys. A91, 145 (1967); H. Arenhövel and J. M. Maison, Nuclear Phys. A147, 305 (1970).
/12/ - T. J. Bowles et al., Phys. Rev. Letters 41, 1095 (1978).
/13/ - A. M. Nathan and R. Moreh, Phys. Letters 91B, 38 (1980).
/14/ - A. Fubini et al., Frascati Reprint LNF-74/62 (1974).
/15/ - N. Powell, Phys. Rev. 75, 32 (1949); V. Petrunkin, Proceedings of the Lebedev Physics Institute, Vol. 41 (1968).
/16/ - G. Matone and D. Prosperi, Frascati Reprint LNF-75/8 (1975).
/17/ - G. Matone and D. Prosperi, Nuovo Cimento 38, 471 (1977).
/18/ - M. Damashek and F. J. Gilman, Phys. Rev. D1, 1319 (1970).
/19/ - F. E. Close, Partons and Quarks, Daresbury Lecture Note, Series No. 12, DNPL/R31 (1973); see also refs. quoted here.
/20/ - H. E. Montgomery, Results on Deep Inelastic Scattering, CERN Report CERN-EP/80-177 (1980).
/21/ - V. J. Gold'anski et al., JETP 38, 1965 (1960); Soviet Phys.-JETP 12, 1223 (1960); Nuclear Phys. 18, 473 (1960).
/22/ - J. Bernaben, T. E. O. Ericson and C. Ferrofontan, Phys. Letters 49B, 381 (1974).
/23/ - P. S. Baranov et al., Phys. Letters 52B, 122 (1974).
/24/ - P. S. Baranov et al., JETP Letters 20, 353 (1974).

/25/ - Proceedings of the Workshop on Intermediate Energy Nuclear Physics with Monochromatic and Polarized Photons, Frascati, July 1980, ed. by G. Matone and S. Stipcich (LNF, 1980).

/26/ - J. Arvieux, Pion-Nucleus Elastic Scattering: what do we learn ?, in the Proceedings of the Eight International Conference on High Energy Physics and Nuclear Structure, Vancouver, August 1979, ed. by D. F. Measday and A. W. Thomas, pag. 353.

/27/ - M. Dillig and M. G. Huber, Excitation of N^*-resonances in Nuclei, Rencontres de Saclay (12-16 Mai, 8-12 September 1975), pag. 111.

/28/ - M. G. Huber and K. Klingenbeck, Subnuclear Excitations, in Proceedings of the 'Workshop on Intermediate Energy Nuclear Physics with Monochromatic and Polarized Photons', Frascati, July 1980, ed. by G. Matone and S. Stipcich (LNF, 1980).

/29/ - M. G. Huber and K. Klingenbeck, Invited paper at the 'International Conference on Hypernuclear and Low Energy Kaon Physics', Jablonna, September 1979.

/30/ - Proposal to move the SLAC hybrid facility into a 20 GeV backscattered Laser beam, SLAC (1978).

/31/ - Experiments at CERN in 1979, CERN, Geneva.

/32/ - Proceedings of the Study of an ep Facility for Europe, ed. by U. Amaldi, Hamburg, 2-3 April 1979 (DESY, 1979).

/33/ - E. Kawai et al., Prog. Theor. Phys. $\underline{60}$, 929 (1978).

/34/ - H. Fritzsch and P. Minkowski, Phys. Letters $\underline{69B}$, 316 (1977).

/35/ - P. Suaya and J. S. Townsend, SLAC-PUB-2190 (1978).

PHOTON SCATTERING

B. Ziegler, Max-Planck-Institut für Chemie (Otto-Hahn-Institut)
Kernphysikalische Arbeitsgruppe, D-6500 Mainz, Germany

1. Introduction

The measurement of angular distributions of elastically scattered photons provides valuable information on nuclear properties. The advantages of this technique compared to charged particle or hadron scattering are known since long ago [1]: The interaction is known exactly and the incoming and outgoing waves are not disturbed by Coulomb or other effects. The measured observables can be interpreted rather directly and unambigously. Photon scattering angular distribution measurements enable the determination of multipole strength distributions in energy. For real photons the momentum transfer to the nucleus is small in the region of the giant resonances and one is restricted to the determination of E1, E2 and M1 transitions. The strength distribution for these multipolarities is not well known, expecially for heavy elements. Therefore, the restriction to the lowest multipole orders is no serious drawback at the present state of the art. Despite these obvious advantages, the number of photon scattering experiments is small. This fact is mainly due to two technical difficulties: One is the preparation of a beam of monochromatic photons with known intensity and the second is the detection of the relatively small number of scattered photons in the presence of certain unavoidable backgrounds.

The different experimental arrangements can be classified according to the manner in which the monochromacity is obtained. Considering only sources with variable photon energies, one finds essentially three types, (i) bremsstrahlung [1,2,3], (ii) positron annihilation in flight [4], and (iii) tagged bremsstrahlung photons [5]. The scattered photons normally have been detected with scintillators, which have mostly been NaI(Tl)-crystals.

When using untagged bremsstrahlung [2,3] the primary photon energy can only be determined from the scattered photon energy, if the energy resolution of the detector is sufficiently good and it is assumed that there is elastic scattering only and no background. In the experiments with monochromatic photons the energy resolution of the detectors was always poorer than the monochromacity of the primary beam. Good resolution of the detector however is essential to improve the discrimination against background counts and inelastically scattered photons.

At the Mainz Linac Laboratory an experiment has been set up to measure the scattering of photons in the energy range 8 to 100 MeV. It consists of a source of variable energy photons from positron annihilation in flight and four NaI(Tl)-detectors installed at different scattering angles. Differential cross sections for the elastic scattering of photons by ^{208}Pb have been measured. An attempt is made to analyse these data and the absorption cross sections [6,7] within a consistent description. The photon energy range considered reaches from about 10 MeV to well above the giant dipole resonance region. It ends however below meson threshold. There, new physical processes become prominent which - in its bearing on photon scattering - will be explored at some time to come. Presently, the program should be, to learn the elementary processes involved below meson threshold and then to proceed to the higher energy domain where the nucleons more and more display their internal structure. In this sense, an attempt is made to include all elementary processes which govern the field from 10 MeV to 100 MeV photon energy.

The nuclear properties involved are giant multipole resonances, sum rule predictions and meson exchange effects, which - if at all - do vary only smoothly along the atomic number scale. Therefore, at the time being, it appears to be best to concentrate on one single nucleus before trying to see individual features. In this sense, for the experiment the ^{208}Pb nucleus was chosen. For this nucleus, and in the energy range of interest, there are total absorption measurements [6,7] and the photon scattering cross sections presented in this paper. For light nuclei, the absorption data [8] are complemented only by a few scattering measurements [9,10], which do not allow a complete analysis yet.

The absorption cross section essentially is taken to calculate the elastic forward (scattering angle $\theta=0°$) scattering cross section. This number is extremely important for the interpretation of scattering cross sections, and can only be obtained from absorption cross sections since scattering under small angles ($\theta<60°$) is obscured experimentally by a number of unavoidable background effects. Absorption of unpolarized radiation and scattering of (polarized) radiation can be regarded as a single combined experiment. It is this combination of absorption and scattering which provides great interpretative strength and unambiguity.

2. The Photon Source

A nearly monoenergetic photon beam of continuously variable energy can be obtained by the annihilation in flight of monoenergetic positrons [11]. The positrons are obtained by bombarding a one radiation length thick, high Z target with an intense beam from an electron accelerator. Bremsstrahlung and subsequent pair production in the target create positrons that have a smooth energy distribution. Such a positron-production target may be installed in a solenoidal magnetic field lens between two accelerator sections. Since in such an installation the positrons will be accelerated from a few MeV to the final energy, a large phase volume of the positrons emitted from the target can be accepted and a ratio of 10^{-3} positrons per incident electron can be reached [12].

In our assembly we use a different arrangement (fig.1), which has the advantage of having little interference with accelerator operation. The positron-production target is installed in the beam switch yard 16 m downstream from the end of the linac. The positrons emitted from the target in foreward direction (solid angle 1.65×10^{-2} steradians) are guided by a transport and energy analyzing system to the annihilation target in the experimental area.

In order to obtain a small emittance for the positrons, the electron beam has to be focused into a narrow spot on the positron-production target. Experimentally it was found, that the highest positron current was reached with an electron beam spot of less than 0.4 mm in diameter. By multiple scattering, the size of the positron source then is approximately 2 mm in diameter. The mean power deposited in the target by the electrons is approximately 1 kW. This will cause no serious cooling problems. However, the high energy density produced in the small target volume hit by the electrons heats the material at the front side of the target during a single linac pulse (100mA, 4µs) by more than 3000 K. Thus, part of the target will inevitably melt during one linac pulse. The positron-production target developed (fig.2) is a "boiling" tantalum target, which is cooled mainly by radiation. The target is a tantalum pellet (4 mm in diameter, 8 mm long) in a graphite container. The pellet is heated above its melting point by the electron beam. When molten the target can restore its shape between beam pulses. Such a target delivers a stable positron current, which is affected only by fluctuations of the incoming electron beam. After a target lifetime of about 200 hours most of the target material is evaporated, as is part of the graphite container, leaving a hole at the beam entrance (fig.2b). In practice several targets are housed in one container, which is surrounded by a water cooled copper jacket (fig.2c). Thus, a used target can easily be replaced by remote control during operation. The target is cheap and mechanically very simple, but has the disadvantage that carbon and tantalum are evaporated onto the surrounding metal pieces.

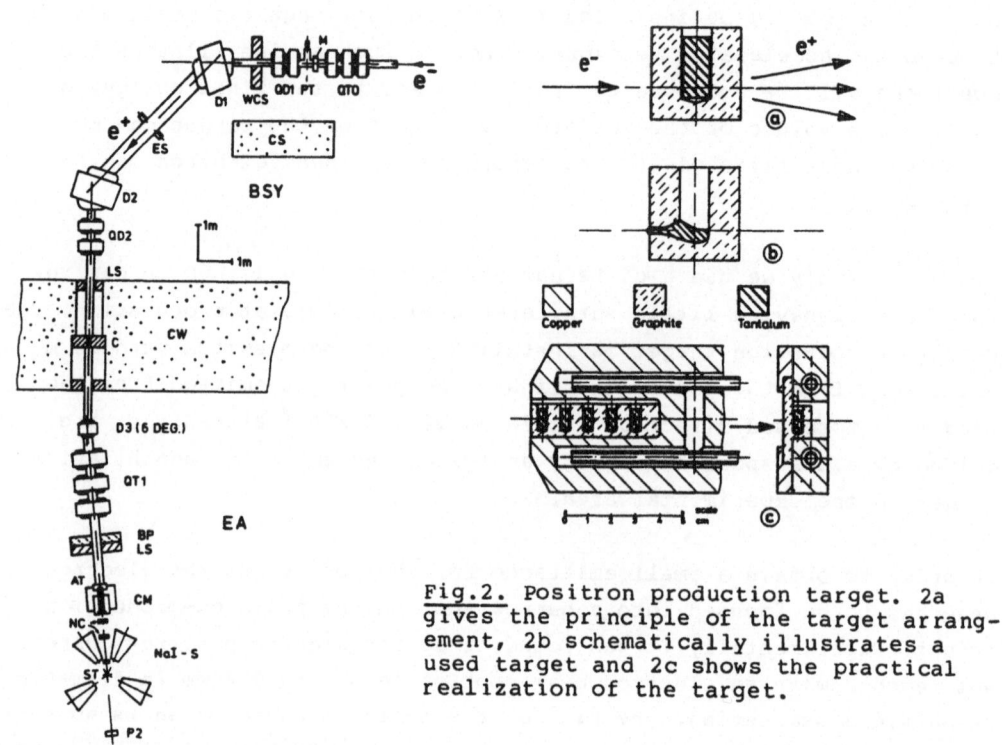

Fig.2. Positron production target. 2a gives the principle of the target arrangement, 2b schematically illustrates a used target and 2c shows the practical realization of the target.

Fig.1. Set up for the production of positrons and their annihilation in flight. The installation in the experimental area, EA, is well shielded from the beam switch yard, BSY, by a 3 m thick heavy concrete wall, CW. The incoming electrons, e^-, are focused by the quadrupole triplet, QTO, onto the positron-production target, PT. M is a beam monitor, QD1 and QD2 are magnetic quadrupole doublets, WCS is a water cooled copper shielding, D1, D2 and D3 are dipole magnets, ES is the energy slit, CS concrete shielding, LS lead shielding, BP borated paraffine shielding, C a collimator, QT1 a magnetic quadrupole triplet, AT the annihilation target, CM a cleaning magnet and a Faraday cup, NC a system of 3 Ni-collimators, ST the scattering target, P2 an ionization chamber, and NaI-S are NaI-spectrometers. For the latter see details in fig.3.

When high energy photons move through matter mainly two processes will occur: Pair annihilation and bremsstrahlung. The process of interest here is the two quanta annihilation of the positron in flight. The cross section of this process is proportional to the atomic number Z of the target material. The cross section of the inevitable bremsstrahlung background is approximately proportional to $Z(Z+1)$. By the difference in the Z-dependence of these two photon production mechanisms one can subtract the bremsstrahlung background using targets of sufficiently different atomic number (e.g. Be and Cu). The same principle is used for the measurement of response functions (shown below in fig.6). The energy flux from these two targets is measured absolutely with a calibrated NBS P2 ionization chamber [13,14]. Since the ionization currents are low (10^{-12} A at 15 MeV), special precautions had to be taken for their integration.

3. The Photon Scattering Arrangement

The details of the photon scattering arrangement are illustrated in fig.3. The scattering target is located 106 cm downstream from the Ni collimator that defines the foreward solid angle for the annihilation photons. The four 10"Øx10" NaI(Tl) spectrometers are installed at distances of 62 cm from the target at scattering angles of $60°$ and $150°$. The size of the collimator in front of the crystal is 15 cm in diameter. It was chosen so that the solid angle for the detection of scattered photons is as large as possible and that the probability of absorbing a scattered photon is always given by the full crystal thickness. The efficiency, E_{sp}, of the spectrometers is then taken to be $E_{sp} = 1 - e^{-\mu \varrho t}$. μ is the mass attenuation coefficient, ϱ and t density and thickness of the NaI crystal. Aluminum absorbers of variable thicknesses may be put into the collimators in order to reduce low energy background from the target. The spectrometers are housed in a 10 cm thick Pb shielding plus additional shielding of borated wax and polyethylene, concrete, iron, and lead to reduce neutron and photon background.

A typical pulse height distribution of scattered radiation (fig.4) shows that most of the analysed events are in the low energy region, whereas the events from the scattered monoenergetic photons appear clearly separated in the tip of the pulse height distribution. In order to prevent too high dead time losses, the lower threshold is set to a relatively high value without loss of information. Since the analyzing

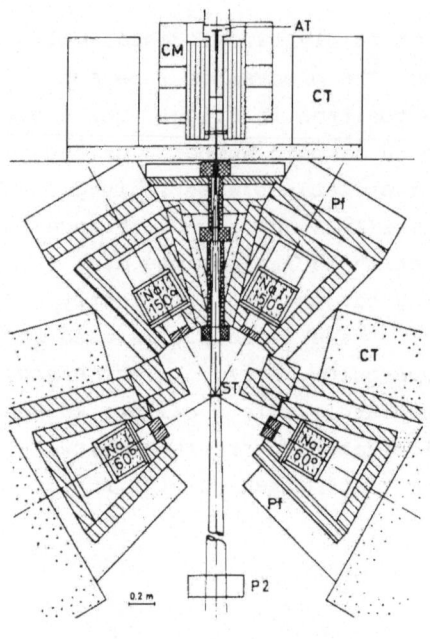

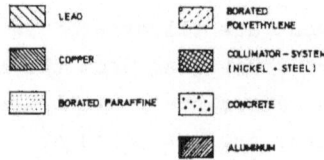

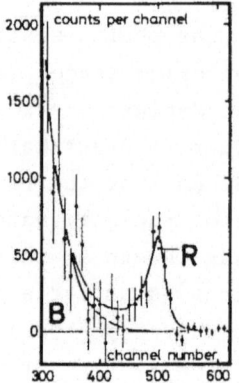

Fig.3. Photon scattering arrangement. Photons scattered from the scattering target, ST, can be simultaneously measured in four NaI(Tl) spectrometers, installed at 60° and 150° to the direction of the incoming photons. Each NaI crystal is housed in a lead castle with a wall thickness of 10 cm (15 cm at the front). The entrance collimators of the spectrometers contain aluminum absorbers of variable thickness. The lead castle is surrounded by borated paraffin Pf and polyethylene. There is additional lead, copper and concrete, CT, shielding. The photon collimation consists of three 10 cm thick nickel collimators with iron tubes in between. AT is the annihilation target, CM a cleaning magnet, and P2 a calibrated ionization chamber.

Fig.4. Pulse height spectrum of 25 MeV photons scattered from ^{208}Pb at 90°. Given is the difference of the pulse height spectra as obtained with a Be- and a Cu-annihilation target. Fitted through the points are the response function, R, of the NaI spectrometer and an arbitrary background function, B. Both functions have been corrected for pile-up.

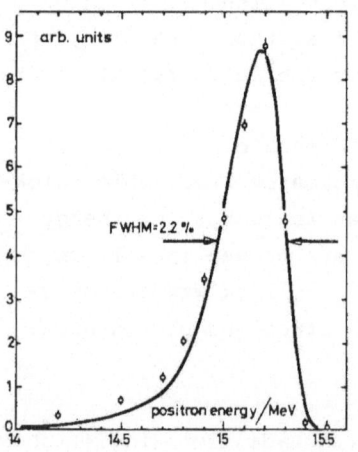

Fig.5. Results of an isochromate measurement (points with error bars), and of a Monte Carlo calculation (full line).

time (≥ 12 μs) is greater than the linac pulse length (4 μs) only one
event per pulse can be analyzed.

An isochromate measurement [15] was carried out using a 1 mm thick Be
annihilation target, a 20 mm thick polystyrene scattering target, and
eleven settings of positron energy from 14.7 to 16 MeV. A relative
normalilzation of the scattered photon numbers was made by dividing them
by the corresponding integrated currents from the P2 ionization chamber
and by the positron energy. The isochromate shape of the annihilation
photon peak was obtained by adding up the scattered photon numbers from
the four NaI-spectrometers for each positron energy. The result is given
in fig.5 together with a curve from a Monte-Carlo calculation. This
calculation takes into account the energy loss and multiple scattering
of the positrons in the annihilation target. The shape of the incident
positron energy spectrum is determined by the setting of the energy
slit, ES. This shape is taken to be rectangular with a width of 1%.
Since the positron source is not an ideal point source, an angular
distribution for the positrons hitting the annihilation target at a
certain point has to be taken into account. The diameter of the beam
defining collimator corresponds to an acceptance foreward angle of
$0.8°$ for the annihilation photons. The FWHM of the annihilation photon
peak is a function of the annihilation target thickness, positron-energy
and energy-spread, and photon collimation. In scattering experiments
this width is smaller than the FWHM of the response function of the
NaI-spectrometer.

The energy calibration was checked by means of the peak energy of the
isochromate (fig.5). Since the mean energy of the annihilation peak
corresponds to the energy to be assigned to measured cross sections,
this energy has to be used for the photon flux determination with the P2
ionization chamber. The calculated difference of this energy and the
positron energy E^+ is less than 0.2%. E^+ is calculated by means of
the magnetic fields in the transport system, which are measured by NMR.

In order to measure the response function of a NaI-spectrometer, the
spectrometer was positioned in the direct beam. The measurements were
taken with two annihilation targets (Be, Cu) and using electron and
positron beams. The targets could rapidly be changed, thereby averaging
over fluctuations of the incident beam. The results are given in fig.6.

The performance of the experimental set up described above shows the
feasebility of nuclear photon scattering measurements with a low duty

factor pulsed positron source in a wide energy range. The limitations
are at the low energy end at about 10 MeV the smallness of the available
positron current and at the high energy end at about 100 MeV the almost
vanishing difference in the counting rates for the two annihilation
targets of different Z. At high energies limitations also may occur due
to excessive piling-up of detector pulses.

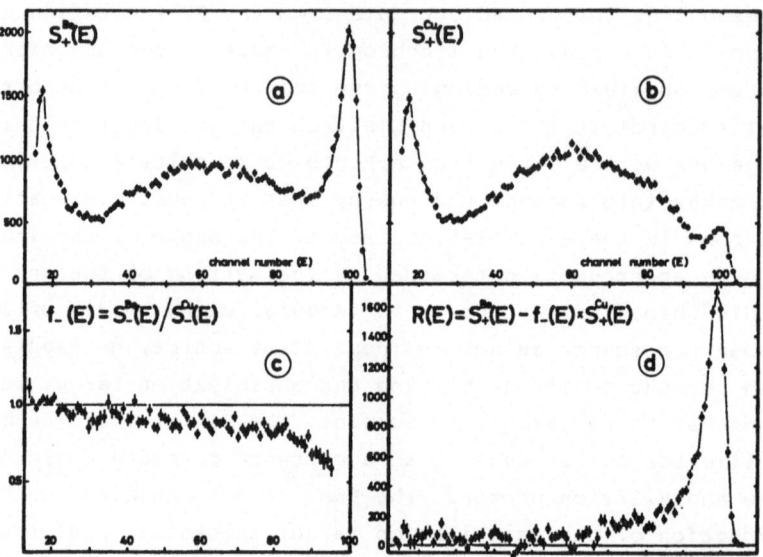

<u>Fig.6.</u> Results of a NaI-spectrometer response function measurement. The
positron energy was 20 MeV. The pulse height distributions a and b were
obtained with a positron beam on a Be- and a Cu-annihilation target
respectively. The ratio of the brems-spectra intensities from an electron beam on a Be- and a Cu-target as a function of photon energy is
shown in c. The response function obtained is given in d.

4. Elastic Scattering Angular Distribution

The two main physical effects determining the angular distributions are the multipole character of resonances and the diffractive scattering pattern of an extended nucleus. In a rather simplified description which however may contain the essential physics involved (see e.g. [16]), one may say, the nucleus looks different for electrons and photons: electrons see the net electric charge, whereas photons see the individual (e_i^2/M_i)'s of the nucleus' constituents. Exchange currents hardly influence elastic electron scattering, at least up to moderate momentum transfers, if one assumes about equal contributions from positive and negative charges, but the presence of charged mesons being exchanged roughly double the scattering cross section of photons. In other words: if the total absorption cross section provides the total integrated strength of the photon-nucleus interaction, then the diffraction pattern of elastic photon scattering, when combined with the absorption measurement, provides an image of the interaction strength density distribution.

Since for a fixed angle θ, photon energy k and momentum transfer q are directly related, the diffraction pattern, usually shown as the scattered intensity as a function of scattering angle, is best seen (and also measured) by the scattering cross section as a function of photon energy for a fixed angle. At higher energies (kR>1; E>30 MeV for ^{208}Pb) the diffraction pattern governs the angular distribution. In fig.7, a form factor $F_\gamma(q)$ was applied to the scattering cross sections as prescribed theoretically [17]. $F_\gamma(q)$ is the result of FOURIER-transforming a FERMI-2-parameter charge distribution. Six different values of the radius-parameter C were chosen. The skin thickness was kept constant equal to .54fm. Clearly a diffraction minimum can be seen the position of which wanders with the half density radius C. The occurence of this minimum is a typical characteristic feature of diffraction and can not easily be camouflaged by resonance interferences. There is therefore a good chance of a quite unambiguous determination of $F_\gamma(q)$ by scattering measurements between 40 MeV and 140 MeV photon energy at one or two angles.

In fig.8, the experimental data and the result of an application of scattering theory [1,17] is plotted. The same set of parameters describe absorption cross sections [6,7] and the new scattering data equally well without the need for an arbitrary adjustment in between.

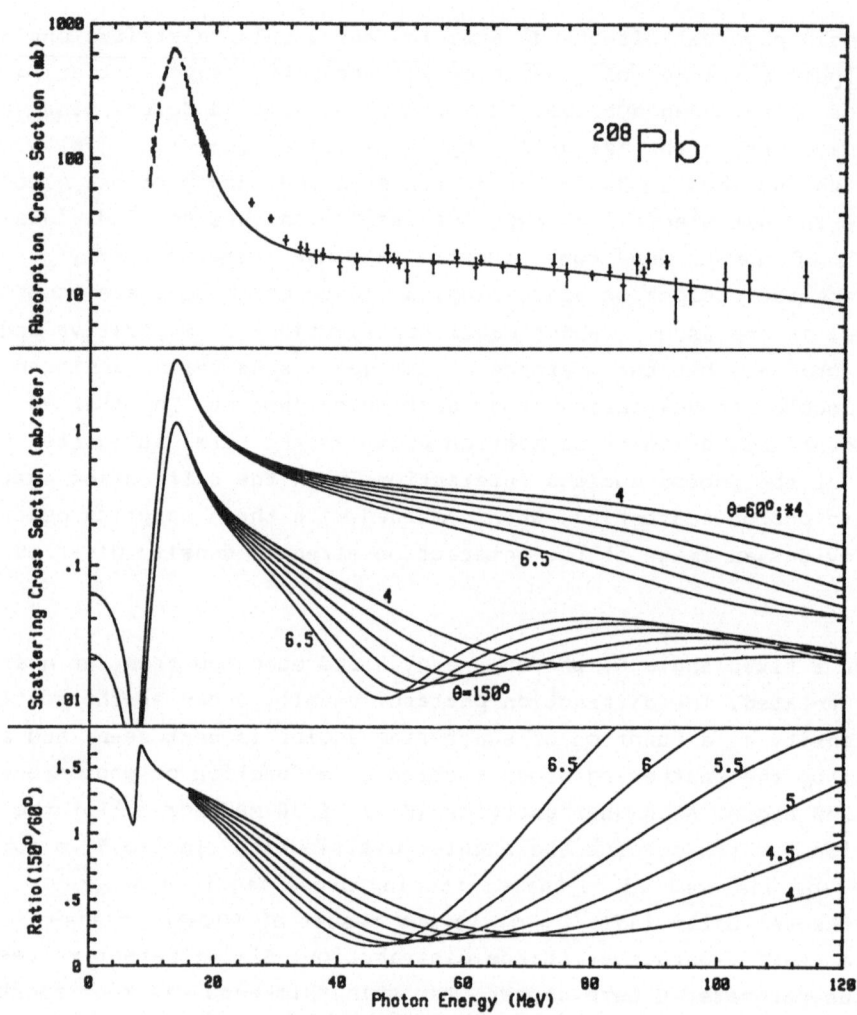

Fig.7. ^{208}Pb, diffraction scattering; the form factor $F_\gamma(q)$ was derived from a FERMI-2-parameter charge distribution with skin thickness t=.54 fm. The half density radius C was varied between 3 fm and 6.5 fm. The electron scattering value is C=6.6 fm. The absorption cross section is assumed to be pure E1 and to be given by two Lorentzians [6,7]. In this figure as well as in the following figures, all scattering cross sections for $\theta=60°$ have been multiplied by 4.

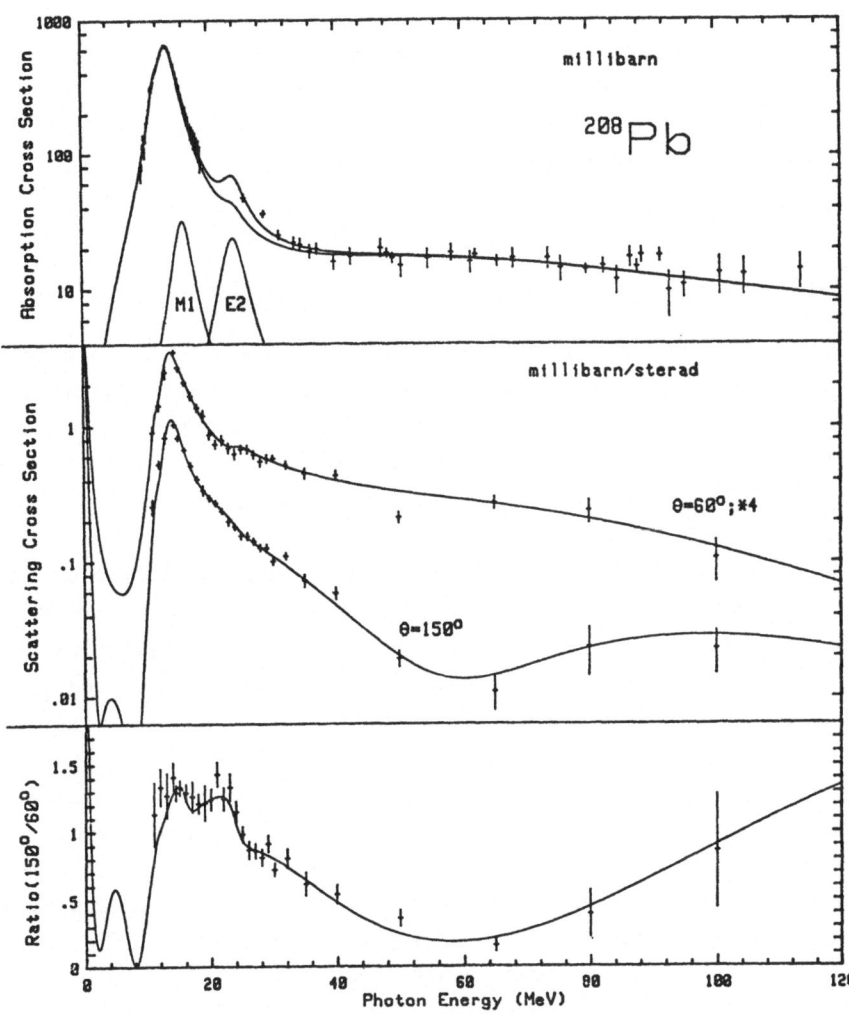

Fig.8. In this figure all experimental information on absorption [6,7] and scattering is summarized. The parameters, describing simultaneously absorption and scattering, are listed in table 1.

D_n in units of the classical dipole sum (3bMeV)

Table 1

	E_n, (MeV)	Γ_n, (MeV)	D_n
E1	11.62	1.3	0.07
	13.5	3.5	1.15
	26±2	8±2	0.05±0.01
	67±3	105±10	0.75±0.02
E2	24.3±0.4	4.5±0.5	0.05±0.015
M1 or E2	16±0.2	5±0.4	0.05±0.015
Sum of all multipoles:			2.12

The form factor giving the best fit is the FOURIER-transform of a
FERMI-2-parameter charge distribution with half-density radius
$C=(4.9\pm.15)$ fm (and $t=.54$ fm). The skin thickness (diffuseness) t is
taken from electron scattering data since the photon data do not justify
yet an independent determination.

5. The Delta-Resonance

The absorption cross section of the resonance part of the π-production
cross section is known experimentally [18,19]. The scattering angular
distribution however is most uncertain, mainly due to the question to
what extent the excitation of the nucleus becomes collective in character (cf.ref.[20]). Two extremes shall be considered here:

a) The transfer of excitation between bound nucleons is fast and strong
enough to set up a collective "giant Delta-resonance". The scattering
cross sections for $60°$ and $150°$, together with the total cross
section of 2 classical sums is plotted as a dashed line in fig.9. The
scattering cross sections in the Delta-resonance region, under these
conditions, are comparable in magnitude with the nuclear giant dipole
region.

b) If in contrast one assumes a Delta-decay fast enough to eliminate the
effects of an internucleonic coupling, then the elementary scattering
amplitudes of the Delta-particles show individually resonance behavior with M1 angular distributions and diffraction structure. In
this case, for the same total absorption cross section, the form
factor $F_\gamma(q)$ reduces the scattering cross section for angles
$\theta>60°$ almost to zero, as shown in fig.9 as the full line.

Compared to present experimental errors, we find the Delta-amplitude
being of negligible influence below 100 MeV photon energy. This result
enables the interpretation of scattering data below 100 MeV photon
energy without complication by more unknowns.

5. Conclusion

In a wide energy range, photo absorption data [6,7] and new scattering
cross sections for ^{208}Pb were described with one and the same set of
parameters. These parameters include the total integrated absorption
strength (taken from absorption cross sections), a partial strength,
width and position for the giant E2-resonance, the same parameters for a
resonance at 16 MeV, which may be M1 in character, and last but not

least, a value for a radius, which can be interpreted as the half density radius of all currents, including the exchange parts, interacting with photons. This latter value turned out to be 4.9 fm, which is considerably smaller than the charge distribution radius as measured by electron scattering (6.6 fm).

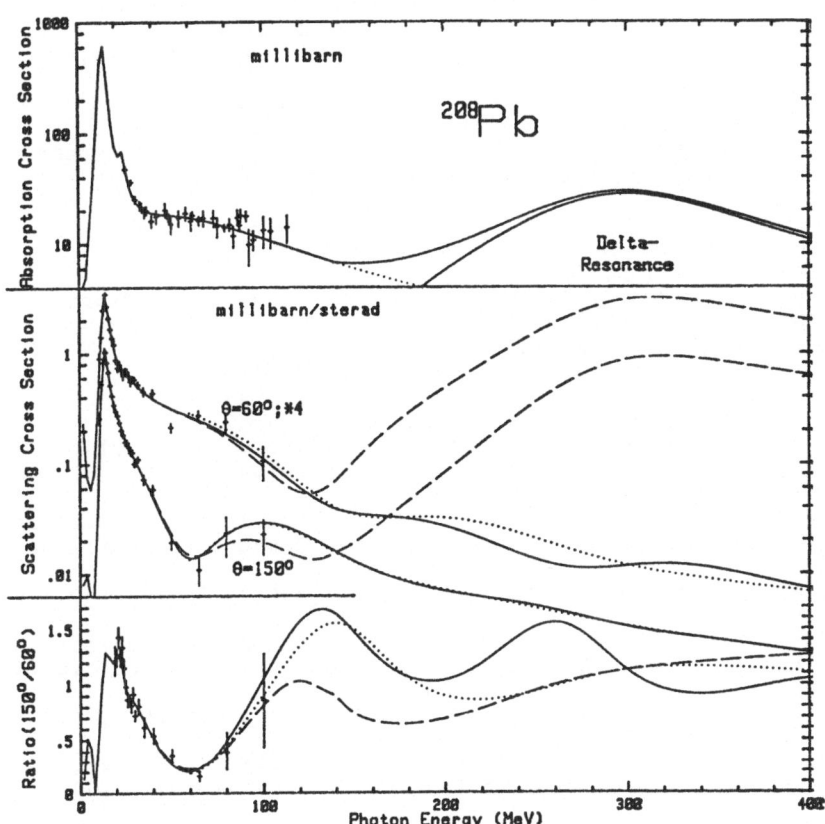

Fig.9. The energy range of fig.8 is extended to 400 MeV, including a M1-resonance with 2 classical dipole strengths at 300 MeV photon energy which imitates the Delta-resonance. The width is taken energy dependent (see expression (1) of ref.[18]), being zero below π-threshold and 140 MeV at resonance energy. The dashed line is scattering with strong internucleon coupling, "giant Delta-resonance": M1-angular distribution. The full line describes scattering by A uncoupled Delta-resonances: M1- and diffraction angular distribution. The dotted line does not contain a Delta-resonance at all. It can hardly be distinguished from the full line. It is assumed that the incoherent scattering by individual nucleons can be seperated out experimentally due to the larger energy shift of scattered photons, since for this elementary process the recoil momentum and also the separation energy of the order of 10 MeV is transferred to a single nucleon.

References

[1] E.G.Fuller and Evans Hayward, Phys.Rev. 101(1955)692
    H.Arenhövel, M.Danos and W.Greiner, Phys.Rev.15(1967)1109
[2] M.Langevin, J.M.Loiseaux and J.M.Maison, Nucl.Phys.54(1964)114
[3] J.Ahrens, H.Borchert, K.-H.Czock, D.Mehlig and B.Ziegler,
    Phys.Lett. 31B(1970)570
[4] G.Tamas, J.Miller, C.Schuhl and C.Tzara, J.Phys.Rad.21(1960)532
[5] J.O'Connel, R.A.Tipler and P.Axel, Phys.Rev. 126(1962)228
    R.M.Laszewski and P.Axel, Bull.Am.Phys.Soc. 22(1977)1022
    P.Axel, R.Starr and L.S.Cardman, Bull.Am.Phys.Soc.22(1977)1022
[6] A.Veyssière, H.Beil, R.Bergère, P.Carlos, and A.Leprêtre,
    Nucl.Phys.A159(1970)561
[7] A.Leprêtre, H.Beil, R.Bergère, P.Carlos, J.Fagot, A.Veyssière,
    J.Ahrens, P.Axel, and U.Kneissl, Phys.Lett.79B(1978)43
[8] J.Ahrens, H.Borchert, K.H.Czock, H.B.Eppler, H.Gimm, H.Gundrum,
    M.Kroening, P.Riehn, G.Sita Ram, A.Zieger, and B.Ziegler
    Nucl.Phys.A251(1975)479
[9] E.Hayward, in Photonuclear Reactions I, ed. S.Costa and
    C.Schaerf, Lecture Notes in Physics, Vol.61 (Springer,
    Heidelberg,1977), p.340
[10] W.R.Dodge, E.Hayward, R.G.Leicht, B.H.Patrick, and R.Starr,
    Phys.Rev.Letters 44(1980)1040
[11] C.Tzara, Compt.Rend. 245(1957)56
[12] A.Veyssière, H.Beil, R.Bergère, P.Carlos, J.Fagot,
    A.Leprêtre and J.Ahrens, Nucl.Instr.165(1979)417
[13] J.S.Pruitt and S.R.Domen, NBS Monograph 48(1962)
[14] H.Burckhart, R.Diehl and B.Ziegler, Nucl.Instr.159(1979)1
[15] G.Audit, N.de Botton, G.Tamas, H.Beil, R.Bergère,
    A.Veyssière, Nucl.Instr. 79 (1970) 203
[16] B.Ziegler, Proceedings Workshop on Intermediate Energy Nuclear
    Physics with Monochromatic and Polarized Photons,Frascati 1980
    ed. G.Matone
[17] P.Christillin and M.Rosa-Clot, Nuovo Cim.43A(1978)172, Physics
    Reports 64(1980)327
[18] B.Ziegler, in Nuclear Physics with Electromagnetic Interactions,
    ed.H.Arenhövel and D.Drechsel, Lecture Notes in Physics, Vol.108
    (Springer, Heidelberg, 1979), p.148
[19] H.Rost, Thesis Bonn University, Bonn-IR-80-10, May 1980, ISSN
    0172-8741
[20] M.Dillig and M.G.Huber, in Interactions Studies in Nuclei, ed.
    H.Jochim and B.Ziegler (North Holland, Amsterdam 1975)p.781

Acknowledgment

The author wishes to acknowledge the participation in this work of
R.Leicht, K.P.Schelhaas, M.Hammen and J.Ahrens.

NEUTRON SPECTROMETRY AND γ-RAY TRANSITIONS

C. Coceva
CNEN, Centro Studi e Ricerche "E. Clementel"
Bologna, Italy

Measurements of radiative transitions between individual compound nucleus states and states at lower energy have two distinctive properties. First, they allow to verify to what extent a collective dipole excitation built on an excited state may be considered to be equal to the same multipole excitation built on the ground state, observed in photo-absorption measurements. Second, they allow a verification of statistical properties of the transition matrix elements through a determination of the distribution function of the radiative widths.

It must be emphasized that, because of the essential complication of compound nucleus states, these experiments lend themselves to a theoretical interpretation only from a statistical point of view; this means that a research work in this field is meaningful only when a high enough number of statistically equivalent compound states can be measured. In other words, these states must have the same spin and parity and, to a good approximation, the same excitation energy. With present techniques, these conditions can be approached only by measuring the spectra of the gamma decays of individual levels excited by resonance neutron capture: the initial states of the radiative transitions are determined by measuring the time-of-flight of the captured neutron, the final states are determined by measuring with a Ge(Li) detector the energy of the emitted gamma-ray, according to the scheme of fig. 1. In this case, for each neutron resonance, the spectrum of the emitted gamma-ray must be measured. We shall refer to this **type** of experiment as to the "discrete resonance capture γ-ray" method. The first problem mentioned before, i.e. the possible dependence of transition probabilities on the final state, can be investigated also with the "average resonance capture γ-ray" method [1]. Here, only one γ-spectrum is measured, corresponding to all neutrons captured in an energy interval containing a high number of resonances (having different spins), enough to average out statistical fluctuations of γ-transition intensities.

In nuclei with $A \gtrsim 60$, that is in those nuclei which can be thought to have a statistical behaviour at an excitation equal to the neutron binding energy, neutron resonances which can be resolved by time-of-flight lie in a very limited energy range, of the order of 0.1 to 10 keV. Correspondently, the low-lying states which can be resolved with Ge(Li) detectors have generally an excitation below 2.5 MeV. As a consequence, the energies of the gamma transitions which can be studied by this method range mostly from 5 to 9 MeV.

An example of the experimental set-up used for discrete resonance measurements is shown in fig. 2. Here, the conditions are indicated in which a measurement was performed by our group at the Geel linear accelerator. The pulsed neutron source can be realized also by means of a fast mechanical chopper placed at a beam-hole of a

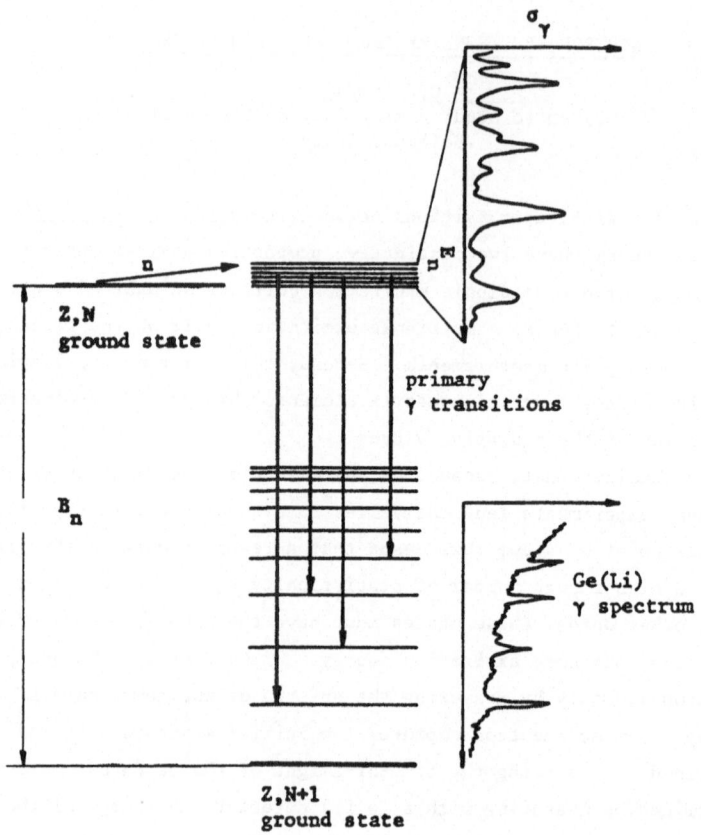

Fig. 1 - Excitation by neutron capture and gamma decay of compound-nucleus states.

high-flux reactor: although an apparatus like this has a much worse timing resolution, part of our information comes from the work done with fast choppers at Brookhaven and Chalk River.

More intense neutron sources are now available, like the same Geel accelerator, which has recently been upgraded, the new Harwell linac and the Oak Ridge linac, which can reach a beam power one order of magnitude higher than in the case illustrated in fig. 2.

However let us see why it is not easy to get substantial improvements. One difficulty comes from the fact that the partial radiative width relative to a particular final state is in general a very small fraction of the total radiative width of the initial state: in fact the branching ratio $<\Gamma_{\gamma i}/\Gamma_\gamma>$ may be of the order of 10^{-3}. As a consequence of this fact and of the characteristics of the response function of Ge(Li) detectors, it happens that out of $10^4 - 10^5$ gamma rays emitted in one resonance and causing a pulse in the detector, perhaps only one may be used for intensity determination of a particular transition. Since, to get a good resolution, each pulse must occupy a rather long detection time, e.g. a few microseconds, it is evident that se-

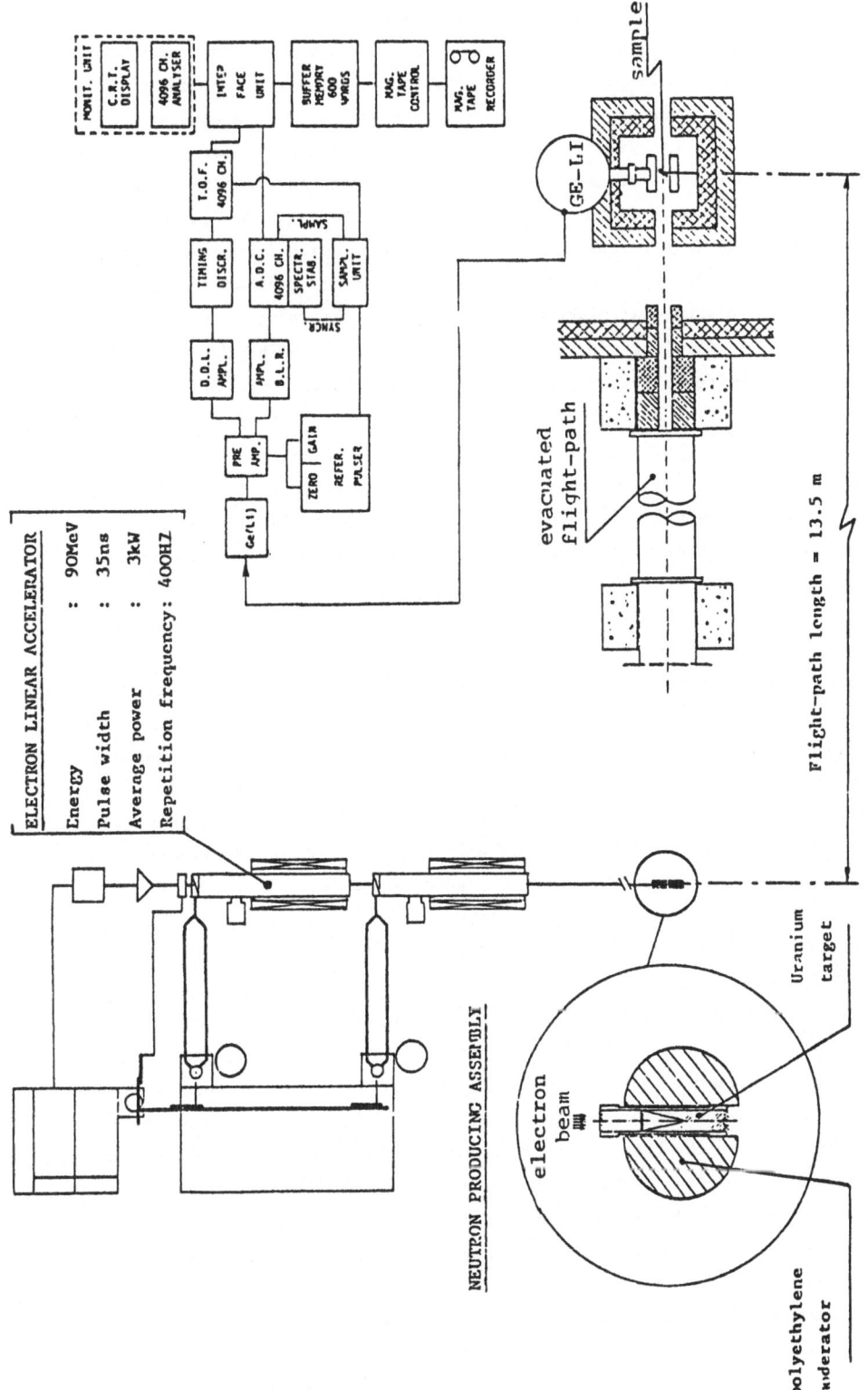

Fig. 2 - Experimental set-up for the measurement of gamma spectra from discrete neutron resonances. The experimental conditions are those of ref. 2.

vere pulse pile-up problems may arise, even when the useful counting rate is very low: a typical value may be of the order of one pulse per hour for a single transition energy. I recall that, in a discrete resonance measurement, typically the intensities of about one thousand γ-transitions are measured simultaneously.

In a measurement, the quality of the gamma spectra becomes worse with increasing neutron energy. Such an effect arises from two concurrent factors: the lower neutron flux per unit energy, which in an apparatus like that of fig. 2 decreases as $E^{-0.9}$, and the larger energy interval per time-of-flight unit, which increases as $E^{3/2}$. Then, in spite of the lower number of neutrons captured per resonance, pulse pile-up problems become more severe at higher energies because the resonances are more closely packed in time. If, for instance, we are in a situation in which the useful number of resonances is limited by pile-up problems, in order to increase this number by a factor 2, we must work on a flight path 2.8 times longer and increase the intensity of the neutron source by a factor 15, approximately. This example illustrates the fact that any improvement is very very expensive. At present, the sensitivity of this kind of measurements is such that, in general, only dipole radiation can be detected; measurements of E2 transitions could be made only in very few favourable cases.

Let us consider all resonances of given spin which can decay by gamma emission of given multipolarity, according to the appropriate selection rules, to a certain final state. From a statistical point of view, the radiative widths corresponding to these transitions may be considered to be sampled from the same population. Their distribution, because the amplitudes are essentially real quantities and because of the random behaviour of the matrix elements, should be a χ^2 function with $\nu=1$ degree of freedom (Porter-Thomas distribution). This behaviour, whose verification we shall see later, gives rise to further difficulties for a check of the distribution and for a determination of the average value of the radiative widths. In fact, as a consequence of the peculiar shape of such a distribution, which has a pole at $\Gamma_{\gamma i}=0$, in a high percentage of resonances the considered transition has a so low intensity that it fails to be observed.

To increase the statistical accuracy, resonances of different spins are usually included in the same statistical sample by taking into account that, according to elementary considerations, the average radiative width should be proportional to the spacing of the states characterized by the same set of quantum numbers as the initial one. However, even in the best discrete resonance measurements performed up to now, the average widths to given final states could be measured with rather low precision, sometimes just enough to distinguish between E1 and M1 multipolarity. As a matter of fact, the best case [2] is that of ^{177}Hf(n,γ) in which the gamma spectra of 38 resonances were measured (see tab. 1).

For this reason, average widths are determined making use of all observed transitions having the same multipolarity, to all resolved final states. A reduction formula must therefore be applied to take in due account the different energies of the

gamma rays. To do so in a correct way, two additional conditions must be verified.
1) The γ-energy dependence of the average transition matrix element must be known. This dependence has to be combined with the E^{2L+1} proportionality factor deriving from the expression of the density of photon states in the phase space (L is the multipole order of the radiation).
2) The average reduced radiative width must not depend on the particular final state considered.

Let us review briefly the state of our experimental knowledge on these two points and also how the measured average widths compare with the expectations.

In the case of electric dipole radiation, the reduction formula is based on the Axel-Brink hypothesis, which states that, if E_x is the excitation of the final state, the transition probability is described by a giant resonance having the same shape as the giant photo-absorption resonance, but with the peak energy displaced upwards by E_x, irrespective of the particular configuration of the final state. A theoretical justification of this hypothesis was given by Rosenzweig [3] in the frame of the hydrodinamical model. But admittedly this is only a gross-feature prediction. In fact, the energy range accessible to this experimental method is well below the peak of the giant dipole resonance (GDR) by one to three times its width, that is where the tail of the Lorentzian shape of the GDR is not expected to describe accurately the photoabsorption cross section. Here, the existence of intermediate structure is a known experimental fact. Here also, Lane and Lynn [4], on the basis of photoreaction data, suggested to apply a reducing factor of the form $\exp a(E_\gamma - E_R)$ to the extrapolated Lorentzian shape (E_R is the peak energy of the GDR).

Data from average resonance experiments [1] show that the energy dependence of the widths is in qualitative agreement with the Axel-Brink hypothesis, but a result [5] obtained in discrete resonance capture in ^{181}Ta, in disagreement with the above hypothesis, demonstrates that detailed experimental checks of the energy behaviour are needed.

As for the absolute values of the reduced widths, a review [2] of experimental data, taken from discrete resonance experiments, concludes that the widths are, on the average, 30% below the predictions deduced from the known parameters of the GDR, in qualitative agreement with the above mentioned suggestion of Lane and Lynn; it concludes also that the A-dependence is well described on the basis of the Axel-Brink hypothesis.

The verification of the independence of the width on particular configurations of the final states is difficult because of the large statistical errors when one has to average only on the resonances, for a fixed final state. However, the clarification of this question deserves an experimental effort in view of the following experimental results concerning deformed nuclei. In the first one [2], the reduced E1 widths of ^{178}Hf (fig. 3) are averaged over final states belonging to rotational bands having the same K quantum number. The different strength of the transitions leading to rotational states with different K values is evident. In the second case [6], the experimental data obtained for ^{174}Yb (fig. 4), although with a worse statistical precision,

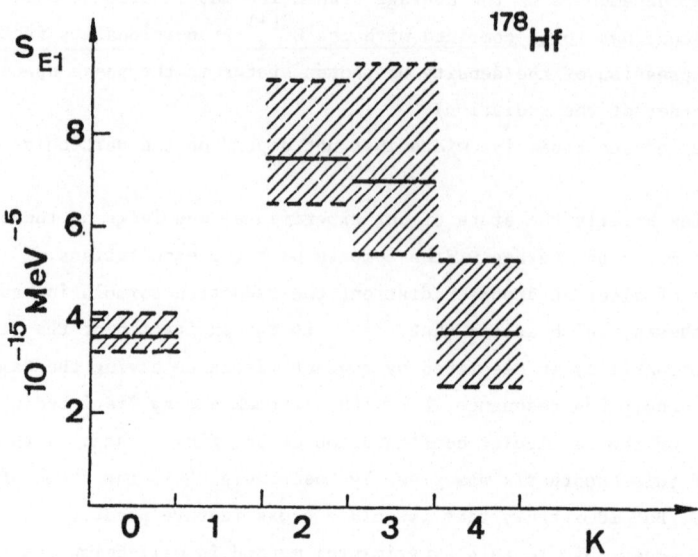

Fig. 3 - Reduced E1 strength $S=<\Gamma_{\gamma ij}/DE_\gamma^5>A^{-8/3}$ for different values of the K quantum number of the final states of discrete resonance gamma-rays in ^{178}Hf. Confidence limits are indicated by shaded areas [2].

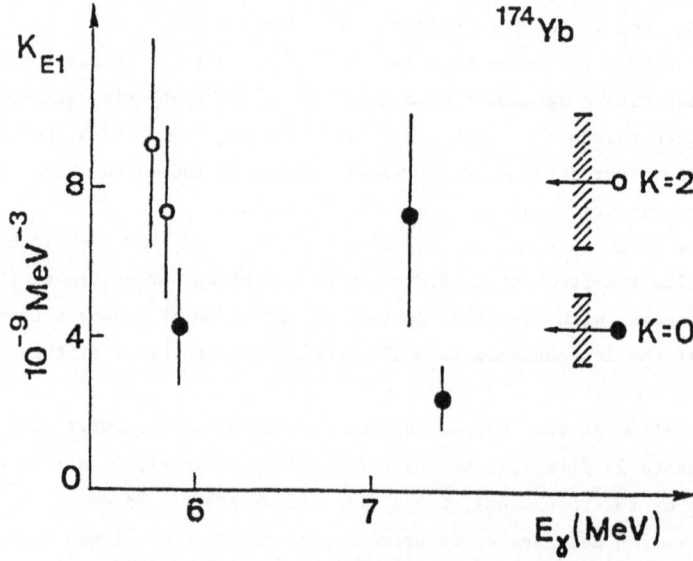

Fig. 4 - Reduced E1 widths $k=<\Gamma_{\gamma ij}/DE_\gamma^3>A^{-2/3}$ for different final states of discrete resonance gamma rays in ^{174}Yb. The average values on three levels of the ground state K=0 band and on two levels of the K=2 γ-vibrational band are indicated by arrows [6].

Table 1

Number of degrees of freedom of the χ^2 function fitting the experimental distribution of reduced E1 and M1 widths.

Nucleus	ν_{E1}	ν_{M1}	References
^{100}Ru		$0.93^{+0.57}_{-0.33}$	Rimawi et al.; Phys. Rev. C9(1974)1978
^{102}Ru		$0.98^{+0.50}_{-0.33}$	
^{106}Pd		$0.94^{+0.26}_{-0.24}$	Coceva et al.; Nucl. Phys. A170(1971)153
^{116}In	$1.42^{+0.14}_{-0.08}$	$1.10^{+0.27}_{-0.09}$	Corvi, Stefanon; Nucl. Phys. A233(1974)185
^{150}Sm	$1.34^{+0.15}_{-0.14}$		Bečvář et al.; Nucl. Phys. A236(1974)198
^{176}Lu	$1.56^{+0.30}_{-0.23}$		Wasson, Chrien; Phys. Rev. C2(1970)675
^{178}Hf	$1.38^{+0.18}_{-0.13}$	$1.15^{+0.60}_{-0.40}$	Stefanon, Corvi; Nucl. Phys. A281(1977)240
^{182}Ta	$1.38-0.11$		Stelts, Browne; Phys. Rev. C16(1977)574
^{196}Pt	$1.35^{+0.43}_{-0.39}$		Samour et al.; Nucl. Phys. A121(1968)1170

show the same behaviour. Stefanon and Corvi [2] suggest that a K-dependence should imply the presence of residual rotational features in compound nucleus resonances. On the other hand, the measured effect might be a consequence of a dependence of the shape of the GDR on particular features of the nuclear surface, which are different for different collective bands. The existence of nuclear surface effects is known for light nuclei; the above results might suggest that these effects are present also in the region of deformed nuclei.

As regards magnetic dipole transitions, due to their lower intensity, the experimental data are more uncertain. The A- and energy-dependence, according to which the widths are usually reduced, are those predicted by Blatt and Weisskopf on the basis of a simple single-particle model. In a recent review [7] of McCullagh, it is estimated that, on the average, the experimental $\overline{B}(M1)$ are enhanced over the Weisskopf B(M1) values by a factor 1.4.

Our knowledge of the energy behaviour of M1 transitions comes essentially from the average resonance capture data of Bollinger [1], which indicate a variation definitely stronger than the E^3 law predicted by the Weisskopf model. Both the enhanced ratio of M1 to E1 strengths, and the energy behaviour suggest that the explored energy range lyes on the low-energy side of a giant M1 resonance. The need for more experimental data, also, for instance, with electron scattering experiments, for an understanding of the giant resonance behaviour of M1 transitions is quite evident.

The measurement of the gamma-decay spectra of discrete neutron resonances is, as I said before, a unique tool for an investigation of the statistical behaviour of transition matrix elements. The experimental distribution of the reduced widths, which is little affected by the uncertainty of the reduction formulae, is usually analysed in terms of number of degrees of freedom ν of the χ^2 function which best reproduces the data. The most significant results are summarized in tab. 1. The consistency of the deviation from the predicted Porter-Thomas distribution in the electric dipole case is quite impressive. It is true that these results must be taken with some caution since a failure in resolving final states or neutron resonances both lead to an overestimate of the number of degrees of freedom. However the same sources of error should be present in M1 transitions which, on the contrary, group nicely around the $\nu=1$ value. Implications of a more uniform distribution than the Porter-Thomas one were discussed by some authors (see for instance ref. 8) and also a theoretical model was proposed by Rosenzweig [9]. However, at this stage, I think that it is not justified to draw any conclusion of physical nature; rather, the E1 data presented in the table should be considered as a challange to experimentalists for their future work.

References

1) L.M. Bollinger and G.E. Thomas, Phys. Rev. C2(1970)1951
2) M. Stefanon and F. Corvi, Nucl. Phys. A281(1977)240
3) N. Rosenzweig, Nucl. Phys. A118(1968)650
4) A.M. Lane and J.E. Lynn, Nucl. Phys. 11(1959)646
5) M.L. Stelts and J.C. Browne, Phys. Rev. C16(1977)574
6) S. Raman and M. Stefanon, private communication
7) C.M. McCullagh, Thesis, State University of New York at Stony Brook, May 1979
8) F. Bečvář, R.E. Chrien and D.A. Wasson, Nucl. Phys. A236(1974)198
9) N. Rosenzweig, Phys. Lett. 6(1963)123

PHOTONUCLEAR PHYSICS WITH SYNCHROTRON RADIATION

W.M.Alberico and A.Molinari

Istituto di Fisica Teorica dell'Università di Torino, Torino, Italy
and
Istituto Nazionale di Fisica Nucleare, Sezione di Torino, Torino, Italy

1. The synchrotron radiation

The properties of the synchrotron radiation (S.R.) associated with LEP, the large electron-positron storage ring to be built at CERN, are listed in table I [1].

Table I

E (GeV)	I (mA)	B (T)	ϵ_c (MeV)	dP/ds (kW/m)	$\left.\frac{d^3 n}{d\phi\, d\psi\, d\epsilon/\epsilon}\right|_{\substack{\psi=0 \\ \epsilon=\epsilon_c}}$ (s mrad^2 %bw)$^{-1}$
86	9.15	.081	0.40	0.56	$\sim 6.9 \times 10^{15}$
130	6.11	.123	1.37	1.96	$\sim 1.3 \times 10^{16}$

Synchrotron radiation from the LEP bending magnets. Energy (E) and current (I) of the electron beam; magnetic field of the bending magnets (B); photon critical energy (ϵ_c) and total power radiated per m orbit length (dP/ds); central brightness of the photon beam radiated per second and mrad^2 of horizontal and vertical angle into a relative bandwidth of 1% at the critical energy.

The critical energy is so defined

$$\epsilon_c = \frac{3}{2} \hbar c \frac{\gamma^3}{\rho} \qquad (1.1)$$

where γ is the relativistic electron factor and the radius of curvature in the LEP bending magnets is ρ = 3544.5 m. Although the photons of the S.R. duly represent a beam of phantastic intensity and collimation and almost complete polarization (linear in the LEP plane), they are too soft for nuclear physics studies.

The application of wiggler magnets, without substantially affecting the performances of LEP for its main purposes, drastically improves the situation, making available polarized photons in the energy range up to 100 MeV (see table II and Fig. 1).

Table II

Wiggler	E (GeV)	I (mA)	B (T)	ϵ_c (MeV)	$\Delta\phi_B$ (mrad)	P_w (kW)	Intensity (s mrad^2 %bw)$^{-1}$
normal magnet	86	9.15	2	9.8	0.35	17.3	$\sim 4 \times 10^{15}$
	130	6.11	2	22.5	0.23	26.3	$\sim 9 \times 10^{15}$
supercond. magnet	86	9.15	4	19.7	1.2	116	$\sim 4 \times 10^{15}$
	130	6.11	4	45.	0.8	177	$\sim 9 \times 10^{15}$

Parameters of the radiation emitted from wiggler magnets: critical energy ϵ_c, horizontal bending angle $\Delta\phi_B$, total power P_w, intensity of photons per second per mrad^2 into 1% bandwidth.

The natural experiments to be performed appear then Compton and Raman nuclear scatterings, the least investigated owing to the poor intensities of the beams available in the past.

High quality, and therefore really useful, experiments should however be done with monochromatic photons, hard to obtain from the continuous S.R. spec-

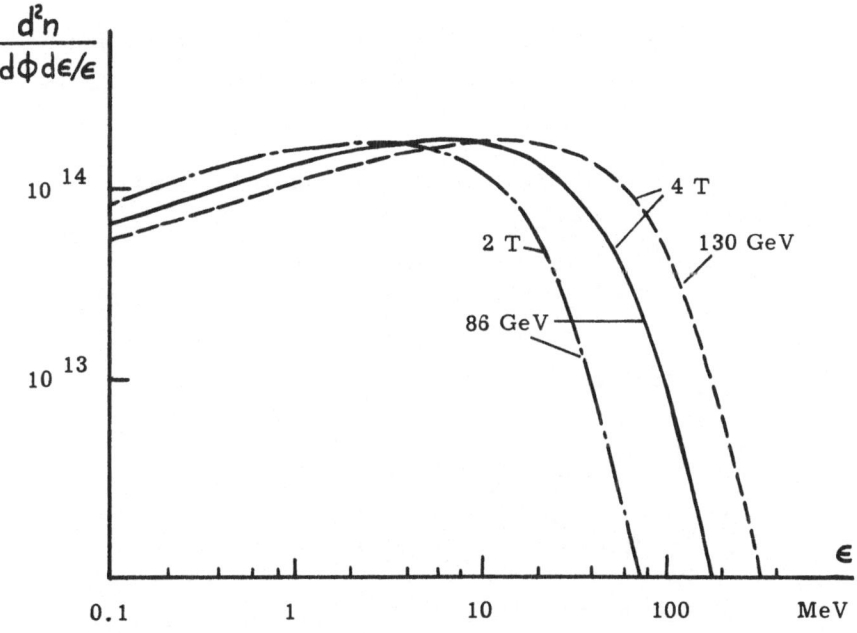

Fig. 1 Spectrum of the radiation from wiggler magnets in LEP.

trum. An undulator partially helps, but the price to be payed in the photon energy is high. A real solution to the problem is provided by the Bragg scattering of the S.R. out of a cristal [(2)] but this method works only in the low energy range, since the intensity of the scattered photons drops as the inverse square of their energy.

Leaving aside a detailed discussion of these difficulties and in the hope that they will be overcome, we deal here shortly with the problem of elastic scattering of photons.

2. The photon-nucleus scattering amplitude

For the interaction Hamiltonian

$$\hat{H}_{int} = -\int d\vec{r}\, j_\mu(\vec{r},t) A^\mu(\vec{r},t) \tag{2.1}$$

where $A_\mu = (\varphi, \vec{A})$ is the four-vector electromagnetic potential and $j_\mu = (\rho, \vec{j}/c)$ the four-vector nuclear current, the scattering matrix, to the lowest order, reads

$$S^{(2)}_{fi} = -\frac{2\pi c^2}{V\hbar\sqrt{\omega\omega'}} \epsilon_\mu \epsilon'_\nu \iint d^4x\, d^4y\, e^{-ik\cdot x} e^{ik'\cdot y} \langle f|T[j^\mu(x) j^\nu(y)]|i\rangle \tag{2.2}$$

$k_\mu = (\vec{k}, \omega/c)$ and $k'_\nu = (\vec{k}', \omega'/c)$ being the four-momenta of the initial and final photons, with polarizations ϵ_μ and ϵ'_ν respectively and V the normalization volume.

The charge-current density, which obeys the continuity equation

$$\frac{\partial}{\partial x_\mu} j_\mu(\vec{r},t) = 0, \tag{2.3}$$

can be expressed, for extended nucleons and neglecting exchange currents, as follows

$$\rho(\vec{r}) = e\sum_{K=1}^{A} \{\tfrac{1}{2} + t_z(K)\} f_E(|\vec{r}-\vec{r}_K|), \tag{2.4a}$$

$$\vec{j}(\vec{r}) = \frac{e}{2}\sum_{K=1}^{A} \{\tfrac{1}{2} + t_z(K)\}[\vec{v}_K f_E(|\vec{r}-\vec{r}_K|) + f_E(|\vec{r}-\vec{r}_K|)\vec{v}_K] +$$

$$+ \sum_{K=1}^{A} \frac{e\hbar}{2M} g_s(K) \vec{\nabla} \times \vec{s}_K f_M(|\vec{r}-\vec{r}_K|) \tag{2.4b}$$

where the spin g_s factor is defined in terms of the corresponding quantities for the proton $g_p = 2\mu_p = 5.59$ and for the neutron $g_n = 2\mu_n = -3.82$

$$g_S = \frac{1}{2}(g_p + g_n) + t_z(g_p - g_n) \qquad (2.5)$$

and

$$f_E(|\vec{r} - \vec{r}_\kappa|) = f_M(|\vec{r} - \vec{r}_\kappa|) = \frac{1}{\pi}\left(\frac{\lambda}{2}\right)^3 e^{-\lambda|\vec{r} - \vec{r}_\kappa|} \qquad (2.6)$$

is the Fourier transform of the nucleon charge and magnetic form factors

$$G_{E_p}(q^2) \approx \frac{G_{M_p}(q^2)}{\mu_p} \approx \frac{G_{M_n}(q^2)}{\mu_n} \approx \frac{1}{[1 + q^2/\lambda^2]^2} \qquad (2.7)$$

which are well reproduced up to $q \approx 25$ fm^{-1} with $\lambda^2 = 18.1$ fm^{-2}.

With a few manipulations (2.2) can be recast into the form

$$S_{fi}^{(2)} = -\frac{2\pi\hbar c^2}{V\sqrt{\omega\omega'}} 2\pi i \delta(E_o + \hbar\omega - E_{N^*} - \hbar\omega')\epsilon_\mu \epsilon'_\nu T^{\mu\nu} \qquad (2.8)$$

with

$$T^{\mu\nu} = \int d\vec{x}\, d\vec{y}\, e^{i\vec{k}\cdot\vec{x}} e^{-i\vec{k}'\cdot\vec{y}} \cdot$$

$$\sum_n \left\{ \frac{\langle N^*|j^\nu(0,\vec{y})|n\rangle\langle n|j^\mu(0,\vec{x})|0\rangle}{E_o - E_n + \hbar\omega + i\Gamma_n/2} + \frac{\langle N^*|j^\mu(0,\vec{x})|n\rangle\langle n|j^\nu(0,\vec{y})|0\rangle}{E_o - E_n - \hbar\omega' + i\Gamma_n/2} \right\} \qquad (2.9)$$

It is graphically shown (terms "a" and "b") in Fig.2.

The scattering amplitude (2.8) is easily shown <u>not</u> to be gauge invariant since

$$k_\mu k'_\nu \int d^4x\, d^4y\, e^{ik\cdot x} e^{-ik'\cdot y} \langle N^*|T[j^\mu(x)j^\nu(y)]|0\rangle =$$

$$= \frac{i}{\hbar} \int d^4x\, d^4y\, e^{+ik\cdot x} e^{-ik'\cdot y} \delta(x_o - y_o) \langle N^*|[\rho(y),[\hat{H},\rho(x)]]|0\rangle \qquad (2.10)$$

and the addition of a counterterm $S^{\mu\nu}$ such that

$$k_\mu k'_\nu (T^{\mu\nu} + S^{\mu\nu}) = 0 \qquad (2.11)$$

is needed to restore gauge invariance. However (2.11) fixes $S^{\mu\nu}$ only in the limit $\omega \to 0$, $\omega' \to 0$ ("low energy theorems").

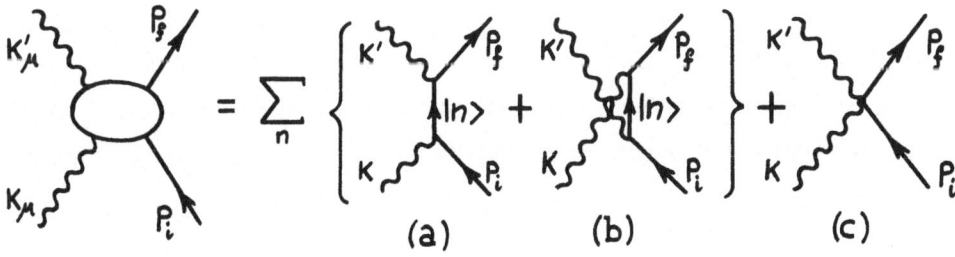

Fig.2 The amplitude for the scattering of a photon out of a nucleus

So, for a nuclear Hamiltonian with ordinary (Wigner) two-body forces, one gets [3]

$$S_{\mu\nu} \equiv S_{\mu\nu}^{kin} = -\frac{e^2}{Mc^2} \delta_{\mu\nu} \langle N^* | \sum_{i=1}^{A} \{\tfrac{1}{2} + t_z(i)\} e^{-i(\vec{K}'-\vec{K})\cdot\vec{r}_i} |0\rangle + \mathcal{O}\left(\frac{\hbar\omega}{Mc^2}, \frac{\hbar\omega'}{Mc^2}\right) \quad (2.12)$$

whereas if also exchange (Majorana) forces are present

$$S_{\mu\nu} = S_{\mu\nu}^{kin} + S_{\mu\nu}^{exch} \quad (2.13)$$

with

$$S_{\mu\nu}^{exch} = -\frac{2e^2}{(\hbar c)^2} \langle N^* | \sum_{i\neq j} \{t_z(i) - t_z(j)\}^2 (r_i - r_j)_\mu (r_i - r_j)_\nu V_M(|\vec{r}_i - \vec{r}_j|) \cdot$$
$$\cdot P_{ij} \exp\{-i(\vec{K}-\vec{K}')\cdot(\vec{r}_i + \vec{r}_j)/2\} |0\rangle + \mathcal{O}\left(\frac{\hbar\omega}{Mc^2}, \frac{\hbar\omega'}{Mc^2}\right) \quad (2.14)$$

Clearly (2.13) stems from the equal time contribution to the time-ordered product (2.2): thus $S_{\mu\nu}$ is equivalently referred to as "gauge" term or "contact" term (compare Fig.2, term "c").

To ascertain its role in the $\omega \to 0$ limit, the corresponding expressions of the direct (resonant) and crossed terms (2.9) of the scattering amplitude are needed and easily shown to be

$$\epsilon_\mu \epsilon'_\nu T^{\mu\nu} = \frac{1}{(\hbar c)^2} \langle 0 | [[\vec{\epsilon}\cdot\vec{D}_g, \hat{H}], \vec{\epsilon}'\cdot\vec{D}_g] | 0 \rangle = \vec{\epsilon}\cdot\vec{\epsilon}' \frac{e^2}{Mc^2} \frac{NZ}{A} \quad (2.15)$$

where

$$\vec{D}_g = e \sum_{i=1}^{A} \{\tfrac{1}{2} + t_z(i)\} \vec{r}_i \quad (2.16)$$

is the dipole operator with coordinates referred to the center of mass of the nucleus.

By combining (2.16) with (2.12) one obtains the zero frequency Thomson amplitude

$$\epsilon_\mu \epsilon'_\nu (T^{\mu\nu} + S^{\mu\nu}) = \vec{\epsilon}\cdot\vec{\epsilon}' \frac{e^2}{Mc^2}\left(\frac{NZ}{A} - Z\right) = -\vec{\epsilon}\cdot\vec{\epsilon}' \frac{(Ze)^2}{AMc^2} \quad (2.17)$$

which gets the dominant contribution from the gauge term. Thus the gauge invariant amplitude (2.17) turns out to be proportional to $(Ze)^2$, a fact of far reaching consequences. Indeed the Born approximation to the scattering amplitude of electrons out of nuclei is proportional to Ze^2 !

3. The structure constants of the nucleus

The next term in the energy expansion of the scattering amplitude, which we consider here in the case of spin zero nuclei without very low lying excited states, provides important informations on the structure of atomic nuclei. Keeping consistently terms up to ω^2 (the linear ones vanish) in the expansion of (2.9) and of (2.12) one gets [4]

$$\epsilon_\mu \epsilon'_\nu (T^{\mu\nu} + S^{\mu\nu}) =$$
$$= \vec{\epsilon} \cdot \vec{\epsilon}' \left\{ \frac{(Ze)^2}{AMc^2} \left[1 - \frac{1}{3} \frac{\omega^2}{c^2} \langle r^2 \rangle \right] - \alpha \frac{\omega^2}{c^2} \right\} - (\vec{\epsilon}' \times \hat{k}') \cdot (\vec{\epsilon} \times \hat{k})(\chi_P + \chi_D) \frac{\omega^2}{c^2} \quad (3.1)$$

where

$$\alpha = 2 \sum_n \frac{|\langle 0 | D_q^z | n \rangle|^2}{E_n - E_0}, \quad (3.2)$$

$$\chi_P = 2 \sum_n \frac{|\langle 0 | \mathcal{M} | n \rangle|^2}{E_n - E_0} \quad (3.3)$$

and

$$\chi_D = -\frac{1}{6} \frac{Ze^2}{Mc^2} \langle r^2 \rangle + \frac{\langle \vec{D}^2 \rangle}{6 A M c^2} \quad (3.4)$$

are the static nuclear electric dipole polarizability, paramagnetic and diamagnetic susceptibilities, respectively, and $\langle r^2 \rangle$ is the mean square radius of the nucleus. In (3.3)

$$\mathcal{M} = i \vec{\epsilon} \cdot \sum_{j=1}^{A} \left\{ \frac{e\hbar}{2Mc} \left[g_s(j) \vec{s}_j + \left(\frac{1}{2} + t_z(j) \right) \vec{\ell}_j \right] \right\} \quad (3.5)$$

is the magnetic dipole operator.

The electric dipole polarizability and the paramagnetic susceptibility can also be obtained through the inverse-square energy weighted sum rules

$$\sigma_{-2}(E1) = \int \sigma_{abs}^{E1}(\omega) \frac{d(\hbar\omega)}{(\hbar\omega)^2} = \frac{2\pi^2}{\hbar c} \alpha \quad (3.6)$$

$$\sigma_{-2}(M1) = \int \sigma_{abs}^{M1}(\omega) \frac{d(\hbar\omega)}{(\hbar\omega)^2} = \frac{2\pi^2}{\hbar c} \chi_P \quad (3.7)$$

For nuclei with $A > 100$, where photoneutron data provide a reliable estimate of the total photoabsorption cross section up to the meson threshold, the experimental value is

$$\sigma_{-2} = (2.7 \pm 0.2) A^{5/3} \; \mu b/MeV \quad (3.8)$$

which implies

$$\alpha \cong 2.7 \, A^{5/3} \, \text{fm}^3 \tag{3.9}$$

if only dipole transitions are involved. For light nuclei α is larger and the dependence upon the mass number more complicated [5].

Although the time-dependent Hartree-Fock theory with realistic interactions (Skyrme forces) accounts for this trend, an independent measure of α would be desirable since in the energy range up to the mesonic threshold multipolarities other than E1 certainly play a role.

The discussion of this section suggests that the natural candidate to accomplish this task is the photon elastic scattering, i.e. Compton scattering at low frequency. By exploiting the differences in the angular distributions and the polarization of the incoming photons, it seems possible to disentangle the electric and the magnetic contributions to the scattering amplitude (3.1), thus providing a measurement of α and $\chi_P + \chi_D$.

The experiment appears difficult partly because the effect is small [6] (for example in ^{16}O at a photon energy of 5 MeV the nuclear electric Rayleigh scattering is only 6.8 % of the Thomson one) and partly because it is not easy to bring under firm control the concomitant Delbrück and Rayleigh atomic processes. The Bragg scattered S.R. of LEP should be "the tool" to carry this experiment out, also because, by simply turning the wiggler, beams of photons linearly polarized parallel and perpendicular to the LEP plane can be obtained.

It is worth pointing out that at present almost nothing is known experimentally on the magnetic susceptibilities. A shell model estimate of χ_P for a spin saturated core with a population n_+ of the single particle level $j = l + 1/2$ and n_- of the level $j = l - 1/2$ reads [7]

$$\chi_P = \frac{2}{3} \left(\frac{e\hbar}{2Mc} \right)^2 (g_\ell - g_s)^2 \frac{n_+ \ell - n_- (\ell+1)}{\Delta (2\ell+1)} \tag{3.10}$$

where Δ is the energy of the spin-orbit splitting and g_l the giromagnetic orbital factor.

Although spin resonances are likely to affect (3.10), nevertheless spin saturated nuclei like ^{4}He, ^{16}O, ^{40}Ca, etc., should be essentially diamagnetic, thus permitting a measurement of χ_D. In this connection it should be noticed that (3.4) is only a rough expression for the diamagnetic susceptibility.

In conclusion we point out that in the expression (3.1) the terms associated with α and χ_p stem entirely from the resonant part of the scattering amplitude, whereas the others get contribution also from the gauge part. How far we can go in utilizing the expression for the contact terms as given by the low energy theorems is a question that needs further investigations.

4. A microscopic model for the gauge terms

Following Arenhövel [8] the gauge term $S_{\mu\nu}$ of the scattering amplitude and, in particular, the contribution of the meson exchange currents (MEC), can be nicely derived from the following expansion for the electromagnetic interaction of the nucleus given up to the second order in the electromagnetic potential

$$\hat{H}_{int} = -\int d\vec{x}\, j^\mu(x) A_\mu(x) + \frac{1}{2}\int d\vec{x}\, d\vec{y}\, A_\mu(x) B^{\mu\nu}(x,y) A_\nu(y), \qquad (4.1)$$

the two-photon operator $B^{\mu\nu}(x,y)$ describing the first order contributions to the two-photon processes, among which the seagull terms.

Starting from a non-relativistic nuclear Hamiltonian

$$\hat{H}_0 = \sum_{i=1}^{A} \frac{\vec{P}_i^2}{2M} + \frac{1}{2}\sum_{i\neq j} V_{ij} \qquad (4.2)$$

one can derive j^μ and $B^{\mu\nu}$ after performing the minimal substitution

$$\vec{P}_j \rightarrow \vec{P}_j - \frac{e_j}{c}\vec{A}(\vec{r}_j)\; ;\quad \hat{H}_0 \rightarrow \hat{H}(A_\mu) \qquad (4.3)$$

where

$$e_j = e\{\tfrac{1}{2} + t_z(j)\} \qquad (4.3')$$

is the charge operator for a single nucleon.

Then the total nuclear current density and the two-photon operator are given, respectively, by

$$j^\mu(x) = -\left.\frac{\delta \hat{H}(A_\mu)}{\delta A_\mu(x)}\right|_{A_\mu=0} \qquad (4.4)$$

$$B^{\nu\mu}(x,y) = B^{\mu\nu}(x,y) = \left.\frac{\delta^2 \hat{H}(A_\mu)}{\delta A_\mu(x)\delta A_\nu(y)}\right| \qquad (4.5)$$

By requiring the gauge invariance of the resulting scattering amplitude, $B^{\mu\nu}$ turns out to be related to the current operator j^μ as follows

$$\frac{\partial}{\partial x^\mu} B^{\mu\nu}(\vec{x},\vec{y}) = -\frac{i}{\hbar c^2}[\varrho(\vec{x}), j^\nu(\vec{y})] \qquad (4.6)$$

which implies $B^{\nu 0} = B^{0\nu} = 0$ and

$$\partial_k \partial_\ell B^{k\ell}(\vec{x},\vec{y}) = \frac{1}{\hbar c^2}[\varrho(\vec{x}),[\hat{H}_0, \varrho(\vec{y})]] \qquad (4.7)$$

analogous to the condition (2.10) leading to the definition of $S^{\mu\nu}$.

The two-photon operator gets contributions from both the kinetic ($B^{kin}_{\mu\nu}$) and the potential energy ($B^{MEC}_{\mu\nu}$), providing that the meson exchange potential, which expresses the nucleon-nucleon interaction, explicitly exhibits the nucleonic and mesonic momenta and isospin variables.

The kinetic term is given (in the radiation gauge) by

$$B^{kin}_{k\ell}(\vec{x},\vec{y}) = \left.\frac{\delta^2 \hat{T}(\vec{A})}{\delta A_k(\vec{x}) \delta A_\ell(\vec{y})}\right|_{\vec{A}=0} \qquad (4.8a)$$

or, explicitly,

$$B^{kin}_{k\ell}(\vec{x},\vec{y}) = \frac{\delta_{k\ell}}{Mc^2} \sum_{i=1}^{A} e^2\left\{\tfrac{1}{2} + t_z(i)\right\}^2 f_E(|\vec{x}-\vec{r}_i|) f_E(|\vec{y}-\vec{r}_i|). \qquad (4.8b)$$

Taking the matrix element of (4.8b) between initial and final nuclear states one gets back essentially (2.12).

Arenhövel has explicitly derived the meson exchange contribution to $B_{\mu\nu}$ for the one pion exchange potential. It is formally defined by

$$B^{MEC}_{k\ell}(\vec{x},\vec{y}) = \left.\frac{\delta^2 \hat{V}^{OPE}(\vec{A})}{\delta A_k(\vec{x}) \delta A_\ell(\vec{y})}\right|_{\vec{A}=0} \qquad (4.9)$$

and splits in four different terms which are illustrated in Fig. 3.

The evaluation of the contribution of these terms to the Compton amplitude appears especially important in the energy range above the nuclear giant resonances, where on one side they are expected to play the dominant role (as demonstrated by the Compton scattering experiment on ^{208}Pb of Ziegler et al.[9]) and on the other the expression for the gauge terms provided by the low energy theorems can be no longer trusted.

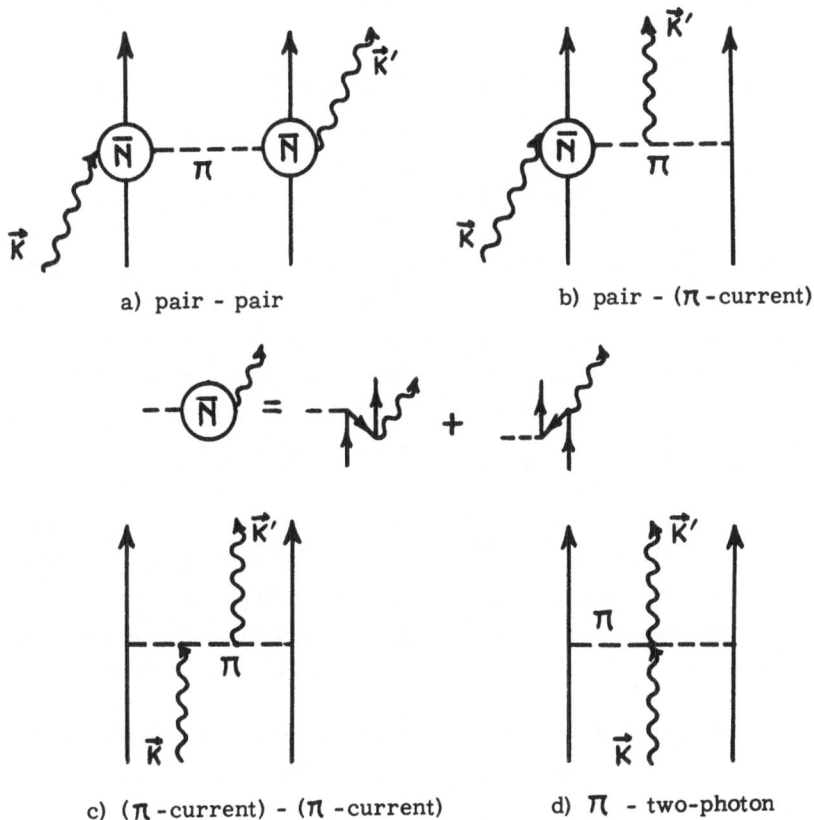

Fig. 3 The various contributions to the π-MEC two-photon scattering amplitude.

References

1) A. Hofmann, Phys. Rep. 64 (1980), 249
2) C. Schultz, private communication
3) P. Christillin and M. Rosa-Clot, Il Nuovo Cimento 28 A (1975), 29
4) T.E.O. Ericson and J. Hufner, Nucl. Phys. B 57 (1973), 604
5) O. Bohigas, Invited talk delivered at the International Conference held in Ames, Iowa, September 1979
6) A. Molinari, Phys. Rep. 64 (1980), 283
7) T.E.O. Ericson, Ann. of the N.Y. Academy of Sciences 257 (1975), 56
8) H. Arenhovel, Z. Physik A 297 (1980), 129
9) R. Leicht, M. Hammen, K.P. Schelhaas and B. Ziegler, to be published.

MEDIUM ENERGY PHYSICS WITH C.W. ELECTRON ACCELERATORS

Dieter Drechsel
Institut für Kernphysik
Johannes Gutenberg-Universität
D-6500 Mainz, Federal Republic of Germany

The past decade has seen an enormous growth of information from photo- and electronuclear investigations. A huge number of nuclei and nuclear levels has been explored in great detail. At the same time, a new level of precision in energy resolution, beam stability and particle detection has been reached, which has made electron accelerators the precision tool in nuclear structure investigations. Such experiments include measurements of ground state charge distributions with an accuracy of about 1 %, rotational levels with an energy resolution of less than 50 keV, magnetization distributions for high spin nuclei, higher giant multipole resonances, quasi-free and deep inelastic electron scattering, photo- and electronuclear reactions and total photonuclear absorption cross sections. The richness of the data and the accuracy of their theoretical interpretation have greatly improved our knowledge about the structure of nuclei and the effective interaction of bound nucleons.

In some cases, however, the analysis has shown phenomena that cannot be explained within the framework of the traditional nuclear A-body system but are possibly connected with subnuclear degrees of freedom. Such effects include, e.g.:

(I) The systematic difficulty to describe nuclear binding energies and charge densities (rms radii, central density) at the same time; in particular the lack of theoretical understanding of the hole in the center of He isotopes[1],

(II) influence of meson exchange currents on the electrodisintegration of the deuteron at high momentum but low energy transfer[2],

(III) the failure of existing theories to describe the forward photodisintegration of the deuteron at essentially all energies[3],

(IV) the substantial ($\sim$ 30 %) reduction of the total nuclear photoabsorption cross section in the $\Delta(3,3)$ resonance region as compared to the absorption on free nucleons [4,5],

(V) the filling-in of the minimum between quasi-free peak and Δ resonance in deep inelastic electron scattering[6] and

(VI) the asymptotic dependence of the form factors for elastic scattering and their relation with the quark or parton content of the wave function[7].

While some of the deviations from the classical pattern have been described qualitatively or, in some cases, even quantitatively with existing models of mesonic and isobaric currents and relativistic effects such as intermediate nucleon-antinucleon pair states, a few of the mentioned effects seem to escape theoretical explanation in a very persistent manner. This has led to a number of speculations that our present interpretation of a nucleus in terms of basically point nucleons and mesons has to be changed in a more radical fashion by introducing explicit effects of nucleon polarization and quark structure at small internucleon separations[8]. Indeed, the growing understanding of nucleon structure in terms of quark bag models[9] and speculations about an effective quark interchange force governing the interaction in the (low energy) confinement phase[10] offer a rich field for new models at small internucleon distances.

The development of (conceivably) completely new pictures of the nucleus are paralleled by the advent of a new generation of high energy and intensity electron accelerators operating with a continuous beam (100% duty cycle, c.w.). Since nuclear cross sections are generally dominated by the effects of traditional nuclear physics, particularly at low energy and momentum transfer, quite special kinematical situations are required to observe really unique signatures of subnuclear degrees of freedom. Such situations are generally described by large transfer of momentum or/and energy, special decay channels, typical resonance energies and scattering angles. The proposed generation of accelerators will be able to cope with these requirements by providing

- a 100% duty cycle to improve the signal/noise ratio by 1 to 2 orders of magnitude as compared to existing accelerators,

- a high current of about 100 µA to keep the counting rates for the extremely small exclusive coincidences at a reasonable level, and

- an energy of about 1 GeV to 2 GeV to explore pions and $\Delta(3,3)$ propagation in nuclei, and, at higher energies, production of ρ, ω, ϕ and K particles in nuclei.

The hierarchy of experiments with virtual and real photons allows to obtain the following information on nuclear structure:

1. Photoabsorption

$$\sigma(\omega) \sim F_T^2(q = \omega, \omega) \qquad (1)$$

The total photoabsorption cross section measures the transverse form factor at the photon point. It determines the energies ω and widths of the resonances.

2. Electron scattering (e,e')

$$\frac{d^2\sigma}{d\Omega_e\, d\varepsilon_e} \sim (V_L\, F_L^2(q,\omega) + V_T\, F_T^2(q,\omega)) \tag{2}$$

Elastic and inelastic electron scattering explores the transverse and longitudinal form factors in the space-like part of the q-ω plane. The variation of momentum transfer q makes it possible to determine the spatial distribution of charges, currents and magnetizations.

3. Photonuclear coincidence experiments (γ,x)

$$\frac{d^2\sigma}{d\Omega_x\, d\varepsilon_x} = \frac{d^2\sigma^T}{d\Omega_x\, d\varepsilon_x} + P\, \frac{d^2\sigma^P}{d\Omega_x\, d\varepsilon_x}\cos 2\phi_x \tag{3}$$

Photonuclear reactions with polarized photons (polarization P) allow to measure the response functions $W_{T,P} \sim (d^2\sigma^{T,P})/(d\Omega_x\, d\varepsilon_x)$, where $W = W(q = \omega, \omega;\, \theta_x, \varepsilon_x)$, for specific decay channels x = γ, p, n, α etc. The "new" structure function W_P can be measured separately due to the specific dependence on the azimuthal angle, $\sim \cos 2\phi_x$.

4. Electronuclear coincidence experiments (e,e'x)

$$\frac{d^4\sigma}{d\Omega_e\, d\varepsilon_e\, d\Omega_x\, d\varepsilon_x} = \Gamma\, \Big(\frac{d^2\sigma^T}{d\Omega_x\, d\varepsilon_x} + P\, \frac{d^2\sigma^L}{d\Omega_x\, d\varepsilon_x}$$
$$+ \sqrt{P(P+1)}\, \frac{d^2\sigma^I}{d\Omega_x\, d\varepsilon_x}\cos\phi_x + P\, \frac{d^2\sigma^P}{d\Omega_x\, d\varepsilon_x}\cos 2\phi_x\Big) \tag{4}$$

The cross section is written in analogy to the photonuclear one, with Γ measuring the flux and P the polarization of the virtual photons produced by the scattered electron. The experiment allows to determine four independent structure functions. The familiar longitudinal (W_L) and transverse (W_T) terms can be separated by varying the polarization of the virtual photons. The "new" transverse polarization term, W_P, and the longitudinal/transverse interference term, W_I, are ear-marked by their dependence on the azimuthal angle, $\sim \cos 2\phi_x$ and $\cos \phi_x$, respectively. The structure functions $W_{T,L,I,P} \sim (d^2\sigma^{T,L,I,P})/(d\Omega_x\, d\varepsilon_x)$, with $W = W(q, \omega;\, \theta_x, \varepsilon_x)$, allow to probe the spatial distribution of the response for any specific decay channel, x. In the limits of small and large final state interaction of the outgoing particle, the response functions may be interpreted by the models of quasi-free[11] and resonance scattering[12], respectively. Both pictures are, of course, limits of the same physical process, as has been shown by Balashov et al.[13] in the case of (e,e'p) and (e,e'n) with a simple shell model including final state interactions. The existing (e,e'p) data of the Saclay[14] and Tokyo[15] groups probe the momentum distribution and the binding energy of the struck protons in their shell model orbits, assuming that the final state interaction is small and may be described by an optical potential. The picture of intermediate resonances, on the other side, applies if the cross section is determined by a few collective resonances like giant multipole resonances in the low energy region, and possibly, coherent collective nuclear resonances in the $\Delta(3,3)$ region. In this case the angular distribution of the emitted particles is determined by the angular momentum of the resonant state. Given the fact that (I) existing low d.c. experiments had to fight a signal to background ratio of about one and (II) coincidence cross sections drop rapidly outside the domain of quasi-free scattering, it is evident that studies of resonance phenomena have to wait for c.w. accelerators. Some preliminary studies in the giant resonance region have been reported by the Stanford[16] and Illinois[17] groups.

5. Triple coincidences $(e,e'x_1x_2)$

Triple coincidences allow to determine the structure functions $W_i = W_i(q, \omega; \theta_1, \varepsilon_1, \theta_2, \varepsilon_2)$ and to explore the nuclear response as function of relative and total energy and angular momentum of a pair of particles. In some cases this will lead to a situation of completely determined kinematics, a completely exclusive experiment. In general, triple coincidences will require careful experiments even with the new generation of accelerators. However, it has been shown recently that experiments of the type $(e,e'\pi^+p)$ are actually quite feasible due to the extremely low background rate with the triples requirement[18], in fact that they might be barely possible with existing accelerators[19]. Typical counting rates have been estimated to be of the order of 0.1 - 1.0/s. Using real photons, Argan et al.[20] were able to make exploratory studies of the processes $^2H(\gamma,\pi p)$ and $^4He(\gamma,\pi p)$ and found intriguing structures, which have led to speculations about dibaryon resonances. The new accelerators will considerably improve the significance level of such experiments and will even allow to extend them to coincidences with the detection of neutral particles. In short, triple coincidences of the type $(e,e'\pi N)$ will provide a "Δ spectrometer" to explore the propagation of the $\Delta(3,3)$ and, possibly, new dibaryon resonances in nuclei in a rather unique way. Quite generally speaking, the new class of experiments will be an excellent tool to study correlations between bound particles in a very direct way. One further exciting experiment will be a systematic study of (γ,np) (or even $(e,e'np)$) with polarized photons in the quasi-deuteron region[21]. With the good energy resolution necessary for an exclusive experiment, e.g. $^{12}C(\gamma,np)^{10}B$, typical triple coincidence rates are some $10^{-2}/s$.

It has been the traditional crux of medium energy physics that aspects of nuclear structure, off-shell ambiguities of the elementary production operator and final state interactions of the struck or produced particles cannot be separated in a clean-cut way. In this situation it is of particular interest that an $(e,e'x)$ experiment yields a total of four independent structure functions corresponding to the different types of polarization of the virtual photon. The combined analysis of all four structure functions will be a very critical test for any theory and help to sort out the three mentioned aspects of medium energy physics, which are almost hopelessly intertwined in a single photonuclear cross section.

In these considerations it should be kept in mind that the "new" structure functions are usually very small, typically of the order of 10% or less of the familiar longitudinal and transverse ones. Moreover, since they are associated with a typical cos ϕ_x and cos $2\phi_x$ dependency, the cross section has to be measured for at least one non-coplanar geometry. However, the structure function W_p (polarized virtual photons!) is expected to be particularly sensitive to subnuclear effects such as isobaric and mesonic currents or, possibly, quark degrees of freedom. As an example, fig. 1 shows the four structure functions for the case of electrodisintegration of the deuteron[22] for a typical kinematical situation outside of the quasi-free peak. We note that the effects of mesonic (MEC) and isobaric (IC) currents are practically negligible for the longitudinal (W_L) and interference (W_I) structure functions but quite important already for transverse virtual photons (W_T). In the case of transversely polarized photons, however, MEC and IC add coherently and even change the sign of the structure function W_p. For incident and final electron energy of 800 and 630 MeV, respectively, a scattering angle θ_e = 22.5° and 110 MeV protons, typical counting rates are of the order of 20/s, i.e. 1-2/s for W_I and W_p[23].

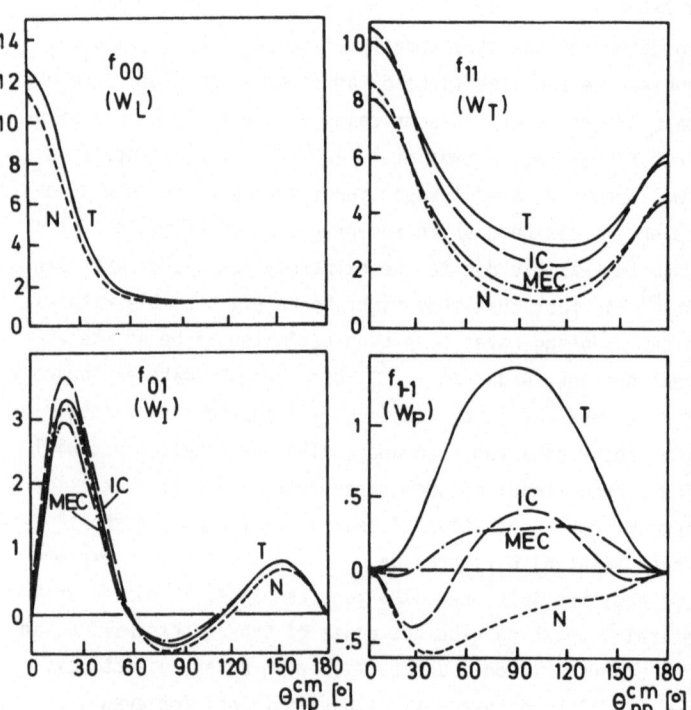

Fig. 1: The four structure functions for d(e,e'p)n as function of proton emission angle. The curves are obtained without (N) and with mesonic and isobaric currents (T). In some cases the effects of mesonic (MEC) and isobaric (IC) currents are shown separately. All structure functions in units of 10^{-3} fm, see also ref. 22.

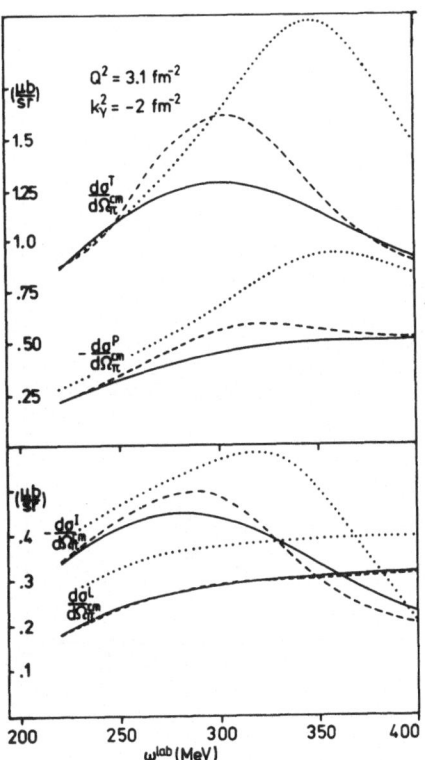

Fig. 2: The four structure functions for ^{3}He(e,e'π^+)^{3}H as function of energy transfer. The kinematical values are momentum transfer to the nucleus, $Q^2 = 3.1$ fm^{-2}, and four-momentum of the virtual photon, $k_\gamma^2 = -2$ fm^{-2}. The dotted curves correspond to electroproduction on free nucleons, the full curves are calculated with an exact treatment of Fermi motion in the production operator, the dashed curves are obtained with the approximation of an average Fermi momentum $<\underline{k}_x> = -1/3\underline{Q}$.

Fig. 2 shows the four "cross sections" for another exclusive process, coherent pion electroproduction on ^{3}He [24]. The binding effects in the production operator are quite substantial, particularly in the case of the transverse cross section due to modifications of the $\Delta(3,3)$-propagator. The approximation of an average Fermi momentum, $<\underline{k}_x> = 1/3\underline{Q}$, leads to an excellent agreement with the exact calculation only in the case of the longitudinal cross section. In this particular experiment one would observe the recoiling triton rather than the pion, leading to an increase of the cross section by about one order of magnitude. With a 50μA, 100% d.c. beam and modest detecting requirements, the maximum cross section in the resonance region corresponds to a counting rate of typically .25/s with a signal to background ratio of about 3. Therefore, a measurement of all four cross sections with at least 5% accuracy would require about 2 hours of beam time for one kinematical value. Further effects to be studied with such a reaction are off-shell ambiguities in the elementary production operator due to the unknown energy-momentum relation of the bound nucleon, modifications of the wave function such as the relative content of S, S' and D states and finally, pion rescattering or $\Delta(3,3)$ propagation. Of course, the latter aspect is of particular interest, because it might possibly lead to resonant states with typical signatures in energy and angular distributions.

Existing data for inclusive electromagnetic processes offer clear evidence for subnuclear effects. Fig. 3 shows the total photonuclear absorption cross section on Be measured by Ziegler et al.[4]. These data have been corroborated by recent neutron

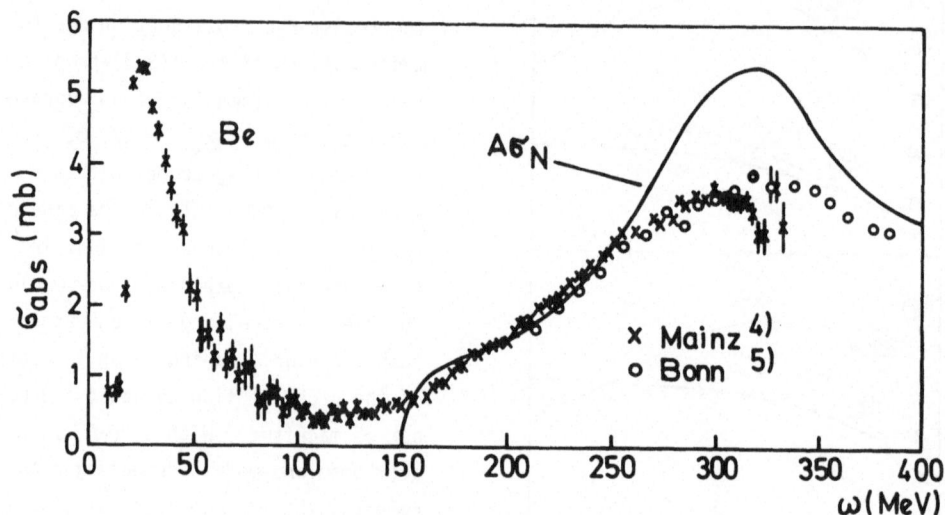

Fig. 3: Total nuclear photoabsorption cross section as function of photon energy. Data from refs. 4 and 5. The curve labeled "$A\sigma_N$" is the cross section of A free nucleons.

photoproduction experiments below 100 MeV[25] and by charged particle photoproduction between 215 and 386 MeV[5], correcting in each case for the unobserved particles. The sum rule integrated up to pion threshold gives about two classical (Thomas Reiche Kuhn) sum rules, mainly due to the existence of tensor forces mediated by the exchange of virtual pions[26]. In the $\Delta(3,3)$ region, on the other hand, an integrated cross section of roughly one classical sum rule is missing in comparison with photoabsorption by free nucleons. In a calculation within the framework of the Δ-hole model, Oset and Weise[27] find a substantial damping but no shift of the resonance, in general agreement with the experimental data. The dominant contribution is due to coupling to two-nucleon continuum channels (true absorption $\Delta h \rightarrow p^2 h^2$). Contrary to processes induced by pions they do not find a shift of the resonance energy to lower values, the differences being due to the different coupling of photons ($\gamma N\Delta \sim \underline{S \times q}$) and pions ($\pi N\Delta \sim \underline{S} \cdot \underline{q}$) to the $N\Delta$ transition spin S. While the total cross section shows little structure and not even a shift in anergy, individual states ($_\Delta A^*$ resonances in the nomenclature of ref. 29) show distinct energy shifts, Δ-hole states with lower multipolarities being pushed downwards, higher multipolarities shifted towards higher energies. As in the schematic model of nuclear giant resonances, a strong energy shift of Δ-hole resonances is also connected with an accumulation of transition strength on the shifted level. Therefore, exclusive coincidence experiments analyzing angular distributions and angular momenta of the involved particles are expected to show energy distributions with considerably more structures than inclusive processes.

The effect of Δ(3,3) resonance broadening has also been seen in deep inelastic electron scattering[6] in the transverse response function above the quasi-free peak. Fig. 4 compares the experimental data at backward angles with theoretical results. A coupling of the Δ to nuclear degrees of freedom in a Δ-hole[29] or nuclear $_\Delta A^*$ model[28] leads to a considerable resonance broadening and filling-in of the minimum. However,

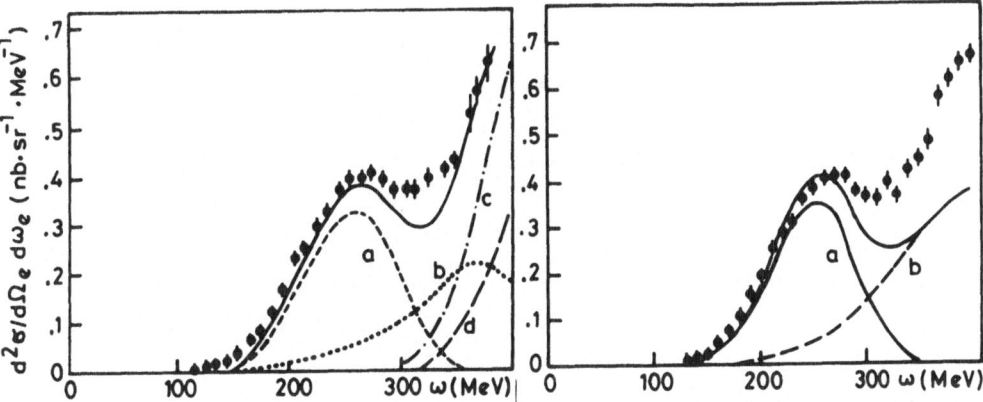

Fig. 4: Deep inelastic electron scattering $^{12}C(e,e')$ as function of energy transfer. The data of ref. 6 (E_e = 480 MeV, θ_e = 145°) are compared with the $_\Delta A^*$ model of ref. 28 (right: a) one nucleon knockout, b) Δ-nucleus resonance) and a calculation of Laget et al.[6] (left: a) one nucleon knockout, b) quasi-deuteron term, c) quasi-free Δ, d) quasi-free Δ and Born terms).

a substantial discrepancy remains. A recent calculation by Laget and Chretien-Marquet including absorption on a quasi-deuteron is able to explain most of the missing cross section. This seems to indicate that more-particle channels are of great importance in the region between quasi-free peak and Δ(3,3) resonance.

We conclude that coincidence experiments with virtual and real (tagged) photons will be a very sensitive tool to explore effects of pion and Δ resonance propagation in nuclei. In particular, exclusive experiments with complete kinematics of the nuclear final state will be invaluable as tests of theoretical models. By measuring angular and energy distributions it will be possible to determine energies and widths of the nuclear and nucleon resonances and the coupling of resonances to many-particle states. Such experiments will test the hypotheses of coherent collective resonances, sharp dibaryon resonances with excitation energies in the range of 200-600 MeV and possible quark content of the nucleon-nucleon wave function at small distances. With an accelerator in the 2 GeV region even more intriguing experiments will become possible, like electroproduction of vector mesons[30], ω, ρ and ϕ, higher nucleon resonances and kaons[31]. In all these experiments a careful planning of the particle detection systems is of utmost importance. It is imperative to take full advantage of all the detailed information which coincidence experiments are able to provide. As a resumé, some typical physical effects and their consequences for accelerators and particle detection systems are summarized as follows:

(I) Signal to background ratio decreases rapidly outside of quasi-free kinematics

High duty-cycle of accelerator (d.c. $\sim$ 100%)

(II) Higher differential cross sections for double and triple coincidences are extremely small, particularly outside of quasi-free kinematics

High beam current ($\sim$ 100 μA); large angular and momentum acceptance of spectrometers (("merit factor" = $(\frac{\Delta\Omega\Delta p}{p})_e (\frac{\Delta\Omega\Delta p}{p})_x$ large)

(III) Typical nuclear level spacing should be resolved, exclusive experiments

Overall energy resolution ≤ 1 MeV, suppresses background in case of sharp resonances in exclusive experiments

(IV) Angular distribution in θ_e, θ_x, ϕ_x should be explored, for both coplanar and non-coplanar kinematics to determine all four structure functions

High mobility and versatility of both spectrometers (e' and x), polar angle θ_x has to be varied about axis of momentum transfer, non-coplanar values of ϕ_x have to be measured

(V) Nuclear and nucleon resonances should be studied using all possible decay channels; neutron momentum distributions should be measured and compared to (e,e'p)

Provisions should be made to integrate counters for neutral particles, e.g., time-of-flight path for neutrons, pair spectrometer for photons, π^0 spectrometer

(VI) Triple coincidences should be studied as crucial test of
  a) $\Delta(3,3)$ propagation, e.g., "collective coherent states" in $(\gamma,p\pi^{\pm})$ and $(e,e'p\pi^{\pm})$
  b) nucleon-nucleon correlations, e.g., "quasi-deuteron effects" in (γ,pn) and $(e,e'pn)$

Appropriate detection devices for triple coincidences, e.g., additional telescope, solid state detectors, neutron time-of-flight path; maximum beam current should be used (extremely small cross section but excellent signal to background ratio due to triples requirement)

References:
1. J.S. Mc Carthy, I. Sick and R.R. Whitney, Phys. Rev. C15 (1977) 1396
2. G.G. Simon et al., Phys. Rev. Lett. 37 (1976) 12
3. P.J. Hughes, A. Zieger, H. Wäffler and B. Ziegler, Nucl. Phys. A267 (1976) 329
   H. Arenhövel and W. Fabian, Nucl. Phys. A282 (1977) 397
4. B. Ziegler, Proc. Int. Conf. on Nuclear physics with electromagnetic interactions, Mainz, 1979, eds. H. Arenhövel and D. Drechsel, Lecture Notes in Physics, Vol. 108 (Springer-Verlag 1979) p. 148
   J. Ahrens, Nucl. Phys. A335 (1980) 67
5. J. Arends et al., Symp. on Perspectives in electro- and photonuclear physics, Saclay, 1980, Short contributed papers, p. 2
6. J. Mougey et al., Phys. Rev. Lett. 41 (1978) 1645
   P. Barreau et al., Saclay Symposium, 1980 (see ref. 5), p.7
7. S. Brodsky and B. Chertok, Phys. Rev. D14 (1976) 3003
   R.G. Arnold, Mainz Conference, 1979, (see ref. 4), p. 76
8. V. Matveev and P. Sorba, Lett. Nuovo Cim. 20 (1977) 435
   A. Aerts et al., Phys. Rev. D17 (1978) 260
   T. Kamae et al., Phys. Rev. Lett. 38 (1977) 468, 471
   M.M. Giannini et al., Phys. Lett. 88B (1979) 13
   E. Hadjimichael and D.P. Saylor, Saclay Symposium, 1980 (see ref. 5), p.32
9. C. DeTar, Nucl. Phys. A335 (1980) 203
   G.E. Brown and M. Rho, Phys. Lett. 82B (1979) 177
10. H.J. Weber, Z. Phys. (1980), to be published
11. M. Gourdin, Nuovo Cim. 21 (1961) 1094
    T. de Forest, Ann. Phys. 45 (1967) 365
12. D. Drechsel and H. Oberall, Phys. Rev. 181 (1969) 1383
13. V.V. Balashov, S.I. Grishanova, N.M. Kabachnik, V.M. Kulikov and N.N. Titarenko, Nucl. Phys. A216 (1973) 574
14. J. Mougey et al., Nucl. Phys. A262 (1976) 461
15. K. Nakamura et al., Nucl. Phys. A268 (1976) 381
16. J.R. Calarco, Mainz Conference, 1979, (see ref. 4), p. 114
17. L.S. Cardman et al., Saclay Symposium, 1980, (see ref. 5), p. 17
18. R.P. Redwine and H.E. Jackson, Contribution to MIT workshop on Future directions of electromagnetic nuclear physics, 1980
19. P. Sargent, Proc. of MIT workshop, June 1977, Conf. Rep. COO-3069-6777, p. 11
20. P.E. Argan et al., Phys. Rev. Lett. 29 (1972) 1191
    G. Tamas, Mainz Conference, 1979, (see ref. 4), p. 363
21. G. Ricco, Workshop on Intermediate energy nuclear physics with monochromatic and polarized photons, Frascati 1980
22. W. Fabian and H. Arenhövel, Nucl. Phys. A314 (1979) 253
23. "Physikalische und technische Aspekte eines 800 MeV-Dauerstrich-Elektronenbeschleunigers", Mainz proposal, Dec. 1978, p. A26
24. L. Tistor and D. Drechsel, preprint KPH 23/80
25. R. Bergère, Mainz Conference, 1979, (see ref. 4), p. 138
26. D. Drechsel, Proc. IVth Seminar on Electromagnetic interactions of nuclei at low and medium energies, Moscow, 1977
27. E. Oset and W. Weise, preprint
28. K. Klingenbeck, M. Dillig and M.G. Huber, Phys. Rev. Lett. 41 (1978) 380
    K. Klingenbeck and M.G. Huber, preprint
29. E. Moniz, Mainz Conference, 1979, (see ref. 4), p. 435
30. J.V. Noble, MIT workshop, 1980, (see ref. 18)
    H. Crannell, MIT workshop, 1980, (see ref. 18)
31. A.M. Bernstein, T.W. Donnelly and G.N. Epstein, MIT workshop, 1980,(see ref.18)

FUTURE DEVELOPMENTS IN PION- AND KAON-NUCLEAR PHYSICS

AND THE EM PROBE: EXAMINING ELECTROMAGNETIC ENTRAILS

J. M. Eisenberg
Department of Physics and Astronomy
Tel Aviv University, Tel Aviv, Israel

The electromagnetic probe enters into pion- and kaon-nuclear physics in three different ways: (1) Mesic and electromagnetic reactions may be used in a complementary way in order to play against each other the various advantages and disadvantages of hadronic and electromagnetic probes - and especially the differences between them - in nuclear studies. (2) Various mixed processes such as photoproduction and its inverse may be used to obtain nuclear information. (3) Features that relate closely to mesic aspects of the nuclear problem, such as meson exchange currents or baryon isobar admixtures in nuclei, may be probed very effectively by purely electromagnetic means. I have been asked to speculate about future directions of such studies, and I shall try to do this by providing some past examples of each of these three uses for electromagnetic interactions and attempting to extrapolate a little into the future. In one or two instances I shall try to elaborate a little on the interaction between the electromagnetic probe and meson-nucleus physics in an effort to underscore that neither should be seen as dominating the other; rather a true parity exists between these tools. (Since the examples brought here are meant only to be illustrative, no attempt whatsoever is made to be complete in coverage or in referencing.)

1. <u>Mesic and electromagnetic reactions as complementary probes</u>

   a. <u>Nuclear neutron radii</u>. As a first - and highly obvious - sample case of the use of hadrons and electrons as complementary probes consider the question of neutron radii in nuclei. As recently as the mid-1970's it was exceedingly difficult to extract convincing information concerning this most basic parameter of the description of the hadronic matter distribution in nuclei, and this long after the charge distribution had begun to be superbly mapped. In the last two or three years however there have become available analyses[1-3] of 1-GeV-range proton scattering on nuclei that yield neutron radii for a variety of nuclei. These analyses aspire to a precision in the radius values of about ±0.07 fm - a truly remarkable stride forward. (Of course these results use as part of their input the proton radii known from electron scattering.) In view of the well-known difficulties in treating hadronic multiple-scattering theoretically, one would like to have mesic results as well and indeed some work[4] in this direction already exists, which even allows for the study of variations due to isotopic effects. (The pion may here offer the advantages of spinlessness and small mass, while the K^+ has a relatively weak interaction with nucleons.)

Naturally the hope here is to continue to analyze scattering data for electrons, protons, and mesons in order to complete the mapping of neutron distributions, and their comparison with proton distributions, that has already commenced.[2]

b. <u>Inelastic scattering: transition densities</u>. The intent here is much the same as in the previous elastic case, namely to compare hadronic with electromagnetic densities. An interesting case in point is pion inelastic scattering to the $J^P = 1^+$, $T = 1$ level at 15.1 MeV. Here the electroexcitation - to be discussed again in a different context below - agrees quite reasonably with theory at least in the forward direction. But the pion excitation[5] is not readily dealt with theoretically,[6] even after an interesting contribution[7] of orbital or convective form arising from the rapid energy variation of the πN amplitude in the 3,3 region is taken into account.[5]

c. <u>Knockout reactions</u>. Here once again part of the motivation for parallel electromagnetic and pionic studies of knockout lies in the possibility of separating charge and hadronic effects. For example one may exploit[8] $^Z_A(\pi^+, \pi^o p)^Z[A-1]$ reactions with the π^o-spectrometer at Los Alamos in order to compare neutron removal with the proton removal of (e,ep). However a separate reason for studying pion-induced knockout may lie in the possibility of exploiting this reaction in order to investigate off-shell features of the pion-nucleon amplitude and distortion effects in the pion wave, especially when it penetrates deeply into the nuclear interior for the knockout of more tightly-bound nucleons. The interplay of these effects produces marked asymmetries about the quasi-free point in theoretical studies[9] of $(\pi, \pi p)$ as opposed to what is expected and seen[10] in (e,ep).

I hope it is clear from the above examples that systematic comparison of mesic and electromagnetic scattering processes has only just begun and offers much interesting work over the next few years. One must also note that the pion and the photon, as bosons, share the ability to be absorbed or produced in nuclear reactions so that one may carry out comparisons of (π,p) - or its inverse - and (γ,p), or of the dominant absorption mechanism $(\pi,2N)$ and $(\gamma,2N)$. These comparisons are made fairly direct by the fact that pions and photons of 150 or 200 MeV total energy carry rather low momentum so that similar kinematic regions are reached in both cases. (Some of the relevant photo-induced processes are discussed here in Session VIII.)

2. <u>Mixed processes containing mesons and electromagnetic interactions</u>

a. <u>Radiative pion absorption</u>. The (π^-_{rest}, γ) process is noteworthy as one that has advanced from cottage industry to assembly line over the last fifteen years. Moreover it bears on some of the same nuclear collective modes that have been popular subjects of study through photonuclear methods, occasionally complementing these through its partial selectivity for spin-isospin vibrations as opposed to isospin vibrations for the photon. The subject has recently been reviewed extensively[11,12] so that it suffices to observe that resolution of $\lesssim 1$ MeV in the nuclear excitation energy is now obtained and the method seems well suited for studying $2^-,1$ excitations in certain cases. Experiments of this sort continue to be pursued vigorously.

    b. <u>Pion photoproduction at threshold</u>. The (γ,π) reaction near threshold and also in the 3,3 region has recently been studied extensively[11-14] with a consequent deepening of our understanding of the pion-nucleus interaction and increase in our confidence in the handling of pion-nucleus effects. Eventually this may lead to further insight into pion off-shell behavior in the nucleus.

    c. <u>Kaon radiative absorption</u>. The success of (π^-_{rest},γ) measurements naturally suggests similar (K^-_{rest},γ) experiments, which would yield information on spin-wave excitations in the hypernucleus. These measurements are made much more difficult by reason of the $\lesssim 300$-MeV photon (though in the early stages of (π^-_{rest},γ) we were also assured of the impossibility of the experiments). Some theoretical estimates of branching ratios for this process exist[15], suggesting about 10^{-3} for producing Λ hyperons and $\lesssim 10^{-4}$ for Σs. It has also been noted[16] that forward (K^-,γ) in flight with 600-MeV kaons produces recoilless Λs of interest for hypernuclear physics. Probably a safe prognostication is that the in-flight measurement is still farther in the future than the one at rest. Last in the context of radiative kaon absorption is the suggestion[17] to measure $d(K^-_{rest},\gamma)\Lambda n$ for information on the final-state Λn interaction of obvious interest for all work on two-baryon systems and hypernuclei.

    Work on the pion-EM processes goes on and we will undoubtedly hear more of it in the near future; kaon-EM reactions seem to remain for the time being a theorist's dream.

3. Electromagnetic probes of hadronic effects

    a. <u>Radiative neutron capture</u>. One must start by noting that one of the earlier - and still one of the more convincing - bits of evidence on the role of meson exchange and Δ-isobars in nuclear processes came from theoretical studies[18] of $n+p \to d+\gamma$ which included such effects and thereby raised the capture cross section by some 10% into agreement with experiment. Subsequently a number of other electromagnetic-nuclear processes - especially electron scattering on A=2 and A=3 systems - have been studied extensively towards this end, with mixed results as reported elsewhere in this conference. The point that I would like to stress is that mesic and isobaric features about which we learn a good deal from hadronic processes may ultimately have their precision tests in electromagnetic reactions. I shall try to clarify this symbiotic relationship further through the two remaining case studies here.

    b. <u>Scalar plus vector (s + v) potentials and electromagnetic effects</u>. A great deal of phenomenological information concerning the interaction of a nucleon in a nucleus is now receiving a uniform systematization in terms of a model[19] in which a relativistic nucleus is assumed to be bound in a combination of scalar potential $U(\underset{\sim}{r})$

and vector potential $V(\underline{r})$ according to the Dirac equation

$$[\underline{\alpha}\cdot\underline{p} + \beta M + \beta U(\underline{r})] u(\underline{r}) = [E - V(\underline{r})] u(\underline{r}),$$

where $\underline{p} = -i\underline{\nabla}$ is the momentum operator, $E = T + M$ is the total relativistic energy for nucleon of mass M described by the spinor $u(\underline{r})$, and $\underline{\alpha}$ and β are Dirac matrices. The scalar potential here takes into account the averaged attraction of the NN force ("σ"-exchange), while the vector part deals with repulsion (ω-exchange). The non-relativistic reduction of this Dirac equation allows one to make contact with a variety of phenomenological analyses. One can account easily for the energy-dependence of the nucleon-nucleus optical potential, spin-orbit features, and so forth. This comparison suggests[19] $U(0) \simeq -420$ MeV and $V(0) \simeq 330$ MeV. If we ignore the action of gradients on the potentials – purely for simplicity of the treatment here – it is easily seen that the Dirac equation is replaced with an equivalent Schroedinger equation

$$[\frac{1}{E+M+U-V} \underline{p}^2 + U + V] f(\underline{r}) = T f(\underline{r}),$$

where $f(\underline{r})$ is the "large" component of $u(\underline{r})$, and we can read off the conventional nonrelativistic potential as $U + V$ (~ -90 MeV for the above numbers) and identify the effective mass as $M^* = \frac{1}{2}(E+M+U-V)$ ($\sim 2M$ in the nonrelativistic limit with weak potentials). Note that for the above numbers $M^* \simeq 565$ MeV $\simeq 0.60 M$ in the central region and for zero kinetic energy T. In other words conventional nuclear potentials are relatively weak because of a large cancellation between scalar attraction and vector repulsion, but in the effective mass, where the absolute values of the potentials add, this model implies very appreciable consequences.

Clearly it is important to reanalyze old results in terms of this s+v model in order to assess its validity. Elsewhere[20] I have argued that sensitivity to the large changes in the effective mass is often less than one would expect, basically because where U and V are large (in the nuclear interior) they are also usually flat so that $\underline{\nabla}U$ and $\underline{\nabla}V$ – the quantities one generally requires – tend to be small.[21] This is the case for a number of electromagnetic situations. Nonetheless effects of M^* may well be expected in looking at the quasi-elastic peak in electron scattering for example. This peak is centered about $k^2/2M^*$ where k is the momentum transfer. Naively one would expect a dramatic rise in the position of the quasi-elastic peak. However most of this effect is vitiated by the energy-dependence of the nucleon optical potential as used in practice: from the above equivalent Schroedinger equation the energy balance for a nucleon in an infinite medium is

$$\frac{p^2}{2M+U-V} = T - (U + V), \qquad T \ll M,$$

or

$$\frac{p^2}{2M} = T - [(U + V)(1 + \xi) - T\xi], \qquad \xi \equiv \frac{U-V}{2M}.$$

The quantity in square brackets on the right-hand side here is the energy-dependent optical potential that is conventionally used in the empirical analyses (making a very clear appearance for example in the deepening of the binding potential of the nucleon for inner shells that is almost universally introduced). It would be interesting to have thorough, systematic, and explicit analyses of (e,ep) in terms of the s+v potential in order to try to sharpen our understanding of this description.

To draw my moral more finely before leaving this topic, I should like to note that a number of the people who have been attracted to the s+v model came to it because it fixes unambiguously the form of the πNN vertex for studying (p,π) reactions.[19] But tests of this important model may well lie in the electromagnetic domain: perhaps again a happy symbiosis.

c. <u>Pion condensation precursors</u>. The rich and intriguing topic of precursor phenomena,[22] or nuclear excitations of pionic character, has been addressed in Session V of this conference, and I shall therefore keep my remarks on it brief, again trying to focus on the mutual sustenance and support that the mesic and the electromagnetic domains can find in it. First a statement of position: whether or not any exotic enhancements are found because of a possible proximity to pion condensation, the elements of the theory are identical to those that enter our entire description of on-shell and off-shell pion behavior in nuclei. Thus we must understand these effects. They are required for a proper understanding of nuclear dynamics. If precursor enhancements do not exist - and well they may not - then either the loosely known parameters of the description must be assigned values that keep us far from the critical point for condensation (e.g., large Landau parameter g') or there is an important and basic flaw in our description.

The main feature of the precursor phenomenon can be stated as an enhancement of the (longitudinal) spin-isospin operator in the nuclear space whereby[23]

$$\underline{\sigma}\cdot\underline{k}\,\vec{\tau} \longrightarrow \underline{\sigma}\cdot\underline{k}\,\vec{\tau}\,[1 + WU]^{-1} \quad ,$$

for momentum transfer to the nuclear system $\underline{k}$. The first major quantity determining the enhancement is

$$W = \frac{k^2}{\omega^2 - k^2 - m^2} + g' \quad ,$$

where ω is the energy transfer to the nuclear system and m is the pion mass. The first term arises here from one-pion exchange, and the second - whose crucial nature for precursor effects has already been hinted at - combines consequences of nucleon correlations for pion propagation (the Lorentz-Lorenz effect, contributing 0.3 to g'?), ρ-meson exchange (contributing another 0.4 or 0.5?), and perhaps other features. The second major quantity is

$$U = \frac{f^2(k)}{m^2}\,[U_N(k,\omega) + 4\,U_\Delta(k,\omega)] \quad ,$$

where f(k) is the πNN vertex function, U_N is the standard Lindhard function (with imaginary part if the nuclear continuum is reached) describing $N^{-1}N$ excitations here and

$$U_\Delta = \frac{8}{9} \frac{\omega_\Delta}{\omega_\Delta + \omega} \frac{1}{\omega_\Delta - \omega - \frac{i}{2}\Gamma(\omega)} \rho, \quad \omega_\Delta \sim 2.3m \quad ,$$

treats $N^{-1}\Delta$ excitations (and allows for an imaginary contribution through the width Γ if the excitation reaches pion-production threshold).

The enhancement factor $[1 + WU]^{-1}$ arises from the summation of bubble graphs of $N^{-1}N$ or $N^{-1}\Delta$ character with π and ρ exchanges and correlation features. On the one hand this factor should make its appearance whenever a nuclear excitation involves the operator $\vec{\sigma} \cdot \vec{k}\ \tau$; on the other, we must seek clean cases of such excitations if we wish to test this matter of precursor enhancements. Many candidates have been put forth, and these were discussed in Session V. Some of them involve inelastic proton scattering[24] or pion reactions.[25] Others are electromagnetic, as in inelastic electron scattering[26], or mixed, as in photoproduction.[27]

Since I am here supposed to discuss things futuristic I shall mention one of the more challenging proposals[28], namely electroproduction of charged pions near threshold. The $\vec{\sigma}\ \tau$ combination dominates here but is not totally longitudinal (∥ $\vec{k}$) and so the enhancement effects are diluted. For $^{12}C(e,e\pi^+)^{12}B_{g.s.}$ with 500-MeV/c incident electrons and 10-MeV outgoing pions the maximal enhancement (near $k \sim 3\ fm^{-1}$) is by a factor of ∼5 for g' = 0.5 and by ∼2 for the more likely g' = 0.7. Unfortunately this is to be applied to a basic differential cross section of $\sim 10^{-41} cm^2/MeV \cdot sr^2$ and requires coincidence measurement: a challenge, as promised, but theorists have a touching confidence in the ability of experimentalists to measure _anything_ if they really want to.

To ease the experimental problem one may exploit the intrinsic preference of the probe for the $\vec{\sigma}\ \tau$ operator and spread consideration to the quasi-elastic peak in the continuum.[25,28,29] The experiment may then be done inclusively insofar as final electron energy and angle are concerned though the pion must be kept near threshold. For 500-MeV/c incident electrons the enhancement factors are then about 2.5 and 1.5 for g' = 0.5 and 0.7 and these are to be applied to a basic inclusive cross section of about $10^{-33} cm^2/MeV$, where some further summation over pion energy is still possible.

Again let me stress my moral. The elements of the precursor phenomenon draw heavily from considerations based on pion-nucleus interactions: pion scattering, the Lorentz-Lorenz effect, the roles of the Δ-isobar and of ρ-meson exchange (see the equations above for W,U, and U_Δ). But the consequences for electromagnetic processes, pure and mixed, are great. The problem is sufficiently intricate that it must be attacked from many directions. I think that there is little doubt that the next few years will see the results of many such efforts - not a few of them of electromagnetic character.

References

1. G. K. Varma and L. Zamick, Phys. Rev. C$\underline{16}$, 308 (1977) and unpublished work.
2. L. Ray, W. R. Coker and G. W. Hoffmann, Phys. Rev. C$\underline{18}$, 2641 (1978).
3. A. Chaumeaux, V. Layly and R. Schaeffer, Ann. Phys. (N.Y.) $\underline{116}$, 247 (1978).
4. R. R. Johnson et al., Phys. Rev. Lett. $\underline{43}$, 844 (1979); J. Alster, in Workshop on Nuclear structure with intermediate-energy probes, H. Baer et al., eds. (Los Alamos Scientific Laboratory, Los Alamos, 1980) p.119.
5. R. J. Peterson et al., Phys. Rev. C$\underline{21}$, 1030 (1980).
6. A. T. Hess and J. M. Eisenberg, Nucl. Phys. A$\underline{241}$, 493 (1975); for a treatment of meson exchange current corrections to the pion excitation, see J. Cohen and J. M. Eisenberg, preprint TAUP 878-80 (Tel Aviv University, 1980).
7. C. Wilkin, Nucl. Phys. A$\underline{220}$, 621 (1974).
8. J. Alster and M. Moinester, private communication.
9. E. Levin and J. M. Eisenberg, preprint TAUP 872-80 (Tel Aviv University, 1980).
10. J. Mougey, Nucl. Phys. A$\underline{335}$, 35 (1980). See also S. Boffi, C. Giusti and F. D. Pacati, Nucl. Phys. A$\underline{336}$, 437 (1980); I wish to thank D. F. Jackson for drawing my attention to this work. See also the report of S. Boffi in these proceedings, as well as those of Session XI.
11. Photopion nuclear physics, P. Stoler, ed. (Plenum, New York, 1979).
12. Nuclear physics with electromagnetic interactions, H. Arenhövel and D. Drechsel, eds. (Springer, Berlin, 1979).
13. F. Tabakin, in ref. 11, p. 301.
14. J. Laget, Nucl. Phys. A$\underline{335}$, 267 (1980) and in these proceedings.
15. G. Ya. Korenman and V. P. Popov, Phys. Lett. $\underline{40}$B, 628 (1972).
16. H. Feshbach, in Meson-nuclear physics--1976, P. D. Barnes et al., eds. (AIP, New York, 1976) p. 521.
17. B. F. Gibson et al., BNL report 18335 (Brookhaven, 1973).
18. D. O. Riska and G. E. Brown, Phys. Lett. $\underline{38}$B, 193 (1972).
19. J. V. Noble, Nucl. Phys. A$\underline{329}$, 354 (1979) and references therein.
20. J. M. Eisenberg, preprint TAUP 871-80 (Tel Aviv University, 1980).
21. An exception may occur in the A=3 systems where central densities are large but sloped. The author is grateful to L. G. Arnold for bringing this to his attention.
22. M. Ericson and J. Delorme, Phys. Lett. $\underline{76}$B, 182 (1978).
23. N. C. Mukhopadhyay, H. Toki and W. Weise, Phys. Lett. $\underline{84}$B, 35 (1979).
24. H. Toki and W. Weise, Phys. Rev. Lett. $\underline{42}$, 1034 (1979).
25. J. M. Eisenberg, Phys. Lett. $\underline{93}$B, 12 (1980).
26. J. Delorme et al., Phys. Lett. $\underline{89}$B, 327 (1980).
27. J. Delorme, in Proc. Workshop on intermediate energy nuclear physics with monochromatic and polarized photons, Frascati, July 1980.
28. J. M. Eisenberg, preprint TAUP 880-80 (Tel Aviv University, 1980).
29. See also W. M. Alberico, M. Ericson and A. Molinari, preprint TH.2813 - CERN (CERN, 1980) and J. M. Eisenberg, preprint TAUP 884-80 (Tel Aviv University,1980).

QUARKS IN NUCLEI

Chun Wa Wong

Department of Physics, University of California, Los Angeles, CA 90024 U.S.A.

Abstract: The quark model of nuclear forces is reviewed to illustrate the new perspective brought to nuclear theory by the quark structure of hadrons. The possible appearance of explicit quark effects is discussed in terms of the production of multiquark states by inelastic electron scattering and of nodal analyses of hadron-hadron phase shifts.

It has been known for some time that nucleons are rather large objects. The experimental proton "charge" radius[1] of 0.88 fm implies a "matter" radius of $\simeq 0.81$ fm, or an equivalent uniform-density radius of 1.05 fm. This hypothetical uniform nucleon matter has a density of $\simeq 0.21$ nucleon/fm^3, as compared to a density of 0.15 nucleon/fm^3 in normal nuclear matter. Thus the nuclear interior may be thought of as roughly 70% nucleon matter and 30% empty space. This is very different from the picture of point nucleons used in "classical" nuclear physics.

The large size of nucleons implies an internal structure which is a consequence of strong interactions. The traditional description of the low-energy or peripheral properties of this internal structure based on virtual mesons has been remarkably successful.[2] I would like to discuss here not this well-established traditional picture, but a much more speculative point of view generated by the more recent development that the internal structure of all strongly-interacting particles has a very simple description based on more fundamental entities called quarks.[3] The quark model has already led to a unified picture of electroweak processes. I would like to discuss a few of its implications in strong-interaction nuclear physics.

There are at least two important issues. We are interested first of all in deciding if the quark description might lead to a better appreciation of the successes and limitations of the usual theories of nuclear properties. The second issue is perhaps more exciting. If there is an underlying quark structure in nuclei, how can we make it betray its presence?

Before I address these questions, I would like to mention briefly the present picture of quark dynamics. Quarks are spin 1/2 fermions of fractional charges (2/3 or -1/3) and baryon number 1/3. They come in different flavors. Five (u,d,s,c,b) have been seen experimentally. The flavors u (for "up") and d (for "down") are of direct interest in low-energy nuclear physics, because they form the isospin SU(2) group which is responsible for the isospin degree of freedom in nuclei.

Each quark of a given flavor comes in three colors, which form the color SU(3) group. Quarks are assumed to interact through the exchange of colors. They do so by the emission and absorption of color-anticolor combinations. There are only eight color-changing combinations called <u>gluons</u> which are described by the eight generators λ^a of SU(3). Thus quark-quark interactions are proportional to $\vec{\lambda}_1 \cdot \vec{\lambda}_2$.

Gluons are assumed to be massless, and to be associated with a gauge, or phase, transformation of the wave function, like photons in QED. Unlike photons, gluons do not commute. Such non-Abelian gauge fields are intrinsically nonlinear, and show many unusual properties. A particularly important property is that they are <u>asymptotically free</u>[4] in the high-frequency limit. That is, the coupling constant vanishes as the momentum of the system goes to infinity, or equivalently as the dimension of the system goes to zero.

The converse situation is also true - the coupling constant becomes very strong at small momenta or large distances. It then becomes more favorable for quarks to bind together into color-singlet bound states, rather than to separate. This binding effect might become so strong that quarks can exist only in such bound states but never separately, at least at zero temperature. If this should happen, the quarks are said to be permanently <u>confined</u>.[5]

This confinement is supposed to take place rather abruptly in the manner of a phase transition as the gluon-quark-quark color coupling constant g^2 increases through a value of $\simeq 2$.[6] Since this coupling increases with distance, this gives the simple picture that quarks interact increasingly strongly up to a certain distance; beyond that they cannot go because of confinement.

The mechanism of confinement is usually associated with gluons rather than quarks. The outside forbidden region is a state of zero color flux. If quarks interact, color flux exists between them. For strong coupling, the most favorable situation is a narrow flux tube of the shortest length, which is the distance r between the interacting quarks. As a result, the interaction energy grows linearly with r:

$$V_{conf}(r_{12}) = \vec{\lambda}_1 \cdot \vec{\lambda}_2 \, k r_{12} . \qquad (1)$$

This confinement potential is expected to vanish rather abruptly as the coupling constant decreases below the critical value mentioned earlier. The flux between charges then spreads out suddenly into the well-known dipole pattern. The quark-quark interaction may then be described perturbatively, the leading term being the octet-gluon exchange potential (8GEP):

$$V_{8GEP}(\vec{r}_{12},\vec{p}_{12}) = \vec{\lambda}_1 \cdot \vec{\lambda}_2 \, (g^2/4) \left(\frac{1}{r_{12}} + f_{BF}(\vec{r}_{12},\vec{p}_{12}) \right) \qquad (2)$$

where f_{BF} is the Breit-Fermi relativistic correction to the Coulomb potential.

Since QCD (quantum chromodynamics) is not well understood yet, it is necessary to rely on phenomenological quark models of hadron structure in order to make educated guesses concerning the possible significance of quark effects in nuclei. Two types of models have been particularly useful in this connection. They are (i) the MIT (Massachusetts Institute of Technology) bag model, and (ii) potential models.

In the MIT bag model,[7,8] quarks and gluons are confined to within the hadron bag by boundary conditions imposed on the bag surface. Inside the bag, quarks interact relatively weakly and perturbatively. The pressure generated by the zero-point motion of quarks and gluons in the bag is counterbalanced by an assumed constant pressure B provided by the bag surface. Thus the energy of a spherical bag of radius R is

$$E(R) = \frac{4\pi}{3} R^3 B + C/R , \qquad (3)$$

where C contains the contributions from the confined quarks and gluons. For n massless quarks in 1s spatial orbitals, C has the simple form

$$C = C_n(S,T) = 2.043n - Z_0 + 0.177 \alpha_s a_n^M(S,T) \qquad (4)$$

for a bag of intrinsic spin S and isospin T. Here 2.043 is the dimensionless eigen-energy of a 1s quark orbit, Z_0 (=1.84) is the adjusted zero-point energy of gluons, $\alpha_s = g^2/4$ (=0.55) is the coupling constant (needed to fit hadron masses), and

$$a_n^M(S,T) = \frac{4}{3} [n(n-6) + S(S+1) + 3T(T+1)] \qquad (5)$$

is the weight of the reduced matrix element (of value 0.177) arising from the color magnetic interaction (proportional to $\vec{\lambda}_i \vec{\sigma}_i \cdot \vec{\lambda}_j \vec{\sigma}_j$) between quarks.

The MIT bag model is basically a shell model. The center of mass (CM) of the system might oscillate inside the bag. Corrections for CM motion requires a refitting of bag parameters to hadron masses. In one such attempt,[9] the end result is to shift ≃ 2 dimensionless units of the kinetic energy appearing in Eq. (4) to the $-Z_0$ term. One advantage of this shift is that the new value of $-Z_0$ (= -0.15) is much closer to the theoretical value[10] of $-Z_0 = 0.51$. The effect of this shift in hadron properties is often quite small, however. Hence I shall ignore it in the following discussion.

Another type of useful model consists of potential models.[3,11] They are characterized by the simultaneous use of a confinement potential for quark confinement, and an octet-gluon exchange potential for mass splitting:

$$V = \sum_{i \neq j} [V_{conf.}(r_{ij}) + V_{8GEP}(\vec{r}_{ij}, \vec{p}_{ij})] . \qquad (6)$$

Typically, no provision is made to distinguish between strong and weak couplings. The color magnetic term in V_{8GEP} is not quite the same as that in the MIT bag model, but it has roughly the same effect, since it is fitted to the same hadron mass splittings. An important property of the $\vec{\lambda}_i \cdot \vec{\lambda}_j$ terms in Eq. (6) which I shall use later is that the expectation value in a color-singlet state, namely

$$\langle \sum_{i \neq j} \vec{\lambda}_i \cdot \vec{\lambda}_j \rangle = \langle \vec{\lambda}^2 - \sum_i \lambda_i^2 \rangle = -n \frac{16}{3} , \qquad (7)$$

is proportional to the quark number n.

We are now in a good position to discuss quark effects in nuclei. One of the best examples of the new perspective afforded by the quark model on old nuclear properties is that on nucleon-nucleon (NN) interactions. This is also an important perspective because NN dynamics is the basic ingredient of nuclear dynamics.

Let us concentrate first at the origin of the NN relative coordinate r, where we assume that the system is in the $(1s)^6$ spatial configuration. Suppose the nucleons do not change size as they collide, as is usually assumed in the resonating-group or generator-coordinate calculations of the potential between colliding composites. Then the MIT bag model gives an energy difference of

$$E_6(R_3) - 2M_3 = -\frac{4\pi}{3} R_3^3 B + [Z_0 + (a_6^M - 2a_3^M)\, 0.097]/R_3, \qquad (8)$$

where R_3 is the nucleon bag radius. To get the NN potential we must subtract out the kinetic energy (KE) of relative motion between the nucleons, because it should appear instead as the KE <u>operator</u> in the scattering equation. By counting degrees of freedom I estimate this to be $\simeq 2.043$ in the dimensionless unit of Eq. (4). Thus the NN potential is

$$V_{bag}(r=0) = -\frac{4\pi}{3} R_3^3 B + (Z_0 - 2.043)/R_3 + V_{pot}(r=0), \qquad (9)$$

where $V_{pot}(r=0) = (a_6^M - 2a_3^M)\, 0.097/R_3$ is 360 (470) MeV for the $^3S_1$ ($^1S_0$) state if $R_3 = 5$ GeV^{-1} is used. The remaining terms in Eq. (9) give -280 MeV, so that the repulsion at r = 0 is roughly 80 (190) MeV in the $^3S_1$ ($^1S_0$) state. This is, of course, only a rough estimate because of the uncertainty in the KE correction, but it does suggest that the NN potential should be much less repulsive than past estimates.[12]

For potential models of quarks, the usual resonating-group method gives rise to a nonlocal potential. The B, Z_0 and KE terms are obviously absent. The contributions from $\vec{\lambda}_i \cdot \vec{\lambda}_j$ quark-quark interactions cancel exactly because of Eq. (7). If we ignore various nonlocality effects, the simplest estimate for the NN potential at r = 0 is the color magnetic repulsion $V_{pot}(r=0)$ of Eq. (9). Indeed, the values shown are in rough agreement with results obtained in more detailed calculations in the literature.[13,14] Thus potential models of quarks give 200-300 MeV more repulsion than the MIT bag model because of basic differences in their dynamical assumptions, especially those concerning quark confinement.

Nucleon bags are actually quite stiff. The compression modulus with respect to R is $3E(R_0)$, or roughly 3 GeV per nucleon. This means that overlapping nucleons at low energies are much more likely to maintain constant densities rather than constant radii. Thus a more realistic picture is that of an adiabatic approximation:

$$V_{Bag}^{adiabatic}(r=0) = M_6' - 2M_3, \qquad (10)$$

where M_6' is calculated like M_6 but with the KE of relative NN motion subtracted out.

The result is -40 (40) MeV for the $^3S_1$ ($^1S_0$) state. This is much weaker than the usual repulsion[12] of 270 (350) MeV in the $^3S_1$ ($^1S_0$) state because of an attraction of $\simeq$ - 310 MeV coming from the subtraction of the spurious KE of relative NN motion. Although the present estimate of this correction is very rough, it should serve to call attention to the problem.

In potential models, the compression modulus is actually a little larger than that for bags. In spite of this, the adiabatic expansion in the radius of the six quark system is much less, leading to an adiabaticity effect in the NN potential of only 20-30 MeV.[15] The reason for this somewhat unexpected result is that the color Coulombic potential, which is present only in potential models, is much more effective in preventing the system from expansion even against the stronger effect of the non-relativistic kinetic energy. It does so because its contribution to dE/dR is positive, when its contributions to E and K_R are both negative.

These considerations suggest that the quark picture of NN potentials is model-dependent: the S-wave NN potentials from the $(1s)^6$ configuration come out to be roughly zero in the MIT bag model, and roughly 400 MeV in potential models.

It has also been realized that $(1s)^6$ is not the only important configuration in the problem. The less symmetrical $(1s)^4 (1p)^2$ is also important because it comes down much lower than expected as a result of strong color interactions involving its hidden-color components.[14,16] In one estimate,[16] the corresponding six-quark state with S = 1, T = 0 comes down to 2.38 GeV in the MIT bag model, and to 2.16 GeV in a potential model. A further decrease is obtained by configuration mixing and channel coupling. The final result for the S-wave NN potentials turns out to be roughly zero for a potential model of quark-quark interactions.[14]

This effect also makes it possible to understand qualitatively the bag results of DeTar[12] which show that the strong repulsion at r = 0 turns rapidly into a strong and sharp attraction with a minimum of -100 to -250 MeV at $\simeq$ 0.7 fm. The initial drop appears to be caused by the increasing admixture of 1p states until an optimal situation is obtained at $\simeq$ 0.7 fm where the mixed-parity single-particle orbitals are roughly 60/40 admixtures of 1s/1p states.

DeTar's results must be corrected for possibly unrealistic features: (1) There might be a significant component of the $(1s)^4(1p)^2$ configuration even at r = 0, so that the NN potential there (after the correction for NN relative KE discussed earlier) might well be attractive (say, -100 MeV). (2) The attraction at $\simeq$ 0.7 fm might have been over-estimated, because the $1p_{3/2}$ eigen-energy in the bag might have been too small. Thus a more likely result for the MIT bag model is that the short-range NN potential from these quark effects might be attractive, vanishing rather smoothly with increasing distance in a gaussian manner. In other words, there is no short-range repulsion at all. This is only a very rough estimate; it must be substantiated by more detailed calculations.

If this picture is accurate, it cannot be complete or realistic, for then there is nothing to prevent nuclei from collapse. We do not know yet what is missing from the theoretical picture. In this connection, the following remarks might be helpful: (1) Since the short-range NN potential appears to depend on the confinement mechanism, which we have not yet understood, it is a little premature to expect a complete quark model of nuclear forces. (2) The spatial $(1s)^4(1p)^2$ configuration would not come down so low if the gluon-quark-quark coupling constant α_s is weaker. Since α_s is deduced from the observed Δ-N mass splitting, other mechanisms must then be proposed for this mass splitting in addition to a weaker 8GEP contribution. (3) It helps to have a smaller bag. These (confinement, interaction strength and bag size) are also questions which are important by themselves.

Given our limited understanding of quark dynamics, we would like to get guidance from experiments if we could. The question is how quark degrees of freedom can be made to stand out.

One possibility involves the confined hidden-color channels of multiquark states. The idea is that their confinement might generate discrete eigenstates which, when coupled to open hadronic channels, might appear as resonances.

Such "color" resonances might be accessible through either production or formation experiments. Typical of these is the inelastic production of dinucleon bag states:[9]

$$e + d \rightarrow e' + d^* \tag{11}$$

One measures the double differential cross section $d^2\sigma/dEdt$. If there is a resonant "signal" with the expected width of $\simeq 50$ MeV, it can be analyzed readily through the energy-integrated differential cross section $(d\sigma/dt)_{inel}$, which has a structure similar to that for elastic scattering, except that it involves a transition density. Estimates of the resulting inelastic form factor $A_{inel}(q^2)$ can readily be made and compared with the elastic form factor $A_{el}(q^2)$. The results[17] are shown in Fig. 1 for (a) different d* size parameters R ($R_o = 0.78$ fm is the expected value), and (b) different admixture amplitudes b of the pure bag state in the deuteron ground state. These results show that the signal is not very strong, and may easily be lost among the Δ and other resonances which might appear at the same energies. Still it might be worthwhile to study whether the signal from this rather compact object might stand out more clearly at large momentum transfers.

An ingenious method for gaining access to strong-interaction dynamics has been proposed by Jaffe and Low.[18] It is based on the idea that as the scattering energy E increases the scattering wave function in an open channel will oscillate up and down, passing through zero recurrently at a certain radius b(E) related to the bag radius R(E) in confined hidden-color channels coupled to it. A discrete set of continuum states can thus be selected; they may conveniently be called <u>nodal states</u> (NS)[19] because they are the states with nodes at b(E).

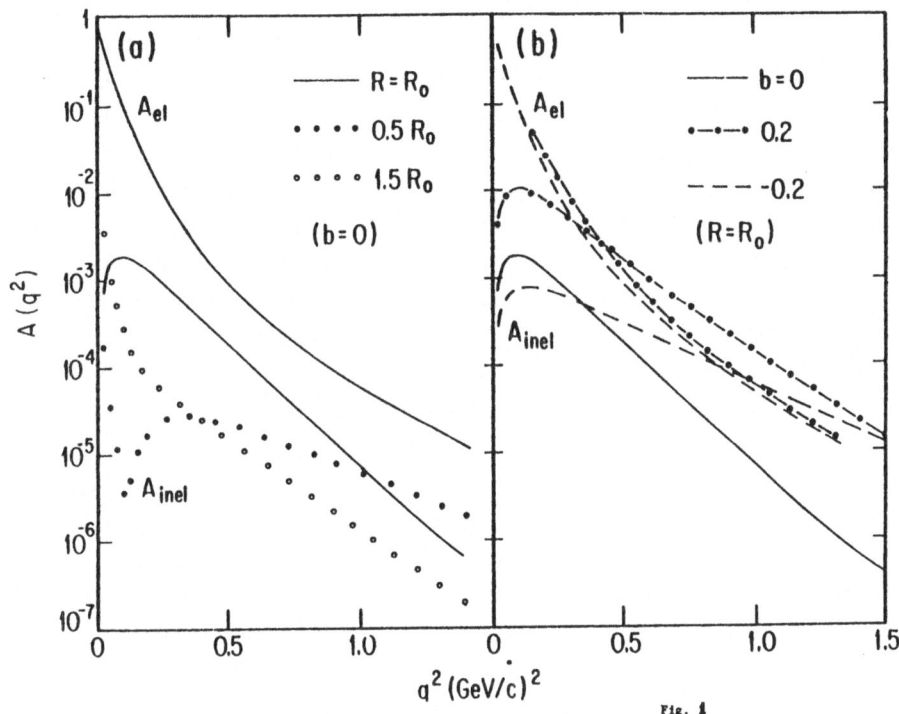

Fig. 1

Like the channel radius of the R-matrix theory, the boundary b(E) separates an internal region from an external region. The internal region is very special in the MIT bag model, because it contains almost all the strong-interaction dynamics. Furthermore, quarks in an MIT bag are individually confined by surface boundary conditions, with the result that the relative wave function in open channels vanishes at a suitably chosen b(E). In other words, the multiquark bag states are just the internal parts of scattering NS, while their masses are essentially identical to those of the corresponding NS.

Given a theory of b(E), NS masses can be deduced from scattering phase shifts δ(E) in an open scattering channel. This is easily seen in S wave scattering, for which the external scattering wave function sin(kr + δ) will start at a node at the boundary b if the nodal function kb + δ is an integral multiple of π. Such an empirical determination of NS mass will be called a <u>nodal analysis</u> of phase shifts. The resulting NS masses can then be compared directly with theoretical masses calculated for multiquark bag states.

It is clear that the choice of the boundary point b is quite crucial to the significance of this comparison. Fortunately, the MIT bag model gives quite specific predictions on this, namely[18,19]

$$b \simeq 1.4R \simeq 8M^{1/3} \text{ GeV}^{-1} \qquad (12)$$

if M is given in GeV. Two other comments might also be relevant here: (1) Coupling to other open channels might shift the NS masses. A preliminary study[19] suggests that the shift is not large when compared to the uncertainties in the input phase shifts, or to the precision needed in the final result. Hence the simple single-channel prescription discussed here will be used in the following. (2) NS are not resonances, but they always appear near sharp resonances where phase shifts are increasing rapidly. Not all NS are associated with resonances, however, since the number of NS is always infinite, while the number of resonance poles may be finite. To compound the complication, not everybody considers far-away resonance poles to be resonances, since the term is also used to denote "physical" resonances with well-defined physical characteristics. Irrespective of the terminology used, nodal analysis is of value by providing a meeting ground between experiment and theory.

In collaboration with K. F. Liu, I have analyzed many hadron-hadron phase shifts for NS.[19,20] The main results are as follows: (1) There is impressive agreement between empirical NS masses and theoretical masses of $q\bar{q}$ and multiquark states in non-exotic meson-meson channels. The best example is found in the I = 0 $\pi\pi$ S-wave channel where the NS masses (theoretical masses in parentheses) are 0.65 (0.65), 0.98 (0.98), 1.17 (1.15), 1.47 (1.45), 1.59 (1.67), 1.76 (1.80), all in GeV. The underlined theoretical masses are those of four-quark states calculated by Jaffe.[21] I consider this agreement to be an important hint that the bag description of quark dynamics and hadron radii may be quite realistic. It does not by itself determine unambiguously whether bags are large or little. (2) All NS associated with the resonances of the Data Table[22] are readily seen. In addition, there are at least the following NS masses (in GeV) not associated with well-established resonances:

(a) In $\pi\pi$ scattering: 1.16 [ℓ = 1, I = 1, ρ'(1.25)?], 1.44(ℓ = 2, I = 0).
(b) In Kπ scattering: 1.29 (ℓ = 1), 1.79 [ℓ = 1; K*'(1.65)?].
(c) In πN scattering: 1.9 [S31; Δ(1.9)?], 2.3(S31), 2.0(P33), 2.4(P33), 1.6[p31; Δ(1.55)?], 2.2(D33), 1.6(D35), 2.4(F35), 2.4?(F37), 2.1(G37), 2.1(G39), 2.5[G39; Δ(2.5)?], 2.9(G39), 2.2[H39; Δ(2.3)?], 2.7 (H39), 2.7(H311), 2.4(I311), 2.8(I311); also 1.9[S11; N(2.1)?], 2.3 (S11), 2.5? (P13), 2.0? (P11), 2.2(P11), 2.2 [D13; N(2.04)?], 2.2 [D15; N(2.1)?], 2.0 [F15; N(2.0)?],2.4(F17), 2.5(G17), 2.5(G19), 2.9 [G19; N(2.8)?], 2.7(H19), 2.2 (H111), 2.7(H111), 2.8(I111).
(d) In $\bar{K}$N scattering: 1.6 [S11; Σ(1.62)?], 2.0? [S11; Σ(2.0)?], 2.1?(P11), 1.7[P13; Σ(1.84)?], 2.1?[P13; Σ(2.08)?], 2.3?(D15); also 1.7[P01; Λ(1.60)?], 1.9[P01; Λ(1.80?], 2.1?(P01), 2.0?(P03), 1.9(D03), 2.0 [F07; Λ(2.02)?].

(3) In exotic meson-meson channels, there is considerable discrepancy between experiment and theory. It is not clear what the problem is.

We are particularly interested in the NN scattering channels. Here the comparison is not conclusive. For example, we find three NS in the pp $^1S_0$ channel, at 1.97, 2.29 and 2.64 GeV using the old Hoshizaki phase shift,[23] as compared to an expected six-quark bag mass of 2.18 GeV. Given the uncertainties in theory, experiment and analysis (i.e. choice of parameters), one is tempted to pick the NS at 2.29 GeV as the bag state. The problems are as follows: (a) OBEP's like that of Ueda and Green[24] already give similar NS masses without having to involve quark degrees of freedom explicitly. (b) The effect of the strong NN potential outside the chosen boundary point b(E) has not been included in the analysis. (c) The effect of inelastic pion channels has also been ignored. Further detailed analyses will be needed to clarify these questions. For completeness, let me give here all the other NS masses (in GeV) based on the old Hoshizaki phase shifts: 2.19, 2.49, 2.85 in $^3P_0$; 2.13, 2.48, 2.79 in $^3P_1$; 2.05, 2.35, 2.71 in $^3P_2$; 2.17, 2.56 in $^1D_2$; 2.30, 2.65 in $^3F_2$; 2.31, 2.68 in $^3F_3$; 2.28 in $^3F_4$; 2.41 and 2.8 in $^1G_4$; also 1.95 in $^3S_1$.

What is increasingly apparent is that there might be very rich structures in NN phase shifts between $\sqrt{s}$ = 2 and 3 GeV which deserve further experimental studies and theoretical analyses.

I must apologize for spending so little time on the experimental aspects of quarks in nuclei, especially those associated with em probes. There are important questions connected with quark signals and meson backgrounds which deserve careful analyses. I am carried away, however, by a preoccupation with <u>nucleon matter</u>, i.e. the stuff inside nucleons, because I find it hard to visualize pion-condensed, Lee-Wick and other interesting high-density nuclear matters without knowing what nucleons are made of. Are they big bags of quarks,[7] or are they full of pions and other virtual mesons surrounding little bags of nucleonic "nuclei"?[2] Is our nucleus made up of little bags swimming in a sea of pions? We believe we understand many mesonic effects in nuclei, but the successes of the quark model in both hadron structure and hadron reactions are also very impressive. (I should also mention the quark-counting rules in reaction cross sections[3] and e.m. form factors,[25] and the quark-counting analysis of elastic proton-proton differential cross sections.[26]) In this connection, the present discussion of short-range nuclear forces and of nodal-state masses is also inconclusive.

REFERENCES:

1. F. Borkowski et al., Nucl. Phys. A222, 269 (1974).
2. M. Rho, invited talk given at the Berkeley Conference, 1980.
3. See, for example, J.J.J. Kokkedee, The Quark Model (Benjamin, New York, 1969).
4. See, for example, A. J. Buras, Rev. Mod. Phys. 52, 199 (1980).
5. See, for example, J. B. Kogut, Rev. Mod. Phys. 51, 659 (1979).
6. See, for example, M. Creutz, Phys. Rev. D 21, 2308 (1980).
7. T. DeGrand et al., Phys. Rev. D 12, 2060 (1975).
8. K. Johnson, Acta Phys. Pol. B6, 865 (1975).
9. C. W. Wong and K. F. Liu, Phys. Rev. Lett. 41, 82 (1978).
10. K. A. Milton, Phys. Rev. D 22, 1441 (1980).
11. See, for example, K. Gottfried, in Proc. 1977 Intl. Symposium on Lepton and Photon Interactions at High Energies, edited by F. Gutbrod (DESY, Hamburg, 1977), p. 667.
12. C. DeTar, Phys. Rev. D 17, 323 (1978); 19, 1451 (1979).
13. D. A. Liberman, Phys. Rev. D 16, 1542 (1977).
14. M. Harvey, Nucl. Phys. A, to be published.
15. C. W. Wong and K. F. Liu, unpublished.
16. I. T. Obukhovsky et al., Phys. Lett. 88B, 231 (1979).
17. C. W. Wong, K. F. Liu and Y. Tzeng, Phys. Rev. C (1980) to be published.
18. R. L. Jaffe and F. E. Low, Phys. Rev. D 19, 2105 (1979).
19. C. W. Wong and K. F. Liu, UCLA preprint, 1980 (unpublished).
20. C. W. Wong, unpublished.
21. R. L. Jaffe, Phys. Rev. D 15, 267, 281 (1977).
22. Particle Data Group, Rev. Mod. Phys. 52, S1 (1980).
23. N. Hoshizaki, Progr. Theor. Phys. 60, 1796 (1978).
24. T. Ueda and A. E. S. Green, Rev. C 18, 337 (1978).
25. See, for example, S. J. Brodsky and B. T. Chertok, Phys. Rev. D 14, 3003 (1976).
26. E. Shrauner, L. Benofy and D. W. Cho, Phys. Rev. 177, 2590 (1969).

SOME ISSUES IN PHOTONUCLEAR PHYSICS

C. Tzara
DPh-N/HE, CEN Saclay, BP 2, 91190 Gif-sur-Yvette, France

Let us consider the simplest processes induced by the electromagnetic interaction, the emission or absorption of one photon. Their cross sections are unambiguously related to definite characteristics of the target, the charge and current form factors associated with the target transition $|a\rangle \to |b\rangle$:

$$\langle b | \int d^3 r \, e^{i \vec{q} \cdot \vec{r}} \rho(r) | a \rangle \quad \text{and} \quad \langle b | \int d^3 r \, e^{i \vec{q} \cdot \vec{r}} \vec{\epsilon} \cdot \vec{j}(r) | a \rangle$$

with $\vec{\epsilon} \cdot \vec{q} = 0$

In reality what is measured is different because higher order e.m. effects are unavoidable :

(i) in photonuclear reactions, the radiative corrections, which are indeed negligible as long as the velocity imparted to the target or the emitted particles is small ;

(ii) in contrast, the corrections to Born approximation in electron scattering :
- distortion of the electron wave,
- two-photon exchange,
- radiative corrections,

are far from negligible and pose eventually difficult problems.

For the purpose of the following discussion, these will be considered as correctly handled, so that the final outcomes of the measurements are really form factors. At high energy, the so-called shadow effect comes into play. Whatever model is invoked to interpret it, the process is a one-photon and, as such, provides a form factor.

It is also known that by equating the form factors to the Fourier transforms of the charge and current operators taken between intrinsic nuclear states, one assumes implicitly that the final nuclear state with total momentum q : $|b, q\rangle$ is identical with $|b, o\rangle \, e^{i \vec{q} \cdot \vec{R}} \, e^{-i\omega t}$, where $\vec{R}$ is the nucleus center-of-mass coordinate. The smaller is the recoiling velocity, $(q/Amc)^2 \ll 1$, the better is the approximation. For heavy nuclei, this condition is insured in practice, because of the limit to momentum transfer set by the smallness of the cross section. For very light nuclei, however, this condition is violated in some existing experiments; some scrutiny is required when analyzing the data in term of spatial densities.

With these reservations one may say that a photo- or electro-reaction cross-section provides a measure of the Fourier transform of a diagonal or a transition density of the target nucleus. Weak interaction gives similarly form factors, but with much less flexibility, whereas the reactions mediated by the strong interaction are testing the properties of the "target + projectile" system.

Such a presentation of the Nuclear Electromagnetic Interaction, which emphasizes its conceptual simplicity, should not mask the complexity, and related difficulties, which arise when a microscopic interpretation of the measured form factors is attempted.

The root of the difficulty is in that the chosen nuclear constituents are always composite systems.

At the simplest level, the description of the nucleus is made in terms of nucleonic degrees of freedom. Their mutual and Electromagnetic Interactions contain phenomenological ingredients: an N-N potential, a model for the charge and current densities.

In a further step, the forms of these interactions are determined from the mesons-nucleon dynamics.

In order to handle processes above the mesons threshold, the meson coordinates must be kept explicitly in the formalism. Still the meson-nucleon dynamics is largely in a phenomenological form. One must accept this unsatisfactory state of affair as long as a fundamental theory of structureless constituents, if any, is not operative.

The first approach noted above is fitted to a large class of properties of most nuclei. The theory of their Electromagnetic Interaction utilizes as much as possible the continuity equation applied to the charge and current density operators:

$$\vec{\nabla}_r \cdot \vec{j}(r) = -i/\hbar c \, [H, \rho(r)]$$

It must not be forgotten that it is an operator equation acting, through the nuclear Hamiltonian H, on the nucleon coordinates. Otherwise one is led to untenable conclusions. For instance, following Foldy, one obtains its formal solution:

$$\vec{j}(r) = \frac{i}{4\pi \hbar c} \vec{\nabla}_r \cdot \int \frac{[H, \rho(r')] \, d^3r'}{|r - r'|}$$

whose long range behaviour is contrary to all expectations. Furthermore, the current matrix element of this irrotational current:

$$- \langle b | \int d^3r \, e^{i\vec{q}\cdot\vec{r}} \, \vec{\epsilon} \cdot \vec{j}(r) | a \rangle$$

would be identically zero, the solenoidal part of the current surviving alone! The valid procedure is to manipulate the operator inside the matrix element in such a way that it splits in two pieces having opposite parities for a given angular momentum, and to use in the electric part the continuity equation. In the version by Foldy [1], the electric matrix element takes the form:

$$- i/\hbar c \, \langle b | \int_0^1 ds \int d^3r \, e^{is\vec{q}\cdot\vec{r}} \, \vec{\epsilon}\cdot\vec{r} \, \rho(r) | a \rangle (E_b - E_a)$$

One notes that the retardation factor is somewhat damped by the integration over Ω. The magnetic one reads :

$$i\langle b|\int_0^1 sds\int d^3r\, e^{is\vec{q}\cdot\vec{r}}(\vec{\epsilon}\times\vec{q})\cdot(\vec{r}\times\vec{j}(r))|a\rangle + i\langle b|\int d^3r\, e^{i\vec{q}\cdot\vec{r}}(\vec{\epsilon}\times\vec{q})\cdot\vec{m}(r)|a\rangle$$

These expressions apply equally to the virtual photon absorption, with $\vec{\epsilon}$ fixed by the electron kinematics.

The predicted value of an electric matrix element depends therefore on the form chosen for $\rho(r)$ in term of nucleon coordinates, on the wave function of the initial and final states and on the computed eigenenergies E_a and E_b (and not on the experimental ones). A striking evidence for its dependence on the nuclear Hamiltonian is provided by the electric dipole sum rule, whose value is porportional to the expectation value of the double commutator in the ground state :

$$\langle a|[\int d^3r(\vec{\epsilon}\cdot\vec{r})\rho(r),[\sum t_i + \sum V_{ij}, \int d^3r\, \vec{\epsilon}\cdot\vec{r}\,\rho(r)]]|a\rangle$$

The computation of a magnetic matrix element requires additional choices, that of the form of $\vec{m}(r)$ and of $\vec{j}(r)$, insofar as the latter is not determined by, but only constrained by the continuity equation.

Is it possible from the analysis of the data on electromagnetic nuclear processes to determine H, $\rho(r)$ and $\vec{j}(r)$ without appealing to the underlying level? Actually the question, in its generality, is far too ambitious. But partial answers do emerge, principally from the consideration of the electric processes and charge form factors.

1. The experimental value of the photon absorption cross section, integrated up to about 140 MeV, reaches approximately 1.8 times the T.R.K. value and 1.4 for the deuteron [2]. This finding is consistent with the following assumptions :

i) The long wavelength limit is a good approximation.
ii) The dipole absorption above 140 MeV contributes negligibly to the integrated cross section.
iii) The charge operator is well approximated by its impulse approximation, non relativistic, one-body form: $\rho_1(r) = e_{1/2}\sum_i \{\rho_s(r-r_i) + \tau_3^i \rho_v(r-r_i)\}$
iv) The NN potential contains an exchange part.

2. The photodisintegration of the deuteron up to $\sim$ 140 MeV confirms that :

i) The process is dominated by the E1 absorption.
ii) The charge density operator $\rho(r)$ is a good approximation of the complete one, $\rho(r)$.

3. The quasi-free charge scattering on the deuteron is compatible with the assumptions :

i) $\rho(r)$ has the one-body form $\rho_1(r)$ complemented with relativistic corrections, necessitated by the momenta reached in this type of experiments ($\sim$ 350 MeV/c) [3].

ii) The wave functions are solutions of a "realistic" NN potential, R.S.C. or Holinde-Machleidt II for instance.

4. The monopole charge form factor of the deuteron, which dominates at moderate momentum transfer the small angle electron scattering, supports the preceding conclusions.

5. The experimental value of the deuteron quadrupole moment $Q = 0.286 \pm 0.0015$ fm^2, considerably more precise than the above-mentioned experimental quantities, is correctly reproduced with $\rho(r) = \rho_1(r)$ (the relativistic terms barely contributes) and a wave function with P_D = 6 to 7 %, with a notable exception, the wave function derived from the H.M. II potential, which yields P_D = 4.3 % and Q_d = 0.287 fm^2 [4].

6. Similar data on the ^{3}He generally confirm the picture obtained from the data on the deuteron, at least at moderate momentum transfers.

7. In heavier nuclei, the precise experimental charge form factors are well predicted again by assuming $\rho(r) = \rho_1(r)$ and wave functions computed by Gogny.

The accuracy of the data concerning electric transverse transitions is only 5 to 10 % and 1 to 5 % for the charge form factors. Within these uncertainties, it seems that the one-body part of the charge operator suffices to predict correctly the experimental data. One notes even that the deuteron quadrupole moment is better reproduced without meson-exchange corrections [5], especially with the H.M. II wave function (which, incidentally, gives a better agreement with the measured cross section for the deuteron photodisintegration at forward angle [6]).

The magnetic nuclear interaction is generally said to be "more" sensitive to the meson-exchange current. In the present phenomenological framework, it means that $\vec{m}(r)$ and $\vec{j}(r)$, or one of these two operators, contain many-body parts. Let us first try the assumption that, like $\rho(r)$, $\vec{m}(r)$ is well approached by its one-body part :

$$\vec{m}_1(r) = \frac{e\hbar}{2Mc} \frac{1}{2} \sum_i \{\mu_S(r-r_i) + \tau_3^i \mu_V(r-r_i)\} \vec{\sigma}_i$$

and, therefore, that the meson-exchange corrections arise in the convective magnetization density. Now, from the exchange character of the NN potential, it ensues that the continuity equation applied to the one-body charge density operator $\rho_1(r)$ is consistent with a current containing a one- and a two-body part :

$$\vec{\nabla}_r \{\vec{j}_1(r) + \vec{j}_2(r)\} = -i/\hbar c \left[\sum t_i + \sum v_{ij}, \rho_1(r)\right]$$

8. The magnetic moment of the deuteron receives from $\vec{m}_1(r)$ the contribution :

$$\mu_1 = (\mu_p + \mu_n)(1 - \tfrac{3}{2} P_D)$$

and from the orbital magnetization associated with the one-body current $\vec{j}_r(r) =$

$$= \frac{e}{2Mc} \sum_i \{\vec{p}_i \, \delta(r-r_i) + \delta(r-r_i) \, \vec{p}_i\} \quad : \quad \mu'_1 = \tfrac{3}{4} P_D$$

In the frame of the impulse approximation, one must add the contribution $\Delta\mu_{rel}$ from the relativistic terms in the one-body current. Its exact value requires the knowledge of the wave function. Assuming for the N-N potential the H.M. II solution, which fixes P_D = 4.3 % and equating the one-body contribution to the experimental value :

$$\mu_1 + \mu_1' + \Delta\mu_{rel} = 0.8574 \quad (N.M.)$$

one gets

$$\Delta\mu_{rel} = -0.0022$$

The H.M. II potential was not included in the systematic analysis made in ref.[5]. But, as a general rule, to smaller P_D correspond smaller corrections to the magnetic moment due to relativistic nucleon motion and to meson-exchange current. This points again towards the adequacy of the H.M. II interaction ; this should be confirmed by the calculation of the various corrections. This N-N interaction should also be used in the estimation of the thermal neutron radiative capture by the proton and of the cross section for the threshold electrodisintegration.

At any event, it is clear that an extensive set of precise data on electric and magnetic nuclear processes is very efficient in discriminating between the possible NN potentials and forms of nuclear currents, provided a comparable precision is reached by the theory. This means calculating the contributions of the N-body currents, which, in the phenomenological framework, are not all determined by the continuity equation. In particular, starting from the one-body charge and spin densities $\rho_1(r)$ and $\vec{m}(r)$, it is only possible to get an information on the part of the current linked to $\rho_1(r)$ by the continuity equation. For instance

$$V_{ij} = c \, \vec{\tau}_i \cdot \vec{\tau}_j \, (\vec{\sigma}_i \cdot \vec{\nabla}_{r_{ij}})(\vec{\sigma}_j \cdot \vec{\nabla}_{r_{ij}}) f(r_{ij})$$

the O.P.E. potential, yields the two-body current identified to the "pair" and the exchanged-meson current. But terms like the many-body parts of the charge and spin density operators and the divergenceless currents are neither determined nor ruled out by the continuity equation.

The methods used in practice for obtain them in a definite form resorts to an underlying theoretical model, generally that of nucleons interacting by exchanging mesons. The calculations proceed by evaluating the contributions of a reasonable set of processes, where the photon is coupled to the different particles considered in the model. This approach is efficient, it is fitted to incorporate the nucleon isobars degrees of freedom, and has the advantage of being applicable to high energy nuclear states, where the mesons appear as real particles [7]. But great care must be exercised to avoid inconsistencies of various type [8,9,10] : violation of gauge invariance due to the selection of a set of processes, use of nuclear wave functions incompatible with the nucleon-mesons dynamics utilized in the calculation of the $\mathcal{E}.\mathcal{M}.$ interaction etc...

The requirement of consistency is sometimes considered as academic because, after all, at the present stage of accuracy of the experimental data, the M.E.C. and other corrections are precisely enough evaluated.

But whenever one faces precise experimental data (the deuteron moments, for example) or peculiar experimental conditions (for instance, far from the nucleon quasi-free kinematics), it is necessary, and perhaps rewarding, to respect the consistency requirement.

The case of complex nuclei (A ⩾ 3) intensifies the difficulties. Applying to them the procedure adapted to the deuteron presupposes that the potential, in the nuclear Hamiltonian, is merely the sum of the free NN potential for all nucleon pairs. But in principle, the NN interaction in a nucleus should differ from the form it assumes in an isolated two-nucleon system. Indeed, the best many-body calculations using a realistic N-N potential fail to reproduce important nuclear characteristics. For the A=3 nucleus, the binding energy and the proton density are not correctly predicted (assuming that a central depression in the ^{3}He proton density truly exists [11]). For heavier nuclei, the predicted density is 40 % above the measured one [12]. These observations seem to indicate that the nuclear Hamiltonian is really more complicated than a simple extrapolation of the two-body Hamiltonian. Therefore those many-body currents which are participating in the continuity equation differ from those derived in the case of the deuteron ; it is tempting to conclude that the corrections to Impulse Approximation are of a different kind than in the two-nucleon case.

This reasoning may be delusive. The fact that in the reduced Hamiltonian the nucleon interaction looks different does not necessarily mean a change in the "full" Hamiltonian from which the reduced one is obtained by a unitary transformation eliminating the mesonic degrees of freedom. In the interest of further developments, it is even recommended to stick to one definite "full" Hamiltonian, valid for any nucleus. In this case, the expressions derived for the many-body currents and already tested on the two-nucleon systems are useful in every nuclei. Of course, the wave functions needed to compute the current matrix elements must be compatible with the full Hamiltonian, or, in other words, the reduced Hamiltonian must originate from the full one.

Apart from the Hamiltonian using the free NN potential, not successful as we saw above, there exists for the heavy nuclei an effective one derived by Gogny and collaborators by fitting various properties of a few spherical nuclei and of the infinite nuclear matter. It is difficult to perceive its link with the more fundamental level. At least is it successful in predicting correctly diagonal and transition charge densities in a number of nuclei [13]. It would be of interest to systematically subject this model to the test of transverse electric and magnetic processes.

As for the A=3 nucleus, attempts are presently made to improve the theory by adding a three-nucleon potential to the Hamiltonian, with mitigated success [14].

One may ask, however, if it is licit to modify the reduced Hamiltonian by simply adding multi-nucleon forces without changing the two-nucleon potential. In the derivation of the effective potential from the original interaction, the nucleus eigenenergies enter explicitly. Its expression depend therefore on the particular nucleus under investigation. The question is whether this dependence is weak or not. A thorough study of the tri-nucleon system in the framework of meson-nucleon field theory should enlighten the point.

The nuclear underlying dynamics, that of nucleons interacting through meson exchange, is nowadays clearly established, and sometimes quantitative accounts have been achieved. Many difficulties remain to be solved, especially in complex nuclei. Is it not too early, then, to attribute apparent failures of the mesons-nucleon description to new degree of freedom?

In order to prove that the exchanged mesons influence the nuclear properties, it was imperative to handle accurately the situation at the level of the Impulse Approximation, as shown by the study of the thermal neutron radiative capture rate on the proton. Similarly, to assert the presence of new degrees of freedom, the predictions of the mesons-nucleon model and the experimental data must be precise enough to demonstrate a significant disagreement.

The claim that a meaningful disagreement exists between predicted and measured cross sections for the deuteron photodisintegration at moderate energies [15] seems premature in view of the uncertainties plaguing the experimental data and the dispersion of the calculated cross sections [7]. The quest for a deeper level of interpretation is quite natural, especially because of the attractiveness of Q.C.D. The question is how to choose among the many possible experimental conditions those which offer the best chance to observe some new effects. A possible tactic is to select processes for which the mesons-nucleon model predict a minimal rate.

[1] L.L. Foldy, Phys. Rev. 92 (1953) 178.
[2] R. Bergère, Nuclear physics with electromagnetic probes, Lecture Notes in Physics 108 (1979) 138 (Springer Verlag).
[3] J. Mougey, Nuclear physics with electromagnetic probes, Lecture Notes in Physics 108 (1979) 124 (Springer Verlag).
[4] K. Holinde, R. Machleidt, Nucl. Phys. A256 (1976) 479.
[5] E. Hadjimichael, Nucl. Phys. A312 (1978) 341.
[6] H. Arenhövel and W. Fabian, Nucl. Phys. A282 (1977) 397.
[7] J.M. Laget, Nucl. Phys. A312 (1978) 265.
[8] P. Stichel and E. Werner, Nucl. Phys. A145 (1970) 257.
[9] M. Gari and H. Hyuga, Z. für Physik A277 (1976) 291.
[10] W. Fabian and H. Arenhövel, Nucl. Phys. A258 (1976) 461.

[11] I. Sick, in Proceedings of Workshop on few body systems and electromagnetic interactions, Frascati (1978).
[12] H. Kümmel, K.H. Luhrmann and J.C. Zabolitsky, Phys. Rep. 36 (1978) C1.
[13] B. Frois, Nuclear physics with electromagnetic probes, Lecture Notes in Physics 108 (1979) 52 (Springer Verlag).
[14] J. Torre, J.J. Benayoun and J. Chauvin (preprint).
[15] E. Hadjimichael and D.P. Saylor, contribution to Saclay Symposium (1980).

NOVEL TECHNIQUES IN REAL PHOTONUCLEAR PHYSICS

LADON collaboration[+]: presented by Carlo Schaerf

The recent results of the LADON project at Frascati have proven that useful yields of monochromatic and polarized photon beams can be obtained by the backward Compton scattering of Laser light against the high energy electron circulating in a storage ring.

In his paper to this conference Gianni Matone has already indicated the concrete possibilities of extending these techniques at higher energies using machines now under construction or in and advanced planning stage. The conclusions summarized in his transparencies indicate that monochromatic and polarized gamma rays with energy up to 800 MeV with an electron storage ring of 5 GeV can be obtained with an energy resolution of 1%. The intensity of such a beam however will be limited to $\sim 10^7$ photons/sec. These estimates indicate the need for future developments in the detectors' field.

Detectors

The intensity of such beams being some orders of magnitude lower than the beams obtained with the more traditional bremsstrahlung technique it is essential to use detectors with a very large solid angle approaching 4π. In this way a gain of 3 orders of magnitude in the solid angle should compensate the comparable loss in beam intensity. In order to be useful for nuclear physics experiments in the energy region of few hundreds MeV the energy resolution of such a detector should be of the order of, or better, than 1% with the hope to reach a part in a thousand.

Unfortunately not a single detector can be designed which will yield, within reasonable cost, the desired characteristics for all particles detected. Therefore we have focused our attention on a ball of NaI (Tl) crystal and a large magnetic spectrometer.

The Crystal Ball

This detector consists of a sphere of NaI (Tl) crystal with a bore along its diameter coincident with the beam axis. The scattering target is located in the center of the phere as indicated in Fig.1. This sphere is divided in 16 equal sectors by planes intersecting on the beam axis. Each sector is subdivided in 7 equal parts each corresponding to a constant interval of $\cos\vartheta$. In this way each of the 112 parts of the crystal covers the same solid angle as viewed from the target.

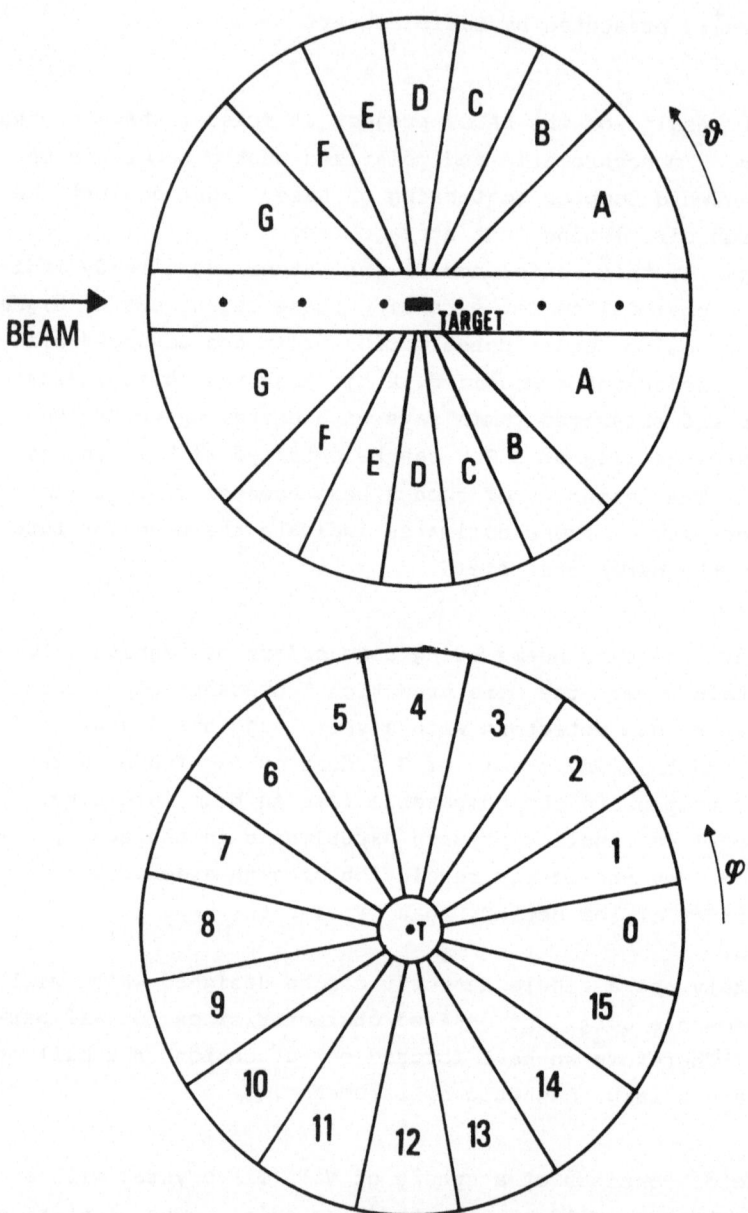

Fig.1 a) Longitudinal cross section of the crystal ball with a plane containing the beam axis. The cuts are made in such a way that for each part
$\Delta \cos \vartheta = 2/7 = 0.286$.
b) Transverse cross section perpendicular to the beam axis. Each sector corresponds to a $\Delta \varphi = \frac{\pi}{8} = 22.5°$

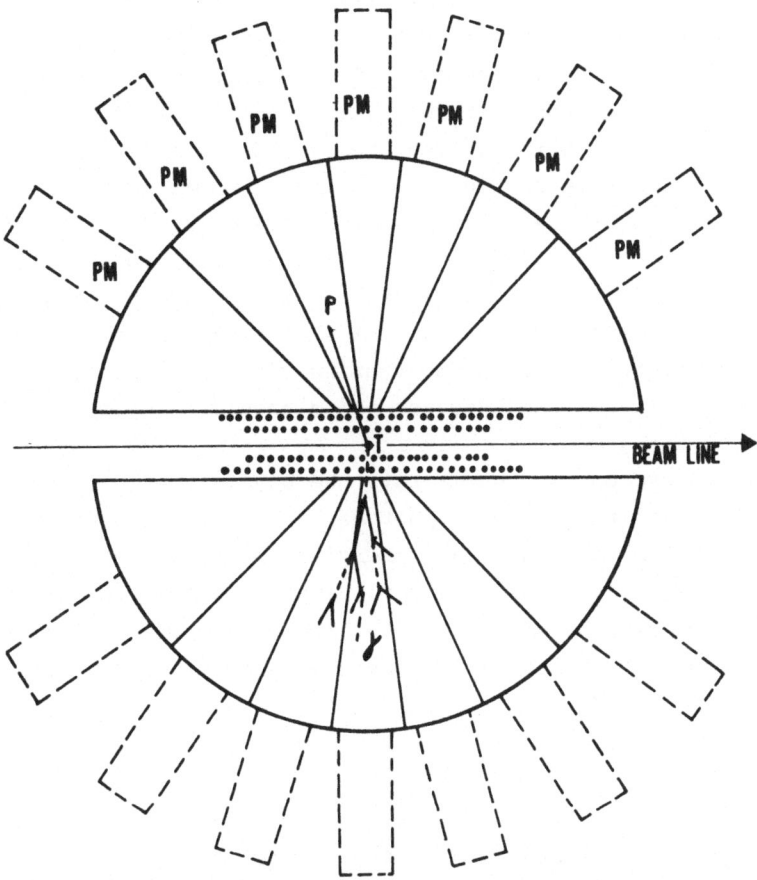

Fig.2 Longitudinal cross section of the crystal ball with an indication of the photomultipliers and the position sensitive detectors around the target. The heavy line starting from the target represents the possible trajectory of a proton, while the broken line and the tree following it give an idea of the development of the electromagnetic shower produced by a gamma ray.

112 photomultipliers each views one part of the phere which is therefore constituted by 112 individual counters. Fig.2 indicates the possible instrumentation of the sphere with plastic or solid state (position sensitive) $\frac{dE}{dx}$ detectors sorrounding the target coaxially with the beam line.

Charged particles like protons emitted from the target travers the $\frac{dE}{dx}$ detectors and enter the NaI (Tl) crystal where they are absorbed releasing their entire energy. The emission angle of the protons are obtained from the position sensitive $\frac{dE}{dx}$ counter and their energy from the light pulse in the crystal ball. In this way an energy resolution of approx. 1-3% should be obtained. Gamma rays scattered in the target travers undetected the solid state detectors and enter the crystal ball where they produce an electromagnetic shower. The energy of the shower can be collected in one or more detector parts. The distribution of the energy amongst the different counters allows a determination of the angles of scattering of the gamma rays. The total light collected in all the detector parts should be proportional to the energy of the gamma ray.

Besides protons and gamma rays the crystal ball is also able to measure π^o detecting its decay photons. In this case the kinematical constraint between the angles and energies of the 2 photons and the energy of the π^o should allow a small improvement on the evaluation of the π^o energy. In conclusion the expected energy and angular resolutions of our crystal ball for energies around 100 MeV is summarized in Table I.

TABLE I

Particle	$\frac{\Delta P}{P}$	$\Delta \cos\vartheta$	$\Delta \varphi$
γ	3-4%	0.15	10°
π^o	2-3%	0.20	20°
P	1-3%	0.10	10°

This crystal ball has already been ordered. Its total cost including detectors, photomiltipliers and coincidence counters, mechanical supports, lead shielding and electronics (high voltage power supplies, ADC, CAMAC modules) should be on the order of $ 1 million.

Magnetic Spectrometers

Magnetic spectrometers with large solid angles approaching 4π usually have a cylindrical symmetry around the beam axis and can be

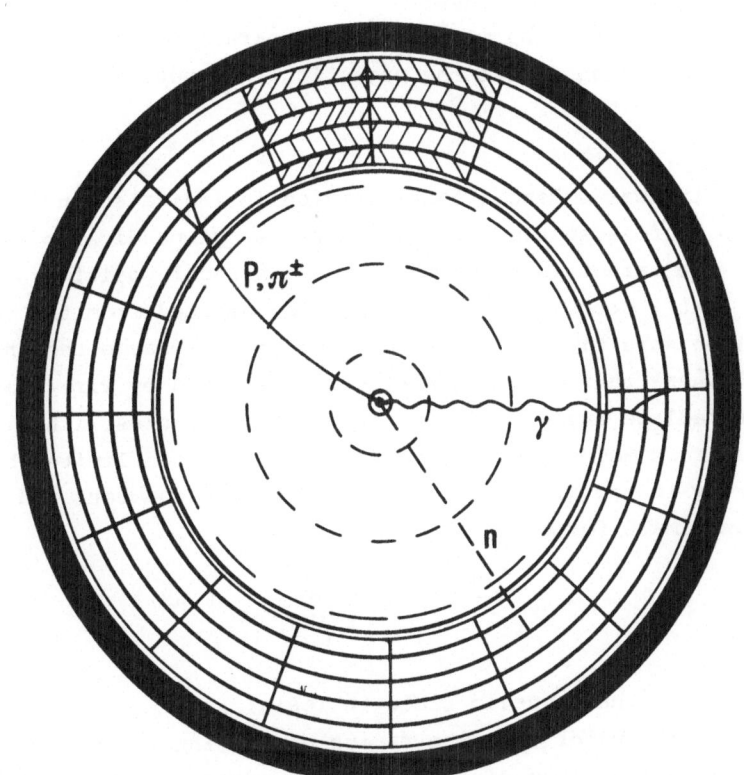

Fig.3 Cross section of a solenoid perpendicular to the beam axis.
The broken circles indicate the position sensitive wire chambers.
The dashed areas the scintillation counters separate by lead foils.

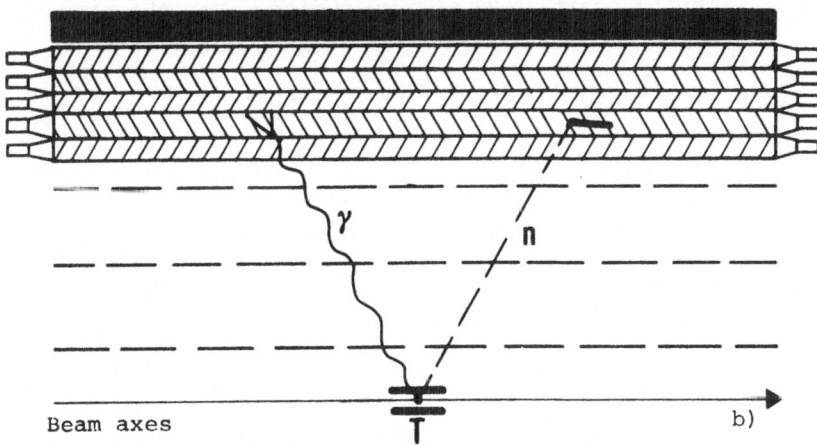

Fig.4 Cross section of the solenoid parallel to the beam axis.

divided mostly into 2 groups:
1) solenoids with the magnetic field parallel to the beam axis
2) toroidal magnet with the magnetic field rotating around the beam axis.

Solenoids

A typical solenoid configuration is indicated in Fig.3 which represents a cross section of the apparatus perpendicular to the beam line. The target is sorrounded by cylindrical position sensitive (solid state) detectors. The particles emitted from the target travers a position sensitive detector and then enter a large region ($\sim$1 m. radius) where their trajectories are identified by wire chambers. At intermediate energies for relativistic particles with a momentum of some hundreds MeV/c, their momentum resolution is limited by multiple scattering in the gas and windows of the wire chambers and is given by the approximate formula:

$$\frac{\Delta P}{P} = \frac{1.5}{B\sqrt{RX_o}} \quad (\text{FWHM})$$

where:

- B is the magnetic field in Tesla;
- R (in meters) is the radius of the inner part of the solenoid where the position sensitive detectors and wire chambers are located;
- X_o (in meters) is the average radiation lenght of the material in the position sensitive region. For the argon-isobutane mixture typical of the wire chambers:

$$X_o = 134 \text{ m}$$

and assuming:

$$B = 1 \text{ Tesla} \quad ; \quad R = 1 \text{ m}$$

$$\frac{\Delta P}{P} = 0.012 \approx 1\% \quad (\text{FWHM})$$

Therefore this apparatus is mostly suited for the detection and measurement of protons and pions (pions cannot be measured with a crystal ball due to their decay and nuclear interaction in the detectors). Neutrons and gamma rays can be detected with this apparatus with the addition of long plastic scintillators separated by lead foils as indicated in Fig.3 and 4. The gamma ray energy can be obtained by the total amount of energy deposited in the plastic scintillators. On the basis of the experience already accumulated with sandwich type shower detectors the

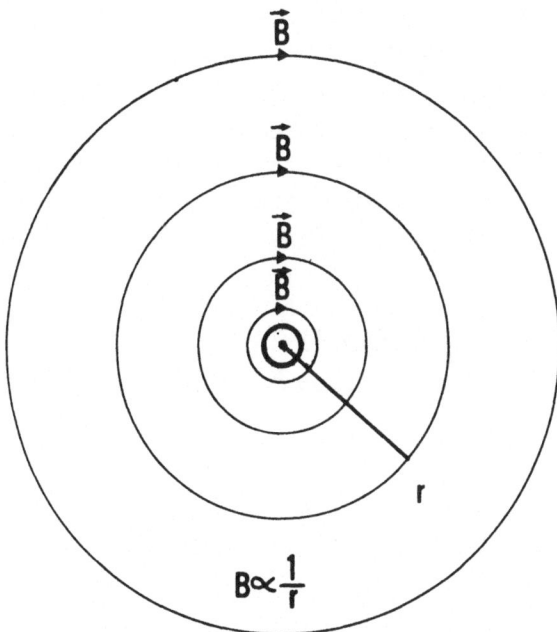

Fig.5 Magnetic field produced by a linear current coaxial with the beam axis. The current enters the figure perpendicular to it.

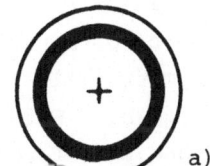 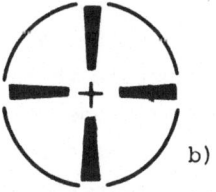

Fig.6 a) Cylindrical conductor coaxial with the beam line, all particles emitted from the target must travers this conductor.
b) Conductors parallel to the beam line but concentrated in certain regions: the particles going through the conductors are lost. The others are unperturbed. Black zones indicate the cross section of the conductor. Heavy line the position sensitive detectors.

energy resolution can be estimated to be not better than 30-50%.
Similarly the energy of the neutrons can be obtained from the time
of flight of the neutron from the target to the detector. The final
result should be $\frac{\Delta En}{En} \sim$ 10-20%.   The emission angle φ for
gamma rays and neutrons is derived from the azymutal position of the
counter in which the particle is detected.   The polar angle
ϑ is obtained by measuring the difference in the arrival time of the
light between the opposite ends of the same scintillator.

Energy and angular resolutions obtainable with such a device are
summarized in Table II.

TABLE II

Particle	$\frac{\Delta p}{p}$	$\Delta \cos\vartheta$	$\Delta \varphi$
π	1%	1-3%	1°
p	1%	1-3%	1°
γ	30-50%	10%	10-20°
n	10%	10%	10-20°

Toroidal Magnets

To improve the energy resolution especially at angles different
from $\vartheta \simeq 90°$ toroidal magnetic fields perpendicular to the planes passing through the beam line are probably preferable. The main advantage
of these configurations is that the momentum of the particle is always
perpendicular to the magnetic field for any value of the emission angles
ϑ and φ. In this way the trajectory of the particle is always contained in the reaction plane.

A field configuration of this type is that given by the law of
Biot and Savart and can be obtained by a very high current running in
a cylindrical conductor parallel to the beam line as indicated in Figs.
5 and 6a. In this case the field outside the conductor is of the
type $\frac{1}{r}$. More sophisticated field configurations can be obtained
using conductors which run parallel to the beam line, symetrically located and with specially studied cross sections as in Fig.6b. With
high enough current, eventually using superconductors, and therefore
with strong magnetic fields it is possible to realize a situation in
which all particles emitted at a given position in the beam line will
return to the beam line at a different position as indicated in in
Fig.7.

In this way the position and the angle of the particles can be

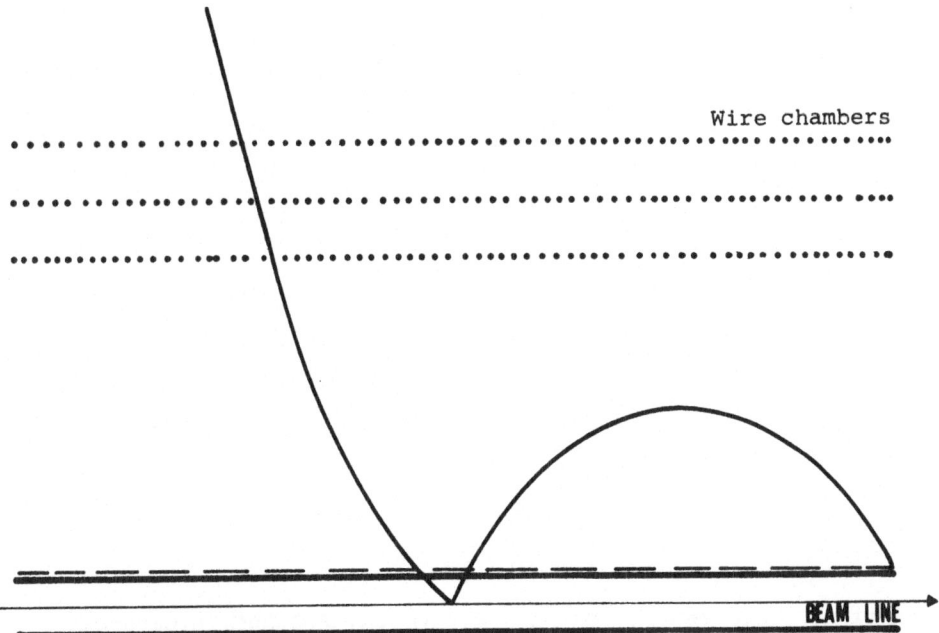

Fig.7 Trajectories of particles in a toroidal field. Particles with high momentum will escape from the high field region and therefore must be detected with large wire chambers. Particles with lower momentum return to the beam axis.

measured only at the 2 positions close to the beam line where the surface covered by the detectors can be small. In all the remaining part of the trajectory the particles do not need to be detected and therefore can travel in vacuum.

It is my pleasure to thank Piero Spillantini for many useful discussions.

+
 L.Federici, G.Giordano, G.Matone, P.Picozza, R.Caloi, L.Casano, M.P.De Pascale, M.Mattioli, E.Poldi, C.Schaerf, P.Pelfer, D.Prosperi, S.Frullani and B.Girolami.

INTERMEDIATE PERSPECTIVES

T.E.O. Ericson
CERN
Genève, Switzerland

The workshop has covered a very wide range of topics connected to nuclear electromagnetic phenomena in one way or another. I will make no attempt to discuss nor the future impact of new technology nor the perspectives of the classical fields of the topic. Let me however say that certain developments in the "classical" areas have been very impressive indeed, as we have heard in the presentation of experimental results concerning charge and magnetization distributions, for example.

Here I will concentrate instead on some areas in flux, for which there is a hot ongoing debate concerning the conceptual basis. The first of these areas concerns the nuclear force. Is the force best understood in a meson picture or in a quark picture? This issue is one of convenience and efficiency, whether conceptual or calculational, and the two approaches are nothing but two extreme approximations to the real situation. The meson description overcultivates the long range aspect of the force, starting from the one-pion exchange and working inwards, running into increasing problems as the range becomes short. The younger partner is the quark picture, in which the short range picture is taken as qualitatively understood and for which the central open problem is to achieve a description of the interaction at intermediate and long range.

More precisely, in the meson picture it is well established by now that the long range nucleon-nucleon force is dominated by one pion exchange, although we heard of some discrepancies in the very high partial waves which are very peripheral. Until very clearly confirmed, our attitude to these as of now should be that they are most likely introduced by the phase shift analysis; anyway, quarks have nothing to do with this effect. In addition to the one-pion exchange, there is by now a nearly model independent understanding of the force as due to 2π exchange with Δ excitation down to a distance of close to 1 fm. The input in this understanding is mainly dispersion relations with physical πN scattering input. Down to such distances one may view the force in this picture as a pion exchange force supplemented by a pionic van-der-Waals force in analogy to molecular forces. This observation is important, since it means that we have here a conceptual understanding of nuclear 3-body forces at similar distances, since the exchanged pions do not have to land on the same nucleon: the mechanism remains identical. Although this many-body force is only a part of nuclear many-body forces, we must expect it to be particularly important. Its introduction has had the further advantage that the nuclear many-body problem now resembles many-body problems in other areas of physics: polarization forces are standard fare there.

In the long range meson exchange approximation the nuclear force is pretty well understood and quantitatively described beyond 1 fm. Inside of 1 fm the description is quite phenomenological. It is based on two observations in that region. First, that ρ and ω are very prominent meson resonances, which have a natural part in a meson approximation and, secondly, that the exchange of exactly the ρ and ω quantum numbers are absolutely needed in the NN force and cannot be obtained naturally from the Δ, N, π picture. One thus "arbitrarily" fixes the mass and coupling constants of such exchanges and in addition, one needs a cut-off of the interaction at some short range. The detailed shape of the interaction is not very well known, and it is in fact possible to use combinations of these parameters in different ways so as to give the same NN amplitude, at least to some extend. It is clearly this part of the NN interaction that one would like to have on a firmer basis, especially the region of the short range cut-off.

Let us now turn to QCD and the description of confined quarks as models for hadrons. This is mainly an area of particle physics with consequences for nuclear physics on a few points only, but it is an extremely interesting field by itself. The whole approach emphasizes very strongly the short range region, where the quarks move as free. Very concrete and rather successful specific models for this situation are the two principal models: the MIT one (big bag) and the Stony Brook one (little bag). The pricipal assumption is that the vacuum pressure confines the quarks, which counteract the inward pressure by their energy; in the little bag this is supplemented by a strong insistence on the chiral symmetry and the importance of the pion field, which provides an additional pressure at the surface. At present the bag models use sharp confinement surfaces, which indicates the early state of developments they still are in, since the corresponding approximations are those used in nuclei 25-30 years ago.

With these assumptions an impressive description has been obtained, in particular of hadron masses but also of magnetic moments and other properties. Partly the successes have been obtained using rather general features of bag models. In the case of masses for example, the colour magnetic mass splitting functions like ordinary hyperfine splittings, so that they depend in each multiplet on only one single dynamical constant, which contains the real model dependence. For other properties like electric form factors there is little success, but this is only to be expected in models with sharp surfaces as we know from nuclei. At this point it would have been very valuable to have a public discussion of the strengths and weaknesses of the two bag-approaches with crucial tests agreed on by both sides. Unfortunately, the two schools continue internal debates with themselves, which is not always the most enlightening procedure for outsiders.

An important general prediction of the bag models is the occurence of dibaryon states in the region below 1 GeV. These states have little to do with deuteron-like resonant states of two nucleons, but are real elementary particle states in their own right and they are rather likely to exist. This has stimulated an intensive search for these states, as we have heard, and whatever the outcome, this search will certainly advance our understanding of the NN interaction considerably.

While dibaryons still concern particle physics mainly, the issue of the NN force is central to nuclear physics. Here the bag models are the first ones which give a direct physical picture of its possible behaviour at short distances. At this meeting practically nothing was discussed about this important issue, so let me review how I feel it now stands.

The program for calculating the force must be first to achieve it qualitatively at short distances from the quark bag. In this region up to about 1 fm, it may have little to do with meson exchanges. As the distance between the nucleons grows, the picture of one and two pion exchange must gradually emerge from the bag model as a natural limit for long range forces.

This program is still in a very early stage. The short range force can in principle be estimated by considering the partial overlap of two bags with their centers some distance apart, because the total energy gives the potential.

The MIT bag has a important problem at this point from the nuclear physics point of view. The bag size in this model is about 1.3 fm, so there is a large overlap of the two bags even at distances of 1.5 to 2 fm. This must be reconciled with an observed NN force, which is nearly completely model independent down to 1.2 fm and only weakly model dependent close to 1 fm. It must in addition be reconciled with the individuality of nucleons inside of nuclei. This situation is uncomfortable for the MIT model, but there are ways out. For example, the quark wave functions are located much more to the center of the bag than the surface, so that the "effective radius" for overlap may be much smaller than the bag radius.

The little chiral bag has resolved part of the force problem simply by assuming the pion field outside via PCAC at the bag surface. This automatically ensrures a correct one-pion force. As for the two-pion force an interesting investigation has just been completed by A.W. Thomas, who shows that πN scattering in a variant of this model can be made consistent with a Chew-Low model for the scattering. Insofar as this can be done, this will automatically give an important part of the intermediate range force correctly, since it only requires correct πN input data on the mass-shell in the Δ region. The little bag has however its own problems in the shorter range region. The issue of the radius of the bag does not yet seem unambiguously settled. It is difficult to see how any reasonably reliable short range force can be obtained before this point is clearly understood. Further, we have heard that in this picture the nucleon would be quite deformed. This is also likely to influence the short range force substantially.

The present picture of the nucleon force from bag models is therefore much more one of hopes and potentialities than one of solid predictions, and much hard work is ahead. Already now we can see that the principal merit of this picture is that of a physically based description of the short range cut-off region. However, it is likely to be a qualitative picture for quite some time.

At this point it may well be asked how well we need to know the NN force in nuclear physics. This question is somewhat imprecise in sofar as various combinations of amplitudes can be considered and in sofar as effective combinations of parameters matter. Roughly speaking we may say, however, that a determination of nuclear levels to about 1 MeV requires at least 5% accuracy in the force.

That the desired precision really is beyond what any model for the force is likely to give for some time, is brought out by the following outrageous example, which is somewhat specious. It is well known that a weakening of the NN force by a few per cent will unbind the deuteron, so there would be no element build-up etc. and the whole world as we know it would cease to exist. Less well known is that if it is strengthened by about 1% the diproton $^2_2$He will bind. If so, the process $p + p \to {}^2_2$He $+ \gamma$ will occur much faster than the weak hydrogen burning in the sun, so that the sun would burn in a few hundred million years. One concludes that our very existence is due to a ±1% gap in the nuclear force strength! The conclusion of this is that we should not expect the quark description of the NN force to provoke any major change in the description of nuclei.

Are there other effects in nuclei, which could be particularly quark sensitive? Clearly as nuclear matter is brought in bulk to very high densities, the possibility of quark matter is a very real one. This is however an issue of the extreme conditions of astrophysics out of normal nuclear physics. Another interesting possibility is the exploration of short distances in high q^2 studies by electron scattering. For hadrons there are very specific predictions of the form factor behaviour as a function of the number of quarks in the system and these are verified experimentally at large q^2. By the same argument nuclei should act like a quark aggregate at "sufficiently" large momentum transfers, and, thus, should be particularly relevant to simple nuclear systems like the deuteron. In fact, the deuteron form factor $F_D(q^2)$ should vary as $|q|^{-10}$ in the quark picture, but only as q^{-2} for point nucleons. The experimental values up to $q^2 = 4(\text{GeV}/c)^2$ indicate nucleons only, provided the nucleon form factors are removed (and these contain quarks of course!). Up to such q^2 nuclei do not yet seem to be in the quark asymptotic region.

Another area of debate is that of pion condensation. This is a field that has made an amazing progress in the last few years. Inspite of the appearance of disagreement, there is in fact a considerable degree of agreement on the main issues, and even more importantly, the crucial experiments have by now largely been identified, although the results are not yet in. The whole development of the debate is therefore very constructive and healthy.

In the early days of discussions of the pion condensation the great issue was whether actual nuclei would have developed condensates or not, or, alternatively, whether one could artificially make nuclear densities so high that a condensation could occur. During this phase there was a certain vagueness about experimental consequences, although it was clear that condensation implied long range spin-isospin correlations in nuclei. The crucial arguments against a developed condensate was that nuclear states of pionic quantum numbers, i.e., unnatural parity and isospin, would be the energetically favoured groundstate. This is not observed.

The decisive advance in the last few years has been the clear realization, that the closeness of a phase transition is signalled by critical or precursor effects of various kinds. The importance of this observation is that the question now becomes "how far is the phase transition from developing"? In addition, it is now clear that coexistence with normal nuclear physics is possible. The following is now agreed on.

The nature of the phase transition is understood in the sense that it can equivalently be considered to be due to the $(\sigma_1 \cdot \sigma_2)(\tau_1 \cdot \tau_2)$ part of the nuclear force or to the pion field. There is therefore no particular difficulty any longer to study it theoretically for finite nuclei in RPA approximation. The onset of the phase transition is sudden, which is why both "particle-hole" and "pion" language can be used interchangeably. The spin-isospin ordering of the system manifests itself by shifting the strength of the spin-isospin operator into a collective state as for the giant dipole state. There will, however, be various such unnatural parity collective states up to a maximum spin.

The consequence of this is that most unnatural parity states are unaffected by the approaching phase transition, and they carry no information about it. It is not enough for a state to have pion quantum numbers to make it pion-like. Only the collective states are relevant to the discussion. The πN coupling is longitudinal. This aspect is crucial for pion-like states, since there is very little coupling to probes, which have transverse coupling like electrons. In the limit of nuclear matter transverse probes are even incapable to excite the longitudinal pionic modes altogether! In view of this poor coupling it may be that the relevant pionic states have been missed altogether so far in present investigations.

A characteristic signature of the pion-like states should be a characteristic enhancement of its form factor for $q \sim q_c \approx (2-3)m_\pi$. This characteristic moment is nearly model independent and represents the periodic spin-isospin wave number of a developed condensate in nuclear matter. It is thus favourable to explore states, which have substantial Fourier components in the region $q = q_c$. The amount of enhancement depends of course on the closeness to the phase transition. It is presently agreed that there are not very large enhancements, but a factor 2-4 may still be possible.

Finally, the pion-like states will be downshifted due to the attraction. This effect is surprisingly small unless one is very close to the phase transition. At the transition the shift of the collective states becomes very important, of course. The upshot of all this is therefore that there now begins to develop a rather clear understanding of the microscopic implications of the closeness of phase transition.

Quite recently an appealing, more global picture has been given of these effects in terms of the spin-isospin response function (see figure).

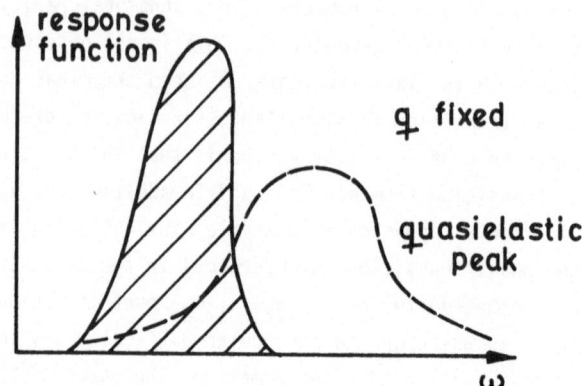

In the absence of a phase transition there is an "unperturbed" quasi-elastic peak of some width for a given momentum transfer. As a phase transition is approached this peak a) narrows (i.e. a collective state absorbs the strength), b) becomes enhanced, for $q \sim q_c$ and finally c) is shifted downward (near transition point). These three effects do not appear simultaneously: the first two develop earlier than the last one (the technical term for these phenomena in phase transition theory is "thermodynamic breaking").

In this perspective, the present discussion now concerns the exact form of the spin-isospin force, in particular in the region near $q = q_c$. This is at present a rather poorly known quantity. Conclusions must largely await additional experimental information. On the experimental side we now know (e,e')-scattering is rather ineffective for these states. Information must be sought from (p,p') or (π,γ) reactions, but the circumstances must even so be favourable in terms of the coupling amplitudes. The ideal thing would be to have experiments only sensitive to longitudinal coupling, but this seems difficult. One possibility is perhaps to use (γ,γ') reactions in the 100 MeV region, since the π^0 pole contributes there as much as the magnetic coupling in magnetic excitations.

All in all, the stage is set for a joint experimental and theoretical clarification of the whole issue within the next couple of years.

Finally, let me make a comment on nuclear exchange currents and the importance of virtual π and Δ in nuclear physics. At present, we see a number of apparently rather successful applications of such virtual processes, the most successful being perhaps the deuteron electrodisintegration. Personally, I have always some feeling that a really concrete picture for these phenomena is still missing. By this I mean a picture such that you can take a given physical transition and say at once: of course, there are big effects of such or such a sign here because of this or that. At present the picture is mainly obtained by calculation of each case.

Now, if we look at electromagnetic processes, we historically approach them by splitting up the amplitudes into magnetic and electric multipoles. This is correct, of course, but initially this was done in the long wave length limit. Now, we do it for $\lambdabar \ll R$. The consequence of this is that we may describe local, near point structures inside the nucleus in terms of high multipoles, and in this way making physics less transparent than necessary.

My suggestion is therefore that it may be useful for such problems to put more emphasis on the local properties. This is not just a gratuitous remark: for axial currents such an identification is possible ("axial locality") and it leads to a close local identification of s- and p-wave pion interactions with the time and space component of the axial current. A special case of such relations is the Goldberger-Treiman relation.

To exemplify what I mean consider the photoproduction of pions on the nucleon from threshold to the resonance:

$$\text{M1 production} \leftrightarrow \text{p wave } \pi$$
$$\text{E1 production} \leftrightarrow \text{s wave } \pi$$

In fact, in both cases there is a very simple relation to other amplitudes. For example, the M1 amplitude is obtained simply by replacing the momentum vector of a π^0 by $(\underaccent{\tilde}{\varepsilon} \times \underaccent{\tilde}{k})$ multiplied by a constant (I am cheating somewhat: there is some pionic E2 in the same region).

How may we use such a picture in a nucleus? First, exchange current phenomena tend to be pion dominated in nuclei for the simple reason that the pion is light, so that e/m_π is large. If we now look at the physical picture of radiation, it always occurs, because charges are accelerated or decelerated. Since the pion is so light, it radiates particularly easily.

In such a situation it would be quite natural to insist on a local E1 or M1 interaction associated with local s and p wave pions. In addition, we would of course have to insist on local gauge invariance. Once we have done this, it would be relatively straightforward to have a similar picture as for the axial current. Such a picture would have an interesting additional bonus. There is no problem in varying the energy ω_γ of the external photon. It would therefore play closely the same rôle as the external pion mass in soft pion theories. We would then have a physical situation in which the "mass variation" can be directly explored down to the soft limit contrary to the usual theoretical constructions for pions.

LIST OF PARTICIPANTS

ADLER, J.-O., University of Lund, Sweden
ALBERICO, W., Università di Torino, Italy
ALTHOFF, K., Universität Bonn, Fed. Rep. of Germany
ARENHÖVEL, H., Universität Mainz, Fed. Rep. of Germany
BELLINI, V., Università di Catania, Italy
BENHAR, O., Istituto Superiore di Sanità, Roma, Italy
BERGERE, R., CEN de Saclay, France
BERNHEIM, M., CEN de Saclay, France
BIZZETI, P.G., Università di Firenze, Italy
BLEULER, K., Universität Bonn, Fed. Rep. of Germany
BOFFI, S., Università di Bologna, Italy
BOHIGAS, O., Université Paris-Sud, Orsay, France
BORTIGNON, P.F., Università di Padova, Italy
BOSCHITZ, E., Universität Karlsruhe, Fed. Rep. of Germany
BOSCO, B., Università di Firenze, Italy
BREGOLA, M., Università di Bologna, Italy
BRUNO, M., Università di Bologna, Italy
CALARCO, J., Stanford University, U.S.A.
CANNATA, F., Università di Bologna, Italy
CARDMAN, L., University of Illinois, U.S.A.
CARLOS, P.-J., CEN de Saclay, France
CAVINATO, M., Università di Bologna, Italy
CENNI, R., Università di Genova, Italy
CHEMTOB, M., CEN de Saclay, France
CIOFI DEGLI ATTI, C., Istituto Superiore di Sanità, Roma, Italy
COCEVA, C., Università di Bologna
COESTER, F., Argonne National Laboratory, U.S.A.
CONCI, C., Kernforschungsanlage Jülich, Fed. Rep. of Germany
CONTE, F., Università di Genova, Italy
D'AGOSTINO, M., Università di Bologna
DAL RI, M., Università di Trento, Italy
DANOS, M., National Bureau of Standards, U.S.A.
DE FOREST, T., IKO Amsterdam, Netherlands
DELFINI, G., Netherlands Energy Research Foundation, Petten, Netherlands
DELLAFIORE, A., Università di Firenze, Italy
DELORME, J., Université de Lyon, France
DE SANCTIS, E., Laboratori Nazionali di Frascati, Italy
DE SWART, J., University of Nijmegen, Netherlands
DE VRIES, C., IKO Amsterdam, Netherlands

DILLON, G., Università di Genova, Italy
DI TORO, M., Università di Catania, Italy
DRECHSEL, D., Universität Mainz, Fed. Rep. of Germany
EISENBERG, J., Tel Aviv University, Israel
ERICSON, T., CERN, Switzerland
FABBRI, F., Università di Bologna, Italy
FALLIEROS, S., Brown University, Providence, U.S.A.
FROIS, B., CEN de Saclay, France
FRULLANI, S., Istituto Superiore di Sanità, Roma, Italy
GARIBALDI, F., Istituto Superiore di Sanità, Roma, Italy
GIACOBBE, P., Università di Bologna, Italy
GIANNINI, M.M., Università di Genova, Italy
GIUSTI, C., Università di Pavia, Italy
GRAMMATICOS, B., CNRS, Strasbourg, France
GUARALDO, C., Laboratori Nazionali di Frascati, Italy
GUIDETTI, M., Politecnico di Torino, Italy
MAMAMOTO, I., Kobenhavns Universitat, Denmark
HØGAASEN, H., University of Oslo, Norway
HOLINDE, K., Universität Bonn, Fed. Rep. of Germany
JURY, J., Trent University, Peterborough, Canada
KHANNA, F., Chalk River Nuclear Laboratories, Canada
KLINGENBECK, K., Universität Erlangen-Nürnberg, Fed. Rep. of Germany
KOWALSKI, MIT, Cambridge, U.S.A.
KREWALD, S., Kernforschungsanlage Jülich, Fed. Rep. of Germany
LAGET, J.M., CEN de Saclay, France
LANDI, G., Università di Firenze, Italy
LEONARDI, R., Università di Trento, Italy
LO JUDICE, N., Università di Napoli, Italy
LONGO, G., Università di Bologna, Italy
LO NIGRO, S., Università di Catania, Italy
LORAZO, B., Université de Montréal, Canada
MACHLEIDT, R., Universität Bonn, Fed. Rep. of Germany
MAINO, G., CNEN, Bologna, Italy
MALAGUTI, Università di Bologna, Italy
MANFREDI, V., Università di Padova, Italy
MARANGONI, M., CNEN, Bologna, Italy
MATONE, G., Laboratori Nazionali di Frascati, Italy
MATTHEWS, J., MIT, Cambridge, U.S.A.
MAXWELL, O., Universität Regensburg, Fed. Rep. of Germany
MENAPACE, E., CNEN, Bologna, Italy
MOLINARI, A., Università di Torino, Italy
MORGENSTERN, J., CEN de Saclay, France

MOSCONI, B., Università di Firenze, Italy
MOTTA, M., CNEN, Bologna, Italy
MOUGEY, J.Y., ILL, Grenoble, France
NICOLESCU, B., Université Paris-Sud, Orsay, France
O'CONNELL, J., National Bureau of Standards, U.S.A.
ORLANDINI, G., Università di Trento, Italy
OWENS, R., University of Glasgow, Great Britain
PACATI, F.D., Università di Pavia, Italy
PACE, E., Istituto Superiore di Sanità, Roma, Italy
PAPPALARDO, G.S., Università di Catania, Italy
PELFER, P.G., Laboratori Nazionali di Frascati, Italy
PICOZZA, P., Laboratori Nazionali di Frascati, Italy
PLATCHKOV, S., CEN de Saclay, France
POMPEI, A., Università di Cagliari, Italy
QUARATI, P., Università di Cagliari, Italy
RAND, R., Stanford University, U.S.A.
REFFO, G., CNEN, Bologna, Italy
RHO, M., CEN de Saclay, France
RICCI, P., Università di Firenze, Italy
RICCO, G., Università di Genova, Italy
RINAT, A., The Weizmann Institute of Science, Israel
ROSATI, S., Università di Pisa, Italy
SALME, G., Istituto Superiore di Sanità, Roma, Italy
SANZONE ARENHÖVEL, M., Max-Planck-Institut für Chemie, Mainz, Fed. Rep. of Germany
SAPORETTI, F., CNEN, Bologna, Italy
SARUIS, A.M., CNEN, Bologna, Italy
SCHAERF, C., Università di Roma, Italy
SCHELHAAS, K.-P., Max-Planck-Institut für Chemie, Mainz, Fed. Rep. of Germany
von der SCHMITT, H., Universität Mainz, Fed. Rep. of Germany
SCHOCH, B., Universität Mainz, Fed. Rep. of Germany
SCHRØDER, B., University of Lund, Sweden
SCHUHL, C., CEN de Saclay, France
SETH, K., Northwestern University, Evanston, U.S.A.
SICK, I., Universität Basel, Switzerland
SIMONINI, R., CNEN, Bologna, Italy
SPRUNG, D., Universität Tübingen, Fed. Rep. of Germany
STEFANON, M., Università di Bologna, Italy
STRUEVE, W., Universität Hannover, Fed. Rep. of Germany
TAMAS, G., CEN de Saclay, France
THIES, H., University of Western Australia, Australia
TORTORA, L., Istituto Superiore di Sanità, Roma, Italy
TRAINI, M., Università di Trento, Italy

TURCHETTI, G., Università di Bologna, Italy
TURCK-CHIEZE, S., CEN de Saclay, France
TZARA, C., CEN de Saclay, France
UGUZZONI, A., Università di Bologna, Italy
VENTO, V., CEN de Saclay, France
VENTURA, A., CNEN, Bologna, Italy
VERONDINI, E., Università di Bologna, Italy
VINH MAU, R., Université Paris-Sud, Orsay, France
VIOLLIER, R., Universität Basel, Switzerland
WAGNER, G., Max-Planck-Institut für Kernphysik, Heidelberg, Fed. Rep. of Germany
WEISE, W., Universität Regensburg, Fed. Rep. of Germany
WIENHARD, K., Universität Giessen, Fed. Rep. of Germany
WONG, C.W., University of California, U.S.A.
YERGIN, P., Rensselaer Polytechnic Institute, Boston, U.S.A.
ZIEGLER, B., Max-Planck-Institut für Chemie, Mainz, Fed. Rep. of Germany
ZUCCHIATTI, A., Università di Genova, Italy

AUTHOR INDEX

Alberico, W.M., 348
Althoff, H., 296
Arenhövel, H., 136

Boffi, S., 186
Bohigas, O., 65
Boschitz, E., 243

Caloi, R., 312
Carlos, P.J., 168
Casano, L., 312
Chemtob, M., 158
Ciofi degli Atti, C., 115
Coceva, C., 339

De Forest Jr., T., 258
De Jager, C.W., 258
Delorme, J., 82
De Pascale, M.P., 312
De Swart, J.J., 196
De Vries, C., 258
De Vries, H., 258
De Witt Huberts, P.K.A., 258
Drechsel, D., 358

Eisenberg, J.M., 368
Ericson, T.E.O., 403

Federici, L., 312
Frois, B., 55
Frullani, S., 312

Giordano, G., 312
Girolami, B., 312
Grammaticos, B., 42

Högaasen, H., 212
Holinde, K., 10

Klingenbeck, K., 102
Koch, J.H., 258
Krewald, S., 31

Jans, E., 258

Laget, J.M., 148
Lapikas, L., 258

Maas, R., 258
Matone, G., 312
Mattioli, M., 312
Molinari, A., 348

Nicolescu, B., 223

O'Connell, J.S., 286
Orlandini, G., 72

Pace, E., 115
Pelfer, P., 312
Picozza, P., 312
Poldi, E., 312
Prosperi, D., 312

Salme, G., 115
Schaerf, C., 312, 393
Schoch, B., 178
Schuhl, C., 277
Sick, I., 125

Tamas, G., 234
Toki, H., 93
Turck-Chieze, S., 251
Tzara, C., 385

Vento, V., 205
Vinh Mau, R., 1

Weise, W., 93
Wong, C.W., 375

Ziegler, B., 325

R. Bass

Nuclear Reactions with Heavy Ions

1980. 176 figures, 31 tables. VIII, 410 pages
(Texts and Monographs in Physics)
ISBN 3-540-09611-6

Contents: Introduction. – Light Scattering Systems. – Quasi-Elastic Scattering from Heavier Target Nuclei. – General Aspects of Nucleon Transfer. – Quasi-Elastic Transfer Reactions. – Deep-Inelastic Scattering and Transfer. – Complete Fusion. – Compound-Nucleus Decay. – Appendices.

The last decade has witnessed an astounding increase in heavy ion research. This book presents – from an experimentalist's point of view – a critical and coherent outline of the results of large scale heavy ion research in the area of low energy nuclear reactions in the 5–10 MeV per nucleon range. Using phenomenological models, the author explains these experimental results, achieving a good balance between a critically selected review and a textbook. This makes it attractive for the advanced student and the specialist alike.

P. Ring, P. Schuck

The Nuclear Many-Body Problem

1980. 171 figures. XVII, 716 pages
(Texts and Monographs in Physics)
ISBN 3-540-09820-8

Contents: The Liquid Drop Model. – The Shell Model. – Rotation and Single-Particle Motion. – Nuclear Forces. – The Hartree-Fock Method. – Pairing Correlations and Suprafluid Nuclei. – The Generalized Single-Particle Model (HFB-Theory). – Harmonic Vibrations. – Boson Expansion Methods. – The Generator Coordinate Method. – Restoration of Broken Symmetries. – The Time Dependent Hartree-Fock Method (TDHF). – Semiclassical Methods in Nuclear Physics. – Addendices. – Bibliography. – Author Index. – Subject Index.

This book, while covering a fair amount of physical observations, stresses the methodology and technical aspects of the different theories presently used in the description of the nucleus. The authors present the more modern theories such as Boson expansions, generator coordinates, time-dependent Hartree-Fock method, and semiclassical models which so far have found only limited mention in textbooks. The book also covers subjects like the liquid drop and the shell model, both presented in a updated version in, for example, rotations and random phase approximation. The full presentation of mathematical details, illustrated by observational data, will help the student fully understand the present views on the nuclear many-body problem.

Springer-Verlag
Berlin
Heidelberg
New York

Lecture Notes in Physics

Vol. 114: Stellar Turbulence. Proceedings, 1979. Edited by D. F. Gray and J. L. Linsky. IX, 308 pages. 1980.

Vol. 115: Modern Trends in the Theory of Condensed Matter. Proceedings, 1979. Edited by A. Pekalski and J. A. Przystawa. IX, 597 pages. 1980.

Vol. 116: Mathematical Problems in Theoretical Physics. Proceedings, 1979. Edited by K. Osterwalder. VIII, 412 pages. 1980.

Vol. 117: Deep-Inelastic and Fusion Reactions with Heavy Ions. Proceedings, 1979. Edited by W. von Oertzen. XIII, 394 pages. 1980.

Vol. 118: Quantum Chromodynamics. Proceedings, 1979. Edited by J. L. Alonso and R. Tarrach. IX, 424 pages. 1980.

Vol. 119: Nuclear Spectroscopy. Proceedings, 1979. Edited by G. F. Bertsch and D. Kurath. VII, 250 pages. 1980.

Vol. 120: Nonlinear Evolution Equations and Dynamical Systems. Proceedings, 1979. Edited by M. Boiti, F. Pempinelli and G. Soliani. VI, 368 pages. 1980.

Vol. 121: F. W. Wiegel, Fluid Flow Through Porous Macromolecular Systems. V, 102 pages. 1980.

Vol. 122: New Developments in Semiconductor Physics. Proceedings, 1979. Edited by F. Beleznay et al. V, 276 pages. 1980.

Vol. 123: D. H. Mayer, The Ruelle-Araki Transfer Operator in Classical Statistical Mechanics. VIII, 154 pages. 1980.

Vol. 124: Gravitational Radiation, Collapsed Objects and Exact Solutions. Proceedings, 1979. Edited by C. Edwards. VI, 487 pages. 1980.

Vol. 125: Nonradial and Nonlinear Stellar Pulsation. Proceedings, 1980. Edited by H. A. Hill and W. A. Dziembowski. VIII, 497 pages. 1980.

Vol. 126: Complex Analysis, Microlocal Calculus and Relativistic Quantum Theory. Proceedings, 1979. Edited by D. Iagolnitzer. VIII, 502 pages. 1980.

Vol. 127: E. Sanchez-Palencia, Non-Homogeneous Media and Vibration Theory. IX, 398 pages. 1980.

Vol. 128: Neutron Spin Echo. Proceedings, 1979. Edited by F. Mezei. VI, 253 pages. 1980.

Vol. 129: Geometrical and Topological Methods in Gauge Theories. Proceedings, 1979. Edited by J. Harnad and S. Shnider. VIII, 155 pages. 1980.

Vol. 130: Mathematical Methods and Applications of Scattering Theory. Proceedings, 1979. Edited by J. A. DeSanto, A. W. Sáenz and W. W. Zachary. XIII, 331 pages. 1980.

Vol. 131: H. C. Fogedby, Theoretical Aspects of Mainly Low Dimensional Magnetic Systems. XI, 163 pages. 1980.

Vol. 132: Systems Far from Equilibrium. Proceedings, 1980. Edited by L. Garrido. XV, 403 pages. 1980.

Vol. 133: Narrow Gap Semiconductors Physics and Applications. Proceedings, 1979. Edited by W. Zawadzki. X, 572 pages. 1980.

Vol. 134: $\gamma\gamma$ Collisions. Proceedings, 1980. Edited by G. Cochard and P. Kessler. XIII, 400 pages. 1980.

Vol. 135: Group Theoretical Methods in Physics. Proceedings, 1980. Edited by K. B. Wolf. XXVI, 629 pages. 1980.

Vol. 136: The Role of Coherent Structures in Modelling Turbulence and Mixing. Proceedings 1980. Edited by J. Jimenez. XIII, 393 pages. 1981.

Vol. 137: From Collective States to Quarks in Nuclei. Edited by H. Arenhövel and A. M. Saruis. VII, 414 pages. 1981.

Selected Issues from
Lecture Notes in Mathematics

Vol. 684: E. E. Rosinger, Distributions and Nonlinear Partial Differential Equations. XI, 146 pages. 1978.

Vol. 690: W. J. J. Rey, Robust Statistical Methods. VI, 128 pages. 1978.

Vol. 691: G. Viennot, Algèbres de Lie Libres et Monoïdes Libres III, 124 pages. 1978.

Vol. 693: Hilbert Space Operators, Proceedings, 1977. Edited by J. M. Bachar Jr. and D. W. Hadwin. VIII, 184 pages. 1978.

Vol. 696: P. J. Feinsilver, Special Functions, Probability Semigroups, and Hamiltonian Flows. VI, 112 pages. 1978.

Vol. 702: Yuri N. Bibikov, Local Theory of Nonlinear Analytic Ordinary Differential Equations. IX, 147 pages. 1979.

Vol. 704: Computing Methods in Applied Sciences and Engineering, 1977, I. Proceedings, 1977. Edited by R. Glowinski and J. L. Lions. VI, 391 pages. 1979.

Vol. 710: Séminaire Bourbaki vol. 1977/78, Exposés 507–524. IV, 328 pages. 1979.

Vol. 711: Asymptotic Analysis. Edited by F. Verhulst. V, 240 pages. 1979.

Vol. 712: Equations Différentielles et Systèmes de Pfaff dans le Champ Complexe. Edité par R. Gérard et J.-P. Ramis. V, 364 pages. 1979.

Vol. 716: M. A. Scheunert, The Theory of Lie Superalgebras. X, 271 pages. 1979.

Vol. 720: E. Dubinsky, The Structure of Nuclear Fréchet Spaces. V, 187 pages. 1979.

Vol. 724: D. Griffeath, Additive and Cancellative Interacting Particle Systems. V, 108 pages. 1979.

Vol. 725: Algèbres d'Opérateurs. Proceedings, 1978. Edité par P. de la Harpe. VII, 309 pages. 1979.

Vol. 726: Y.-C. Wong, Schwartz Spaces, Nuclear Spaces and Tensor Products. VI, 418 pages. 1979.

Vol. 727: Y. Saito, Spectral Representations for Schrödinger Operators With Long-Range Potentials. V, 149 pages. 1979.

Vol. 728: Non-Commutative Harmonic Analysis. Proceedings, 1978. Edited by J. Carmona and M. Vergne. V, 244 pages. 1979.

Vol. 729: Ergodic Theory. Proceedings 1978. Edited by M. Denker and K. Jacobs. XII, 209 pages. 1979.

Vol. 730: Functional Differential Equations and Approximation of Fixed Points. Proceedings, 1978. Edited by H.-O. Peitgen and H.-O. Walther. XV, 503 pages. 1979.

Vol. 731: Y. Nakagami and M. Takesaki, Duality for Crossed Products of von Neumann Algebras. IX, 139 pages. 1979.

Vol. 733: F. Bloom, Modern Differential Geometric Techniques in the Theory of Continuous Distributions of Dislocations. XII, 206 pages. 1979.

Vol. 735: B. Aupetit, Propriétés Spectrales des Algèbres de Banach. XII, 192 pages. 1979.

Vol. 738: P. E. Conner, Differentiable Periodic Maps. 2nd edition, IV, 181 pages. 1979.

Vol. 742: K. Clancey, Seminormal Operators. VII, 125 pages. 1979.

Vol. 755: Global Analysis. Proceedings, 1978. Edited by M. Grmela and J. E. Marsden. VII, 377 pages. 1979.

Vol. 756: H. O. Cordes, Elliptic Pseudo-Differential Operators – An Abstract Theory. IX, 331 pages. 1979.

Vol. 760: H.-O. Georgii, Canonical Gibbs Measures. VIII, 190 pages. 1979.

Vol. 762: D. H. Sattinger, Group Theoretic Methods in Bifurcation Theory. V, 241 pages. 1979.

Vol. 765: Padé Approximation and its Applications. Proceedings, 1979. Edited by L. Wuytack. VI, 392 pages. 1979.

Vol. 766: T. tom Dieck, Transformation Groups and Representation Theory. VIII, 309 pages. 1979.

Vol. 771: Approximation Methods for Navier-Stokes Problems. Proceedings, 1979. Edited by R. Rautmann. XVI, 581 pages. 1980.

Vol. 773: Numerical Analysis. Proceedings, 1979. Edited by G. A. Watson. X, 184 pages. 1980.

Vol. 775: Geometric Methods in Mathematical Physics. Proceedings, 1979. Edited by G. Kaiser and J. E. Marsden. VII, 257 pages. 1980.

Vol. 779: Euclidean Harmonic Analysis. Proceedings, 1979. Edited by J. J. Benedetto. III, 177 pages. 1980.

Vol. 780: L. Schwartz, Semi-Martingales sur des Variétés, et Martingales Conformes sur des Variétés Analytiques Complexes. XV, 132 pages. 1980.

Vol. 782: Bifurcation and Nonlinear Eigenvalue Problems. Proceedings, 1978. Edited by C. Bardos, J. M. Lasry and M. Schatzman. VIII, 296 pages. 1980.

Vol. 783: A. Dinghas, Wertverteilung meromorpher Funktionen in ein- und mehrfach zusammenhängenden Gebieten. Edited by R. Nevanlinna and C. Andreian Cazacu. XIII, 145 pages. 1980.

Vol. 786: I. J. Maddox, Infinite Matrices of Operators. V, 122 pages. 1980.

Vol. 787: Potential Theory, Copenhagen 1979. Proceedings, 1979. Edited by C. Berg, G. Forst and B. Fuglede. VIII, 319 pages. 1980.

Vol. 791: K. W. Bauer and S. Ruscheweyh, Differential Operators for Partial Differential Equations and Function Theoretic Applications. V, 258 pages. 1980.

Vol. 792: Geometry and Differential Geometry. Proceedings, 1979. Edited by R. Artzy and I. Vaisman. VI, 443 pages. 1980.

Vol. 793: J. Renault, A Groupoid Approach to C*-Algebras. III, 160 pages. 1980.

Vol. 798: Analytic Functions, Kozubnik 1979. Proceedings. Edited by J. Ławrynowicz. X, 476 pages. 1980.

Vol. 799: Functional Differential Equations and Bifurcation. Proceedings 1979. Edited by A. F Izé. XXII, 409 pages. 1980.

Vol. 801: K. Floret, Weakly Compact Sets. VII, 123 pages. 1980.

Vol. 802: J. Bair, R. Fourneau, Etude Géometrique des Espaces Vectoriels II. VII, 283 pages. 1980.

Vol. 804: M. Matsuda, First Order Algebraic Differential Equations. VII, 111 pages. 1980.

Vol. 805: O. Kowalski, Generalized Symmetric Spaces. XII, 187 pages. 1980.

Vol. 807: Fonctions de Plusieurs Variables Complexes IV. Proceedings, 1979. Edited by F. Norguet. IX, 198 pages. 1980.

Vol. 810: Geometrical Approaches to Differential Equations. Proceedings 1979. Edited by R. Martini. VII, 339 pages. 1980.

Vol. 816: L. Stoica, Local Operators and Markov Processes. VIII, 104 pages. 1980.

Vol. 819: Global Theory of Dynamical Systems. Proceedings, 1979. Edited by Z. Nitecki and C. Robinson. IX, 499 pages. 1980.